FIRE INVESTIGATION

Russell K. Chandler

DELMAR
CENGAGE Learning™

Australia • Brazil • Japan • Korea • Mexico • Singapore • Spain • United Kingdom • United States

Fire Investigation
Russell K. Chandler

Vice President, Career and Professional Editorial: Dave Garza

Director of Learning Solutions: Sandy Clark

Product Development Manager: Janet Maker

Managing Editor: Larry Main

Senior Product Manager: Jennifer Starr

Editorial Assistant: Amy Wetsel

Vice President, Career and Professional Marketing: Jennifer McAvey

Marketing Director: Deborah S. Yarnell

Senior Marketing Manager: Erin Coffin

Marketing Coordinator: Shanna Gibbs

Production Director: Wendy Troeger

Production Manager: Mark Bernard

Senior Content Project Manager: Jennifer Hanley

Art Director: Benj Gleeksman

Technology Project Manager: Christopher Catalina

For product information and technology assistance, contact us at
Professional Group Cengage Learning Customer & Sales Support, 1-800-354-9706

For permission to use material from this text or product, submit all requests online at **cengage.com/permissions.**
Further permissions questions can be e-mailed to **permissionrequest@cengage.com.**

Library of Congress Control Number: 2008940736

ISBN-13: 978-1-4180-0960-1

ISBN-10: 1-4180-0960-1

Delmar
5 Maxwell Drive
Clifton Park, NY 12065-2919
USA

Cengage Learning products are represented in Canada by Nelson Education, Ltd.

For your lifelong learning solutions, visit **delmar.cengage.com**

Visit our corporate website at **cengage.com.**

Printed in Canada
1 2 3 4 5 XX 11 10 09

Special thanks to my wife.
As we have dedicated our lives to each other, I dedicate this book to her.

CONTENTS

PREFACE

INTENT OF THIS BOOK

This book is designed for both the fire science student enrolled in a college course on fire investigation as well as the firefighter and the fire officer who need to know the basics of fire investigation to be successful on the job. A reader-friendly narrative walks the learner through the necessary information to achieve a complete understanding of the fire investigation process—from initial response to the final report and testifying in court. Comprehensive in nature, this book covers the requirements and guidelines outlined in NFPA Standard 1033, *Standard for Professional Qualifications for Fire Investigator* (2009 edition) and correlates to the *Fire Investigator I and II* course outcomes created by the Fire and Emergency Services Higher Education (FESHE) committee at the National Fire Academy in efforts to standardize training in colleges across the United States.

How This Book Achieves These Goals

Every fire that goes uninvestigated is a lost opportunity to prevent future fires. It is an economic fact that there are not enough formally trained full-time fire investigators to look at each and every fire in any given jurisdiction. For this reason, many departments rely on the fire officer in charge of the incident to collect information on the incident at the very least to enable them to complete the fire run report hopefully using the National Fire Incident Reporting System.

The fire officer or firefighter does not need the full breadth of fire investigative knowledge to make an accurate determination of many of the smaller, less damaging incidents. Many fires have an obvious area of origin as well as an identifiable cause. In addition to the fire officer or firefighter recognizing the origin and cause of these uncomplicated fires, it is important for them to also know when to call for a full-time investigator to do a more in-depth investigation.

Investigating the fire and making an accurate determination of the area of origin and cause, such as when the fire officer identifies that the fire started as a result of a malfunctioning furnace, is the first step to preventing future fires. The assigned fire investigator may realize that this is not the first such incident and then notices that there is a frequency of such fires associated with one furnace repair company. A detailed investigation using scientific methodology may further identify that the local contractor not doing a proper maintenance and inspection on the furnaces was a direct cause of these fires. Actions can then be taken to stop the contractor or educate the contractor, thus preventing future incidents.

On a wider scope, the identification of a specific model of an electrical appliance that has malfunctioned on multiple occasions, overheating and igniting nearby combustibles, can also be identified. Accurate investigations identifying this problem could result in a national recall of that product—preventing future fires.

Then, in the case of an incendiary fire, identifying the perpetrator with subsequent conviction and incarceration will most likely prevent that person from setting future fires. Just as important, the arrest and conviction of the arsonist can be a deterrence to others who may be considering committing similar crimes.

The first line of defense is the firefighter and fire officer, who respond to the fire scene, extinguish the fire, and collect the data on how the fire started. By educating these fire suppression personnel, we enhance our chances of preventing fires.

However, the text does not stop with the responsibility of initial investigators, the firefighter and fire officer (suppression forces). The text goes further to explain the actions necessary for the more in-depth investigation conducted by an assigned, full-time investigator. This is twofold: first to help fire officers understand what else may need to be accomplished in the more in-depth investigation and what they can do to assist. The second reason is to educate new full-time assigned fire investigators on their duties and how suppression forces can assist them in their endeavor to seek the area of origin, fire cause, and person or thing responsible for the incident.

WHY I WROTE THIS BOOK

I came to the realization early in my career of the necessity of using fire personnel to enhance the investigative process. On my first investigation as a Fire Marshal, I was met by 10 firefighters beaming with pride because they had removed every bit of debris from the structure and had also used water hoses and water vacuums to clean off the floor. I was presented with the cleanest burnt house I had ever seen.

Countless hours were spent creating or adapting one program after another that could educate our firefighters in conducting a preliminary investigation. Everything from 6-hour to 24-hour programs were piloted with some success.

Later, as an adjunct instructor at the community college it was obvious that this was an excellent opportunity to train our fire suppression personnel on conducting those basic fire investigations and to educate them a little further so that they could then become a valuable asset to the assigned fire investigator when they arrive on the scene. Over the 20 years of teaching, many outstanding texts were used in this endeavor, but none that were dedicated to this topic with that target audience and that could also be easily adapted to a college curriculum. Thus, the creation of this textbook.

Of course, the primary motivation for writing about conducting an investigation is that the investigation itself is the first step in preventing future fires. However, my teachings and this project have two other motivations. It boils down to numbers and ethics. First, as an average there are only two to four full-time investigators in any given jurisdiction assigned to investigating fires. But there may be as many as 200 to 400 firefighters (volunteer, career, or a combination of volunteer and career) in that same locality. A desire to reach all the fire service was a major drive in writing this text.

The next and very important motivation in the creation of this text is ethics—a desire to have everyone believe, as most investigators do, that in our investigations we are seekers of truth not only in the criminal cases but also in the accidental cases. The use of scientific methodology in the process of conducting the investigation ensures that there will be accurate determinations as to the cause of the fire. The victims deserve to know what happened. They do not need to be burdened with conjecture or inaccurate summations that could result in misguided or misplaced guilt.

HOW TO USE THIS BOOK

Both the method in which the chapters are organized, as well as the manner in which the design and topics within those chapters are structured are intended to assist the reader in learning the critical concepts. Like investigations, learning is a process, and one must therefore learn and approach the subject with a logical progression of thought. It is for this reason that I recommend you follow the sequence of the chapters in the order that they are presented in this book:

- **Chapters 1–4 (National Standards for Fire Investigators; Safety; Role of First Responders; The Fire Investigator)** introduce you to the role of the fire investigator in the context of incident response. These chapters outline the basic information that you need to know to understand your responsibilities and operate as a fire investigator at the scene.
- **Chapters 5–8 (Legal Issues and the Right to Be There; Spoliation; Science, Methodology, and Fire Behavior; Physical Evidence Collection and Preservation)** provide you with the principles necessary to conduct an effective investigation. Each chapter brings to light the duties and responsibilities of the investigator and explains their importance.
- **Chapters 9–13 (Sources of Ignition; Electricity and Fire; Patterns: Burn and Smoke; Examining the Scene and Finding the Origin; Fire Cause)** break down and explore the potential causes of a fire to lead you to make the appropriate analysis when provided with different types of evidence and clues.
- **Chapters 14–16 (Vehicle Fires; Fatal Fires and Fire Injuries; Arson)** highlight specific types of fires and the characteristics of each. These chapters explain the uniqueness of these fires and the important clues that are necessary for accurate analysis and reporting in these situations.
- **Chapters 17–23 (Investigative Resources; Documenting the Scene; Interviews; Human Behavior; Sources of Information; The Expert Witness; The Final Report, Testifying in Court, and PIO)** wrap up the final, yet important, details that are also critical in a fire investigation. Beyond the investigation of the structure there is the important aspect of gathering information, reporting out on the findings, and (in some cases)

actually testifying in court on an incident. These chapters illustrate how to accomplish these tasks and provide practical advice on how to do so successfully.

FEATURES OF THIS BOOK

This book is designed to provide the necessary information on the topic of fire investigation, while presenting it in a practical, straightforward approach that is conducive to learning. For these reasons, the following features were added to the book to enhance learning:

- **Standard Guidelines and Requirements** are outlined, including coverage of NFPA Standard 1033, *Standard for Professional Qualifications for Fire Investigator,* 2009 Edition and the National Fire Academy FESHE course outcomes for *Fire Investigator I and II.* A chapter explaining the role of these standards and of the investigator at a fire scene is also included at the beginning of the book.
- **Safety** is emphasized throughout the book, including a separate chapter that outlines various potential dangers an investigator may face even *after* the fire is extinguished so that you can stay safe on the job.
- **Case studies** open each chapter by giving an example of an actual incident that explains the importance of good fire investigation tactics. Each case study relates to the information contained within that chapter and highlights important lessons learned from the incident. Throughout the text are additional real-life examples germane to the topic at hand. The names and unit numbers in the retelling have been changed to protect the innocent.
- **Incident photos** from past investigations illustrate various clues to look for when conducting the investigation and help you understand the importance of preserving and collecting evidence as well as documenting findings.
- **Advice on interviewing witnesses and testifying in court** is also included in chapters in the book to better prepare you for the job of the fire investigator. Practical advice on how to identify certain traits of human behavior within your investigation and how to prepare for an appearance in court in cases of suspected arson highlights the professional skills required of a successful investigator.
- **Discussion questions** in addition to the *review questions* at the end of each chapter allow for classroom discussions to develop critical thinking skills as well as best practices and methodologies for approaching a fire scene investigation.

AN IMPORTANT NOTE ABOUT THIS BOOK

It is important to understand that before you begin your lessons in fire investigation that local history has dictated the title of who enforces the laws in that community. Titles range from constable, deputy sheriff, to police officer. It is the same in the fire investigative community but with an additional twist: local history also dictates which agency conducts in-depth fire investigations. This can be the local sheriff's office, police department, or the local fire department. Respective titles can vary between sheriff's investigator, police detective, or police inspector from the law enforcement agency or the investigator can be the fire prevention officer or fire marshal from the fire department. This variance from locality to locality and from state to state makes it difficult to come up with the one title that would identify the person with the full-time responsibility to conduct the fire investigation.

The national trend toward a uniform title presently seems to be *fire marshal.* This is evident in the fact that there is a National State Fire Marshal Association and the National Fire Protection Association has created a new standard, NFPA 1037, *Standard for Professional Qualifications of Fire Marshal.*

Being partial toward the term *fire marshal* would not have been in the best interest to the wide use of this text. Being sensitive to the various titles and agencies involved, I chose to use the term *assigned investigator* to represent the person in that locality who is responsible for conducting more complicated, detailed investigations. The instructor can then further identify to the students the proper agency and associated titles as to who that investigator will be in each jurisdiction.

SUPPLEMENTS TO THIS BOOK

This book is accompanied by handy resources to assist both the student and the instructor in learning and delivering the content presented in these chapters.

e.resource CD-ROM

The e.resource CD-ROM offers instructors the necessary tools to prepare for classroom presentation and evaluation of student knowledge of the content. This CD includes the following features:

- **Lesson Plans and Answers to Questions** provide chapter outlines and information for classroom instruction and discussion. Lesson Plans include correlations to the accompanying *PowerPoint presentations* and provide a cohesive presentation.
- **PowerPoint Presentations,** including full-color photos as well as graphics from the book, keep instructors organized for classroom presentations. Instructors may utilize the PowerPoint presentations as is or may wish to insert additional information, including images from the accompanying *Image Library.*
- **Test banks** in ExamView 6.0 provide questions for each chapter in the book. The flexible format of each exam allows instructors to add, delete, or edit existing questions to meet the needs of their students. Instructors may also create their own exams from existing questions to suit their needs.

- An **Image Library** includes full-color photos as well as graphics from the book. Instructors may add to the existing *PowerPoint presentations* on this CD or create their own presentations utilizing this art.
- **NFPA Correlation Guides** are also included on this CD and correlate the content of NFPA Standard 1033 (2009 edition) and NFPA Guide 921 (2008 edition) to the chapters and pages in this book. Instructors can gauge student knowledge of specific requirements and recommendations by focusing and reviewing content as identified in the book.
- The **FESHE Correlation Grid** outlines the course outcomes developed by the National Fire Academy FESHE committee for *Fire Investigator I and II* and correlates these outcomes to the content in this book. By following this recommended course outline, instructors can help standardize training for this course.

Online Companion

An Online Companion complete with full-color photos provides an opportunity to enhance learning in the classroom. Instructors may wish to insert photos into existing *PowerPoint presentations*, or they may wish to direct their students to review important clues and evidence for upcoming exams. This FREE Online Companion includes photos from the book as well as additional photos and materials for further review. Simply visit our Web site at *www.delmarfire.cengage.com* and click "Online Companions" to be directed to this resource.

ABOUT THE AUTHOR

I always teach my students that "we are the sum of our experiences." My experiences actually started before I got involved in public safety. My grandfather, who taught me my core values, also gave me an opportunity to become a licensed electrician and plumber before I even got my driver's license. In addition to wiring homes and businesses, we did repairs on electrical household appliances. Little did I know at that time how valuable this would be in my investigative ventures.

My career path has taken me from being a career firefighter, to a full-time safety manager and Fire Chief at a major amusement park, to becoming the first Fire Marshal in my new home of Hanover County, Virginia. I was then lured by the fiscal rewards of being a fire investigator in the private sector, primarily serving the insurance industry. This certainly added to the sum of my experiences, seeing an entirely different side and motivation toward completing the fire investigation. Not a bad side by any means, it was certainly a different aspect of the investigative world where I took great pride in our accomplishments and found a whole new world of colleagues and peers.

In 1990, I was hired by the Virginia Department of Fire Programs as their Manager of Investigations and Inspections. Within a few years, I was able to redesign that entire division, creating the Virginia Fire Marshal Academy. The academy delivers national certification level training for investigations, inspections, and public fire and life safety education. One vital school the academy provides to Virginia's fire marshals is the Law Enforcement School for Fire Marshals. This school enables local investigators to get their police powers from an academy dedicated to make them investigators rather than the other law enforcement academies dedicated to turning out new police patrol officers.

Presently, I serve as the Branch Chief of the Virginia Department of Fire Programs. My team now consists of the Program Chiefs for Airport Rescue Firefighter, Heavy Technical Rescue, Incident Management Systems, Course Development and Quality Assurance, and the new Chief for the Virginia Fire Marshal Academy. In my current position, I am still involved in many aspects of educating and serving Virginia's fire investigators.

During these past 40 years, I expanded my horizons by completing my associate's degree in fire science, with a major in investigations. I also picked up my professional qualifications in Firefighting, Fire Officer, Fire Instructor, Fire Inspector, Fire Investigator, and Public Education. Part-time employment included teaching for the State Fire Training Agency and the local community college, where I taught *investigations*, *inspections*, *fire service law*, and other fire-related courses. Today I am currently teaching the *Forensics of Fire Investigations* course at the Virginia Commonwealth University, School of Forensics.

I have been fortunate to work with an outstanding team of peers in the validation of IFSTA manuals, including the second and third editions of the *Introduction of Fire Origin and Cause* and the first edition of *Fire Investigator*. I was chosen to work on the Technical Working Group on Fire/Arson Scene Investigation in their creation of the *Fire and Arson Scene Evidence, A Guide for Public Safety Personnel*, which was a research report published by the United States Department of Justice, Office of Justice Programs, National Institute of Justice.

One of the greatest rewards and an unbelievable education has been to serve on the National Fire Protection Association, NFPA 921 Technical Committee, working on *the Guide for Fire and Explosion Investigations* text. For more than 16 years, I have served on this committee working alongside some of the greatest minds in the fire investigative field. These meetings can be rewarding, frustrating, infuriating at times, but regardless, always educational.

Through most of this time I was able to maintain a role in the volunteer fire service, working from firefighter through each and every rank until earning the position of Chief in my volunteer fire company. This opportunity allowed me to experience the aspect of completing the fire incident reports first hand and to see how difficult it could be at times and to know the value of the role of the firefighter and fire officer in making that determination of fire cause.

I hope that in sharing the sum of my experiences, volunteering, working, and teaching, I provide the motivation for the reader to conduct the investigation as a seeker of truth as well as provide the knowledge and skills necessary to come up with the accurate determination of the fire origin and cause.

ACKNOWLEDGMENTS

We, the author and the publisher, are proud to acknowledge the following reviewers who carefully reviewed the content within this book as it was in development and who offered insightful advice for its final form:

Eddie Bain
Investigation, Prevention and Public Education Program Director
University of Illinois
Illinois Fire Service Institute
Champaign, IL

Thomas Blair
Fire Science Technology Instructor
Weatherford College
Weatherford, TX

Steve Malley
Fire Academy Director
Weatherford College Regional Fire Academy
Weatherford, TX

Dr. James Nardozzi
Program Manager, Criminal Justice Department
Post University
Waterbury, CT

Marion B. Oliver
Director of Fire Protection Technology
Midland College
Midland, TX

Lt. Tom Peal
Fire Investigator/Instructor
Beckley Fire Department
Mountain State University
West Virginia

This book was a cooperative effort with Bobby Bailey writing Chapter 14, "Vehicle Fires," and Karl Mercer writing Chapter 19, "Interviews." Bobby has served both as a police officer and fire marshal and presently has taken on the reins of the Virginia Fire Marshal Academy as their Chief, and Karl Mercer, a retired police sergeant, was instrumental in the growth of the Virginia Fire Marshal Academy and now serves as the Deputy Chief.

Technical editing of the entire text was conducted by David Smith of Associated Fire Consultants in Tuscan, Arizona. Dave is a former police officer and prior president of the International Association of Arson Investigators. He has served on and chaired validation committees for the International Fire Service Training Association and is long-standing member of the National Fire Protection Association, Technical Committee on Fire Investigations (NFPA 921, *Guide for Fire and Explosion Investigations*).

My sincere and heartfelt appreciation goes to Senior Product Manager Jennifer Starr for her perseverance, prodding, and encouragement. Without her invaluable assistance this text might never have been written. It is just as clear that were it not for the encouragement of Billy Shelton, Executive Director of Fire Programs, I would never have considered this project. I also offer my appreciation to Dave Diamantes, author extraordinaire, whose advice I took and accepted for this assignment.

There are many others who offered assistance, guidance, support, and a needed bit of enthusiasm from time to time. Special thanks goes to the Henrico Division of Fire, Henrico Division of Police, Hanover Circuit Court, Hanover Fire & EMS, and my colleagues at the Virginia Department of Fire Programs.

I gratefully acknowledge the assistance of the staff, cadre, and students of the Virginia Fire Marshal Academy; and the assistance from Virginia State Police, Chief Arson Investigator, Lieutenant Ira Matney and First Sergeant Kelly Graham, along with their Special Agents, in setting up photo opportunities during their training events.

Special thanks goes to my wife, who not only excused me from many home chores, but also gave me encouragement and assistance during the creation of this text.

NOTE TO READERS

The sharing of ideas and input into the content of this text are always welcome. If you have any ideas for improvements, comments on content, or ideas of what you might want to see in future editions, please contact me at my e-mail address: Vafirelaw@aol.com.

FIRE AND EMERGENCY SERVICES HIGHER EDUCATION (FESHE)

In June 2001, the U.S. Fire Administration hosted the third annual Fire and Emergency Services Higher Education Conference at the National Fire Academy campus in Emmitsburg, Maryland. Attendees from state and local fire service training agencies, as well as colleges and universities with fire-related degree programs, attended the conference and participated in work groups. Among the significant outcomes of the working groups was the development of standard titles, outcomes, and descriptions for six core associate-level courses for the model fire science curriculum that had been developed by the group the previous year. The six core courses are *Principles of Emergency Services, Fire Protection Systems, Fire Behavior and Combustion, Fire Protection Hydraulics and Water Supply, Building Construction for Fire Protection,* and *Fire Prevention.*[1]

In addition, the committee also developed similar outlines for other recommended courses offered in the fire science programs. These courses included *Introduction to Fire and Emergency Services Administration, Occupational Health and Safety, Legal Aspects of the Emergency Services, Hazardous Materials Chemistry, Strategy and Tactics, Principles of Fire and Emergency Services Safety and Survival, Fire Investigation I,* and *Fire Investigation II.*

FESHE CONTENT AREA COMPARISON

The following tables provide a comparison of the course outcomes and the FESHE content areas with the content of this text.

[1] Since the establishment of these courses in 2000–2001, these courses are continually reviewed and updated by the FESHE committee on an annual basis to meet the changing needs of the fire service. The FESHE outline presented in this book reflects this most recent edition of the course outline and outcomes at the time of printing. For current information on this course, as well as other FESHE-related courses, please visit *http://www.usfa.dhs.gov/nfa/higher_ed/feshe/feshe_model.shtm.*

FIRE AND EMERGENCY SERVICES HIGHER EDUCATION (FESHE) COURSE CORRELATION GRID

Name:	Fire Investigation I
Course Description:	This course is intended to provide the student with the fundamentals and technical knowledge needed for proper fire scene interpretations, including recognizing and conducting origin and cause, preservation of evidence and documentation, scene security, motives of the firesetter, and types of fire causes.
Prerequisite:	Principles of Emergency Services, Building Construction for Fire Protection, Fire Behavior and Combustion, or instructor approval

Outcomes:		***Fire Investigation* Chapter Reference**
	14. Identify and explain the responsibilities of the fire department from a firefighter's perspective when responding to the scene of a fire, including the possibility of incendiary devices often encountered.	3
	15. Define criminal law and explain the constitutional amendments (Fourth, Fifth, Sixth, Eighth, Fourteenth) as they apply to fire investigations.	5
	16. Analyze the precedents set by constitutional law case studies that have affected fire investigations.	5
	17. Define and explain the common terms used in fire investigations.	Chapter key terms; Glossary
	18. Describe the basic elements of fire dynamics and how they affect cause determination including fire behavior, characteristics of fuels, and methods of heat transfer.	7
	19. Analyze the relationship of building construction on fire investigations including types of construction, construction, and finish materials.	Prerequisite; 11
	20. Evaluate fire protection systems and building services and discuss how their installation affects the ignition of fires in buildings.	9
	21. Discuss the basic principles of electricity.	10
	22. Explain the role of the fire investigator in recognizing health and safety concerns including potential hazardous materials awareness.	2
	23. Describe fire scene investigations and the process of conducting investigations using the scientific method.	7
	24. Explain how an investigator determines the point of origin in a room.	12
	25. Identify the types of fire causes and differentiate between accidental and incendiary causes.	13

(Continues)

Outcomes:		*Fire Investigation* Chapter Reference
	26. Describe and explain the basic procedures used for investigating vehicle fires.	14
	27. Identify the characteristics of arson and common motives of the firesetter.	16
	28. Identify and analyze the causes involved in line of duty firefighter deaths related to structural and wildland firefighting, training and research and the reduction of emergency risks and accidents.	Prerequisite; 2 and 15

Course Outline:		*Fire Investigation* Chapter Reference
	I. Emergency Responder Responsibilities and Observations	
	A. Responsibilities of the Fire Department	3
	B. Responsibilities of the Firefighter	3
	C. Responsibilities of the Fire Officer	3
	D. Observations When Approaching the Scene	3
	E. Observations Upon Arrival	3
	F. Observations During Firefighting Operations	3
	G. Identification of Incendiary Devices	3 and 16
	II. Constitutional Law	
	A. Criminal Law	5
	B. Constitutional Amendments	5
	III. Case Studies	
	A. *Michigan v. Tyler*	5
	B. *Michigan v. Clifford*	5
	C. *Daubert* Decision	22
	D. *Benfield* Decision	22
	E. *Kuhmo/Carmichael* Decision	22
	IV. Fire Investigations Terminology	
	A. Terms as They Apply to Structural Fires	11 and 12
	B. Terms as They Apply to Vehicle Fires	14
	C. Other Common Investigative Terms	Glossary
	V. Basic Elements of Fire Dynamics	
	A. Ignition	7
	B. Heat Transfer	7
	C. Flame Spread	7

(Continues)

Course Outline:		*Fire Investigation* Chapter Reference
	D. Burning Rate	7
	E. Fire Plumes	7
	F. Fire Analysis	7
	VI. Building Construction	
	A. Types of Construction	Prerequisite
	B. Building Materials	Prerequisite; 11 and 12
	C. Building Components	Prerequisite; 11 and 12
	VII. Fire Protection Systems	
	A. Extinguishment Systems	Prerequisite
	B. Detection Systems	Prerequisite
	C. Signaling Systems	Prerequisite
	D. Other Building Services	Prerequisite
	VIII. Basic Principles of Electricity	
	A. Basic Electricity	10
	B. Wiring Systems	10
	C. Common Electrical Systems	10
	IX. Health and Safety	
	A. Methods of Identification	2
	B. Common Causes of Accidents	2
	C. Common Causes of Injuries	2
	X. Fire Scene Investigations	
	A. Examining the Fire Scene	12
	B. Securing the Fire Scene	8
	C. Documenting the Fire Scene	18
	D. Evidence Collection and Preservation	8
	E. Exterior Examination	12
	XI. Determining Point of Origin	
	A. Interior Examination	12
	B. Area of Origin	12
	C. Fire Patterns	12
	D. Other Indicators	12
	E. Scene Reconstruction	12
	F. Point of Origin	12
	XII. Types of Fire Causes	
	A. Accidental	13
	B. Natural	13

(Continues)

Course Outline:		*Fire Investigation* Chapter Reference
	C. Incendiary	13
	D. Undetermined	13
	XIII. Vehicle Fires	
	A. Examination of Scene	14
	B. Examination of Exterior	14
	C. Examination of Driver and Passenger Areas	14
	D. Examination of Engine Compartment	14
	E. Examination of Fuel System	14
	F. Examination of Electrical System	14
	XIV. Firesetters	
	A. Characteristics of Arson	16
	B. Common Motives	16

FIRE AND EMERGENCY SERVICES HIGHER EDUCATION (FESHE) COURSE CORRELATION GRID		
Name:	Fire Investigation II	
Course Description:	This course is intended to provide the student with advanced technical knowledge on rule of law, fire scene analysis, fire behavior, evidence collection and preservation, scene documentation, case preparation, and testifying.	
Prerequisite:	Fire Investigation I	
Outcomes:		***Fire Investigation*** **Chapter Reference**
	1. Explain the rule of law as it pertains to arrest, search and seizure procedures, and their application to fire investigations.	5
	2. Recognize and interpret fire scenes common to various types of fires.	13
	3. Describe the chemistry of combustion and the relationship of atoms, elements, compounds, and organic compounds on fire.	7
	4. Explain the nature and behavior of fire including the effects of heat.	7
	5. Explain and identify the combustion properties of liquids, gases, and solid fuels.	7
	6. Identify and explain electrical causes of fires.	10
	7. List and explain the procedures for lifting fingerprints, evidence collection, and preservation.	17
	8. List and identify the make-up and use of incendiary devices, explosives, and bombs.	17
	9. List the procedures for documenting fire scenes, including sketching, photography, and report writing.	18
	10. Analyze fire-related deaths and injuries and describe methods of documentation.	15
	11. Identify the techniques for interviewing and questioning suspects and subjects.	19
	12. Explain the role of the fire investigator in courtroom proceedings including courtroom demeanor and testifying.	19
	13. Identify and list the sources and technology available for fire investigations.	17
	14. Identify and analyze the causes involved in the line of duty firefighter deaths related to structural and wildland firefighting, training, and research and the reduction of emergency risks and accidents.	15

(Continues)

Course Outline:		*Fire Investigation* Chapter Reference
	I. Rule of Law	
	A. Arrest Procedures	5, by reference
	B. Search and Seizure	5
	C. Warrant Searches	5, by reference
	II. Interpretations of Fire Scenes	
	A. Structure Fires	11
	B. Vehicle Fires	14
	C. Ship Fires	15
	D. Explosions	17
	E. Wildland Fires	17
	F. Hazardous Materials Fires	17
	III. Chemistry of Combustion	
	A. Atoms	7, by reference
	B. Elements	7, by reference
	C. Compounds	7, by reference
	D. Organic Compounds	7, by reference
	IV. Behavior of Fire	
	A. Heat	7
	B. Flame Plumes	7
	C. Sequence of a Room Fire	7
	D. Effects of Environmental Conditions	7
	V. Combustion Properties	
	A. Liquids	7
	B. Gases	7
	C. Solids	7
	VI. Electrical Causes of Fires	
	A. Wiring Systems	10
	B. Ignition Sources	10
	C. Investigation of Fires	10
	VII. Collection of Evidence	
	A. Photography Procedures	18
	B. Sketching Procedures and Techniques	18
	C. Fingerprint Lifting and Collection Techniques	8
	D. Preservation of Evidence	8
	VIII. Incendiary Systems	
	A. Basic Incendiary Devices	17, by reference
	B. Explosives	17, by reference
	C. Bombs	17, by reference

(Continues)

Course Outline:		*Fire Investigation* Chapter Reference
	IX. Documentation of Fire Scene	
	A. Sketches	18
	B. Photographs	18
	C. Incident Reports	3, by reference
	D. Log Sheets	18
	E. Investigation Report	23
	F. Chain of Custody	8
	X. Investigation of Fire-related Deaths and Injuries	
	A. Homicide Fire Investigation	15
	B. Scene Security	2, 3, 5 and 15
	C. Scene Examination and Search	15
	D. Scene Documentation	15 and 18
	E. Autopsy Report	15, by reference
	XI. Interview Techniques	
	A. Interviewing	19
	B. Questioning	19
	C. Advising of Rights	19
	D. Exceptions to the Rule	19
	E. Waiver of Rights	19
	XII. Courtroom Demeanor	
	A. Court Procedures	23
	B. Pre-trial Preparation	23
	C. Trial Exhibits	23
	D. Physical Appearance	23
	E. Testifying	23
	F. Court Decisions	5
	XIII. Court Decisions	
	A. *Daubert* Decision	5
	B. *Benfield* Decision	5
	C. *Kuhmo/Carmichael* Decision	5
	XIV. Sources of Information	
	A. Local	17
	B. State	17
	C. Federal	17
	D. Website	17

FIREFIGHTER SAFETY SECTION

FESHE OBJECTIVE 28 [Fire Investigator I]: Identify and analyze the causes involved in line of duty firefighter deaths related to structural and wildland firefighting, training, and research and the reduction of emergency risks and accidents.

FESHE OBJECTIVE 14 [Fire Investigator II]: Identify and analyze the causes involved in the line of duty firefighter deaths related to structural and wildland firefighting, training, and research and the reduction of emergency risks and accidents.

Each year, approximately 100 firefighters are killed and 100,000 are injured. The United States Fire Administration (USFA), the National Fire Protection Association (NFPA), and the International Association of Firefighters (IAFF) maintain statistics on these deaths and injuries. Collectively, these studies indicate the following:

Sudden heart attack is the most common cause of death of firefighters killed on the job. Heart attacks claim anywhere from 40 to 50 percent of the firefighters who die in the line of duty each year. A 10-year study by the NFPA, covering 1995 to 2004, found that 440 of the 1,006 line-of-duty deaths (43.7 percent) were related to sudden cardiac events, while many other deaths were the result of stress-related conditions such as strokes and aneurysms. Most of these victims had known heart diseases or had conditions that could have been diagnosed had the members undergone mandatory fitness-for-duty medical examinations, as recommended by the NFPA.

Not surprisingly, the most likely place for a firefighter to be killed or injured is at an emergency scene. Fireground deaths account for between 28 and 40 percent of firefighter deaths each year, with nonfire emergencies accounting for an additional 8 to 10 percent.

Approximately 25 to 35 percent of the firefighters who die each year succumb while responding to an alarm or returning from an alarm. Vehicle accidents cause many of these deaths. Research has identified tanker rollover accidents as accounting for a statistically significant number of firefighter fatalities over the years. Training deaths commonly account for between 10 and 20 percent of firefighter fatalities annually.

Age has been shown to be a risk factor for firefighters. NFPA studies have concluded that a firefighter over the age of 60 is three times more likely to die than a firefighter who is between 40 and 49. Older firefighters are more likely to die from heart-related causes, while younger firefighters are more likely to die from trauma.

While firefighter line-of-duty deaths and injuries can never be eliminated completely, fire departments can achieve positive results by adopting comprehensive occupational safety and health programs. These programs address a wide range of firefighter activities, including:

- Periodic medical examinations
- Physical fitness programs

- Health and wellness programs
- Accident prevention programs
- Effective training programs
- Acquisition of proper equipment
- Proper equipment maintenance programs
- Enforcement of operational procedures
- Risk management programs

The single most important measure that fire departments and firefighters can take to reduce the death and injury rate is to implement NFPA 1500, *Standard on Occupational Safety and Health Programs.**

For the fire investigator, there is no standard specifically addressing safety in this role. However, this is not to say that there are no recommendations for their safety. The same safety and health issues and occupational preventative programs would apply to the fire investigator as it applies to the firefighter.

Fire investigators with police powers share in the high stress and working conditions of their brothers in the local police force. This includes the hazard of doing interviews and searching out the criminal in the carrying out of their duties. Under such circumstances it would be best served for the fire investigators with police powers to also take advantage of any occupational safety and health programs provided to other police officers in that jurisdiction.

Utmost to all concerned is the fact that the most important person on that fireground, at that investigative scene, or in the search for the suspect is you. You, the public servant, have an obligation to take care of yourself, to put your safety first, and to make sure you make it home safely to be with your family and to provide for their welfare.

SOURCES

LeBlanc, Paul, and Rita Fahy, "U.S. Firefighter Fatalities for 2004," *NFPA Journal,* July/August 2005.

LeBlanc, Paul, and Rita Fahy, "U.S. Firefighter Fatalities for 2003," *NFPA Journal,* July/August 2004.

LeBlanc, Paul, and Rita Fahy, "U.S. Firefighter Fatalities for 2002," *NFPA Journal,* July/August 2003.

National Institute for Occupational Safety and Health, *Fire Fighter Fatality Investigation and Prevention Program—Annual Report: 200* (Atlanta, GA: NIOSH).

United States Fire Administration, *Firefighter Fatalities in the United States in 2003*, FA-283, Item # 9-0981 (Emmitsburg, MD: USFA).

United States Fire Administration, *Firefighter Fatalities: A Retrospective Study 1990–2000*, FA-220, Item # 9-0437 (Emmitsburg, MD: USFA).

*Portions of this section are created in part by J. Curtis Varone, author of *Legal Considerations for Fire and Emergency Services* (Clifton Park, NY: Delmar, Cengage Learning, 2007).

NATIONAL STANDARDS FOR FIRE INVESTIGATORS

Learning Objectives

Upon completion of this chapter, you should be able to:

- Understand the history of fire and why we investigate fire origin and cause.
- Describe and understand the National Consensus Standards and National Fire Protection Association's role in the fire service.
- Understand the requisite knowledge necessary before starting investigative training.
- Describe the difference between NFPA 1033, *Standard for Professional Qualifications for Fire Investigator,* and NFPA 921, *Guide for Fire and Explosion Investigations.*

CASE STUDY

As the first fire marshal to be hired by this jurisdiction, I faced a daunting task to be everything to everybody. There was neither state nor local fire code, so one had to be written by a committee of local business leaders and fire officials. There were no consistent public education programs for the public schools, so programs had to be created that could address the need without taxing limited time and resources. In addition, there had been no fire investigations to speak of other than an arrest or two by the sheriff's office in the preceding years.

A code was written, public education programs flourished with the assistance of volunteers, and over the next several years an average of 27 felony arrests for arson occurred. Pride dictates that I report that all arrests lead to felony convictions with the exception of one, and this was not a result of awe-inspiring talent so much as an overwhelming case of ethics: The thought of putting handcuffs on an innocent person led us to take to court only those cases that could be won.

As required by law, every fire marshal must provide a report to the county board of supervisors each year. That first report along with a board member's historical perspective led to a lesson in history, or at least in making new history. I read the report, and at the end of the report I read a list of each arrest and successful conviction. With only two more cases to report, one of the members of the board of supervisors interrupted me. Standing up, the board member looked incredulous and challenged me, stating: "Sir, before you came along there was no arson problem in our county." Silence followed, absolute silence, for as much as 20 to 30 seconds. Finally, another member of the board thanked the Fire Department for our fine report, and business of the evening continued.

Over the years the retelling of the events of that night has led to many chuckles. This was not an attack on an individual but a lesson in history. For years and years, there had been no arson arrests; thus, with no arrests, arson fires were not occurring. So, I guess you could say that history was made that first year.

INTRODUCTION

In this nation, the Constitution and Bill of Rights dictate the law of the land. As needs arise, legislatures create new laws to address various issues. The U.S. Supreme Court, using contested lower court decisions, interprets and defines the laws and how they are to be enforced across the country. Granted, this is an overly simplistic view, but it does get complicated and intricate because some specific court cases dictate many of the actions of the fire investigator.

There is no absolute way to ensure that the fire investigator's actions stay within various rules established by the courts, but the creation of the national consensus standard is the first step to address those concerns as well as many other fire investigator issues. Today the **National Fire Protection Association (NFPA)** provides a specific **standard** on fire investigator competencies called NFPA 1033, *Standard for Professional Qualifications for Fire Investigator.*

National Fire Protection Association (NFPA)
a private, nonprofit organization dedicated to reducing the occurrence of fires and other hazards

standards
model providing minimum mandatory training requirement to meet a professional competency in a form that can be referenced or adopted into regulations or law

■ Note
A professional qualification standard is a document designed to give the minimum qualifications for an individual to obtain to meet a job description.

A professional qualification standard is a document designed to give the minimum qualifications for an individual to obtain to meet a job description such as firefighter, fire officer, instructor, or in this case, investigator. An authoritative body whose expertise includes both technical as well as educational experts promulgates these documents. Professional qualification documents contain two primary features: prerequisite knowledge and prerequisite skills. These items are measurable, enabling the student or a training agency to evaluate training progress. The professional qualifications are minimum standards. It is the responsibility of individuals to constantly improve their knowledge and skills, taking them beyond these minimum expectations.

Many documents, publications, and texts on the market today are good resources for the investigator but are not necessarily standards. One such document is the NFPA 921, *Guide for Fire and Explosion Investigations.* Notice that this is called a *guide* and not a *standard.* The intent of this document is to provide recommendations for the safe and systematic investigation of a fire or explosion incident. However, it is not intended to be adopted as a mandate as are NFPA standards.

Regardless of the fact that NFPA 921 is a guide, it is a vital tool used to ensure the credibility of investigators who will be rendering an opinion in a fire case. As a guide, not every aspect of the NFPA 921 text need be followed in each and every case. But when a process is used contrary to the guide, the investigator should ensure that another authoritative document can substantiate that the process is scientifically valid and proper. This is important to establish the fact that the investigator is adhering to a standard of care, being reasonable and prudent in the application of science to the findings of the fire's origin and cause and the identification of the person or thing that was responsible for the incident.

To understand where we are today in fire investigations we need to see where we have been. The history of humankind is peppered with tragic events, many of which were catastrophic fires. These events have shaped our cities through safer construction and have shaped our laws, codes, and regulations.

HISTORY

■ Note
We need to learn from our failures or we are doomed to repeat them again.

History is not something that is just nice to know. History explains how something came to be or how we arrived at where we are today. History can also explain a failure in the past. We need to learn from our failures or we are doomed to repeat them again. In the fire service, like in so many other occupations, we have to embrace our past so that we can improve and enhance our future. From time to time in this text, I discuss the history of the fire service or history of fire disasters. In this chapter, we discuss how our history has led to the development of professional standards for the fire service, including fire investigators.

Accounts of destructive fires come from many nations, but in particular from European countries, especially England. London's history is marked repeatedly with devastating fires. This does not mean that London was any more dangerous than other cities were, but that more historical records have survived through the centuries to give a snapshot of the city's history.

The London Fire of 1212

On July 11, 1212, a fire destroyed a large part of London, including part of the famed London Bridge. The bridge construction was primarily of stone, but many structures placed upon the bridge were wood and had combustible contents. High winds also allowed the fire to skip across the River Thames, igniting structures on the other side and spreading the destruction even farther. Depending on the source, there were anywhere from 1,000 to 12,000 fatalities. History most frequently documents that there were 3,000 deaths.[1]

While the city still smoldered, officials met to see what could be done to make sure this did not happen again. To successfully accomplish this they needed to know what happened—essentially how the fire started. What factors led to the explosive spread of the fire? Most important, what led to the massive loss of life? These questions were answered by taking down eyewitness accounts, knowing about life at that time, and applying common sense. Although fire investigators as we know them today did not exist then, certain individuals used common sense and deductive reasoning to come up with fundamental answers to each of the questions.

Before the fire in 1212, London had established building codes that required stone construction and limited the use of combustible materials. However, citizens found the cost of importing stone from the countryside prohibitive. Furthermore, the London officials failed to enforce the provisions of the code, so Londoners continued to use wood primarily in building construction. Thatch roofs were also popular. The thick bundles of straw used in construction of these roofs, as shown in **Figure 1-1,** could easily be ignited and provided a tremendous fuel load that allowed a fire to spread quickly. A more comprehensive code was adopted following the 1212 fire reemphasizing the encouragement of stone construction among other things to slow down the spread of fire.

The Great Fire of London in 1666

More than 450 years later London had grown considerably. The year was 1666 and again fire ravaged the city, this time destroying more than four-fifths of the city's structures. Even though the fire was more extensive than the fire in 1212, the loss of life was significantly less. History only recorded six fatalities. As in earlier fires, documents give varying death statistics. There could have been more deaths but this will never be known. This low loss of life may have been a direct result of lessons learned from the London fire in 1212 as well as other

Figure 1-1 *Typical thatch roof showing the volume of combustible material which would easily allow a fire to spread more easily.*

fires over the years. The slow speed of the fire allowed most citizens not only to escape, but escape with their most valuable possessions. Some of this may be attributed to widening of roads and alleyways in comparison to the roads in the year 1212. At least some of the structures did meet the required building codes.

But this was not to say that London was safe. London Alderman Daniel Baker predicted in 1659 that the city would burn. In April of 1665, King Charles II approached the mayor of London with his concerns about London burning. Both of their issues dealt with their concern that the entire city would most likely be destroyed by fire. They based this belief on the vast wood construction used throughout the city. There were some thatched roofs still in the city, but most believe that did not play a big role in this fire. However, vast quantities of lumber, coal, heather, and flax were stored within the city limits.

jetties
projecting or overhanging parts of a building

Prior to 1666, the city was growing so fast that land became a rare commodity. Building owners searched for methods to get more square footage out of buildings placed on their limited parcels of land. One way to accomplish this was to make taller and taller buildings. To increase the square footage on each upper floor the builder would sometimes create overhangs on each of the upper floors. The parts of buildings that jutted out were referred to as **jetties**. **Figure 1-2** shows a typical jetty. Sometimes they would only extend a few inches and other times they could be measured in feet. There was no standard design, the jetty could just extend on the second floor and rise straight up for subsequent floors, and other times each floor would have its own jetty. Consider a six-story building with only a 6- or 7-inch jetty on each floor. It could extend over the street by as much as 3 feet or more when you get to the roof eaves. With the potential of this happening on both

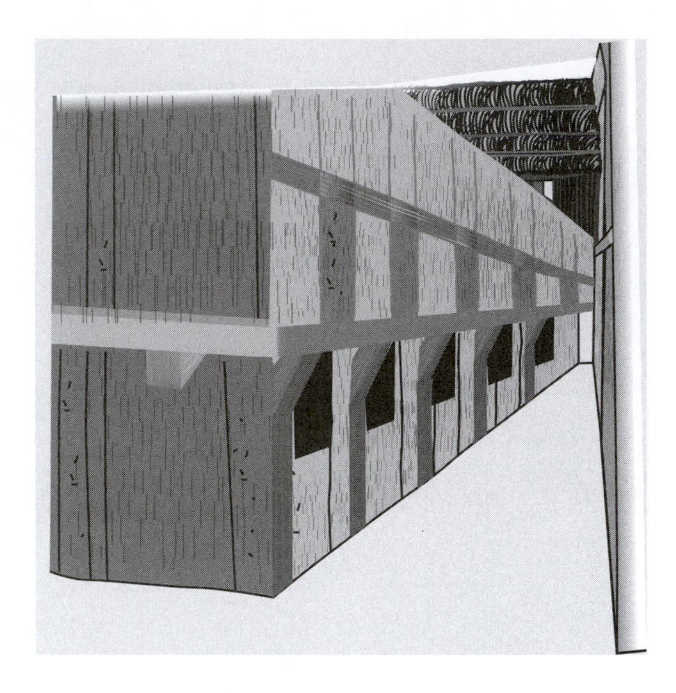

Figure 1-2 *Typical London structure with a jetty, which narrowed the gap between buildings allowing fire to spread more easily.*

sides of the street, any protection from fire spread provided by wider streets or alleys would be negated as the result of jetties lessening the gap between structures and thus allowing fire to easily jump from one side of the street to the other.[2]

Early Investigations

Once again, authorities were pressured to come up with measures to prevent future fires; yet they needed to know how fires started and to know how to prevent them. With no such profession as fire investigator in these early days, it became the responsibility of the community leadership to make decisions. For larger fires, sometimes research committees were formed to study the issue and come up with both blame and recommendations for changes to prevent further occurrences.

Statistically, these committees and other authorities were not always on the mark for placing blame accurately. All too often, they were overly influenced by the politics of the day. In the case of the fire of 1666, the research committee blamed the papists for starting the fire and for the subsequent destruction of London. *Papist* was a derogatory name for people who belonged to the Roman Catholic Church, the implication being that they would be more faithful to Rome than to England.

The rumor that the papists started the fire can be partially justified in light of the confession of a simple-minded French watchmaker named Robert Hubert. He first stated that he was an agent of the Pope and that he started the London fire in Westminster. He changed his story later to say that he started the fire on Pudding

Lane adjacent to a bakery. The fire started on Sunday, September 2, and extinguished on Wednesday, September 5. Mr. Hubert was hanged at Tyburn Gallows on September 28. This proved to be a gross mistake when it was later discovered that he did not arrive in London until two days after the fire had started.

The fire did in fact start at the bakery on Pudding Lane. This building was both a shop and residence owned and operated by Thomas Farynor, baker to King Charles II. It is now believed that Farynor failed to extinguish the cooking fires in the ovens the night before the great fire. Embers from the oven fires ignited nearby stacked firewood, which led to the destruction of both the shop and the house. Of interest is the fact that a law existed requiring that all fires be extinguished at night.

Note It has become obvious that we need individuals with specialized knowledge in science and technology to make accurate determinations on the origin and causes of fires.

For centuries, people depended primarily on speculative common sense to discover how and why fires start. However, in the last 50 years it has become obvious that we need individuals with specialized knowledge in science and technology to make accurate determinations on the origin and causes of fires. This has brought about the birth of the fire investigation profession. One link still exists between today's fire investigators and those seeking answers in earlier centuries: We still rely on the use of common sense as a tool to determine the cause and spread of fire.

Fires in America

The history of fire in the United States did not differ much from that in European countries. The Jamestown settlement suffered many small fires and a devastating fire that burned all the cabins in January 1608. Even though English laws required stone construction, the first American settlements did not have the resources or ability to build stone forts and structures. The materials available for settlers to use were wood and thatch. Chimneys were made of wood and coated with mud. Failure to maintain these chimneys led to many fires. As more and more brick masons arrived in the colonies and as the communities developed, more brick structures were built. However, not many settlers had the funds to afford this type of construction, and trees were abundant. So, wood remained the primary building material.

As the colonies grew, so did towns and cities. Just like in Europe, regulations, rules, and laws were established for fire safety. As in the days of Rome, fire watches were established, with individuals walking the streets at night keeping a watchful eye out for uncontrolled fires. Their duty was to wake the populace to come and fight the fire. If the fire was too close and out of control, citizens could assist others in saving personal belongings from the fire or, in the worse case, simply flee the fire.

Development of Fire Sprinkler Systems

It was not until the early 1800s that people began to install pipes to deliver water into a structure to put out fires. This practice started in 1806, in England, with an invention by John Carry in which water simply flowed out of holes in the pipes. This device did help control fires but resulted in just as much water damage as fire damage. Sprinkler heads developed in the mid-1800s improved the efficiency of dispersing water as it came out of pipes. In the late 1800s, Henry Parmelee and Frederick Grinnel developed a sprinkler head that activated individually only when it was affected by heat. Although sprinkler systems have gone through many changes to meet specific hazards, the basic sprinkler head is remarkably similar to the concept first developed by Grinnel; essentially the same systems are used today throughout the world.

Development of Electricity

About this same time, Thomas Alva Edison patented the incandescent light bulb and became a pioneer for electrical safety. When exploring which type of current to use—alternating current (AC) versus direct current (DC)—Edison chose DC because it posed fewer hazards. In contrast, George Westinghouse, another inventor and entrepreneur, was impressed with new developments in transformers and the induction motor, both of which relate to AC electricity. He concluded that AC would have more potential for electrical distribution.

The battle was on. In 1882, Edison built a steam-driven power station in New York City that was capable of providing enough power to light up 7,200 lamps. Not to be outdone, Westinghouse set up a multiple-voltage AC power system in Great Barrington, Massachusetts. This AC system produced 500 volts, stepped up to 3,000 volts for transmission, and stepped back down to 100 volts at different locations to power electric light bulbs. Westinghouse proved his point about the versatility of AC power.

An inventor and engineer Franklin Leonard Pope assisted Westinghouse in the development of the AC system. Even though he was working against Edison's idea of a world powered by DC, he was a former collaborator with Edison. They met early in their careers when they both worked in the telegraph industry, and they eventually formed a cooperation called Pope, Edison and Company in 1869. It was through their efforts that the stock ticker tape machine came into wide use in larger cities. Their partnership only lasted a short while, and they then went on their different ways.

In 1885 in Great Barrington, the Westinghouse AC lighting project was running while being refined. Pope ran a 2,100-volt overhead line into his basement workshop where he was experimenting with some well-used converters, devices that would take the voltage down to a more usable 52 volts. On a stormy October evening, the lights in Pope's home dimmed. He went to the basement to check his AC electrical converter. There he found that the window adjacent to the converter had blown open and rain had entered the basement, wetting both the converter and the basement floor. As he reached to close the window, he came in contact with the converter and, a victim of his own device, was electrocuted at the age of 55. An investigation into the incident showed that the primary feed had shorted out on the casing of the converter, creating the hazard that lead to Pope's death. Ironically, this event proved that Edison's concerns about the dangers of AC energy were correct.

Regardless, as you know, the world adopted the use of AC power. Edison's own electric company eventually became the General Electric Company and followed the pack, converting its power stations to AC energy.

The Adoption of Fire Safety Standards

Fire sprinkler systems and electrical systems were not consistent from one location to another in their design, configuration, or installation. The 1890s was a time of technological growth when new ideas emerged at a surprising rate. Along with this growth in technology came challenges to fire safety. Both industry in general and the insurance industry recognized the problem and stepped forward to work toward a common solution.

Electrical Code Several different electrical codes were in use, but they were neither consistent nor compatible. To solve this problem an assembly of key players met in New York on March 18, 1896. They represented the insurance industry, electrical engineers, electric companies, manufacturers, electrical associations as well as the telegraph and telephone industries. Representatives from Germany and England were also present.

National Electrical Code (NEC)
a consensus code for the safe installation of electrical components and appurtenances

During their organizational meeting, they decided to call themselves the Joint Conference of Electrical and Allied Interest [now known as the **National Electrical Code (NEC)**]. The primary topic for this meeting was the need for consistency and the virtues of adopting a common code for the nation and internationally. The group created a working committee chaired by Professor Francis B. Crocker of Columbia University. This committee chose what they felt to be the most suitable criteria from the various existing electrical codes and wrote a first draft. To obtain input from all areas of the electrical industry, copies were then sent to 1,200 individuals in North America and Europe.

In 1897, after three meetings of the Joint Conference, they unanimously approved the National Electrical Code. As a testament to the quality of the NEC, the National Board of Fire Underwriters, which had its own version of an electrical code, adopted the new code, issuing it as the NEC of 1897.

Automatic Sprinkler Code In some ways, the insurance industry was no different in 1895 than it is today: fiercely competitive. It was a sign of the times that insurance companies were willing to come together to seek solutions in regard to the fire sprinkler industry. Representatives from Factory Mutual Fire Insurance Company, New England Fire Insurance Exchange, Factory Insurance Association, and Boston Board of Fire Underwriters met with the Underwriters Bureau of New England in Boston. Also at this meeting was Frederick Grinnell of the Providence Steam and Gas Pipe Company. It is important to note that the Factory Insurance Association eventually became Industrial Risk Insurers (IRI), and the Providence Steam and Gas Pipe Company became Grinnell Fire Protection, both big players in today's fire safety arena along with Factory Mutual Fire Insurance Company (FM).

This initial meeting was to address the problems involved with the growing number of sprinkler installations. Around Boston and outlying areas, there were more than nine different codes, all vastly different from each other. With no single standard for installation or design, it was becoming difficult for local plumbers to install systems properly. More important, the group thought it was critical for new sprinkler systems to be tested to make sure they could do as they proposed, control fires. At this meeting, it was obvious that interested stakeholders all desired a single set of rules for the installation of fire sprinkler systems. Another meeting in December that included more insurance representatives proved that not only was one code needed, but a national association should be formed to oversee and administer the standards.

On March 18 and 19, 1896, members of the group met in New York City and released a document titled *Report of Committee on Automatic Sprinkler Protection*.

Formation of the National Fire Protection Association

The sprinkler document outlined a set of rules addressing fire sprinkler installation. More important, the group appointed a separate committee to create guidelines and a recommendation on an association that would administer the rules that were being established. On November 6, 1896, at the offices of the New York Board of Fire Underwriters, a group including representatives from the various insurance companies who were present at the 1895 meeting considered the rules established for sprinkler installation. Of particular significance, they also reviewed the articles presented for the creation of a new association. Article 1 established that the new association was to be called the National Fire Protection Association.

What started out more than 100 years ago as a small meeting of minds to find a fire safety solution has developed into the leading association addressing a wide range of fire safety issues. The NFPA has more than 75,000 members representing almost 100 nations. It produces 300 standards and guides, including NFPA 70, *National Electrical Code*, and NFPA 13, *Installation of Sprinkler Systems*.

■ Note
The NFPA has more than 75,000 members representing almost 100 nations.

So, what does this have to do with fire investigations? First and foremost, investigators determining fire origin and cause must also consider whether the failure to comply with established codes and standards has had any impact on the fire's cause, spread, or increase in the level of damage. It is important to know whether proper compliance with code would have prevented the fire, in which case this knowledge is a good tool to use in advocating for code enforcement and an opportunity to sell others on the benefit of the code. It is also important to know whether the code did not address the specifics that lead to the fire, in which case the next question is about whether the code needs to be changed to

■ Note **Needless to say, a good fire investigator should be knowledgeable of both fire and building codes.**

prevent similar fires in the future. Needless to say, a good fire investigator should be knowledgeable of both fire and building codes.

Second, the birth of the NFPA is relevant to fire investigators because this is the same entity that now maintains the Professional Qualifications Project. This project provides NFPA 1033, *Standard for Professional Qualifications for Fire Investigator.*

PROFESSIONAL QUALIFICATIONS

National Professional Qualifications Board the entity created to oversee the application of professional standards by training entities to ensure that they meet the professional qualifications

In 1972, the **National Professional Qualifications Board (NPQB)** came into existence. A group known as the Joint Council of National Fire Service Organizations created this organization. The organization was made up of representatives from the International Association of Fire Chiefs (IAFC), the Metropolitan Fire Chief's Committee of the IAFC, the International Association of Firefighters (IAFF), the International Association of Black Professional Firefighters (IABPF), National Volunteer Fire Council (NVFC), International Fire Service Training Association (IFSTA), International Society of Fire Service Instructors (ISFSI), Fire Marshals Association of North America (FMANA), International Association of Arson Investigators (IAAI), International Municipal Signal Association, and the NFPA, which was selected to serve as secretary to this board.

The desire of the group was to create a board that could administer accreditation, providing accountability and compliance to a set of professional qualification standards for the uniformed fire service. The Joint Council asked the NFPA to create the standards that would be administered by the NPQB. NFPA established four committees made up of experts on the various topics to develop professional qualification standards to meet firefighter, fire officer, fire service instructor, and fire inspector and fire investigator duties. The latter was to be a combination document that eventually addressed fire inspector, fire investigator, and fire prevention education officer. This committee met from 1973 through 1977 with the NFPA, adopting their document as the first edition of the NFPA 1031 standard.

Recognizing that the standards needed to be more comprehensive and that many fire investigators did all three duties, the NFPA divided the standard into three separate documents in 1986, creating NFPA 1031, *Standard for Professional Qualifications for Fire Inspector and Plan Examiner,* NFPA 1035, *Standard for Professional Qualifications for Public Fire and Life Safety Educator,* and NFPA 1033, *Standard for Professional Qualifications for Fire Investigator.* New committee members were assigned to each document, and they went forward and developed their own specialized standard to address the necessary competencies each discipline needed to be credible and accountable for that level of knowledge and skill.

Until this time, there was a career path for each standard: Before you could certify as a fire officer, you had to have firefighter certification. Likewise, before you could become an investigator or inspector, you had to be a certified firefighter.

The Joint Council recognized that there would be qualified individuals who would not necessarily be firefighters, so they decided to have an independent path allowing civilian entry into the investigator field. It should be noted that civilian entry implies individuals such as police officers or other investigator-type individuals who were not necessarily employed by a fire department. The new stand-alone professional standard for fire investigator was adopted by the National Fire Protection Association in 1987.

This was not the end of changes in the development of professional standards. In 1990, the NFPA took over the responsibility of appointing committees and the development of professional qualifications. Its first task was to establish the Professional Qualifications Correlating Committee. This group identified the need for the various professional qualifications and was involved in the appointment of personnel on these committees. Groups interested in the standards-making system for fire investigators can include representatives from the insurance industry, private investigators, federal law enforcement, state and local fire investigators, attorneys, and members of private industry. Each group has fair representation on the committee, ensuring that any one specific group does not have an unfair advantage in slanting the qualifications one way or another. As you can see, keeping a balance is important.[3]

Note

NFPA creates the standards. The role of the NPQB, more commonly known as ProBoard, is to evaluate entities to see whether they meet the standard.

NFPA creates the standards. The role of the NPQB, more commonly known as ProBoard, is to evaluate entities to see whether they meet the standard. Entities usually assessed are state training agencies. ProBoard sends in a team to assess the agency to observe how it develops and packages training programs and to validate that the entities meet the NFPA professional qualification standard. ProBoard reviews the delivery techniques, the material used in the training, the associated paperwork to ensure credibility, and proper tracking of individuals' training. It also reviews the testing process, both written and practical. When an entity, such as a state training agency, has been deemed to have met a specific NFPA standard by the ProBoard, students taking those programs can be nationally certified. Each program matching an NFPA professional qualification standard that meets the ProBoard measure of quality is then allowed to have students who complete the program successfully apply for national registry with the National Professional Qualifications Board.

To ensure that there were no critical doubts as to how the standards were developed, the NFPA adopted the guidelines established by the American National Standards Institute. This system ensures fairness, consistency, and credibility in the development of professional standards.

American National Standards Institute (ANSI)
coordinates the creation, development, promulgation, and use of voluntary consensus standards and guidelines

American Society for Testing and Materials (ASTM)
a trusted source for technical standards for materials, products, systems, and services

STANDARDS

The primary sources of consensus standards used by the fire service are the **National Fire Protection Association (NFPA), American National Standards Institute (ANSI),** and **American Society for Testing and Materials (ASTM).**

■ **Note**
A *consensus standard* means that the majority of those creating the standard are in agreement as to its content and meaning.

fairness test
a process adopted by the American National Standards Institute that requires openness, balance with a lack of dominance, consideration of all views and objections, an appeals process, and audit requirements; these requirements produce a level playing field for all involved

A *consensus standard* means that the majority of those creating the standard are in agreement as to its content and meaning. ANSI acts as the administrator and coordinator of voluntary consensus standards. Its goal is to promote openness, balance, consensus, and due process in the adoption and creation of standards. NFPA, as a member of ANSI, complies with these standards, ensuring that they meet the **fairness test** both here in the United States and around the world.

ASTM was created in 1898 in an effort to standardize the manufacture of steel for the railroad industry. Trains were getting larger and carrying heavier loads, and existing steel rails were failing. Some rail companies even imported steel from Britain to ensure quality. However, a scientific method of testing was developed, the rail industry accepted the standard, and the manufacturing of steel became consistent and had better quality assurance. ASTM went on to create standards for cement and concrete. Through a consensus system, it developed methods of testing these products that led to better products and safer structures. It was only a matter of time before ASTM entered the fire field.

Some standards for fire safety were established as early as 1904. Science was used to improve fire safety, and ASTM assisted with creating a standard that enhanced products and their uniform production. For example, Halon for fire extinguishing systems met the ASTM standards in 1958, and each subsequent product for fire extinguishing systems complies with an associated ASTM standard. ASTM was involved in forensics as early as 1970, and today almost every test used in forensic labs and in the field is associated with ASTM standards. Even though the average fire investigator in the field may not be familiar with the specifics of that standard, forensic procedures and processes are associated with ASTM.

The best way to understand today's standards is to know where they come from and how they evolved. The most important thing to know is that you can be involved in the process to make changes and improve the standards. To see how this is possible, it is important to understand the anatomy of an NFPA standard.

Anatomy of an NFPA Standard

All NFPA standards essentially follow the same format and order. When you first open the text, you see a brief outline of the specific standard that includes its name and dates associated with its adoption. The date the standard was issued for approval by the Technical Correlating Committee and the Standards Council is listed. The next important date listed is the standard's effective date and the fact that it supersedes all previous editions.

Also in the front of the document is a history of the development of the standard. The document lists the members of the Technical Correlating Committee on Professional Qualifications. Next is a list of individuals sitting on the standard's specific technical committee along with a description of the committee's scope.

The primary text of the document gives the specific requisite skills, requisite knowledge, and general information about the specific standard. Following is Annex A, which contains material that helps explain the specifics of the standard, clarifying the more complex topics. However, the annex is not part of the requirement itself.

job performance requirements
a statement that describes a specific job task, items necessary to complete the task, measurable and observable outcomes of the task

Annex B usually is the section titled "Explanation of the Standard and Concepts of **Job Performance Requirements**" (JPRs). The JPRs represent the knowledge and skills necessary for a specific profession such as fire investigator. Annex B explains any unfamiliar terms, such as *requisite knowledge*, in detail along with the performance outcomes of the position and how they are evaluated.

The contents of Annex C and Annex D vary. For example, the fire inspector Annex C contains sample job descriptions, but in the fire investigator standard, Annex C is the informational references. Annex D does not exist in the NFPA 1033, but in NFPA 1031 Annex D contains an outstanding tool, a table showing how different sections of the standard match various JPRs.

Following the annex sections are two small columns that show the sequence of events necessary for a committee to get a document published. One column gives the coding and definitions of committee members' classification. One important aspect of the creation of an NFPA document is the fact that no one specialty group has dominance over the creation of the document. Attempts are made to have the make-up of the committee evenly divided between the various groups such as users, manufacturers, insurance industry personnel, and special experts. This is done to ensure the creation of a fair and impartial document that will serve the entire fire service.

The last few pages of NFPA professional qualification standards or guides are an invitation for anyone to submit a proposal to the NFPA regarding these documents. All are encouraged to submit a change or an addition to a document that might enhance, improve, or correct it. The first opportunity is when NFPA does a call for proposals to amend existing documents. NFPA also entertains any proposals for creation of new documents as well. Each committee is tasked with preparing updates to the standard or guide, including any properly prepared proposals from outside the committee. When the document is ready and has received two-thirds approval via a written ballot by the committee, the document is printed and sent out as a **Report on Proposals (ROP).** The ROP is sent to every member of the NFPA and to any interested party upon request so that everyone has an opportunity to review what is proposed as the new, updated document.

Report on Proposals (ROP)
a report published by NFPA and sent to its membership and other interested parties to solicit feedback on proposed documents

Properly written proposals are compiled and forwarded to the appropriate committee for action. The committee reviews all proposals and acts upon each one. They can reject the proposal but must give a reason why. They can accept in principle, which means that they agree with the proposal but still feel the committee work is more appropriate, or accept only a part of the proposal. They can accept the submitted proposal, where the proposal is put in place in its entirety. Again, individuals do not have to be a member of the NFPA to make a change to any NFPA document.

After committee approval (two-thirds), the committee report on actions taken on all submitted proposals is published as a supplementary report called a **Report on Comments (ROC).** As before, it is sent to all NFPA members and any interested parties upon request. The public has one more opportunity to make a change on a document: If a proposal was not received favorably by the committee, an individual can petition to be heard at the annual meeting prior to membership vote on the document.. The last step is for the NFPA membership present at their annual meeting to vote on the document.

Report on Comments (ROC)
a report created by NFPA showing the committee actions taken on each proposal

What may appear as a complex, confusing process at first is in fact what makes the NFPA standards and guides so successful. This process of **peer review** gives the documents their credibility and strength. The next time you hear someone complain about a part of an NFPA standard or guide, remind that person that he or she has an opportunity to submit a proposal to change the error.

peer review
a process of review of written documents by persons in similar fields or professions in which anyone can make comments for changes and improvements, making the document more reliable and credible

NFPA 1033, Standard for Professional Qualifications for Fire Investigator The NFPA 1033, *Standard for Professional Qualifications for Fire Investigator,* was approved by the NFPA membership and was approved as an American National Standard. At the beginning of this document is the scope and purpose of the document. The scope identifies job performance requirements and the purpose of NFPA 1033 is to specify minimum qualifications resulting in a professional level of performance for investigators in both the private and public sectors. Simply said, the document addresses performance requirements and the qualifications to meet those requirements.

Note
Purpose of NFPA 1033 is to specify minimum qualifications resulting in a professional level of performance for investigators in both the private and public sectors.

Other general administrative requirements require that the investigator must be at least 18 years of age and a high school graduate or equivalent. Sometimes the obvious must be stated. An investigator must be able to testify in court and the individual must be credible if he or she is to sway a jury to see testimony as accurate and factual.

Background Investigation The **Authority Having Jurisdiction (AHJ)** should conduct a detailed background investigation on each perspective investigator. *AHJ* has many definitions depending on its context. In this case, the AHJ is that entity that has the legal authority to employ and empower individuals to conduct fire investigations within the political boundaries of that entity. This can be the elected officials of a city, county, or town, or it could be those appointed and given this authority from a political board or council, such as the county administrator or the fire chief.

Authority Having Jurisdiction (AHJ)
the governing body of any political entity who has the authority to adopt and enforce codes or laws

A full background investigation is required to ensure the candidate has no prior convictions (misdemeanors or felonies), no outstanding warrants, and no restraining orders. The background investigation is necessary even if the fire investigator candidate comes from the firefighter ranks. In years past, many fire departments did not do detailed background investigations for firefighters. In today's environment, even the best investigator who has done an outstanding job can have his or her expertise questioned if the person does not have their personal life in order. A fire investigator must be of good character and cannot have any convictions of moral turpitude, which are crimes contrary to justice, honesty, and good morals. This is an important requirement when you consider that this individual will take an oath in court to give expert testimony. If there is anything in the person's past that would give a jury reason not to believe the testimony, an arsonist could win the case and return to the street.

Continuing Education Completing a basic course in fire investigation that meets the NFPA 1033, *Standard for Professional Qualifications for Fire Investigator* is only the first step to becoming a fire investigator. The NFPA 1033 standard states that an individual must continually keep current on new court decisions, changes in codes, science, and methodology by attending additional training courses throughout the investigator's entire career.

International Association of Arson Investigators (IAAI)
a private, nonprofit organization dedicated to the professional development of fire and explosion investigators through training and research of new technology

National Association of Fire Investigators (NAFI)
an organization whose primary purpose is to increase knowledge and improve skills of persons involved in fire and explosion investigations

Groups such as the **International Association of Arson Investigators (IAAI)** and the **National Association of Fire Investigators (NAFI)** host a variety of seminars that address almost all aspects of the fire investigation field. State or province associations such as chapters of the IAAI also put on training events. These events are usually good at identifying training topics that address local problems or issues. In many areas, local and regional associations also sponsor training.

Some private entities such as SIRCHIE host training to address the proper use of their evidence collection kits, specialized equipment, and supplies. NFPA puts on seminars on systems and code applications that can also be of benefit to the fire investigator who is pursuing continuing education. Even fireworks manufacturers give seminars on shooting fireworks and proper storage.

If the AHJ or certifying agency requires a minimum number of training hours during a given cycle, such as 40 hours in 24 months, it should be remembered that this is just the minimum requirements. In addition to organized training, investigators can enhance their knowledge and expertise by reading periodicals on fire investigations, taking Internet courses, or doing research on recent court cases. With the Internet and information technology available today, the learning opportunities are limitless.

By attending organized training events, especially those put on locally and regionally, the investigator complies with another section of the standard: Maintaining a liaison with fellow investigators and investigative associations is also an NFPA 1033 mandate. It is clearly recognized that interaction between professionals creates an atmosphere in which the exchange of ideas and concepts can take place.

requisite knowledge
basic knowledge a person must possess to perform an assigned task

Requisite Knowledge and Skills The **requisite knowledge** criteria in the standard makes it clear what knowledge is necessary to carry out the duties and tasks of the fire investigator. This is the minimum knowledge and understanding an individual must acquire to be qualified to carry out that task. It enables educators to design training programs that give clear and concise guides for the lesson plans. In addition, certifying agency can be sure which skills and knowledge must be included in the test process. (See **Figure 1-3.**)

4.2.4 Interpret burn patterns, given standard equipment and tools and some structural or content remains, so that each individual pattern is evaluated with respect to the burning characteristics of the material involved.

(A) Requisite Knowledge. Fire development and the interrelationship of heat release rate, form, and ignitibility of materials.

(B) Requisite Skills. Interpret the effects of burning characteristics on different types of materials.

Figure 1-3 *Sample of a NFPA 1033 Requisite Knowledge and Requisite Skill.*

Source: Reprinted with permission from NFPA 1033–2009: *Standard for Professional Qualifications for Fire Investigator,* Copyright © 2009, National Fire Protection Association, Quincy, MA. This reprinted material is not the complete and official position of the NFPA on the referenced subject, which is represented only by the Standard in its entirety.

requisite skills
the skills a person must have to perform an assigned task

It is also important to know that an individual can physically carry out a skill to accomplish a given task. The **requisite skills** as they are listed are not specific instructions, but they identify that an individual must have appropriate communication skills, both oral and written. Other examples are that the individual must be able to operate a piece of equipment or employ a specific technique. For example, the fire investigator must be able to use a 35-mm camera, as shown in **Figure 1-4.**

These requisite knowledge and skills requirements provide a level of measurement that allows an individual to demonstrate his or her ability to meet the minimum qualifications for the position of fire investigator.

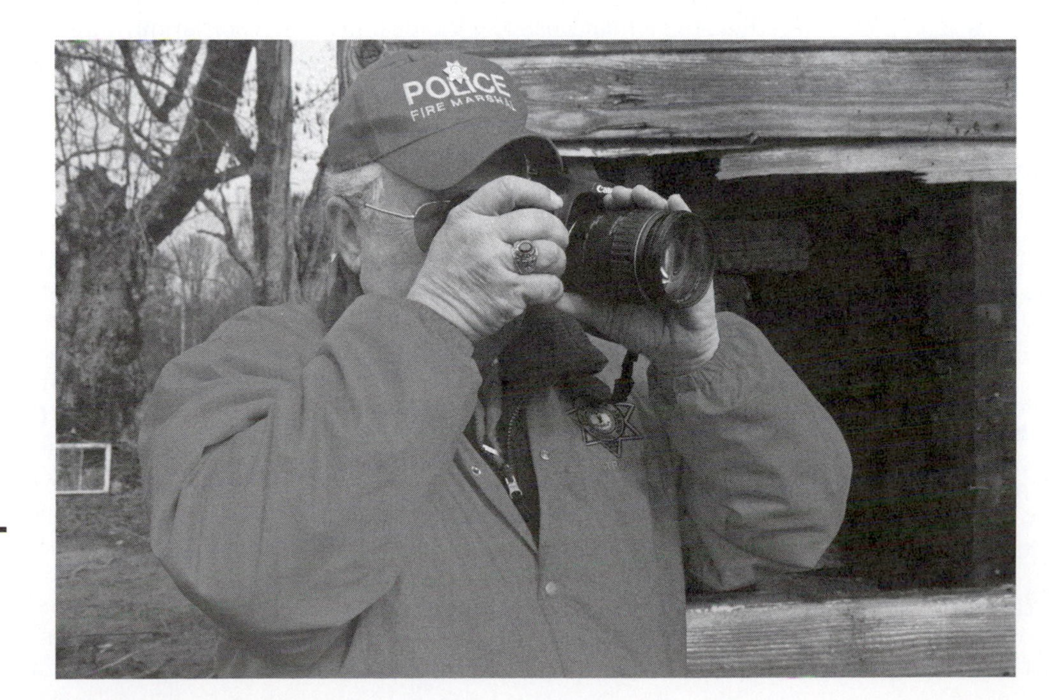

Figure 1-4
Requisite skills include ability to use a 35 mm type camera.

Additional Knowledge and Skills In addition to the life skills that an individual brings to the job, specific levels of training and general knowledge should be obtained to be an efficient and productive investigator. Much of this knowledge can be obtained in formal training events such as college-level physics and chemistry courses. Although the NFPA 1033 standard does not specifically mandate college classes, it does require that the individual be knowledgeable in chemistry and physics.

State and federal regulations may also require the investigator to have hazardous materials training. At the very least, the NFPA 1033 standard requires investigators to take a Hazardous Materials Awareness training course and understand the standard as per NFPA 472, *Standard for Competence of Responders to Hazardous Materials/Weapons of Mass Destruction Incidents.* The federal mandate requires investigators to be certified at a certain level to work in a hazardous atmosphere, which describes many fire scenes.

■ Note
A fire investigator cannot accurately make a determination of the fire cause unless the individual understands building construction and building systems.

A fire investigator cannot accurately make a determination of the fire cause unless the individual understands building construction and building systems. In some cases, these systems must be eliminated as the potential heat or energy source that was involved in the ignition sequence. Unless the investigator understands these systems, it would be unlikely that he or she could effectively interpret whether they were or were not involved.

Local community colleges or state training agencies usually offer specialized classes in building construction, which may include topics on building systems. It is imperative for the student investigator to endeavor to obtain this training to increase his or her level of knowledge. Another way to increase general knowledge is to develop a relationship with local heating, ventilation, and air conditioning (HVAC) and utility contractors. Some will welcome the opportunity to pass on knowledge and skills. This relationship can also be handy should the investigator ever need the help of these individuals in the future with an investigative incident.

Fire Marshal
the predominant title used to describe government employees who have the primary responsibility to investigate fire and explosions, enforce a fire code, and provide public fire and life safety education

NFPA 1037, *Standard for Professional Qualifications for Fire Marshal*
a standard outlining the minimum professional qualifications for fire marshals or equivalent positions

NFPA 1037, Standard for Professional Qualifications for Fire Marshal In the United States, the title **Fire Marshal** is becoming the title of choice for persons with fire prevention–type duties, but there is no clear definition of what a fire marshal is and what they do. To complicate this further, the fire marshal's duties can vary from one jurisdiction to the next, even within the same state. In some jurisdictions, fire marshals conduct fire prevention inspections for code enforcement, perform full fire investigations, and handle public education events. In neighboring jurisdictions, the fire marshal can be responsible for code enforcement only. In some areas, investigative duties of the fire marshal can be limited to conducting the origin and cause scene work, and then turning the case over to the local police if further investigation is warranted, whereas other jurisdictions give their fire marshals full police powers to investigate all fires and handle any criminal action associated with the fires.

The creation of the **NFPA 1037, *Standard for Professional Qualifications for Fire Marshal,*** is a first step in providing a tool for the future. A professional standard may not solve this inconsistency, but it is the first step in giving AHJs something to use to define what its fire marshals can or should be doing. The standard is designed to be used in whole or in part at the discretion of the AHJ and citizens. This professional qualification standard was created in 2006 and adopted in 2007, so it is a new tool for most localities and states.

The scope of this standard is to identify a professional level of performance for a fire marshal position. The document provides definitions of minimum qualifications and job performance requirements for fire marshal duties.

Some of the general requirements for a fire marshal are the ability to accurately and efficiently create correspondence and maintain the appropriate files in accordance with local policies. Also required is the ability to create and efficiently maintain a budget, establish a strategic plan, and manage all available resources including personnel.

Specific topics include items such as conducting risk analysis with solutions, maintaining a good relationship with community groups, and creating media communication strategies. Dealing with the media is critical to enable the office to disseminate public safety information such as seasonal safety.

The standard addresses inspections programs, plans review, public education, and investigations. It concentrates on general leadership roles in each of these areas because the skills to accomplish each type of task is already addressed in other standards, such as NFPA 1033, *Standard for Professional Qualifications for Fire Investigator.*

Importance of Professional Qualifications for Investigators

Specialists in fire investigations are employed in most government entities in the United States today. Fire investigators at the federal level exist in several agencies including the Bureau of Alcohol, Tobacco, Firearms and Explosives (BATF), the Federal Bureau of Investigation (FBI), and the Department of Defense (DoD) in each branch of the military. The Department of Homeland Security has several investigators including individuals in the U.S. Coast Guard. States employ fire investigators in their state police divisions or state fire marshal's offices, and most major municipalities assign investigative authority to some of their own personnel.

With so many investigators, there is a justifiable concern that laws will be inconsistently and inappropriately applied and simple concepts of fundamental science will be improperly applied. Regretfully, some court cases resulting in wrongful convictions have proved this point. Therefore, it is critical that we take steps necessary to ensure that all fire investigators are knowledgeable about all federal and state statutes and associated case law. It is also critical that investigators are trained to apply basic fundamental concepts of science and forensics properly.

How can we be sure that federal, state, and municipal investigators are knowledgeable on all statutes and case law? How can we be sure that investigators are current on the science of fire investigations? How can we be sure that the level of investigator professionalism will be consistent? The first step is to create a standard that lays out the minimum knowledge and skills necessary to carry out a fire investigation. This is exactly what has been done in the NFPA 1033, *Standard for Professional Qualifications for Fire Investigator.*

A standard shows us a level of knowledge and skills, but it does not necessarily show someone how to investigate a fire effectively and efficiently. There are many ways to accomplish this task, and many texts have been written on the subject over the years.

Remember that only a small percentage of all fires are actually investigated by someone assigned full time to an investigative office. More often than not, the fire officer in charge of a scene is responsible for completing the fire report. If using the National Fire Incident Reporting System, or a state version of this software, the report will include an entire section on the origin and cause of the fire. This makes the first responder also the first investigator on the scene and responsible for the initial analysis. Therefore, these standards should be of interest to any fire officer or firefighter who will take charge of a fire scene.

STANDARDS VERSUS GUIDES

A dictionary defines *standard* as something that is established by an authority or something to serve as a model. A standard is something that is uniform such as a rule or a measure of quality.[4] NFPA **Standards** are designed to be used as models or minimum qualifications. If an AHJ (a governmental entity) so desires, it can adopt an NFPA standard and mandate compliance, for example, adopting NFPA 1, *Uniform Fire Code.* The AHJ could be a state agency that adopts a professional qualification such as NFPA 1033, *Standard for Professional Qualifications for Fire Investigator.* This standard then becomes the minimum mandated knowledge and skills necessary for an individual to be certified as a fire investigator at the state level.

■ Note

A guide is something that can lead or direct but not necessarily mandate.

guide
a nonmandatory document giving advice and recommendations on how to carry out a task

NFPA 921, Guide for Fire and Explosion Investigations
a document designed to assist individuals investigate fire and explosion incidents in a systematic and efficient manner

In contrast, a **guide** is something that can lead or direct but not necessarily mandate. In **NFPA 921, *Guide for Fire and Explosion Investigations,*** it states that the document is to be used as a guide, giving recommendations for the safe and systematic investigation of fires. This implies that it is not necessarily the only way to investigate a fire, but just one way to do so.

Many entities and courts might use NFPA 921, *Guide for Fire and Explosion Investigations,* as the benchmark for challenging cases in court. As a guide, this document recommends actions to take at a fire scene that in all probability will help the investigator come up with a sound conclusion in the investigation. The NFPA 921 document also clearly states that deviating from the procedures recommended in the text is not necessarily wrong or inferior. If deviating from the text, the investigator simply needs to be prepared to justify his or her actions. This justification needs to be sufficient so that a jury could understand the actions.

For example, the text recommends that the flow of the fire scene examination should be from the least damaged area to the most damaged area. This procedure is great for teaching new investigators and as a way for experienced investigators to remain consistent in all their scene exams. However, an investigator may decide to start a scene exam from the top and work down or do a clockwise search of the scene at each level, starting at the lowest level of the scene. This procedure, although different from what is recommended, should not imply that the fire investigation was not systematic or thorough.

NFPA 921, GUIDE ON FIRE AND EXPLOSION INVESTIGATIONS

Every investigator and investigative student must be familiar with NFPA 921 in its entirety. The NFPA 921 text was not designed to be adopted by an AHJ, but was instead designed as a guide. The NFPA 921 committee made a great effort in writing the scope and purpose of the document to make sure that readers understand it is a guide. The intent of the document is to give the reader a method to follow for performing a systematic and effective investigation.

■ Note **The NFPA 921 committee made a great effort in writing the scope and purpose of the document to make sure that readers understand it is a guide.**

There are approximately 28 members on the NFPA 921 committee, with an additional 20 alternate members. Everyone is involved in composing, debating, recomposing, and putting forward proposed text for a consensus vote. The professionalism and wealth of knowledge present at any given meeting are phenomenal. The cross-section of backgrounds of committee members is one of the features that make this document so successful. Scientists, engineers, lawyers, and field fire investigators from both the private and public sectors help to keep the document in line with the needs of the reader.

To ensure drafting of the best document possible the committee occasionally seeks outside specialty experts who work in areas such as electrical fires or wildfire investigations. As the meetings move across the country to various meeting locations, guests and visitors attend the meetings. They are invited to give opinions and suggestions, but only committee members can vote.

NFPA documents have international impact; thus, the NFPA encourages people from other nations to contribute and serve on committees. Presently, an investigator with the London Fire Brigade has contributed valuable input to NFPA 921. Knowledge of investigative issues used in the United Kingdom has given great insight to the committee. Although some issues are quite unique to that nation, such as automotive systems and electrical circuits, as a basic rule, concepts, procedures, and issues encountered by investigators in the United States are no different from those encountered in other nations.

■ Note **The NFPA 921 committee incorporates an international flavor on legal sections as often as possible.**

Several representatives from Canada have also sat in on committees. They experience very similar problems with potential hazards, but they have a different legal system. The NFPA 921 committee incorporates an international flavor on legal sections as often as possible. Although references to U.S. case law and the U.S. Constitution are made, in most cases similar legislative rules apply in other developed nations. Regardless, investigators must be familiar with national and local laws to ensure proper compliance with legal procedures.

However, NFPA 921 is just one document and cannot give you all the tools you need to become a successful fire investigator. You must learn your strengths and weaknesses, and from there choose texts and learning experiences to improve your ongoing education. It is critical to make sure that you are using up-to-date documents and materials in your education. For example, an earlier version of NFPA 921 dedicated sections of its content to correcting misconceptions that were created by earlier textbooks and articles. Basic concepts, such as spalling, crazing, and annealed springs, which are discussed later, were portrayed as incendiary indicators in the earlier documents. All of these indicators can occur in an accidental fire today. NFPA 921 not only pointed out these misconceptions, but later issues strive to include good science and technology to educate the reader properly. Thus, documents such as NFPA 921, which is continually updated and changed as technology and the environment changes, are necessary tools to use in conducting a fire investigation.

The best tool, however, that investigators can use in conducting the simplest investigation or examining complex fire scenes is *common sense*. There are many definitions of this term, but the one that fits best in these circumstances is “the ability to decide what is sound and prudent.” If something defies logic, trust your instinct and delve into the topic further by referring to the NFPA 921 text. One of two things will happen: You will learn something new, adding to the sum of your experiences, or you will learn that the information is worthless and of little use. Either way, you will have benefited.

Science

good science
science that can be verified by independent means that meet accepted norms of the scientific community

junk science
untested scientific theory or faulty scientific data sometimes used to promote a private agenda or theory

crazing
glass with small cracks that go in all directions where the cracks do not go all the way through the glass

One goal of the NFPA 921 committee is to include **good science** in the document. Good science is science that is reproducible and based on common truths and that comes from scientific knowledge. Of course, the term *science* can mean many things and covers a broad spectrum of knowledge, from mathematics to paleontology. However, when conducting fire investigations it is essential that investigators use sound scientific methodology and avoid **junk science.** Use of the term *junk science* has become universal as a result of the Supreme Court ruling on the *Daubert v. Merrell Dow Pharmaceuticals* case. Junk science references testimony that is based on faulty or unsound principles that have not been tested or accepted in the scientific or investigative communities. Junk science is also sometimes referred to as bad science.

To ensure that you are getting up-to-date scientific knowledge always look at the date of the text or reference material you are using. One reason to do so, for example, is that several older texts indicate that the appearance of crazed glass can be an indication of the presence of an accelerant. It was contended that the rapid heat buildup would create small cracks in the glass, as shown in **Figure 1-5.** However, this is an incorrect assumption. An artist who uses a kiln to make pottery can tell you that spraying water on the surface of hot glazed pottery, similar in characteristics to glass, creates a **crazing** effect, creating small, fine cracks in the surface

Figure 1-5 *Small piece of glass from a screen door showing the intricate cracks in the glass commonly referred to as crazing. This is an indication of rapid cooling, such as would occur with the application of water on hot glass.*

■ **Note**
This is not to say that older texts have no value—only that you need to be knowledgeable on what is good science and what is not.

■ **Note**
Investigators should also develop a network with others in the investigative field because these other individuals will become valuable resources who are able to share their knowledge and skills, which leads to everyone becoming better investigators.

of the item, yet no accelerants were used to heat the glass in the kiln. Common sense should prompt you to challenge the conclusion that the fire was incendiary because the glass was crazed. You should look for further knowledge on the topic. A quick search on the Internet can point you to up-to-date articles on controlled laboratory testing using good science.

An article by John Lentini of Applied Technical Services titled "Behavior of Glass at Elevated Temperatures" shows that the heating of glass alone does not cause crazing.[5] In fact, it is the rapid cooling of glass that creates the crazing effect. The logical interpretation is that fire suppression water could easily create the crazing of glass, not accelerants.

This is not to say that older texts have no value—only that you need to be knowledgeable on what is good science and what is not. As the NFPA 1033 standard mandates, investigators should further their education through classes and seminars, which should increase their scientific knowledge. Investigators should also develop a network with others in the investigative field because these other individuals will become valuable resources who are able to share their knowledge and skills, which leads to everyone becoming better investigators.

ETHICS

It is essential that every investigator, both public and private, possess strong ethics. The International Association of Arson Investigators (IAAI) has a code of ethics that best describes what investigators must know and do to be ethical in carrying out their duties. (See **Figure 1-6.**)

International Association of Arson Investigators

Code of Ethics

I. I will, as an arson investigator, regard myself as a member of an important and honorable profession.

II. I will conduct both my personal and official life so as to inspire the confidence of the public.

III. I will not use my profession and my position of trust for personal advantage or profit.

IV. I will regard my fellow investigators with the same standards as a hold for myself. I will never betray a confidence nor otherwise jeopardize their investigation.

V. I will regard it my duty to know my work thoroughly. It is my further duty to avail myself of every opportunity to learn more about my profession.

VI. I will avoid alliances with those whose goals are inconsistent with an honest and unbiased investigation.

VII. I will make no claim to professional qualifications which I do not possess.

VIII. I will share all publicity equally with my fellow investigators, whether such publicity is favorable or unfavorable.

IX. I will be loyal to my superiors, to my subordinates, and to the organization I represent.

X. I will bear in mind always that I am a truth-seeker not a case maker; that it is more important to protect the innocent than to convict the guilty]

Figure 1-6 *The International Association of Arson Investigator's (IAAI) Code of Ethics is an excellent reminder on why we investigate fires. (Courtesy of International Association of Arson Investigators.)*

Note
They must be truth seekers and not case makers.

The last article of the IAAI *Code of Ethics* sums up the most important task assigned to fire investigators: They must be truth seekers and not case makers. This means that investigators keep an open mind and do not go forward into the investigation with any preconceived ideas about the case.

A very difficult aspect of the investigation is knowing that most likely someone else will also look at the same scene, and based on their expertise, they will examine the evidence and come up with an independent area of origin and fire cause. It is critical that steps be taken to preserve all evidence, even indicators that were discounted or discarded, so that other investigators can come up with their own conclusions.

Sometimes the findings of both investigators match, and sometimes they do not. Regardless, it is an ethical investigator who keeps an open mind. Investigators must judge the findings of the investigation and not judge the other individual just because they disagree professionally.

SUMMARY

Knowing the history of fire prevention is critical to preventing future disasters. The fire service must learn from its mistakes to prevent repeating them. This was true through history in both the United Kingdom and the United States where laws and codes to address fire safety issues had to be created to keep the general public safe from the ravages of fires. Even in colonial times, our predecessors had to examine fires that occurred and apply common sense to keep from having similar fires. They were our earliest investigators.

As technology advanced, so did our ability to see what was going right as well as what was going wrong. For years, electricity has been a leading cause of fires in the United States. If Thomas Edison had gotten his way and we were now using direct current, no doubt there would be fewer fires and fewer accidental electrocutions. Nevertheless, even Edison had to concede that alternating current was the only technology that allowed electricity to be delivered to so many locations at such a low cost.

In the process of advancements, we have found that we needed national consensus standards. With great foresight, the designers of the electric code saw the need for a national entity to oversee the code process, which was the birth of the National Fire Protection Association. Through the peer-review NFPA and ANSI process, we have a means of creating professional qualifications for uniformed fire service personnel. The creation of NFPA 1033, *Standard for Professional Qualifications for Fire Investigator,* has instilled professionalism in the fire investigative field. The NFPA 1033 standard provides a list of requisite skills and knowledge necessary to carry out the tasks of fire investigator.

Anyone can submit a recommendation to make changes to any NFPA document. The process is allowed and encouraged by the NFPA and the individual committees. In the back of all NFPA standards and guides, you can find the form for submitting proposals on NFPA technical committee documents, a completed sample form, and directions for submittal. The most important tool for success in making a change is to make sure that the form is filled out completely, listing the verbiage in question and a complete recommendation for replacement along with justification of why the reader wants to make that change.

In addition to professional qualification standards, which are developed to be adopted by the Authority Having Jurisdiction, the NFPA also develops guides that are specifically to give direction and guidance. NFPA 921, *Guide for Fire and Explosion Investigations* is an outstanding document that provides assistance in conducting safe and systematic fire and explosion investigations. The guidelines are not intended to be followed in every case, but any deviation from the NFPA 921 text need only be justified.

KEY TERMS

American National Standards Institute (ANSI) Coordinates the creation, development, promulgation, and use of voluntary consensus standards and guidelines.

American Society for Testing and Materials (ASTM) A trusted source for technical standards for materials, products, systems, and services.

Authority Having Jurisdiction (AHJ) The governing body of any political entity who has the authority to adopt and enforce codes or laws.

Crazing Glass with small cracks that go in all directions where the cracks do not go all the way through the glass.

Fairness test A process adopted by the American National Standards Institute that requires openness, balance with a lack of dominance, consideration of all views and objections, an appeals process, and audit requirements. These requirements produce a level playing field for all involved.

Fire Marshal The predominant title used to describe government employees who have the primary responsibility to investigate fire and explosions, enforce a fire code, and provide public fire and life safety education.

NFPA 1037, *Standard for Professional Qualifications for Fire Marshal* A standard outlining the minimum professional qualifications for fire marshals or equivalent positions.

Good science Science that can be verified by independent means that meet accepted norms of the scientific community.

Guide A nonmandatory document giving advice and recommendations on how to carry out a task.

NFPA 921, *Guide for Fire and Explosion Investigations* A document designed to assist individuals investigate fire and explosion incidents in a systematic and efficient manner.

International Association of Arson Investigators (IAAI) A private, nonprofit organization dedicated to the professional development of fire and explosion investigators through training and research of new technology.

Jetties Projecting or overhanging parts of a building.

Job performance requirements A statement that describes a specific job task, items necessary to complete the task, measurable and observable outcomes of the task.

Junk science Untested scientific theory or faulty scientific data sometimes used to promote a private agenda or theory.

National Association of Fire Investigators (NAFI) An organization whose primary purpose is to increase knowledge and improve skills of persons involved in fire and explosion investigations.

National Electrical Code (NEC) A consensus code for the safe installation of electrical components and appurtenances.

National Fire Protection Association (NFPA) A private, nonprofit organization dedicated to reducing the occurrence of fires and other hazards.

National Professional Qualifications Board The entity created to oversee the application of professional standards by training entities to ensure that they meet the professional qualifications.

Peer review A process of review of written documents by persons in similar fields or professions in which anyone can make comments for changes and improvements, making the document more reliable and credible.

Report on Comments (ROC) A report created by NFPA showing the committee actions taken on each proposal.

Report on Proposals (ROP) A report published by NFPA and sent to its membership and other interested parties to solicit feedback on proposed documents.

Requisite knowledge Basic knowledge a person must possess to perform an assigned task.

Requisite skills The skills a person must have to perform an assigned task.

Standards Model providing minimum mandatory training requirement to meet a professional competency in a form that can be referenced or adopted into regulations or law.

REVIEW QUESTIONS

1. What is the role of NFPA in the standards-making process?
2. What is the meaning of consensus standards and why are they important?
3. What is the difference between a standard and a guide?
4. Describe the Professional Qualifications System.
5. Define *requisite knowledge* and *requisite skills*.
6. What can you as an individual do to improve consensus standards?
7. If an investigator has already proven his or her abilities by being a nationally certified investigator, why should this person be required to obtain recertification training?
8. What is the role of ANSI in the standards-making process?
9. Name two national professional organizations that provide investigative training on a regular basis for their membership.
10. What is meant by the term *junk science*?

DISCUSSION QUESTIONS

1. Why should NFPA 1033 dictate that the minimum qualifications of an investigator be a high school or equivalent diploma and that the person be at least 18 years of age?
2. What does it mean to be a seeker of truth? Why is this so important?

ACTIVITIES

1. Go to the local library and ask for any documents on fires that have occurred in the past in your jurisdiction. Research the documents and consider whether anything that happened in the incidents could repeat today. What recommendations would you make to prevent a repeat of the events?
2. Examine the current NFPA 921, *Guide for Fire and Explosion Investigations,* text and decide whether you agree completely with the contents. If you find something you feel could be improved, complete the form titled "Form for Proposals on Technical Committee Documents" in the back of the text and forward your recommendation to NFPA to propose your change to the document.

NOTES

1. Neil Hanson, *The Great Fire of London* (New York: John Wiley and Sons, 2002).
2. Hanson, *Great Fire.*
3. National Fire Protection Association, *NFPA 1033, Standard for Professional Qualifications for Fire Investigator* (Quincy, MA: NFPA, 2003).
4. *Webster's New World College Dictionary*, 2nd ed. (New York: Simon & Schuster, 1984).
5. John Lentini, "Behavior of Glass at Elevated Temperatures," *Journal of Forensic Sciences* 37, no. 5 (September 1992).

SAFETY

Learning Objectives

Upon completion of this chapter, you should be able to:

- Describe the various hazards found on the fire or explosion scene.
- Describe the protective measures that must be taken on the fire/explosion scene using accepted safety practices.
- Describe the protective equipment available to the fire investigator and regulations requiring use of such equipment.
- Describe the necessity of safety training for each person involved in the investigation of a fire or explosion.

CASE STUDY

Because of limited funding and resources, most fire scenes are assigned only one investigator. For larger losses of $1 or $2 million, there are usually multiple investigators on the scene. Investigators can be from other entities, such as the insurance industry, or in the more organized areas, from neighboring jurisdictions in the form of a regional fire investigation team.

However, that is not what happens on a day-to-day basis. Many fires are looked at after hours and on weekends. The engine company will stay on the scene most of the time for a suspicious fire. But, if the fire officer wants a second opinion on an accidental fire scene, the engine could be dispatched to another call in the meantime or back in quarters loading hose to get back in service. The point is that even though we would like there to be more than one person on any given scene, this does not always happen.

Engine 6 had reported a fire that was accidental but wanted the fire investigator to look at the wiring in the kitchen. It was after hours, but in the summer months the sun was still shining and there was plenty of ambient light. Following safety protocols, the investigator put on full turnout gear and had a flash light but had left his portable radio in the charger in his vehicle. The investigator had to chuckle when he looked at the scene because the roof was gone and the debris in the rooms had been almost completely removed. This made a complete investigation more difficult, but he was there just to look at some wiring in question.

On the floor there was a lot of charring; stepping into the structure, he placed a tentative foot on the floor. It felt stable and he knew there had been firefighters on this same floor only hours earlier. Had they thought the floor was unsafe he thought they would have advised as such. Herein lies the investigator's first error: assuming that the floor was safe using negative logic. Like so many others in his field, he felt sure there would be telltale signs if the floor was about to give way: the second error in judgment.

Only four steps into the structure the investigator stopped to take a photo of the hall. With no warning whatsoever, no warning crack, no telltale softness in the flooring, the floor gave way and down went the investigator. Although there was only a 3-foot crawl space, that much of a drop could cause quite a jarring and even some minor injuries. However, along with poor judgment came fateful placement of the feet—one foot on one side of a floor joist and one foot on the other side of the floor joist. The investigator went straight down. There was no damage to the floor joist and it gave no indication of give as the investigator came in contact.

Fortunately, the investigator suffered only minor physical injuries. But he will never forget the long crawl back to the car to get his radio and call for assistance. Neither will his fellow members of the fire service let him forget for some time to come.

INTRODUCTION

For years, the fire service has stressed that the most important point to remember at the fire scene is firefighter safety. The same is true for the fire investigator. Not thinking of your own safety may very well put others in danger unnecessarily.

Taking care of yourself will also allow you to be around to take care of your family. The first step toward safety is being aware of all federal, state, and local regulations as they pertain to the fire or explosion scene. The investigator must also be aware of the safety protocols of the fire department. If your agency or department does not have safety protocols, the investigator should be diligent in creating such a document for the safety of all personnel on the investigative team.

Knowledge of federal and state regulations dealing with labor laws and the proper use of safety equipment, including the fit testing of self-contained breathing apparatus (SCBA) masks and respirators, is crucial to the protection of investigators and the success of the mission. As an investigator, you must be proficient with the use and operation of atmosphere multigas meters, which provide immediate readings of oxygen in the atmosphere and the presence of unwanted gases. Wearing, maintaining, and caring properly for protective clothing are important to your safety, health, and well-being.

Proper building construction knowledge is essential in determining structural stability. Investigators must be cognizant of special hazards such as slippery conditions, sharp objects, and electrical and water hazards. Even unusual hazards such as territorial pets and the occasional snake can create a challenge for you!

FEDERAL, STATE, AND LOCAL REGULATIONS

Note
You must be aware of many rules and regulations and follow them diligently.

the Act
Occupational Safety and Health Act

OSHA
Occupational Safety and Health Administration; a federal agency created by the Occupational Safety and Health Act whose function is to ensure safety in the workplace

You must be aware of many rules and regulations and follow them diligently. They range from your agency's rules on safety to those requirements outlined in state and federal regulations. In 1970, Congress enacted the Occupational Safety and Health Act, commonly known as **the Act.** The Act empowered the U.S. Department of Labor to create a new agency, the Occupational Safety and Health Administration (**OSHA**), to administer and enforce the provisions of the Act. A stipulation of the Act allows states (and territories) to come up with their own job safety and health plan, allowing the state, in contrast to federal authorities, to enforce these regulations. The state plan must incorporate the requirements under the Act, and, if so desired, the state may also include more stringent standards or delve into areas not covered. To date, about half of the states have successfully presented their own plans. This makes it more important than ever before for you to know who the enforcing entity is within your state.[1]

Part of the federal requirements is the minimum training necessary for any individual to attain before being allowed to work in a hazardous atmosphere. Without this training, an investigator may not be allowed legally to enter a fire scene to conduct the fire investigation. Contact your state or federal enforcer of OSHA regulations for more information about the minimum safety training requirements and where to obtain this training.

Your own department may have locally mandated safety standards. These rules may be enforced by your department's safety officer or by an officer with other titles such as quality assurance. Other agencies within your local jurisdiction, such as the **Building Officials Office,** may have certain safety requirements. Within this office are the individuals who approve building permits and review plans. These are valuable skills in determining whether a structure is safe. Other individuals within this office are electrical and plumbing inspectors. The latter can assist with water issues, and the electrical inspector can be an added resource for safety. Other resources may be found in the Department of Personnel, which may have a division titled Loss Prevention. *Loss prevention* is exactly what it implies: this office seeks means to prevent accidents and as such could be beneficial to evaluating a fire scene.

Building Officials Office
empowered by the Authority Having Jurisdiction (AHJ) to enforce the provisions of a state or local building code, electrical code, plumbing code, and so forth; in some instances empowered to enforce the fire code

Other state and federal agencies have rules and regulations specific to a certain industry or situation. For example, the Bureau of Mines has its own enforcement regulations separate from the Act's requirements. Similar to OSHA, the state may have its own enforcement authority. Be sure to do your research on all safety requirements well in advance so that you can be in compliance prior to pulling onto the scene of the emergency incident.

PROTECTING THE INVESTIGATOR

Many regulations require that investigators wear protective clothing for the environment in which they will be working. Depending on the hazard, protective clothing may be something as simple as a **Tyvek coverall** and filter mask, as shown in **Figure 2-1,** to fully encapsulated suits with self-contained breathing apparatus. For first responder investigators, the most common protective clothing to be worn is that as specified in the most recent issue of National Fire Protection Association (NFPA) 1971, *Standard on Protective Ensemble for Structural Fire Fighting.*[2] This clothing protects and covers the investigator from head to toe, consisting of boots, pants, coat, hood, helmet, and gloves, as shown in **Figure 2-2.** This section discusses several aspects of protecting investigators while on the scene of a fire investigation.

Tyvek coveralls
coveralls made out of Tyvek, which the manufacturer DuPont describes as lightweight, strong, vapor-permeable, water-resistant, and chemical-resistant material that resists tears, punctures, and abrasions

Respiratory Protection

While in a hazardous atmosphere, investigators should wear suppression quality **self-contained breathing apparatus (SCBA).** Before using any SCBA, investigators must have taken a physical to ensure that they are physically fit enough to don and work in such equipment. Before wearing any face piece, you must be fit tested to ensure a proper seal can be maintained while you wear the mask. Once fit tested, you should receive documented training on the proper use and care of such equipment before being allowed to wear the SCBA in a hazardous atmosphere.

self-contained breathing apparatus (SCBA)
a delivery system of breathable air from a compressed cylinder

Figure 2-1
Investigators may be required to wear protective clothing, from basic Tyvek coveralls to fully encapsulated suits with self-contained breathing apparatus.

■ Note
When the fire is out and the structure has been properly ventilated, the atmosphere throughout the structure should be tested to ensure the presence of oxygen.

multigas detector
a portable device for reading the levels of oxygen and hazardous gases

canister
a cylindrical attachment for a filter mask, through which air is drawn removing airborne contaminates

When the fire is out and the structure has been properly ventilated, the atmosphere throughout the structure should be tested to ensure the presence of oxygen. If possible, the atmosphere should also be tested for carbon monoxide and other potential toxic gases as well as for the presence of any explosive gases through the use of a **multigas detector,** as shown in **Figure 2-3.** Even though the atmosphere may indicate proper oxygen levels exist and there is no presence of carbon monoxide, the scene is not necessarily safe. Airborne particulates stirred up by the act of conducting the fire scene examination are bits of dust that can be toxic and carcinogenic. When you or the suppression safety officer has determined the atmosphere to be safe, the use of SCBA can be curtailed in favor of a light and easy-to-use respirator. However, be sure to make continual checks on oxygen levels and the presence of toxic gases diligently while at the scene.

Respirators must be **canister** type, in contrast to paper mask filters, which are not acceptable. The masks are sized so that the wearer can attain a proper fit that ensures that all breathable air is pulled in through the filter canister. No mask should be worn until the wearer has been fit tested for an accurate fit. Because the vast majority of airborne particulates are carbon you should check with the respirator manufacturer to ascertain which filter is the safest to use in that atmosphere. There may be other contaminates, such as asbestos, present as

Figure 2-2 *Fire investigator suited out in full turnout gear with self-contained breathing apparatus.*

respirators
a mask (full or half face) designed to cover the mouth and nose, allowing the wearer to breathe through filters attached to the mask so as to prevent the inhalation of dangerous substances usually in the form of dusts or airborne particulates

well. When deciding which canister to wear, investigators should choose the filter providing the most protection as recommended by the manufacturer.

The choice of half or full face mask, as shown in **Figure 2-4,** is up to the investigator. If wearing a half mask, the investigator must also wear **nonvented goggles** to prevent the absorption of contaminates through the sclera of the eye. Both must be worn at all times while within the hazardous atmosphere. A common problem in the use of any eye protection is the fogging up of the lenses of the goggles or the face piece of full masks. Fogging reduces visibility, creating an additional safety issue by limiting your ability to see clearly. Numerous chemicals on the market coat the lenses, preventing fogging. Failure to take steps necessary to stop fogging can further endanger you from unseen hazards.

Figure 2-3 *Fire investigator using a multigas detector. The most important test is for the proper availability of oxygen in the atmosphere and the absence of dangerous gases such as carbon monoxide.*

Figure 2-4 *An investigator may wear either full-face or half-face canister masks. If wearing a half-face mask, the investigator must wear nonvented goggles.*

goggles, nonvented eyewear that makes a tight seal with the face, preventing the introduction of any airborne particulates; *nonvented* implies that there are no vent holes so that the glasses are airtight

■ **Note**
Wear latex gloves under work gloves.

Protective Clothing

In the search for evidence, you may handle fire debris more than the fire suppression personnel on the scene do. During the investigation, you will handle fire debris in the process of sifting and collecting evidence. In handling this debris, some chemicals can be absorbed into your gloves, transferring contaminates, such as flammable liquids, to the inner surface of the glove and allowing them to be absorbed by your skin. To prevent this, wear latex gloves under work gloves, as shown in **Figure 2-5.** Some investigators even take the extra precaution of wearing two or more layers of latex gloves under their leather gloves. When the fire is under control and suppression forces have completed all overhaul, the fire scene is then considered to be a **cold scene.** When this is the case, you can wear lighter leather gloves to provide more dexterity.

Nails, glass, and metal shards create puncture hazards. Investigators should invest in proper footwear. **Steel toe boots** as well as **steel sole boots** can be a preliminary defense from puncture wounds. Steel soles are boots with a thin steel plate embedded in the sole of the boot; this is not to be confused with steel shank, which provides better support. Of course, the first line of defense is to watch where you step. Even steel sole boots are capable of being punctured under certain conditions. Proper fitting boots are important as well. Uncomfortable boots

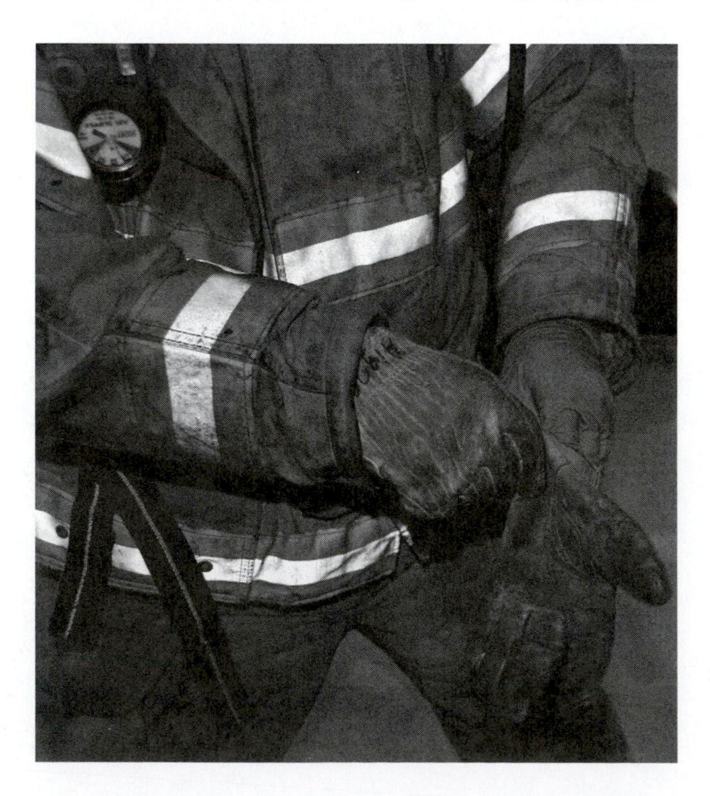

Figure 2-5
Protection from absorbing chemicals through the skin can be provided by using latex or plastic gloves under leather work gloves.

can be a considerable distraction while examining and digging a fire scene. Proper fitting boots allow the investigator to concentrate on the job at hand.

During warm weather, you may prefer to take off your heavy fire suppression clothing (turnout gear). In such cases, protect yourself by using coveralls. A safe alternative is a disposal coverall made out of materials such as Tyvek that can be removed and properly discarded. If not using a disposable product, you must clean the garment properly between each use. Never take these garments home to be washed in a household washer and dryer because doing so puts others in danger from the residue left behind in the appliances, which can contaminate personal items in the future.

The same goes for fire protection gear: wash and dry gear according to the manufacturer's specifications. Never place fire gear in any appliance that will be used for other clothing. Most fire departments have commercial washers and dryers dedicated to cleaning protective clothing. Should these machines not be available, wash the gear with a hose and soft scrub brush, using soap recommended by the manufacturer, and then rinse thoroughly. Once cleaned, your turnout gear or coveralls should be dried in dryers that are dedicated for turnout gear. Some departments spread out their gear on hose dryer racks, using gentle heat, and others simply hang it up to dry naturally. Again, be sure to check with the gear manufacturer before using any heat to dry the turnout gear. NFPA 1851, *Standard on Selection, Care, and Maintenance of Protective Ensembles for Structural Fire Fighting and Proximity Fire Fighting*, offers guidance on the maintenance and cleaning of firefighting gear.

Head protection must always be worn. When at a fire scene for any appreciable amount of time, investigators should consider switching from a suppression-type helmet to an industrial hard hat such as seen in **Figure 2-6.** The lighter hard hat can provide protection to the head and reduce fatigue and stress on your neck. Wear lighter headgear only when the hazards on the scene do not dictate the wearing of heavier protection such as a suppression helmet.

cold scenes
the remains of a fire scene after the fire has been completely extinguished; although the debris may still be warm, there is little or no chance of reignition; cold scenes may be days old

■ Note
Never place fire gear in any appliance that will be used for other clothing.

■ Note
Be sure to check with the gear manufacturer before using any heat to dry the turnout gear.

steel toe boots
work boots designed with a steel protective cap covering the toes to prevent crushing blows that would otherwise injure the wearer's toes

steel sole boots
boots with a lightweight steel plate that provides a barrier between the foot and the surface being walked on; the intent is to keep the wearer safe from punctures through the boot

SITE SAFETY ASSESSMENT

Fire investigators must perform a site safety survey assessment at every fire or explosion scene. Ask yourself: Is the structure safe to enter? Recognition of structural stability should be a skill you acquire. It is recommended that investigators become proficient in the subject of building construction. This knowledge is just as crucial in scene assessment as it is when making a proper determination of the fire area of origin and cause. Should you not possess this knowledge, seek classes in building construction available from state agency training institutions as well as through local colleges and universities. Of course, there are other ways you can be proficient in this knowledge such as having a background or experience in building construction or engineering.

Figure 2-6
Switching from a heavier fire suppression helmet to a lighter hard hat will not only be more comfortable but will reduce the investigator's fatigue.

Construction Features and Hazards

■ Note
By knowing construction features, investigators can recognize hazards that could make the structure unsound.

By knowing construction features, investigators can recognize hazards that could make the structure unsound. Suppression water and weather can affect a fire scene and the stability of the structure. You must take into account the added weight of water when doing your initial and ongoing scene assessments. For example, a flat roof may well be able to handle standing water under normal circumstances. But if the structure is weakened by the fire or explosion, the additional weight of water could push the stability of the remaining structural elements beyond their capability, causing failure. Likewise, snow and ice on any roof can add sufficient weight enough to cause structural failure.

When a roof is not present, the rain or snow can collect on the floor and be absorbed by contents and debris, creating more weight and causing collapse.

In winter conditions with temperatures below freezing, investigators must consider that suppression water that would normally flow back out of the structure could freeze inside, adding more weight than that experienced in the spring or summer.

Unsafe Chimney Leads to Loss of Life

What may be assumed as a safe condition may not always be the case. In New York, a fire department investigator was looking at a fire scene that was 5 days old. The fire involved an older residence that had been converted into three apartments. The fire was an exterior attack with deck guns and was brought under control in only an hour. But the investigation was delayed because of subzero temperatures.

On the day of the investigation, the assigned fire investigator was accompanied by a private investigator, an insurance adjuster, an electrical consultant, and the victim. The insurance adjuster was downstairs assessing damages while the other four stayed in the attic to find the origin and cause of the fire. There was considerable damage to the roof, leaving the chimney free standing, rising about 13 feet above the attic floor. This chimney had direct fire exposure, had been struck by water streams from the deck guns operating on the day of the fire, and since the fire had been exposed to 54-mph wind gusts. This along with freezing and thawing temperatures was eroding the structural integrity of the chimney.

About two and a half hours into the investigation, the group had worked their way closer to the chimney. Of interest is that the three working the investigation had commented to each other that they had seen the chimney swaying slightly in the wind gusts. There was some concern because the private investigator had actually pushed and pulled on the stabilizing bar connected to the chimney, but the chimney did not move. That along with the fact that it was still free standing for 5 days gave them a false sense of safety.

■ Note

Of interest is that the three working the investigation had commented to each other that they had seen the chimney swaying slightly in the wind gusts.

The department investigator was 8 feet from the base of the chimney, directly in line with the chimney. The others were off to one side or the other. No one knew there was a problem until they heard the chimney hit the attic floor, crushing the fire department investigator. They tried in vain to move the chimney off the victim, but to no avail. To summon help one of the others had to run next door to call 9-1-1. At about the same time, another investigator arrived on the scene and two other chiefs arrived shortly thereafter. With five of them, they were able to move the chimney off the victim. CPR was initiated and he was transported. He died later at the hospital from his injuries.

The section of the chimney that landed on the victim measured 98 inches high, 38 inches wide, and 17 inches deep, and weighed between 2,500 and 3,000 pounds. At the time of the incident, the winds were between 33 and 36 mph. The construction of the chimney was bricks and mortar only, no flue liner, and there appeared to be two types of brick near the roofline, indicating that the chimney had been added onto sometime in the past. All these issues contributed to the failure of the chimney.

Figure 2-7
A freestanding chimney may give the appearance of stability, but the mortar may be damaged to the point that a gust of wind can cause it to topple.

The presence of any chimney, freestanding or still attached to structural elements, is a potential hazard. Even those with mass that give the appearance of stability, such as shown in **Figure 2-7,** can be unstable enough to topple, putting anyone in proximity in danger. All chimneys should be evaluated for their potential as a hazard, and should it be necessary, topple the chimney, by safe means, to reduce the likelihood of it presenting a danger to the investigator or those visiting the scene well after the emergency is over.

The National Institute for Occupational Safety and Health (NIOSH) is a federal agency responsible for conducting research and coming up with prevention recommendations to prevent future work-related accidents. Their May 14, 1999, report, number F99-06, made the following recommendation:

> Fire departments should conduct an assessment of the stability and safety of the structure, e.g., roofs, ceilings, partitions, load-bearing walls, floors, and chimneys before entering damaged, e.g., by fire or water, structures for the purpose of investigation.[3]

Not all fire investigators and first responders will have the necessary skills to make a fair assessment of the scene, so qualified experts may be necessary.

Should there be any question as to the stability of the structure, most jurisdictions have personnel within the Building Officials Office who can assist in making a safety assessment. Structural engineers may also be available to make such assessments for a fee. To make a scene safe, there are occasions when heavy equipment may need to be brought in to remove walls and overhangs. When this is done, be sure to photograph the area adequately prior to disturbing the scene for future use in making your determination of origin and cause. Photographs are also beneficial for others who may be working at the scene after the structure has been altered.

Of course, the use of heavy equipment is a last resort to make the structure safe because as a result some evidence may be inadvertently contaminated. Steps can be taken to avoid such contamination: (1) choose the location for equipment placement for minimum exposure to the debris and the scene; (2) prior to the arrival of the heavy equipment thoroughly examine the placement area for potential evidence; (3) clean any part of the equipment that will be entering the scene, either in direct contact or hovering over the scene, and test it using a hydrocarbon detector to ensure that contaminates are not being brought into the scene.

The stability of the structure is an ongoing concern and should be monitored constantly. (See **Figure 2-8**.) If resources are available, maintain the safety officer position used during the suppression phase of the event.

Figure 2-8
Damaged structural components may be unstable, capable of failure at any time.

■ **Note**
Before entering the scene it is critical that investigators follow their department's directive on the ICS.

Incident Command System (ICS)
an on-scene, standardized, all hazard management process

ENTERING THE STRUCTURE

Before entering the scene it is critical that investigators follow their department's directive on the **Incident Command System (ICS).** Investigators should report to the incident commander (IC) or the IC aide and request permission to enter the building, advising where they need to go and who will be accompanying them on their initial survey of the scene. In some jurisdictions, fire investigators have concurrent jurisdiction with the incident commander. Regardless of statutory authority, safety issues dictate that investigators follow the existing ICS. Likewise, being accompanied by a fellow investigator or a firefighter is important to ensure assistance is available should a dangerous situation present.

All Incident Command Systems should include accountability tags of some type, two examples of which are provided in **Figure 2-9.** Leaving your first

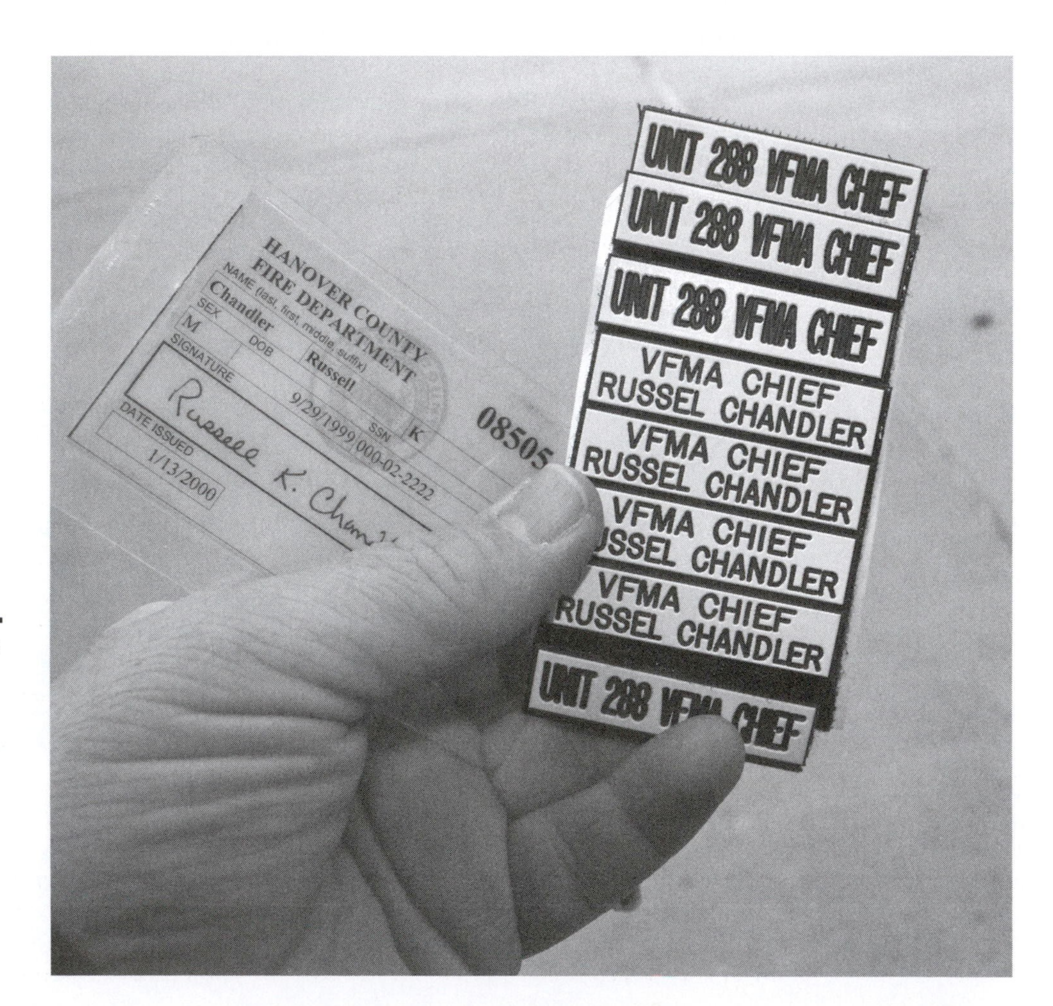

Figure 2-9 *There are many versions of accountability tags. A common type is the plastic tag with the name and unit number attached with Velcro so that the tag can be attached to the ICS accountability board.*

tagged
in reference to an electrical disconnect switch, the action of placing a tag on the device warning others not to turn the handle to the on position due to a safety concern

lockout
in reference to an electrical disconnect box, when it is in the open position with no current flowing and a lock is in place to keep the circuit from being energized until the lock is removed

Note
Check to ensure that extension cords or heavier wires are not leading from a neighboring structure into the fire scene.

multimeter
an electronic device used to measure AC/DC voltage, current, resistance, capacitance, and frequency; also referred to as a multitester

voltage detector
tester or probe that either glows or emits a sound when placed in proximity of an energized electric circuit (electrical wires or cords)

accountability tag with the IC serves to remind Command that you are on the scene. If the scene is large enough or tall enough to require separate command sectors, additional accountability tags should be left with sector commanders. Be sure to pick up those tags when leaving that sector.

When entering a burning building, always keep track of your location and establish at least two routes of escape at any given time. Always carry a strong, reliable hand light whether it is day or night. The light may be necessary to observe items such as burn patterns or evidence, but it is just as valuable a tool to observe tripping hazards and sharp objects that could be risks to you. You can also use the light to discover other hazards such as exposed electrical wires that are energized.

Electricity

When possible the source of electricity to the structure should be completely disconnected, not just at the breaker or fuse panel, but at the street, whether it is overhead service or underground. The power company should disconnect electrical service because its personnel have the tools and necessary equipment to make the disconnect safely. Although it may have been common practice in the past to pull meters, most departments avoid such actions because they are a potential hazard. Investigators should examine overhead power lines especially when they are near the damaged structure. Have the lines been damaged, or are they hanging closer to the structure now, creating a potential hazard?

Even if electricity has been disconnected, all lines should be considered "hot" until each circuit can be verified not to be energized. Before entering the building, ascertain the status of all utilities. Electrical meters or disconnects should be **tagged** and locked out using a **lockout** device such as the one shown in **Figure 2-10** to ensure they stay unenergized.

When using ladders in or around overhead power lines, use extreme caution so as not to come into contact with the wires. Even if the wires appear to be insulated, that does not negate the potential for energy leakage that could cause electrocution should your ladder come in contact or get near these wires.

When doing a safety assessment of electrical hazards, investigators must also consider alternate sources of electricity other than that provided by the power company. Check to ensure that extension cords or heavier wires are not leading from a neighboring structure into the fire scene. Such practices of borrowing or stealing electricity are not common but are possible.

As a final measure, even if you've been assured that the electricity has been disconnected, be familiar with the operation of a simple electric **multimeter.** These devices, as shown in **Figure 2-11,** can verify the presence of electricity or lack thereof when properly used. Some other simple devices commonly referred to as **voltage detectors** give an audible or visual signal when the device is put in

Figure 2-10 *A lockout device can accommodate many different locks for each individual working on or near that circuit. The disconnect device will have a tag advising that the circuit is locked out and deenergized.*

close proximity of an energized electrical cord or wire. These are great devices and good tools but should not serve as a replacement or be used solely instead of a voltmeter. As with all tools, investigators must properly maintain and test their equipment on a regular basis.

Slippery Surfaces and Sharp Objects

Slippery surfaces can be present, not just from ice but from water, chemicals, or even soot on certain surfaces. If the investigator is attempting to make these areas safe, no additional materials should be introduced to the scene that could

Figure 2-11 *Every investigator must be familiar with the use of a typical multimeter.*

cause contamination or the appearance of contamination. Instead of adding salt or sand, consider marking the area so that others are aware of the hazard and finding an alternative path to avoid the slippery area whenever possible.

Sharp objects are of a particular concern. All debris contains a multitude of objects such as metal shards, glass, nails, and staples that can easily penetrate most clothing if the wearer does not take precautions. Wearing protective equipment is only one measure to take to prevent injuries from sharp objects. Wearing proper protection such as gloves and boots is not enough; to prevent injuries you must also be cautious of where you step and where you reach with your hands.

Rehabilitation and Switching to Respirators

Investigators must follow all local policies regarding the amount of time they spend breathing compressed air or using respirators and the necessary time to spend in rehabilitation. Rest and rehydration can be important at an emergency scene even for the fire investigator not just for physical well-being but to refresh the mind to allow clear thinking. Monitoring vital signs and maintaining proper body temperature are also part of the rehabilitation process. It

Safety

Remember, before removing the SCBA you should test the atmosphere for proper oxygen levels of at least 16 percent, preferably 20 percent, and the carbon monoxide (CO) readings should be under the recommended levels established by your department.

carbon monoxide (CO) a colorless, odorless gas usually produced as a by-product of a fire

is important that fire investigators follow NFPA 1584, *Standard on the Rehabilitation Process for Members During Emergency Operations and Training Exercises,* which references the recommended procedures to be followed for incident scene rehabilitation. Even if investigating a cold scene, investigators should always have sufficient water for rehydration and snacks for energy. They should also always be sure to clean up or at least wash their hands and face before handling foods to prevent contamination.

Remember, before removing the SCBA you should test the atmosphere for proper oxygen levels of at least 16 percent, preferably 20 percent, and the **carbon monoxide (CO)** readings should be under the recommended levels established by your department. Most publications indicate that you can develop headaches from as low as 40 parts per million. The threshold for safety should be much lower.

Carbon monoxide is a colorless, odorless gas that acts as an anesthetic. In higher concentrations it becomes an asphyxiant. CO combines with the hemoglobin in blood to form carboxyhemoglobin. Hemoglobin is the part of the blood that carries oxygen throughout your body; essentially, the hemoglobin prefers CO 250 more times than oxygen.[4] With no or limited oxygen being carried by the hemoglobin, insufficient oxygen reaches your brain and muscles including your heart. The effect is narcosis, a condition in which you enter a stupor and eventually lose consciousness. The effects of CO poisoning are all too often not recognized in time.

Every investigator should have access to a multigas detector. These units can be small, constant monitoring units attached to your clothing while you are working on the scene or can be larger models with which you can check the atmosphere on a regular basis.

If a scene safety officer is on site, this person may be monitoring atmospheric gas levels, giving the all clear for removal of SCBA when it is appropriate. However, once the suppression personnel have completed their task at the scene and have left the premises, investigators should continue to monitor the atmosphere to ensure that they are working in a safe environment.

If the scene is an industrial or mercantile property, extra precautions should be taken to prevent exposure to dangerous chemicals. A hazardous response team can make an assessment as to the presence of such hazards and can advise you as to when it is safe to enter the scene.

Other Hazards

Holes in floors are not always obvious. They can be masked by loose debris or standing water. Always use caution when stepping into any standing water because it may be much deeper than you expected and may hide sharp edges that could cause injuries. Holes or weakened floors could lead to a fast entry into the floor, basement, or crawl space below, with serious consequences. Holes in floors need to be identified and marked or rendered safe.

Figure 2-12 *A large aggressive dog can give obvious signs to cause you to approach cautiously, but even a scared, small, timid dog can cause a serious injury.*

■ Note
Even when they appear timid, a family pet can turn and cause serious injuries.

Standing water in basements should never be entered. Unknown electrical sources could lead to electrocution hazards. Hidden debris under the water could constitute tripping hazards, and even basement floors could be multi-level or contain cisterns, sump pump wells, or other deeper openings. Portable pumps are available in most departments or through a local equipment rental company so that you can remove the water for a safer examination of the basement area. Be sure to test any devices to be put into the water to ensure that they do not have any trace residue of hydrocarbons that could contaminate your scene.

Animals can present special hazards. Even when they appear timid, a family pet can turn and cause serious injuries. Cats rarely pose a threat unless you decide to detain them when they do not want to be handled. Dogs, regardless of their size, can be a hazard. (See **Figure 2-12.**) Treat them with caution. When a loose dog appears in a threatening manner, it may be defending its domain and perceiving you as an intruder. Thinking you can "calm the savage beast" may be a flaw in your plan. Pay careful attention even to distraught dogs that appear to be safely tethered.

If family members are not available to assist, the locality may have animal control officers to assist for both your protection and the protection of the animal. Another tool that may work from time to time will be to offer dog or cat

Dog Determination

On one occasion, an investigator pulled onto the property of a fire that had occurred the night before. No one was at the property, but the investigator could hear a barking dog in the back yard, which was the location of most of the fire damage. The investigator's systematic process of checking the exterior of the structure first revealed that the backyard was not fenced in. However, a Rottweiler mix was chained to a doghouse in the center of the yard. For the record, Rottweilers average between 80 and 100 pounds or more.

What appeared comical at first was then seen as reassuring. The doghouse was made from 2- by 6-inch treated boards and looked large enough to house half a dozen dogs. Around the dog's neck was a heavy, thick, 2-inch-wide collar with chrome spikes. In the D-ring of the collar was a climbing carabiner that attached a chain to the collar. It was a short chain, only about 8 feet long, but looked heavy enough to use with a boat anchor. Then, a large eyebolt and base attached the chain to the doghouse using four large bolts. As large as the chain seemed, the dog had no problem pulling it taut as the investigator approached the house. But there seemed no way this dog could escape his tether.

As with any repetitive noise, the investigator eventually blocked out the barking and went to work sifting through the scene. After finding the fire area of origin, he was kneeling down looking at different items searching for the fire cause. The investigator eventually realized that the barking was louder and looked up. There was the dog, about 2 feet away, snarling and snapping, looking the investigator right in the eye. What originally was a 20-foot barrier between the dog and the fire scene was now only a couple of feet. The dog had dragged the dog house across the yard, leaving a furrow in its wake. It did not take long for the investigator to decide that the final cause of the fire could be determined at a later date.

snacks as appropriate. A snack cannot be depended on to make friends with worried or aggressive animals, but a barking dog can be quite distracting at times, and snacks have been known to at least satisfy dogs enough so that the barking will stop or at least lessen.

Depending on where you work, snakes, spiders, and insects can be a hazard. Knowledge and recognition of poisonous snakes known to the area may be beneficial. Spider bites, although not usually deadly, can cause serious sickness and damage. If poisonous spiders are known to be in the area, you may want to take precautions to prevent unnoticed exposure to spiders such as covering all exposed skin, sealing clothing at the ankles and wrists, and ensuring that your shirt stays tucked in. The same precaution can be taken in southern states that have fire ants or other swarming insects. A can of wasp/hornet spray can be a valuable tool to remove nests that may be a potential hazard. However, because these products may contain petroleum distillate, it is very important to use extreme caution not to contaminate the scene in a way that jeopardizes an accurate determination of the fire cause.

■ **Note** **The federal government, in particular the Bureau of Alcohol, Tobacco, Firearms and Explosives (ATF), provides outstanding training events on terrorism, explosives, and the postblast scene.**

EXPLOSIVES AND TERRORISM

No safety chapter is complete without mention of explosive scenes and terrorism. Investigators should be well aware of all safety concerns at any emergency scene involving explosives. Issues with residues, hazardous materials, and secondary devices are all very real and potentially deadly. A multitude of training events are available to the local, state, and federal fire investigators who address explosives, their detection, and the handling of explosive debris. The federal government, in particular the Bureau of Alcohol, Tobacco, Firearms and Explosives (ATF), provides outstanding training events on terrorism, explosives, and the postblast scene. These events usually include hands-on experiences with primary and secondary devices.

SUMMARY

Upon receiving an assignment, safety should be your primary concern. Investigators tend to start their size-up of the situation on their way to the scene. Considering safety conditions and steps to make a scene safe can be a part of your thought process while responding. But safety has to start long before the assignment is received. Investigators should have strong knowledge of all laws, regulations, and standards to be met while carrying out their assignments. Knowledge of building construction and structural integrity is essential and must be learned to carry out the initial safety assessment. And, of course, investigators must be familiar with hazards they may encounter and all the tools at their disposal to work within these hazardous environments.

Most of all, investigators must have a proper frame of mind to keep safety in the forefront while completing the assessment, carrying out the fire scene examination, and even when cleaning up after the scene examination. They must always remember that they are the most important persons, and in taking care of themselves, they will be able to take care of others.

KEY TERMS

The Act Occupational Safety and Health Act.

Building Officials Office Empowered by the Authority Having Jurisdiction (AHJ) to enforce the provisions of a state or local building code, electrical code, plumbing code, and so forth. In some instances empowered to enforce the fire code.

Canister A manufactured metal or plastic cylindrical device filled with filtering material. Canisters are designed to be attached to a filter mask with an airtight seal, forcing all air entering the mask to pass through the filtering medium, essentially removing hazardous airborne particulates.

Carbon monoxide (CO) A colorless, odorless gas usually produced as a by-product of a fire.

Cold scenes The remains of a fire scene after the fire has been completely extinguished. Although the debris may still be warm, there is little or no chance of reignition. Cold scenes may be days old.

Goggles, nonvented Eyewear that makes a tight seal with the face, preventing the introduction of any airborne particulates. *Nonvented* implies that there are no vent holes so that the glasses are airtight.

Incident Command System (ICS) An on-scene, standardized, all hazard management process.

Lockout In reference to an electrical disconnect box, when it is in the open position with no current flowing and a lock is in place to keep the circuit from being energized until the lock is removed.

Multigas detector A portable device for reading the levels of oxygen and hazardous gases.

Multimeter An electronic device used to measure AC/DC voltage, current, resistance, capacitance, and frequency. Also referred to as a multitester.

OSHA Occupational Safety and Health Administration; a federal agency created by the Occupational Safety and Health Act whose function is to ensure safety in the workplace.

Respirators A mask (full or half face) designed to cover the mouth and nose, allowing the wearer to breathe through filters attached to the mask so as to prevent the inhalation of dangerous substances usually in the form of dusts or airborne particulates.

Self-contained breathing apparatus (SCBA) A delivery system of breathable air from a compressed cylinder.

Steel sole boots Boots with a lightweight steel plate that provides a barrier between the foot and the surface being walked on. The intent is to keep the wearer safe from punctures through the boot.

Steel toe boots Work boots designed with a steel protective cap covering the toes to prevent crushing blows that would otherwise injure the wearer's toes.

Tagged In reference to an electrical disconnect switch, the action of placing a tag on the device warning others not to turn the handle to the on position due to a safety concern.

Tyvek coveralls Coveralls made out of Tyvek, which the manufacturer DuPont describes as lightweight, strong, vapor-permeable, water-resistant, and chemical-resistant material that resists tears, punctures, and abrasions.

Voltage detector Tester or probe that either glows or emits a sound when placed in proximity of an energized electric circuit (electrical wires or cords).

REVIEW QUESTIONS

1. List and describe safety protective clothing used by the fire investigator.
2. When clothing is exposed to fire scene debris, where should it be washed and why?
3. What needs to be done to ensure that all electrical circuits are not energized?
4. Why is standing water a concern on a floor or in a basement?
5. What type of protective clothing should be worn by the fire investigator when entering a structure still under suppression activities?
6. Why is it important to follow national guidelines on incident scene rehabilitation?
7. Why should proper eye protection be nonvented?
8. It is recommended that the investigator wear latex gloves under leather gloves. Why?

9. The presence of some animals may constitute a hazard to the investigator. What are some measures that could be taken to render the condition safe?
10. What is the benefit of the investigator reporting to the officer in charge (OIC) upon arrival at the emergency scene?

DISCUSSION QUESTIONS

1. What resources are available to assist the investigator in making the scene safety assessment?
2. How could the investigator render slippery surfaces safe without introducing any foreign material that may contaminate the scene?
3. Holes in floors constitute one of the most common hazards to be encountered by the fire investigator. How could the investigator render such holes safe without introducing something onto the scene that would contaminate the scene and affect any evidence collected?
4. List and explain the reason for using all the safety equipment that should either be carried by the fire investigator or made available to investigators at the fire scene.
5. Discuss who might be available in your jurisdiction to give opinions as to the stability of a structure. What are some creative ways to get help from local businesses in this endeavor?

ACTIVITIES

1. Research and list training opportunities for investigator safety that are available locally and at the state and federal levels. Remember, they do not have to be specific to firefighting.
2. Research the labor safety laws in your state and locality. List the agencies responsible for enforcing the Occupational Safety and Health Act and the contact person for your locality.

NOTES

1. Occupational Safety and Health Act (OSHA). http://www.dol.gov/dol/location.htm.
2. NFPA 1971, *Standard on Protective Ensembles for Structural Fire Fighting and Proximity Fire Fighting,* http://www.nfpa.org/assets/files/PDF/CodesStandards/TIAErrataFI/TIA1971-07-2.pdf.
3. Richard W. Braddee, "Fire Investigator Dies After Being Struck by a Chimney That Collapsed During an Origin and Cause Fire Investigation—New York," May 14, 1999, NIOSH Report F99-06, http://www.cdc.gov/niosh/fire/reports/face9906.html.
4. Arthur E. Cote, ed. *Fire Protection Handbook,* 20th ed. (Quincy, MA: National Fire Protection Association, 2008), 6–14.

ROLE OF FIRST RESPONDERS

Learning Objectives

Upon completion of this chapter, you should be able to:

- Describe who is a first responder.
- Describe the role of first responders as they approach the fire or explosion scene.
- Describe the role of the first responder as the potential first investigator on the scene.
- Describe the value of first responders' observations at the fire or explosion scene.

CASE STUDY

In one incident, it was 3 in the morning when an alarm came in that the local drinking establishment had flames shooting through the roof. It was not far from the local fire station, and in only minutes both the engine and the ladder truck arrived on the scene, parking in front of the property. As a result, the only exit to the street from the parking lot was blocked. At this time in the morning, there seemed to be no reason why this would be a problem.

The local rescue squad was setting up a rehabilitation site where the firefighters coming out of the building after depleting an air bottle could sit down, rest, and take in fluids. During this setup, one of the emergency medical technicians (EMTs) noticed a car parked next to the Dumpster and could see that someone was slumped down in the driver's seat but was looking around as the firefighters were going about their business.

The EMT reported this unusual situation to the fire marshal who had arrived on the scene. Immediately, the fire marshal approached the car and found the individual just as described. The person in the vehicle stated that he came out of the bar and was too drunk to drive, so he fell asleep on the seat. This seemed a valid explanation and was a commendable decision.

But the fire marshal still had his doubts—the driver's body language was classic for being deceitful—so he asked permission to look in the trunk of the car. The driver said it was not his car and he did not have a trunk key. The fire marshal called the dispatcher to have the license plates run, which revealed that the Department of Motor Vehicles listed the car as belonging to the driver.

About this time a local police patrol officer walked over to see what was happening. Wanting to make sure that the driver understood the questions, the local police officer gave the driver a sobriety test. He passed 100 percent; there was no notice of alcohol on the driver's breath and all other indications were that he fully understood the questions he was being asked.

In the meantime, the fire was still burning, but firefighters told the fire marshal that the back door of the building had been forced and was open upon their arrival. They also had found other items that they felt were incendiary indicators. The fire marshal took a quick look and confirmed that the back door had in fact been forced. With the fire still raging, little could be done inside at that time.

The fire marshal did not ask any more questions about the fire but did ask the driver again if he could look inside the trunk. The driver clearly stated that if he (the fire marshal) could open the trunk, he was welcome to look in it. The fire marshal warned the driver that the lock and the trunk itself would be damaged by forcing it open, but the driver stood firm and said to go right ahead.

The truck company lieutenant was standing nearby. Having heard the conversation he had already picked out his best man and he was standing at the back of the vehicle with his forcible entry tool and a flat-head axe. Anyone in the fire service knows that it can take several strikes with a tool to open a trunk, but this was not to be the case. One strike and the trunk popped open.

(Continued)

(Continued)

Sometimes it is good to listen to your instincts. The trunk contained two cases of beer labeled with a delivery sticker for the bar that was burning. Next to them were six metal coin boxes (the type from vending machines and games), and almost all were half full of quarters. Sitting on top of the coin boxes was the drawer for a cash register, still holding coins and paper money.

As the fire marshal placed the cuffs on the driver and read him his rights, the driver just stood there shaking his head. It was later confirmed that the serial numbers on the coin boxes matched the video games and pool table in the bar. The bar owner identified the cash drawer; it even contained the owner's business card with some figures written on the back. Confirmation was obtained that the beer in question was delivered to the bar and it was not sold or given away by staff.

Some may wonder why the individual was so brazen about allowing the search of the trunk. There is more to this than can be covered in one chapter, let alone one book. But it would surprise the average individual how many times a case can be solved just by asking the right question.

The fire marshal used his knowledge of human behavior, and then followed policies set up by his department. Local police assisted in administering a sobriety test to make sure the individual was not under the influence and could understand what was being asked. He was not questioned about the fire scene and neither was he physically detained; he could have walked away at any time. But most important, the observations made by first responders helped to solve this case.

INTRODUCTION

first responders any public safety responders who may be or may have the potential of being first on the scene as the result of being sent there by the Emergency Communications Department

The term **first responders** is just as it implies: the first public safety providers dispatched or to arrive on the emergency scene. First responders can be patrol law enforcement (police) officers, **emergency medical technicians (EMTs),** or firefighters. They are the public safety officials first to observe the scene, the surrounding area, as well as the people in and around the incident. Their observations can be critical to the ultimate disposition of the investigation. (See **Figure 3-1.**)

■ Note **First responders can be patrol law enforcement (police) officers, emergency medical technicians (EMTs), or firefighters.**

Responders have their own responsibilities and will do an individual size-up as they approach the scene. As they exit their vehicles, they will be assessing injuries as well as observing damages and road blockages that will hamper the arrival of additional resources. They will be thinking of safety issues, and if they are arriving at an explosion scene, the thought of a secondary device will not be far from their minds.

They will also be making observations that may be of value for the fire investigator. The importance of these observations may not be obvious at first; however, a thorough interview by the investigator may assist in bringing to light some valuable data.

Figure 3-1 *First responders arriving at the scene make many valuable observations in the course of their duties.*

emergency medical technicians
individuals with specialized medical skills, trained to offer basic first aid in the field under emergency circumstances, and then to assist in the transporting of the patient to medical facilities if necessary

The emergency communications dispatcher is usually the first public safety person to collect any information about the incident. For this reason, we include them in the category of first responder even though they do not travel to the scene. They, like the emergency medical personnel, are not in a position to investigate the fire, but they can be a valuable asset in this process. The patrol officer, too, is not likely to be the primary investigator in most jurisdictions, but by the very nature of the occupation they may recognize the value of what they see and will readily present this information to the proper authorities.

The firefighter, on the other hand, is both a first responder and can be the first investigator on the scene. In this text, we group fire officers and firefighters in the same category. Their rank does not establish their role so much as their assigned responsibilities. The person who fills out the fire report is the one who should collect information necessary to complete the reporting task.

In this chapter, we discuss the role of each first responder and how that role can affect the fire investigation. We discuss the fact that some first responders, especially fire service personnel, have a tremendous impact on the fire scene. They may also be the investigator of record documenting the fire scene, suppression, and their determination of the fire origin and cause of the fire in the National Fire Incident Report System.

■ **Note**
Usually, the first public safety personnel involved in an incident are those in the Emergency Communications Center (Dispatch).

FIRST RESPONDERS' ROLE IN AN INVESTIGATION

An incident starts with the first report of the incident. Usually, the first public safety personnel involved in an incident are those in the **Emergency Communications Center (Dispatch).** They, in turn, broadcast via radio the information necessary to send the appropriate units for the situation at hand. Those being sent are the first responders, the very first personnel to respond to the scene to handle the emergency situation. They may know information that will be beneficial to the investigation of the incident, including observations they make while en route to the scene or comments they heard as they were packing up to leave.

Emergency Communications Center (Dispatch)
sometimes referred to as the E 9-1-1 center; a place where calls from the public for emergency assistance are received and whose staff (dispatchers) alert and send the appropriate emergency unit(s) to handle the situation

dispatcher
a person who works in an Emergency Communications Center (E 9-1-1 center), whose duties are to receive calls from the public for emergency assistance, and then to send the appropriate emergency unit(s) to handle the situation

Although the hierarchy of each type of responder may be different from jurisdiction to jurisdiction, their overall operational tasks remain somewhat consistent across the country. We explore these consistencies in the following subsections.

Emergency Dispatchers

Dispatchers are not typically considered first responders because they do not physically go to the scene. However, when considering the investigation of an incident, dispatchers are usually the first emergency personnel to have information about the fire or explosion scene (see **Figure 3-2**). Most modern Emergency Communications Centers not only have their radios, telephones, and proper protocols, they also have recording equipment that documents all telephone and radio traffic.

Most Emergency Communication Centers have 24-hour recorders. In older systems, there are 24-hour tapes on reels that are saved on the shelf for 7 to 30 days, or even longer. For these systems, someone must take the tapes off the machine and rotate the oldest tape back onto the machine for re-recording. The data on the older tape are erased as the newer data are recorded. Most facilities have enough spare tapes to enable them to take a few out of the rotation to be saved for an investigation or court proceedings.

Emergency Communications Centers with newer systems record on various media with sufficient data memory to capture 30 days or more with no need to physically change media devices each day. As would be expected, the new

Figure 3-2 *The dispatcher can obtain valuable information from those reporting the emergency.*

media have better quality and is capable of discerning which information came off which frequency or which phone line. If data must be kept, the information can be recorded onto a variety of devices for long-term storage. The most important aspect is that the assigned fire investigator must be familiar with the dispatch center, its capabilities, and its resources. The investigator must also make sure to request communications data before they are erased.

■ **Note**
The assigned fire investigator must be familiar with the dispatch center, its capabilities, and its resources.

These recorders usually apply a time stamp to the data, and the dispatch center should be able to transfer data to a smaller recording device. Time stamping is when recordings are stamped with the exact time of their creation. Time stamps help in searching at a later date and assist with future investigations.

The dispatching staff must also be trained to recognize that their interaction on the phone (9-1-1 lines) and the radio may provide crucial evidence about the incident. Dispatchers have developed keen listening skills and can provide the investigator with insights that can be beneficial before the investigator even listens to the tapes. The dispatcher can also be a valuable resource on what is heard on the tapes. Dispatchers can recognize stray sounds that the recording equipment makes or that are unique to the communications center in contrast to a sound that is unusual.

Do not discount a dispatcher's life experiences. Dispatchers have been known to recognize specific sounds, such as coins dropping into a toll booth basket, the distinctive beeping of a French fry machine in a fast-food restaurant, and the characteristic noise of a water fountain in a park. Call takers in the dispatcher center have also had years of experience dealing with people reporting emergencies on the phone. They may be able to provide insight about the attitude or demeanor of the person reporting the emergency. Many of these bits of information are not necessarily something you will use in court on your case, but they can help lead you in the right direction so that you can get the case to court if necessary.

Emergency Medical Personnel

An ambulance or rescue squad is called to the scene of a fire or explosion for two primary reasons. The first is that there may be a reported injury, and the second is that they need to be on standby at the scene in case of an injury to a first responder or citizen. The emergency medical technician (EMT) is trained to make assessments, observe patients, and look for the method of injury as a means of diagnosis. The method of injury or any comment made by a patient may be of importance in the investigation.

■ Note
The method of injury or any comment made by a patient may be of importance in the investigation.

The method of injury could be multiple cuts from flying glass. The incident may have been an explosive scene in the early morning hours in a remote industrial site. The fact that the injuries were from flying glass would be of interest to the investigator because it can indicate that patient may be an eye witness or the person who set the explosive device, getting hurt from a premature detonation.

Burn injuries are definitely of interest to the fire investigator. If the patient with the burn injuries was inside a business that was closed at the time and supposedly unoccupied, the burns may be a result of setting the fire. The first responder, and investigator, however, should never make rash assumptions. The person with burn injuries might also have been a bystander who saw the fire and tried to extinguish the flames.

Even if arriving as a standby unit, EMTs may observe suspicious actions of a person in the crowd. When something stands out as being unusual to an EMT, it should be of concern to the fire investigator. For example, consider a bystander who seems overly exuberant and willing to assist the firefighters and shows an uncanny joy about the scene. Assumptions can be dangerous because this person might be absolutely innocent and this may be a normal behavior for him or her. But when more than one first responder reports seemingly unusual behavior, this information can be verified or corroborated and of value to the investigation.

When EMTs approach the scene, they begin assessing the number of victims and deciding whether they need more resources. If there are no victims, EMTs may be assigned to setting up a rehabilitation site (rehab) where the firefighters can receive fluids, heat, or cooling depending on the situation and have their vital signs checked before assuming another suppression task. For this, the EMT crew seeks a location convenient to the scene but out of the way.

On a fire scene, EMTs may observe where the smoke was originating, and smoke or flame color may be of interest, but be cautioned that it is not an accurate determination of the first fuel ignited or a fire cause. EMTs might notice the

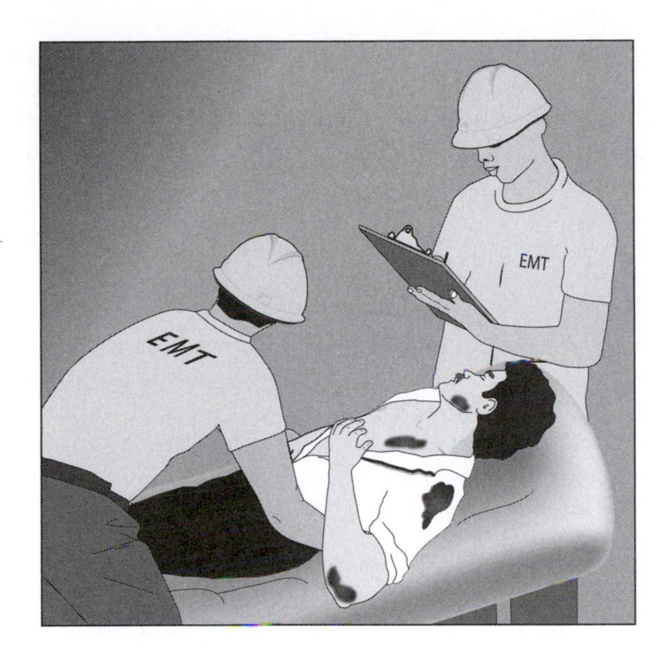

Figure 3-3 *While assessing patients from the fire or explosion scene, EMTs might hear and observe valuable information.*

direction of travel of the smoke, which indicates the wind direction; this may be of interest, for example, if the smoke was blowing from the west to the east yet the fire traveled from the east to the west.

EMTs may also observe victims or bystanders (see **Figure 3-3**). The attire of the building's occupants should match the time of day. If a residence is burning, for example, it would be unusual to see the family fully dressed at 2 in the morning. Regarding bystanders, what is their attire and is it normal to have them at the scene? EMTs might also observe vehicles or pedestrians leaving the scene. Most people are drawn to accident or emergency scenes; to run from or speed away in a vehicle would not be a normal behavior and would warrant investigation.

■ **Note**

Clothing may contain residue of ignitable liquids or chemicals. This in itself, however, does not indicate criminal-related behavior.

On a fire or explosion scene, the role of the EMT in the investigation, in addition to their observations, is to preserve any evidence found on victims. Clothing may contain residue of ignitable liquids or chemicals. This in itself, however, does not indicate criminal-related behavior. It may be an accidental fire started as the result of the victim using gasoline as a cleaning fluid. In all situations, it is critical to make proper and accurate observations but to avoid speculation—just report the facts.

■ **Note**

Should there be an odor on the clothing, the clothing should be protected from contamination.

Should there be an odor on the clothing, the clothing should be protected from contamination and turned over to a fire investigator as soon as possible. Any other items that may be potential evidence should also be protected and saved. If an investigator is not readily available and there are no clean metal cans or clean glass canning type (Mason) jars to hold these items, the last resort is to place them in a clean plastic bag that seals. EMTs should contact an investigator immediately to come collect this evidence.

If victims are transported to the emergency department of a hospital, the medical personnel can recover additional evidence. Parts from an explosive device can be embedded in the body of the victim or perpetrator. If the investigator has not arrived at the emergency room, an EMT can stay with the patient and bag that evidence as well. As soon as practical, all the evidence should be turned over directly to the investigator; to maintain the chain of custody, the evidence should never leave the EMT's control until turned over. In years past, investigators' efforts have been frustrated because no chain of custody for evidence existed.

Emergency medical personnel should be trained and educated on all aspects of their role in assisting in investigations, including collecting, preserving, and protecting evidence. Their help is valuable for all types of investigations. It is just as important to solve the accidental fire scene as it is to solve the criminal case. By solving all investigations on fire and explosion scenes, we can go a long way to prevent similar incidents from occurring in the future.

The Police Patrol Officer

Because police officers are on patrol, they are frequently the first to arrive on the scene of a fire or explosion. On these occasions, law enforcement first responders are in a position to make many important observations should the location turn out to be a crime scene. As trained observers they can interpret what they see and can even take further actions should they feel it is necessary.

A good example is when a police officer observes someone walking or running away from a fire scene. Most citizens move toward a fire scene out of curiosity. The officer, if time allows, can approach the fleeing individual to investigate. (See **Figure 3-4.**) They can verify the individual's identification and conduct a brief interview to ascertain why the person was in the area and why the person was leaving. Simply leaving a scene does not make a person guilty of a crime. A person may have a good reason for leaving the area. On the other hand, that person may have been the one to set the fire and may want to distance him- or herself from the scene before being noticed. When a police officer obtains the person's identification information, the scene investigator may be able to interview the person later. If the person is an employee or ex-employee of the business burning, the individual would want to stay at the scene and not leave. This would make the person's actions suspicious and worthy of future investigation.

Depending on the situation, the police officer may not have time to take any such action. Injured victims might need immediate attention. The officer may need to move a crowd away from the incident scene to protect them and to protect the scene because it might be a crime scene. The scene could be blocked with bystanders' vehicles, and the officer might need to clear traffic and parked vehicles to make room for arriving fire apparatus. Even with all these other duties, a police officer can most likely describe suspicious individuals or even call in the information for another patrol officer to investigate.

Figure 3-4 *The police officer on the scene can interview or detain suspicious individuals of interest.*

■ Note

Just because someone appears to be helpful does not make him or her guilty of a crime. On the other hand, the motivation of the person setting the fire might be to show the fire department that this person would be a good candidate for a firefighting job.

Directing traffic away from the scene is an important function of the police officer from a suppression aspect because it ensures that the fire apparatus can lay lines and get into the best position for pumping and spotting ladders. Sometimes even the police officer's own vehicle can get in the way of fire apparatus placement. As they direct traffic and people away from harm's way, officers might notice an overly enthusiastic citizen who appears eager to assist. It is not unusual to even see these individuals pull lines off the apparatus or drag hose behind the firefighters. If the officer moves the individual away from the scene, he or she should also obtain the person's identification information so that the investigator can interview this individual at a later time. Obtaining identification can easily be done as a guise of wanting to thank the individual for the assistance at a later date. Again, just because someone appears to be helpful does not make him or her guilty of a crime. On the other hand, the motivation of the person setting the fire might be to show the fire department that this person would be a good candidate for a firefighting job.

barricade tape
wide, brightly colored tape with a clear message that will prevent entry onto the property

In addition to keeping people clear of the scene for their own safety, it is imperative that the officer consider the security of the scene a priority as well. When possible, the police officer should start placing **barricade tape** around the scene. (See **Figure 3-5**.) This tape is usually yellow with wording such as *Fire Line* or *Police Line* and also includes the wording *DO NOT CROSS,* which is usually understood by the public. As additional resources arrive, they can assist in ensuring that only public safety personnel approach and enter the scene.

Figure 3-5 *Police or fire personnel can cordon off and protect the fire scene by using barricade tape.*

As the fire is placed under control and fewer resources are needed, firefighters can assist with security of the scene.

In the absence of a fire investigator, it can be the responsibility of the police officer to protect the scene or to collect obvious evidence to prevent its destruction or disappearance.

The Firefighter

Like the police officer and the EMT, the firefighter enters strategy planning mode from the minute the alarm is broadcast over the radio. Considerations about the time of day, weather conditions, the location of the fire, whether it is in a commercial district or residential area are going through firefighters' minds as they speculate about what they may encounter when they arrive on the scene.

■ Note
The presence of obstacles en route to or at the scene can trigger a suspicion that someone didn't want the firefighters to put a quick stop to the fire.

As they approach the scene, they will be following through with their strategy of where to lay the lines for water supply, where to place any aerial apparatus and pumpers to enable them to carry out their assigned task, and whether they will need additional resources. During this process, they are also observing the scene.

While carrying out their suppression duties firefighters can also make observations that are important to the investigation. For example, the presence of obstacles en route to or at the scene can trigger a suspicion that someone didn't want the firefighters to put a quick stop to the fire. This is not to be confused with usual problems with traffic or cluttered lawns or hallways; however, should a downed tree block a driveway or road, the firefighter or officer should note whether it is a natural break or whether there are saw marks on the trunk. Even if the tree fell from high winds, an arsonist might have waited until a large storm was approaching the area before setting the fire. A word of caution: Just because there are storm-created obstacles does not mean the fire was incendiary. The culmination of *all* the facts determines the origin and cause of a fire, not individual circumstances.

■ Note
The culmination of *all* the facts determines the origin and cause of a fire.

Other observations made by firefighters upon arrival are the condition of the doors and windows. Were they locked or unlocked, open or closed? Locked doors are not unusual, but it is unusual to find doors nailed shut. This may be a security measure, or it may be a stalling tactic used by an arsonist to hinder firefighters from getting to the fire. On the other hand, windows nailed shut may be a security measure for the occupants. It is critical that firefighters ascertain door and window conditions by simply trying to open them before forcing. Basic firefighting training requires the firefighter to try the door knob before using forcible entry tools. The point is to prevent unnecessary damage and to keep from masking prior attempts by perpetrators who may have forced the door before the firefighter's arrival. (See **Figure 3-6.**)

casement windows
windows with hinges on the side that usually open with a rotating crank assembly

The condition of the windows is important. Were they forced prior to the firefighter's arrival, indicating an illegal entry? Were they left open on an extremely cold day? This could indicate the arsonist's desire to ensure sufficient air to feed the fire. For example, in one residential fire the **casement windows** were cranked open, and then the arsonist removed the handles in an effort to keep anyone from closing the windows—more important, it raised a red flag because this was not a typical behavior of the occupants of such a structure. A firefighter noticed this strange situation and reported it to the officer in charge (OIC) of the scene, who in turn advised the arriving fire investigator.

■ Note
Nothing is more interesting than to have the homeowner state that he was eating breakfast and just barely escaped with his life only to have a firefighter observe that the kitchen cabinets and refrigerator were completely empty.

Firefighter first responders also usually notice the type, quantity, and arrangement of the structure's contents. Firefighters should find the living room contents to consist of couches, chairs, and entertainment equipment such as stereos or televisions. Bedrooms should have beds, bureaus, and sufficient clothing to show they were occupied. Kitchens should have cooking and eating utensils along with sufficient quantities of foods, spices, and condiments. Nothing is more interesting than to have the homeowner state that he was eating breakfast and just barely escaped with his life only to have a firefighter observe that the kitchen cabinets and refrigerator were completely empty.

The arrangement of the structure's contents is of concern. In one instance, firefighters made their initial entry into a real estate office and found piles of magazines, books, and papers in the middle of the floor making a path of combustibles leading from one room to another. The obvious intent was to create a trailer to spread the fire. When the firefighters realized what they had discovered, they radioed the OIC, who immediately started securing the area as a crime scene.

■ Note
An unusually colored flame can indicate the presence of a chemical not usual to that type of structure.

Firefighters also observe the fire itself. After a few years of firefighting experience, a firefighter can recognize when a fire has unusual burning characteristics. Situations in which the fire burns, extinguishes, and then reignites may be an indication of a petroleum accelerant. An unusually colored flame can indicate

Figure 3-6 *Tool mark impressions must be protected. Unnecessary damage from suppression personnel could mask these imprints.*

the presence of a chemical not usual to that type of structure. Of course, obvious facts are of interest as well—was the structure burning from the outside or was the fire contained in the interior?

FIRST RESPONDER RESPONSIBILITIES

incendiary
the willful and intentional setting of a fire, with malice

The one ultimate responsibility of first responders is to call for assistance should they not be able to make an accurate determination of the origin and cause of the fire. If they suspect for any reason that the fire may be **incendiary**—intentionally set with malice—or if the fire should involve any criminal activity, they should call for the assigned investigator.

Before the arrival of the investigator, there are actions that should be taken to ensure the protection of any evidence and establish a safer environment. The first responder can also start looking at various systems to ascertain their status, such as fire alarm systems, sprinkler systems, or intrusion alarms. The documentation of suppression and individual actions as well as the observations of fire personnel will be necessary. The sooner these steps can be taken, the more accurate will be the results.

■ Note
The scene must be secured not only from outsiders but from other first responders as well.

Securing the Scene

As soon as practical, the scene, which might include the building and surrounding area, must be secured to prevent unauthorized entry. Securing the scene is done regardless of whether there are indications that the fire is incendiary in nature or accidental. It is just as important to protect the scene of an accidental fire so that evidence can be secured and preserved for a civil trial that may help prevent future fires.

sheetrock
often called drywall, gypsum, or wallboard; a crumbly material called gypsum sandwiched between two layers of thick paper; gypsum is fire resistive

The scene must be secured not only from outsiders but from other first responders as well. Once the scene is secure, entrance should be limited to those who need to carry out assigned tasks such as overhaul and safety. Needless walking around inside the fire scene can cause unnecessary damage to unknown evidence; create additional damage to the structure, such as leaning against **sheetrock** and causing it to crumble, which could destroy a fire pattern (see **Figure 3-7**); or create further contamination of the scene from fire boots.

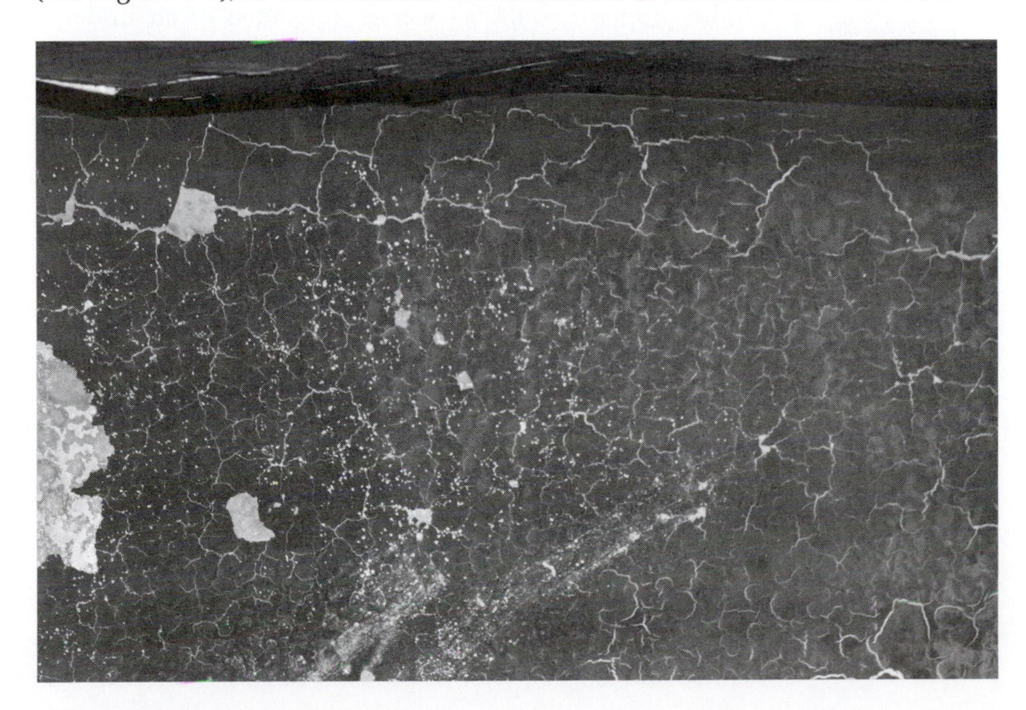

Figure 3-7 *Sheetrock walls show burn patterns but become fragile as a result of heat and can be easily damaged from fire suppression streams or movement of debris during an initial investigation.*

The area outside may contain evidence such as tire tracks, footprints, or discarded items such as cigarettes or empty or not-so-empty containers that may have contained an accelerant. Various methods are used to protect this evidence. One is to place an empty bucket upside down over the evidence to protect it from damage. Interestingly enough, it is sometimes difficult to prevent the most innocent firefighter from lifting the bucket and taking a peak underneath.

The method to secure the scene needs to be obvious to the general public. Both police and fire apparatus should carry barricade tape to mark the area that is off limits to all clearly, protecting the scene and any potential evidence. This is also a safety factor because it also limits entry into an area that may be dangerous. Natural items around the scene can be used to secure the tape: Trees, signposts, mailboxes, and even bushes are used to support the tape, keeping it up enough to be seen and send the message for all to keep out.

Ascertaining Status of Electrical Service

First responders should notice the status of electricity to the structure. Were there lights on in the structure when they arrived? Granted, during the day this is hard to determine. Just as important is the fact that there may have been no electricity to the structure. When conducting size-up, a fire officer may notice that the structure is older and the electric company's meter is missing from the meter base, which now has a blank cover. (See **Figure 3-8.**)

Of course it is always wise to check for secondary sources of electricity. The property could have been upgraded to an underground service, and the meter was relocated somewhere else on the structure. A thorough examination is important to ascertain the status of electricity for safety and for the eventual investigation of the fire.

■ Note
Some people steal electricity from their neighbors.

An additional note on electricity: Some people steal electricity from their neighbors. Investigators have found extension cords leading to an outside outlet on a neighbor's home and the wire hidden with leaves or dirt. The amount of electricity coming from that electrical branch may be limited, but it can still be a hazard and a source of heat for the ignition sequence. This type of theft typically works on neighbors who do not usually walk around the outside of their homes, such as the elderly or those who go on frequent or long trips away from home.

It is important to note whether there is no electricity to the structure prior to the fire. This eliminates most electrical appliances, electrical distribution system, and auxiliary devices from being involved in the ignition sequence. But the first responder conducting the preliminary investigation or the assigned investigator must still examine and document all appliances to make certain they are not involved with the fire ignition.

Dealing with Alarm Systems

Both commercial and residential structures can have fire alarms as well as intrusion alarms. First responders can verify from the structure's occupants or

Figure 3-8 *Just because the electrical company meter is missing does not necessarily mean that there is no electricity in the structure. Look for other potential sources.*

owners the status of these alarm systems. Each system will most likely have a local control box within the structure that may provide valuable information. A local alarm system provides an alarm only within or on the outside of the structure. A monitored alarm system sends a signal to a remote location where it is supervised. Should an activation of a detector or sensor occur in a monitored system, the owner of the system and the local authorities are notified for a response depending on the type of alarm. Local police or fire may have experienced previous activation of this system that can indicate whether the system was faulty or experienced previous problems that may be related to the emergency at hand. The most obvious fact that should be noted by first responders is whether they heard anything that would indicate that the alarm had activated such as a siren, horn, or klaxon.

Figure 3-9 *When the OIC obtains information that may be valuable to the incident, it is always best to take complete notes.*

Documenting and Making Sense of Observations

When first responders arrive on the scene and interact with individuals such as occupants, owners, witnesses, or neighbors, it is important that they remember what they see and hear. Some statements may be crucial to the investigation. If an observation is unusual enough, the first responder should write down any comment so that it can more easily be recalled at a later time. (See **Figure 3-9.**) An investigation is a process of seeking the truth. Statements showing innocence are just as important as those giving an indication of guilt.

■ Note
Statements showing innocence are just as important as those giving an indication of guilt.

First responders should briefly document for future recall all activities they are involved in. This includes when an EMT administers first aid to a victim, a police officer moves around to the rear of the structure to check for activity, and a firefighter forcibly enters the structure or places fans for ventilation.

The observations of different first responders may conflict. For example, the observation of smoke and flame color varies with the stages of the fire. These observations may be helpful to the investigator to provide insight on what was burning; it is important, however, for the first responder to document when he or she made the specific observation. It is unreasonable to expect first responders to constantly look at their watches. However, there are other methods of documenting time; for example, should they report black smoke as they pulled onto the scene, their arrival time will be documented by Emergency

Communications Center (Dispatch). Knowing the time is important in order to compare observations. One person could report brown or gray smoke, another black, and a third person could report white. What may seem like a conflict in observations is easily explained in doing a time line comparison. For example, consider a barn fire in a rural area. Several first responders report that the smoke was white; another says black, and a third reports that the smoke was brown. After an in-depth interview, the investigators put together a time line that explains the differences. The local patrol deputy was the first on the scene. As she arrived, she observed brown and gray smoke. About 8 minutes later, firefighters on the first fire engine arriving on the scene reported black smoke coming from the roof. The rescue squad arrived about 6 minutes after the first engine, and the EMTs reported seeing white smoke.

Each can be logically explained. The deputy saw ordinary combustibles burning, such as the hay and wood inside the barn, which give off gray and brown smoke. When the first engine arrived, the fire had progressed from inside the barn to the tar shingle roof, which gives off black smoke when burning. By the time the EMTs arrived, the fire company had started putting water on the fire and, as expected, steam was observed, which is what was described as white smoke. The time line explains the differences in observations. But even better, after knowing the time line, the observation of the deputy points toward the fire starting on the interior of the barn. This simplistic example points to the importance of conducting good in-depth interviews and the value of the observations of first responders.

building official
the person responsible for the enforcement of the local and state building code; the building official's office is responsible for issuing construction permits and certificates of occupancy as a means of ensuring compliance with the code

Documenting Other Suppression Activities

From the beginning of the incident all the way through the overhaul and cleanup, safety issues must be monitored and abated, and these actions should be documented. This process can include shutting off all utilities; protecting an exposed, downed electrical line; and diverting water from the structure to lessen the weight. Additional resources may be available to determine structural stability, such as the local **building official.** The expertise and experience of the building official are valuable resources for first responders.

Note
The expertise and experience of the building official are valuable resources for first responders.

Documenting Any Holes in the Floor During the suppression of the fire, the fire crews may have noticed the existence of holes in the floor that created a serious hazard for the firefighters and investigators. These can be created by flame impingement or over a longer period of time from smoldering embers under the debris. They can also be created by the arsonist of a means to slow down suppression operations.

These holes can also be created by the firefighters during suppression activities such as when they cut holes in the floor to drain off water. The weight of water can lead to further unstable conditions, thus the need to remove water as soon as practical.

■ **Note**
Because attack lines and ventilation can change the direction of fire travel, initiating these activities is important to document and relay to the investigating officer.

Documenting Fire Suppression Ventilation Activities Because attack lines and ventilation can change the direction of fire travel, initiating these activities is important to document and relay to the investigating officer. Firefighters use ventilation to remove smoke and combustible gases from a structure. Sometimes when ventilation is improperly implemented, it can actually change the direction of the fire travel, pulling the fire into areas that would not logically have burned.

Documenting Delay in Fire Suppression Investigators are also interested in any problems with water supply that may have delayed suppression activities and that could explain additional fire damage. Firefighters also need to use caution with their fire streams because careless application of fire streams can damage evidence—although preserving evidence should always be a third consideration with rescue and suppression as the primary concerns. However, an innocent application of a straight stream of water down an undamaged aisle in a store, for instance, can knock items off shelves and walls, making it look as though vandals caused the destruction.

The OIC can communicate these issues to the investigator as well as provide details on any water supply problem to assess whether it is important to the investigation. Debris inside a hydrant or a standpipe system may have been the work of juveniles with nothing better to do. But an arsonist could also plug the water supply to keep the firefighters from extinguishing an intentionally set fire.

■ **Note**
It is extremely desirable that all fire departments participate in the National Fire Incident Reporting System (NFIRS).

Documenting Hazardous Materials Issues Hazardous materials can be found in every structure. The discovery of additional hazards from chemical, biological, or other sources must be reported to the investigator for safety purposes. It is important for the investigator to assess whether these hazards are common to the structure and its use or not. Today, investigators must address additional concerns about illegal activities that involve hazardous materials such as drug labs in homes, motel rooms, and abandoned structures.

Documenting Potential Terrorist Indicators Since September 11, 2001, fire investigators must consider terrorism, domestic and foreign, as a motive for the fire and explosion scene. First responders and fire investigators must be suspicious of finding secondary devices. Some such devices are aimed at the first responders to kill or maim them as part of the overall objective.

National Fire Incident Reporting System (NFIRS)
a national computer-based reporting system that fire departments use to report fires and other incidents to which they respond; this is a uniform system to collect data for both local and national use

Completing the Fire Incident Report

The OIC has an equal interest in many of the issues that concern the fire investigator. Each department may have different reporting criteria and procedures. But it is extremely desirable that all fire departments participate in the **National Fire Incident Reporting System (NFIRS).** The data collected from these reports have immense value for many aspects of the fire service. At the conclusion of the event, the OIC, or someone the OIC designates, must fill out the fire report. The report documents items such as the first material ignited, the heat source that provided the energy to start the fire, and what brought the heat source together with the first material ignited.

By using the NFIRS, first responders can investigate and report their findings on the fire area or origin and cause. These data can be used to educate the public on preventing future fires, to combat arson through education, and to justify fiscal allocations for investigative resources. These statistics can also be used to justify changes to the fire code and taking a dangerous product off the market in the form of recalls. Another benefit of contributing to the NFIRS is identifying suppression fiscal needs, training, and resource allocations, which all can justify future purchases of apparatus and equipment.

SUMMARY

First responders are public safety personnel who are dispatched to an emergency scene by an Emergency Communications Center on behalf of the government or under contract to the government. First responders are emergency medical personnel, patrol police officers, and firefighters. In many instances, the fire officer in charge, the OIC's designee, or the patrol police officer on the scene may be the first investigator to work the scene and determine a plausible fire origin and cause. When the fire scene is more complex or when there are indications that the fire may be incendiary in nature, the OIC contacts a fire investigator to conduct a detailed investigation. In many situations, the OIC, or the OIC's designee, is also responsible for making sure that the fire report is accurately and completely filled out.

First responders are critical to the overall success of fire investigations. Their observations upon arrival and in handling the emergency provide a wealth of knowledge that can be put to use in the final analysis of the scene. They may be the first to recognize evidence and to take steps to protect that evidence. They hear comments and see actions of individuals that may give insight as to what happened on the scene prior to the arrival of first responders. They might even observe unusual behavior of individuals at the scene that could point to the perpetrator should the incident turn into a crime scene.

KEY TERMS

Barricade tape Wide, brightly colored tape with a clear message that will prevent entry onto the property.

Building official The person responsible for the enforcement of the local and state building code. The building official's office is responsible for issuing construction permits and certificates of occupancy as a means of ensuring compliance with the code.

Casement windows Windows with hinges on the side that usually open with a rotating crank assembly.

Dispatcher A person who works in an Emergency Communications Center (E 9-1-1 centers), whose duties are to receive calls from the public for emergency assistance, and then to send the appropriate emergency unit(s) to handle the situation.

Emergency Communications Center (Dispatch) Sometimes referred to as the E 9-1-1 center; a place where calls from the public for emergency assistance are received and whose staff (dispatchers) alert and send the appropriate emergency unit(s) to handle the situation.

Emergency medical technicians Individuals with specialized medical skills, trained to offer basic first aid in the field under emergency circumstances, and then to assist in the transporting of the patient to medical facilities if necessary.

First responders Any public safety responders who may be or may have the potential of being first on the scene as the result of being sent there by the Emergency Communications Department.

Incendiary The willful and intentionally setting of a fire, with malice.

National Fire Incident Reporting System (NFIRS) A national computer-based reporting system that fire departments use to report fires and other incidents to which they respond. This is a uniform system to collect data for both local and national use.

Sheetrock Often called drywall, gypsum, or wallboard; a crumbly material called gypsum sandwiched between two layers of thick paper. Gypsum is fire resistive.

REVIEW QUESTIONS

1. List the role responsibilities of EMT first responders and how these first responders can contribute to the fire investigation.
2. List the role responsibilities of patrol police officer first responders and how they may contribute to the fire investigation.
3. What is the importance of finding all windows open in a burning structure in freezing weather?
4. Why should the first responder observe the contents of the structure? Provide an example of a situation that would be unusual or suspicious.
5. When the fire is under control and essentially put out, why is it important to keep firefighters from entering the structure unless necessary to accomplish an assigned task or mission? Provide examples.
6. Why is it important for first responders to write down as soon as practical any comments made by witnesses, victims, neighbors, and others on the scene?
7. Why does the fire investigator want to know about the use of attack lines and ventilation?
8. Why might a fire investigator consider the dispatcher as a first responder?
9. Why is it important that the fire investigator be advised of any delay in applying water to the fire when first responders arrived on the scene?
10. Does the fact that windows are nailed shut prove that the fire was incendiary in nature? Explain your answer.

DISCUSSION QUESTIONS

1. Of what significance is it that the building on fire is known to have an audible fire alarm system, but on arrival the first responders hear no audible alarm?
2. If you have prior fire or police experience, what have you observed that you now know may be of interest to a fire investigator?
3. There is no doubt that the fire is incendiary in nature. The homeowner has been mean and behaved badly toward the first responders. Even though you suspect that he may have been involved in setting the fire, you hear him utter a comment that may indicate his innocence. Should you report this to the investigator and why?

ACTIVITIES

1. Using your imagination, list the sounds that a dispatcher may hear in the background of a telephone call that could be of importance to a fire investigator. Compare your list with the lists of other students or readers of this text.
2. Using your imagination, list the type of things that an emergency medical responder may hear or see that could be of importance to the investigator. Compare your list with the lists of other students or readers of this text.

THE FIRE INVESTIGATOR

Learning Objectives

Upon completion of this chapter, you should be able to:

- Describe who has the responsibility to investigate fires.
- Describe the different types of responsibilities assigned to local, state, and federal authorities.
- Describe the role of the insurance industry in the investigation of fire origin, cause, and responsibility.
- Describe various research entities and their involvement in fire investigations.
- Describe the importance of maintaining relationships with each of these resources.

CASE STUDY

A homeowner refueled his kerosene heater in his living room from a 2-gallon container he had just filled at a local service station. He did everything right when fueling and starting the unit. However, when he started the heater the flame from the wick was higher than he expected and eventually the flame escaped the unit and ignited nearby combustibles. The fire department was notified, and fortunately, the fire was contained to the living room.

The engine company officer recognized how the fire started but realized that it warranted further investigation, so he called for the duty fire marshal, investigator, to respond to the scene. Upon arrival, the engine company officer apprised the fire marshal of the facts surrounding the case. After looking at the scene, the fire marshal concurred with the officer in charge (OIC) that the kerosene fuel must have been mixed with gasoline. Examining the contents of the 2-gallon container, she realized that the color was not red, which is the color of the dye in kerosene sold for heaters in that area. To confirm their suspicions, the fire marshal took a sample to send to the laboratory for testing.

The owner advised that he had gone to the gas station on the corner that evening to get fuel. The homeowner was very emphatic that he did not accidentally purchase gasoline and that the container in question had only contained kerosene since he bought it new last year.

A trip to the local service station paid off. The fire marshal spoke with the station manager who stated that no problems had been reported to him about his products. He also mentioned that they had received a shipment of fuel that very afternoon. The fire marshal purchased 1 gallon of kerosene, putting it in a new 1-gallon container. There was a red tint, but it was very slight. The service station manager shut down the pump immediately when he recognized that the fuel being dispensed was not kerosene but gasoline. The lab later confirmed that the fuel from the pump was gasoline. The delivery person had placed 1,500 gallons of gasoline in the wrong tank. All the markings on the ground were correct; it was just human error by a new employee. Fortunately, that was his first and only delivery for that company.

An effort to track down all persons who purchased K-1 kerosene from that station that afternoon was under way. Regretfully, two more small fires burned that evening from the contaminated kerosene, but none thereafter. Thanks to rapid response from the fuel station once the manager realized there was a problem, cooperation from each of the local news stations, and a good investigation with prompt follow-up, other disasters were prevented.

As you can see, the steps of a fire investigation can prevent future adverse events. In this case, the fire department took further steps by providing information for local newspaper articles and by conducting speaking engagements to further educate the public on kerosene safety.

INTRODUCTION

Regardless of who investigates the fire, the person assigned the investigative task must be completely familiar with all related laws, ordinances, and regulations pertaining to the investigator's responsibilities in the investigation. The old saying "Ignorance of the law is no excuse" not only applies to citizens, but to investigators as well. Part of fire investigator responsibility is to know the legal limitations and responsibilities in carrying out the investigation.

Fire investigators come from two different domains, the government and the private sector. Both types of investigators essentially perform the same job, they both should have the same expertise, and they both should know their assigned duties. Although the focus of their tasks may vary from time to time, they are all seekers of truth. The investigation responsibilities vary as well. Investigators could be assigned to conduct only a scene investigation with no interviews. In contrast, another investigator might be assigned to work the criminal aspect of the case based on the findings from a previous scene investigation. The purpose of the investigation could even be limited to examining whether the insurance carrier has a liability before it expends more funds on a detailed investigation.

There is yet another type of investigation, one based solely on research. These investigations can vary from digging fire scenes, compiling information from previous investigations, to reconstructing a fire scenario in a laboratory to document the results. These efforts can be done on behalf of the government or a private entity.

What is unique to the field of fire investigations is that investigators can testify in court and give their expert opinion on where the fire originated and how the fire started. Their opinion is based on the evidence they obtain during their investigation. Differing opinions will exist, and two colleagues could very well find that they are testifying for opposing sides in a court of law. It is human nature not to want someone to contradict your theories and opinions. Regardless, you have to strive not to take disagreement personally and to recognize your peers' expertise and respect their opinions, even though you may strongly disagree with them. The professional fire investigator must understand that there will be differences of opinion and that, at the end of the day, they are all colleagues.

This chapter examines the role of fire investigators from both the private and public sectors. It describes the role of local, state, and federal investigators and discusses the duties each may perform. It is important for both first responder and assigned investigators to know the available resources and their capabilities.

FIRE INVESTIGATOR VERSUS ARSON INVESTIGATOR

Simple terminology can make the difference in perceptions. If someone investigates fire scenes to make an accurate determination of the fire origin and cause, that person is a *fire investigator*. For a multitude of good reasons, the public fire investigator has become known as an *arson investigator,* printing this designation on their business cards, coveralls, and even their vehicles. For the most part, these individuals do work arsons and they are arson experts. However, there is good reason to consider using the terms *fire marshal* and *fire investigator* to describe this public official. To limit the title to *arson investigator* may imply that the individual looks only for arson; as you will learn, this can jeopardize the credibility of the investigation that is being conducted.

■ Note
Fires are considered accidental in nature until proven otherwise.

■ Note
Finding the cause of the fire is for the good of the people.

In most states, fires are considered accidental in nature until proven otherwise. This is important for many reasons. Foremost is the fact that incident is not a crime until evidence can be shown that the fire was not accidental in nature. Nevertheless, the scene can still legally be protected and treated as if it is a crime scene as long as it is under the control of the fire investigator or fire department. This is based on the fact that finding the cause of the fire is for the good of the people. Several rules, exceptions, and decisions about scene security are discussed later in this book.

Consider another factor. To classify a fire as incendiary, accidental causes must be ruled out. To do this, the investigator must look at all aspects of the fire. The label *arson investigator* implies that the investigator looks only at arsons or searches for arson evidence. Worse yet, it implies that the investigator has a predetermined mindset that the fire was caused by **arson.**

arson
intentionally and willfully setting a fire with malice

Using the term *arson investigator* in other situations can be just as damaging. It is bad enough that a family loses its home to the ravages of fire. Then, there is even more pain because parked in front of their former home is a government vehicle with the bold letters *Arson Unit* or *Arson Task Force* printed on its side. The neighbors might think this unfortunate, innocent family set their own home on fire. As public servants, we have committed an injustice in using this label indiscriminately.

A person who investigates the cause of a fire is a *fire* investigator. This person might specialize in arson, accidental fires, or even environmental crimes. Nevertheless, he or she is a fire investigator first. As such, when you appear in court using the title *fire investigator* can imply that you looked at the fire—the entire fire—before concluding the cause of the fire.

public fire investigator
an investigator working for the government representing the locality, state, or federal government; usually denotes someone whose job requirements include the determination of the area of origin and cause of fires and explosions

PUBLIC FIRE INVESTIGATORS

The **public fire investigator** works for a government entity at the local level (town, city, or county) or at the state level (working for a state fire marshal or the state police). In some circumstances, public fire investigators may work for the state's Department of Insurance. At the federal level, several entities work on fire investigations, the most prominent of which is the **Bureau of Alcohol, Tobacco, Firearms, and Explosives (ATF),** which has dedicated National Response Teams (NRTs) prepositioned across the entire United States. Also

Bureau of Alcohol, Tobacco, Firearms, and Explosives (ATF)
a federal agency with fire investigative authority; it operates a National Response Team, which can provide almost every resource necessary at a fire scene

within the federal system other law enforcement agencies are involved in arson investigations, including the Federal Bureau of Investigation, the U.S. Forest Service, and all branches of the Department of Defense.

The requirement of the government to investigate the cause of the fire is usually mandated through a local, state, or federal ordinance, legislated laws, or implied mandate. This can be a direct authorization clearly stating who is to investigate or it can be implied in general phrases. Many of the city and county charters created in the 1700s and the early 1800s may state or imply that it is the responsibility of that government to provide for the protection and safety of its citizens. As such, it gives the authority to the local leaders, the Authority Having Jurisdiction (AHJ), to create and maintain police and fire departments. The next logical step is to assign the duties of the various departments including who will investigate fires and determine who was responsible for the incident or the cause of the incident.

Investigations: Police Department versus Fire Department

The decision of which department is to investigate fires usually rests with the AHJ. History and politics have a hand to play in this as well. *What has been, will be* is a phrase describing the fact that because the fire (or police) department has always investigated fires, it shall also in the future. However, the AHJ can be seen supporting this issue, not by vote, but by delegation of fiscal resources to continue that operation. If the budget allocates a lump sum of funding for new arson tools to the fire department, there is no doubt the fire department has fire investigative responsibilities.

■ Note

Every law enforcement agency has the responsibility to enforce the law, all the laws.

Every law enforcement agency has the responsibility to enforce the law, all the laws. Even if, officially, the fire department has the investigation unit and a police detective uncovers sufficient information about an individual in an arson case, then the police detective should make the arrest.

Both police and fire department have a good reason for wanting the fire investigation unit within their department. Fire departments want to know all the details about the cause of fires to establish training programs to prevent future fires. Police departments use as one measure of success the closing of investigations with successful convictions, generally referred to as the *clearance rate*. The more criminal cases solved, the higher the clearance rate.

However, when the different departments and individuals work together, everyone benefits. Sometimes this appears in the form of simple courtesies, such as the fire marshal makes an arrest for arson, and then uses the sheriff's department to make reports to the National Crime Information Center (NCIC). This way, the sheriff gets credit for the arrest and clearing a criminal case.

The most cooperative concept, one from which everyone benefits, is the task force or strike force concept. Depending on your location, the terms can mean different things. Primarily, for this unit, both the fire and police departments dedicate two members each to serve on this team. There might then be a tendency for the two fire employees to work the fire scenes and the two police employees to do the follow-up and make the arrest. Sometimes with small units, using a task force is not always successful because it has the appearance that the fire department personnel do all the digging and dirty work and the police department personnel do the interviews. Strong leadership can still make the task force unit work by creative assignments so that each understands the other's duties.

Another type of team concept that has been quite successful is where each individual is crossed-trained so that all four team members have the same fire and law enforcement training. As with any group, each individual brings the sum of their own experiences to the team, adding that much more to the team's potential success. Each individual is responsible for his or her own case, taking the case from the scene examination to the final report. Other team members are there to assist, especially for the larger incidents, and to listen to as well as share ideas and concepts.

■ Note
Each individual brings the sum of their own experiences to the team.

Investigators: Police versus Fire

There is an ongoing debate as to who makes a better fire investigator—a police officer or a firefighter. Police officers need to learn all aspects of the dynamics of fire chemistry, the impact of firefighting techniques, and other factors that can affect a fire and its resulting burn pattern. Firefighters need to learn legal principles, interview techniques, and basic police practices. What really makes a good investigator is the individual, his or her capabilities and desire to be meticulous, thorough, and consistently systematic in the investigation. Such a person might work for either a fire or a police agency.

■ Note
What really makes a good investigator is the individual, his or her capabilities and desire to be meticulous, thorough, and consistently systematic in the investigation.

What we have established is that conducting a fire investigation requires the skills found in both the fire service and law enforcement. There are several ways to find the needed skills regardless of who does the job. Although the candidate from the fire service might lack what may be commonly referred to as street smarts, a ride-along program in which the fire department member rides along with both a patrol officer and a detective squad can help fill in those educational gaps.

The same is true for the local police investigator, who may never have entered a burning building to experience how a well-laid fire stream can change the direction of fire travel and cause an unusual burn pattern. A ride-along with a local engine company can go a long way in helping educate a police investigator about the technical aspects of a fire scene.

Both individuals will need some basic training for their own protection and to protect others. Fire personnel will definitely need a training program to teach them basic law enforcement skills. They may not need an extensive course to cover traffic or drug laws, but all other legal aspects need to be on their training agenda. The training program for the police personnel would not necessarily be a full firefighter training program. Police officers do not need to know how to pack hose, but they need extensive training on self-contained breathing apparatus and fireground safety. Building construction education and fire behavior training are necessary for them to be able to make accurate determinations of fire origin and cause.

Some states have mandatory training specifically for all fire investigators. This training program might address most of the needs for both police and fire. Most fire investigator training programs follow NFPA 1033, *Standard for Professional Qualifications for Fire Investigator*. The police personnel attending will also need some prerequisite knowledge in various fire topics to be on a level playing field once they show up for the fire investigator training program.

■ Note
Most fire investigator training programs follow NFPA 1033, *Standard for Professional Qualifications for Fire Investigator*.

Local Fire Investigators

As discussed in the previous chapter, the first fire investigator to arrive at most scenes is the fire officer in charge or someone that officer designates to determine the origin and cause of the fire. Should the cause not be obvious or should the facts surrounding the incident indicate that the fire was maliciously set, it is the duty of the first responder to contact a full-time fire investigator as the department's procedures dictate. That person will most likely be the local public fire investigator, frequently titled fire marshal.

■ Note
The term *fire marshal* is the predominate title associated with those in the fire investigative field.

The term *fire marshal* is the predominate title associated with those in the fire investigative field, but it is not the only associated title. Today a fire marshal can be an inspector, an investigator, or even a public educator. However, investigators, inspectors, and public educators exist that do not carry the title of fire marshal.

Some jurisdictions call all their investigators fire marshal where others use that title for the person in charge of the fire investigative office. (See **Figure 4-1.**) The staff then carry the title of deputy or assistant fire marshal. This is strictly a local issue that matches the rank structure with their peers in the suppression side of the department. It should also be mentioned that not all fire marshals report to the fire chief. Less frequently, fire marshal offices are located within the Building Department, within a police department, or as a stand-alone agency reporting to the head of that locality.

Not every locality has enough investigators to handle every fire investigation in that jurisdiction. First, it is not necessary to have that many investigators because the area of origin and the cause of many fires are obvious and an experienced fire officer or senior firefighter can easily make the official determination.

Second, it is just not cost effective to have a full-time investigator on standby for every fire that occurs. The average investigator works a 40-hour shift. This is only for an average of 48 weeks a year, considering holidays and annual leave. Usually, investigators take at least a week's worth of training, every year, which reduces their working time to 47 weeks. A simple fire investigation can take 3 to 4 hours to conduct including travel, looking at the scene, doing brief interviews, and then writing up the report. If this was all investigators ever did, an

Figure 4-1 *The title fire marshal is becoming the predominate title for those enforcing fire codes and investigating fires.*

investigator would average only 11 fires a week for 47 weeks; the most fires to be investigated would be 517 incidents.

■ Note
An average fire investigation can take as long as 2 or 3 days to finish with scene work, interviews, delivering evidence to the lab, and writing the final report.

However, this is not the case. An average fire investigation can take as long as 2 or 3 days to finish with scene work, interviews, delivering evidence to the lab, and writing the final report. (See **Figure 4-2.**) Then, there are the complex fires that require multiple interviews, doing background investigations, following leads, and putting the case together. Should there be an arrest, the suspect has to be processed, including all of the associated paperwork. Then, there are the meetings with the local prosecutor and the detailed report. Any follow-up work prior to trial must be completed and trial preparation takes time as well. None of this includes daily operations issues such as maintaining equipment or vehicles or attending meetings. It is not unusual for an investigator to only average 120 to 140 investigations a year.

Assigned Duties of Fire Investigators The duties of the local fire investigator vary according to the objectives assigned by their departments. Some fire investigators working for the fire department are assigned to conduct only fire scene examinations, interviewing only those who may be at the scene at the time of their visit. Their only assignment is to report the area of origin of the fire and the cause, if possible. After this has been completed, their assignment is finished. Should they discover that the fire was incendiary in nature, they can turn their case over

Figure 4-2 *Some fire scenes are so large that it is not practical to think that the investigation can be finished in only a matter of hours.*

to local law enforcement personnel such as a police or sheriff's department to pursue the criminal aspect of the incident. Should the local police apprehend a suspect, the scene fire investigator would then testify in court as to the fire's area of origin and cause along with any other evidence he or she legally discovered in pursuit of the limited assignment.

Other investigators could be assigned to do the scene inspection and limited off-site interviews up to the point of making a determination of the fire's area of origin and cause. Again, if the fire turns out to be incendiary, the case is turned over to the local law enforcement agency.

The third type of investigative assignment is of the investigator who has police powers to look at the scene, conduct interviews, and make an arrest for arson. (See **Figure 4-3.**) The fourth type of fire investigator has full police powers—the same as any other law enforcement officer—and carries out the investigation from the initial fire scene examination to the apprehension of suspects and conviction. The full police powers grant this fire investigator the authority to handle and make arrests for other crimes associated with the arson. Even with full police powers, though, if the case involves illegal drug activity, for example, it is always best to team up with drug experts for the fire investigator. The same is true for any other type of criminal activity associated with a fire.

■ Note
The public fire investigator has an obligation to the citizens to investigate all fires occurring in his or her jurisdiction, not just arson.

The public fire investigator has an obligation to the citizens to investigate all fires occurring in his or her jurisdiction, not just arson. In fact, arson is a very

Figure 4-3 *Some investigators, conduct in-depth interviews with those involved with the incident.*

small percentage of the total number of fires. Through the efficient investigation of accidental fires, the investigator may have the opportunity to have defective products taken off the market. For example, with enough documentation on a faulty product, the Consumer Product Safety Commission can negotiate with a manufacturer to take that product off the market and conduct a recall. Or the fire investigator can produce public training events that can reduce such fires and save lives. Even on the local level, the fire investigator can act to eliminate a hazard, protecting others from similar events.

People make mistakes and fires occur. By investigating these fires, efforts can be made to prevent future fires. The public is educated each year on seasonal safety through the media as to the hazards of candles, combustible Halloween costumes, and Christmas trees not adequately watered. This is all in an effort to protect the public based on actual events, not because the fire service thinks it might happen.

The assigned duties of the investigator may also mean making the arrest of an individual who, with sufficient probable cause, is identified as the perpetrator of the crime. Of course, this should only be done when authorized by the jurisdiction and even then only after the investigator has received the proper training. This training is usually based on that state's minimum training standards and is usually provided in state-run or state-supported law enforcement schools. (See **Figure 4-4.**)

Figure 4-4 *A student from Virginia Fire Marshal Academy, Law Enforcement School for Fire Marshals doing a take-down procedure and arresting the suspect (role player). Cadre staff observing the procedure.*

State Fire Investigators

■ Note **Most state investigators experience a heavy workload, no different from their peers working other types of investigations.**

■ Note **The only thing consistent about fire investigations at the state level is that there is no one clear model that every state follows.**

Most state fire investigators operate the same as the local fire investigators. In some states, they are referred to as fire marshals either in their own department or within the state police. In other states, they are labeled investigators or special agents performing multiple types of investigations, with fire being just one category. More often than not, they cover a wide geographic area, sometimes three or more local jurisdictions. Most state investigators experience a heavy workload, no different from their peers working other types of investigations. Those affiliated with a state police agency generally concentrate on arson. This is understandable in consideration of the mission of their parent agency.

The only thing consistent about fire investigations at the state level is that there is no one clear model that every state follows. Because of differences in each state government, differences in their mission statement, and where fire investigators stand in an organizational chart it is almost impossible to create a model to match all circumstances.

Other state agencies have investigative authority too. Some of the state parks and state forest services have their own police who have investigative authority. The forester of some states assigns investigative duties to certain members of its staff. In other instances, these investigators have police powers as the law in that state dictates. As a rule, the park service investigator is limited to working cases on park property. The forestry investigator works on fires occurring in state forests, usually on both state property and on private land.

Federal Fire Investigators

Several federal agencies have personnel dedicated to the fire investigation field. Few are called fire marshal except at individual facilities such as the Pentagon and some military bases. These individuals are dedicated to cover their own facility, and, other than providing assistance to other facilities, they do not venture beyond government-owned or -leased property.

The Bureau of Alcohol, Tobacco, Firearms, and Explosives (ATF) has the resources and mission to conduct fire investigations on a national level. In addition to having special agents with fire investigation experience, this agency has set up a National Response Team with specialized equipment and personnel. This team, upon arrival at any given incident can immediately start working with special agents doing the scene examination while another part of the team initiates the interview process. They have their own lab personnel and when necessary can bring in their K-9 units to search for accelerants.

If the scene should present a unique problem, the Fire Research Laboratory is capable of doing reconstructive studies on the fire scene; the lab can re-create entire rooms, creating a scenario to test a theory and document a burn to study its results to be compared with the actual incident as well as other fire investigation issues. As part of its research in creating the Fire Research Laboratory, the ATF hosted the International Conference of Forensic Science where concepts and ideas from around the world were a focus of the event. Fire research and forensic laboratories from other nations shared information about their existing programs and lab layouts and also what they felt would be enhancements to any new facility. After much research, the new ATF laboratory was built, incorporating many of the recommendations from other nations. Part of the ATF lab strategic goal is to conduct internationally recognized research to enhance knowledge of fires, fire scene reconstruction, and fire investigations. Nevertheless, for the purposes of investigating the fire scene they have the ability to re-create entire rooms, creating a scenario to test a theory and document a burn to study its results to be compared with the actual incident.

■ **Note**

Part of the ATF lab strategic goal is to conduct internationally recognized research to enhance knowledge of fires, fire scene reconstruction, and fire investigations.

The **Federal Bureau of Investigation (FBI)** has also worked arson cases on many occasions. It has done so with its wealth of experts working white-collar crimes, organized crimes, and other illegal activities. The FBI has worked cooperatively with the ATF and other federal law enforcement agencies to obtain arrests and successful convictions. It too has a lab and has been instrumental in refuting some mistruths and misconceptions about the science of fire investigations.

Federal Bureau of Investigation
the nation's law enforcement agency; it is available to assist on major fire incidents; in addition to an outstanding crime lab, the FBI has one of the best training academies in the world

Both the FBI and the ATF have provided some outstanding training for fire investigators. The FBI Academy has provided training events for local and state fire investigators at its facility in Quantico. The ATF provides multiple types of training from white-collar arson to advanced arson training at its training facilities in Glynco, Georgia, and Fort A. P. Hill in Virginia.

Although the National Fire Academy (NFA) does not provide investigators, it does provide outstanding training for both firefighters and fire investigators at its facility in Emmetsburg, Maryland. It offers a 2-day course on arson detection for the first responder and a 2-week fire investigation course. It offers many other courses of interest to the fire officer and the fire investigator. The arson detection class is a hand-off course, allowing it to be taught by the states.

Other investigative agencies look at fire scenes as well. The Federal Aviation Administration (FAA) has a responsibility in the investigation of all incidents involving aircraft. Its largest role is in the compilation of statistics and ensuring that corrective action is taken on any hazards uncovered as a result of the investigation. However, it does respond to the scene of major incidents and works closely with the National Transportation Safety Board (NTSB). Usually, the lead at the actual incident scene is a member of the NTSB team.

The National Transportation Safety Board is an independent federal agency that has the responsibility of investigating accidents involving highway, rail, marine, pipelines, and aviation. It also can investigate the release of a hazardous waste as the result of a transportation incident. When each incident occurs, the NTSB sends a Go Team consisting of experts in that particular transportation mode. As a result of its investigations it has produced many reports that have lead to improvements in many aspects of transportation in the United States as well as in other nations around the world.

PRIVATE SECTOR INVESTIGATORS

■ Note
The largest self-insurer is the federal government.

The insurance industry is the largest contributor to the field of fire investigation in the private sector. Some larger companies have opted to be self-insured, setting aside special funding in case of loss from fires or lawsuits. The largest self-insurer is the federal government. To help understand the private sector, you need to be knowledgeable about the insurance industry and its terminology.

Insurance Policies and Subrogation

policyholder
person who owns the insurance policy, possibly the property owner or someone who has an interest in the property

risk
the item covered by the insurance policy; the item that may be lost

In an investigation, the public fire investigator does not always provide sufficient detailed information that is beneficial to the insurance industry. For this reason, insurers must use their own investigators who are either their own employees or they must hire fire investigators from the private sector.

The motive for the insurance industry to investigate fires is based on funding. Let's first establish some basic terms that are specific to the insurance industry: Businesses and homeowners take out fire insurance policies and pay annual premiums. They are then considered the insured, or the **policyholder.** That which is insured, the business or home, is considered the **risk.** The contents of the structure are also considered the risk, if stated as such in the policy. Another part of the policy may cover business interruption, or in the case of a residence, it may cover the cost of temporary housing should that be necessary.

The insurance company collects the premiums, putting so much aside to cover losses as they occur. Part of the premium pays for the operation of the insurance company such as employee salaries, rent, taxes, and so forth. A good year for the insurance companies is when there are few losses, allowing them to take some of the premium proceeds and expand the company, potentially increasing profits. Do not forget that insurance companies exist to make a profit.

A bad year for the insurance companies is one like 1992. Hurricane Andrew slammed into Homestead, Florida, essentially wiping out the city and surrounding area. Several insurance companies were quite concerned that they would not have enough reserve funds to pay all the claims submitted by home and business owners. To pay for these large losses, the premium rates had to increase. The hurricane in Florida resulted in rate increases for home and business owners across the entire country.

The same theory applies to fire losses. For example, a fire occurs at Jones Business Machines, which has a fire and liability policy with the ABC Insurance Company. The loss is $2.5 million for the building alone. ABC Insurance hires a fire investigator, who reports that the fire started in the business next door, Acme Tool Manufacturer. The fire was the result of an Acme employee improperly using a heating appliance. The XYZ Insurance Company covers Acme Tool Manufacturer.

loss
object of value that was destroyed

ABC Insurance Company pays Jones Business Machines the limits of the policy. Because the **loss** exceeded the policy limits, Jones Business Machines gets the full amount of the policy. XYZ Insurance Company pays Acme the limits of its policy as well. The two businesses are now on the way to recovery.

Because the fire started next door, some may feel that Jones Business Machines could file suit against its neighbor. As part of the insurance policy, ABC Insurance Company has a right to do just that. In paying the Jones Business Machines, ABC Insurance Company has the right to essentially step into the business's shoes and file a suit to recover the funds the insurance company paid out in the insurance claim. This process is called **subrogation,** the substitution of one person in the place of another with reference to a lawful claim.[1]

subrogation
the substitution of one person in the place of another with reference to a lawful claim

Subrogation can result from any loss. A fire could occur as the result of a faulty or poorly designed appliance. The insurance company, after paying the claim, could file suit against the appliance manufacturer to recover its expended funds. Remember that subrogation is a means of recovering funds, not for making a profit.

■ **Note**
The average insurance policy is not there to replace absolutely everything.

There is one basic premise that many do not realize about insurance. The average insurance policy is not there to replace absolutely everything. Individuals are not supposed to make a profit on a claim. Having a $200,000 home and buying a $400,000 policy will only get you $200,000 if the home burns. If you had four $100,000 policies on the building, each policy pays a part of the loss, probably $50,000 each to the maximum total of the property value. The way most insurance policy applications are written, the insured must divulge the existence of other polices.

■ Note
Insurance policies are there to protect people from their own mistakes.

■ Note
Larger insurance firms sometimes maintain an investigative staff in a unit commonly referred to as a Special Investigations Unit (SIU).

Special Investigations Unit (SIU)
a division usually within an insurance company that oversees or conducts investigations involving a loss

claims managers
within almost every insurance company is a division assigned to evaluate submitted insurance claims to assess the value of the loss and, if justified, to write the checks necessary to settle the claims

private investigator
an individual involved in providing professional investigations of fires and explosions on a contract basis or as an employee of a private enterprise or organization

Insurance policies are there to protect people from their own mistakes. A policyholder could be responsible for the cause of the fire such as dropping a cigarette onto a sofa, where it falls between the cushion and the armrest. The actions by the policyholder caused the fire, but the insurance company still pays the claim. Exceptions exist when the insured creates an extra hazard that is unusual, such as storing gasoline in 55-gallon drums in the basement next to an oil-fired furnace. This is not to say that the insurance company will not pay the claim, but it certainly may have the right to refuse to pay under such circumstances.

Insurance Investigations

Insurance companies might conduct their own fire investigations for several reasons. Not all jurisdictions in the United States have a local, readily available, qualified fire investigator. Thus, insurance companies may have to rely on their own staff or hire fire investigators to look at a fire scene. The local fire investigator, after determining that a fire is accidental, may not go into any more depth with the investigation, sometimes because the investigator's heavy workload might require him or her to move on to the next case, which may be criminal in nature. Therefore, an insurance company might hire its own investigators to delve into a fire scene if it feels that there was a product failure that may justify subrogation. An entire industry has developed to assist insurance companies with their investigations.

Larger insurance firms sometimes maintain an investigative staff in a unit commonly referred to as a **Special Investigations Unit (SIU).** These individuals specialize in all forms of insurance losses, with fraud a primary concern. Some units hire individuals with special skills such as retired law enforcement or retired fire investigators. Other SIU agents may have insurance experience such as working the claims aspect of the industry.

SIU personnel can work fire scenes themselves or hire personnel from the private sector, such as private fire investigators or engineers with expertise in fire origin and cause, to do so. When SIU personnel work a fire scene, and if they believe that fraud is involved, they may still hire an outside expert to back up their findings and lend their case more credibility should it end up in court.

Instead of SUIs, some insurance companies have **claims managers,** who may hire specialists to do fire investigations. Some insurance companies don't even handle their own claims but hire independent claims management companies to process their claims. These companies may hire the experts, and some of these companies have investigators on staff to either review an investigative file or work the scene themselves.

These **private investigators** or engineers provide reports to the claims department on their findings. Depending on the situation, the claims manager may ask for a verbal report only. If the preliminary report indicates that the fire is accidental with no suspicious situations, the case may be closed. Sometimes claims managers want only a brief written report, and on other occasions they want a detailed report with photographs, video, and interviews. When the brief report is requested and the preliminary information indicates that the fire was incendiary, the case can be expanded to include a more detailed report.

■ Note
Private fire investigator specialists sometimes come from the fire service or law enforcement.

Private fire investigator specialists sometimes come from the fire service or law enforcement. Some work on their days off or leave their public safety job to join the more lucrative private sector. Other private sector investigators are individuals with specialized training such as engineering or forensic degrees. They can work independently or for private companies, both large and small; a few insurance companies even have special agents (investigators and engineers) covering almost every state.

The insurance industry does not typically seek criminal convictions but will cooperate with officials when the situation justifies this action. What insurance companies do want to do is identify issues involving fraud, in particular, when individuals knowingly falsify information in their policy application or when they falsify information on their insurance claims. When these falsifications take place, the common action is to deny the claim, which is usually justified by the language in the insurance policy. Some insurance companies will also take legal action.

■ Note
To falsify information for personal gain is fraud.

fraud
a false statement of fact or intentionally misleading statement made to cause someone to give up something of value

To falsify information for personal gain is **fraud.** The illegal claims placed against insurance companies range from staging false car accidents to collect on damages to falsely collecting workers' compensation monies when fully capable of working. Fraud could be lying on an insurance application when asked if you had previous insurance claims or lying about whether you have other insurance policies on the same property.

■ Note
Even an accidental fire scene could turn into a fraud case.

■ Note
For example, all claims require the policyholder to submit a written proof of loss.

proof of loss
formal statement made by the insurance policyholder to the insurance company validating the loss; if the policy covers contents, this statement lists all contents in the loss at the time of the incident

Even an accidental fire scene could turn into a fraud case. For example, all claims require the policyholder to submit a written **proof of loss** listing all items lost in the fire. For whatever reason, a family decides to inflate the content losses on the insurance claims after a house fire. A dozen shirts purchased at a discount retailer are listed at twice their value. A fabric coat could be listed as a leather or a fur coat. The family heirloom half-carat diamond becomes a full carat. Sometimes the inflation can be small, but all too often greed takes over and this misreporting becomes fraud, a major felony. Although they did not set the fire, they have broken the law. Sometimes fraud is hard to prove because the receipts for purchases burn up in the fire as well. However, there are ways to sift a fire scene that can show enough deceit in the inflated proof of loss to allow the insurance company to deny the entire claim, for both the house and contents.

Sifting a Fire Scene A description of the process of sifting a fire scene is included in this section because it is more often a tool used by the private sector than by the public sector. Even when it is a private sector project, the public sector has either assisted or the off-duty firefighters have been hired to assist. Thus, first responders should be aware of the procedures so that they can be more valuable assets in the process.

Figure 4-5 *Sifted piles next to piles of discarded objects are a clear indication that the area has been sifted.*

sifting
process of using a screen to separate out small particles so that larger particles can be examined

When there is reason to believe that a fire scene may involve an inflated claim, investigators can set up a **sifting** process. Sifting immediately after the fire is best, but there usually is no indication of fraud until after the proof of loss is submitted. If the scene has remained reasonably undisturbed during this time, investigators can sift the scene, comparing the items in the proof of loss with what is found on the scene. (See **Figure 4-5.**)

In a structure with major damage where most of the house is on the ground, a limited number of items can be identified. However, specific items can be found that will support the proof of loss or refute its validity. Remember, seeking the truth should be the overall motivation of the fire investigator.

■ Note
Remember, seeking the truth should be the overall motivation of the fire investigator.

Items such as power tools, hand tools, metal features on sporting equipment, kitchen appliances, metal lamps, and even traces of fiberglass equipment can sometimes be identified. Parts of a CRT (picture tube) television can survive and can verify its existence in the house at the time or can validate the size of the TV. Some household items, such as the antique clock as those shown in **Figure 4-6,** can survive sufficiently enough to be identified.

The sifting process involves using screens, as pictured in **Figure 4-7.** They can be commercially made or constructed from 2 × 4s and wire mesh from the local hardware store. The openings should be small enough to capture what is expected to be found in the debris.

Figure 4-6 *Household items with metal parts that usually survive are items such as older bathroom scales, printers, grandfather clocks, and portable heaters.*

Figure 4-7 *Sifting screens being used on a fire scene. Investigators shovel debris in the screens and other investigators rock the screens back and forth so that smaller particles drop to the ground and the investigator can examine larger items.*

Safety is the key in this process, and all personnel must be wearing respiratory and eye protection because of the carbon and other hazardous particles that float in the air as a result of this process. Care must be taken to provide facilities for sifters to clean up when taking breaks for food and fluid so they do not ingest particles that have settled on their hands, face, and clothing.

One person should be in charge of sifting to oversee the entire operation and examine all evidence found. During sifting, the scene should be documented using photographs before, during, and following to show items found.

Self-Insured

Some companies are so large and have sufficient fiscal resources that they will not take out insurance policies but keep a cash reserve to handle any loss they may experience. Most do not have investigators on hand but have a loss prevention division that handles safety issues and conducts preliminary investigations on any damages the company experiences.

Some companies may have an insurance policy but will have anywhere from a $100,000 to a $1 million deductible. They operate in a similar manner as self-insureds except when experiencing a major loss. As mentioned before, the largest self-insured entity is the federal government, but it has resources to investigate fires and to examine its losses to seek restitution when necessary.

Research, Independent, and Other Investigators

The National Fire Protection Association has been conducting on-site fire investigations since the 1940s through its Fire Investigations Division. As a private entity, the NFPA has no legal authority to take the place of the local investigator and neither is that its purpose for existence. However, it is there to provide technical support upon request and to collect data that can provide lessons for the fire protection community.

■ Note
NFPA investigations are conducted to document fires of a highly technical educational interest.

NFPA investigations are conducted to document fires of a highly technical educational interest. The analysis of these events may also be of importance to the various technical programs and committees both within NFPA and as data to be used by third parties interested in fire research.

The detailed data collected as part of this process are vital to both the private and public sectors. These reports are available free of charge to NFPA members and for a small fee for nonmembers. Areas documented in these fire investigative reports include the following:

- Details on fire ignition, growth, and development
- Contributions and impact from building construction as well as interior finishes and contents
- Action, or lack of action, of fire detection and suppression systems
- Smoke movement and control

National Institute of Science and Technology (NIST)
the Fire Research Division consists of several groups that can conduct research on fire incidents

- Human reaction related to the fire scene; how did people respond to the event
- Firefighting and rescue response and actions
- Loss of life and injuries as a result of the fire

The **National Institute of Standards and Technology (NIST)** is a government entity that is viewed as a third party in any investigation. NIST is not involved in regulatory enforcement and does not issue codes or standards. Furthermore, federal law (NCST Act) does not allow any NIST reports to be used in lawsuits for damages. NIST employees cannot serve as expert witnesses in court proceedings. For these reasons, NIST can concentrate on the research value of its actions.

Note NIST can concentrate on the research value of its actions.

NIST staff members have an enormous amount of expertise and are experts in construction, engineering, fire systems, materials, manufacturing engineering, and electronics to name a few. They are allowed to conduct basic and applied fire research, assisting both governmental and private entities upon request.

Note One of NIST's primary tasks is to discover why buildings fail in fires and other disasters.

One of NIST's primary tasks is to discover why buildings fail in fires and other disasters. They have been instrumental in almost every investigation of major disaster in recent years, providing valuable answers on how to prevent or limit future damages from similar incidents. Incidents from the World Trade Center and Pentagon terrorist attacks in 2001, to the Station Nightclub fire in Rhode Island and the terrorist bombing of the Murrah Federal Building in Oklahoma City in 1995 were all investigated by NIST.

INVESTIGATIVE NETWORKING

Note It is recognized that one valuable tool in training events in addition to the speakers and instructors is the opportunity for the students to interact with peers.

Every investigator must maintain a certain level of ongoing investigative training. Most certification entities require that this training be documented and in a classroom environment. It is recognized that one valuable tool in training events in addition to the speakers and instructors is the opportunity for the students to interact with peers. Networking is an extremely valuable tool for sharing ideas and concepts.

International Association of Special Investigation Units (IASIU)
a nonprofit international organization with local chapters dedicated to combat insurance fraud through training, awareness, and legislation

This process is made valuable when professional entities such as the **International Association of Special Investigation Units (IASIU)** put on training for their investigative membership. The International Association of Arson Investigators is another organization that offers both international and national training events. Moreover, both organizations have chapters in most states that put on local training on a regular basis for their members and guests. Because both associations have members serving on committees to set up the training events, there is a good probability that they will address the needs of their membership.

Many other smaller local associations gather on a regional basis monthly or quarterly. These groups discuss local issues that are common to their neighboring jurisdictions.

Fire and Arson Associations

The Central Virginia Fire and Arson Association meets on a monthly basis. The typical meeting covers some administrative issues for the association, and then provides a 2- or 3-hour training event taught by a guest or one of the members. Following that, there is a session where each jurisdiction briefly describes issues dealing with code enforcement or recent fire investigations that it is dealing with. At one meeting, one jurisdiction brought up an issue it had recently with trash cans in service station restrooms being set on fire. Another jurisdiction gave its report and stated that they too had fires in ladies restrooms in two service stations. Smugly, a third jurisdiction reported similar fires and that it had made an arrest the previous day. Although the suspect could not be linked to the other incidents sufficiently to make further charges, it was interesting that all such service station fires ceased.

Sharing information can be quite beneficial. Criminals do not stay within certain jurisdictional boundaries. Local investigators can tear down those boundaries as well by meeting regularly with their peers to discuss recent cases, problems, and issues.

SUMMARY

Who investigates fires—police or fire departments—is usually dictated by history. There is no clear-cut answer to who makes a better investigator, a police officer or a firefighter. Success depends on the individual's abilities and not the choice of employer. Regardless, those assigned to investigative duties must be given clear dictates as to how to handle and how far to delve into the investigation of the incident.

Fire investigators come from the ranks of police and fire departments and also may come from an independent office of the fire marshal at either the local or the state level. Various other state agencies have fire investigators as well. The state department of forestry sometimes has dedicated staff to investigate forest fires. Even state parks departments with their own police force may have individuals with specific expertise to investigate fires.

The insurance industry has a need for its own investigators to look at fire scenes. In some cases, the report from the local investigator is sufficient. However, when the insurance companies want to follow up on possible fraud aspect of the incident, the local government may not have those resources to assist, which is why some insurance companies have their own Special Investigation Units or hire outside, private investigators to look at their scenes and carry out investigations to the insurance companies' specifications.

Research is the key to helping us understand the scientific aspect of fire investigations. Independent research conducted by agencies such as the National Institute of Standards and Technology or private corporations such as the National Fire Protection Association provides insight into all aspects of the fire, structural problems as well as how humans will react.

All investigators—public and private, from police and fire departments, engineers and researchers—should network to ensure the growth of the collective knowledge and abilities. Several associations have training events or regular meetings to nurture this exchange of knowledge and information.

KEY TERMS

Arson Intentionally and willfully setting a fire with malice.

Bureau of Alcohol, Tobacco, Firearms, and Explosives (ATF) A federal agency with fire investigative authority; it operates a National Response Team, which can provide almost every resource necessary at a fire scene.

Claims managers Within almost every insurance company is a division assigned to evaluate submitted insurance claims to assess the value of the loss and, if justified, to write the checks necessary to settle the claims.

Federal Bureau of Investigation The nation's law enforcement agency; it is available to assist on major fire incidents. In addition to an outstanding crime lab, the FBI has one of the best training academies in the world.

Fraud A false statement of fact or intentionally misleading statement made to cause someone to give up something of value.

International Association of Special Investigation Units (IASIU) A nonprofit international organization with local chapters dedicated to combat insurance fraud through training, awareness, and legislation.

Loss Object of value that was destroyed.

National Institute of Science and Technology (NIST) Founded in 1901, a nonregulatory federal agency in the U.S. Commerce Department's Technology Administration. Its mission is to promote U.S. innovation and industrial competitiveness by advancing measurement science, standards, and technology in ways that enhance economic security and improve people's quality of life. The Fire Research Division consists of several groups that can conduct research on fire incidents.

Policyholder Person who owns the insurance policy, possibly the property owner or someone who has an interest in the property.

Private investigator (for fire and explosion): An individual involved in providing professional investigations of fires and explosions on a contract basis or as an employee of a private enterprise or organization.

Proof of loss Formal statement made by the insurance policyholder to the insurance company validating the loss. If the policy covers contents, this statement lists all contents in the loss at the time of the incident.

Public fire investigator An investigator working for the government representing the locality, state, or federal government. Usually denotes someone whose job requirements include the determination of the area of origin and cause of fires and explosions.

Risk The item covered by the insurance policy; the item that may be lost.

Sifting Process of using a screen to separate out small particles so that larger particles can be examined.

Special Investigations Unit (SIU) A division usually within an insurance company that oversees or conducts investigations involving a loss.

Subrogation The substitution of one person in the place of another with reference to a lawful claim.

REVIEW QUESTIONS

1. What misconceptions are associated with using the title *arson investigator* on coveralls or vehicles?
2. List four federal government agencies that have fire investigation responsibilities.
3. Who decides which department has fire investigative authority at the local level?
4. Who makes a better fire investigator, someone from the fire department or the police department? Explain and justify your answer.
5. Describe the various duties (type of fire investigation) that may be required of the local fire investigator.
6. Describe the various types of reports that the insurance company may require from the fire investigators it hires.
7. Which two federal law enforcement agencies have laboratories to assist in fire investigations and in research on types of fire incidents?
8. Define *subrogation*.
9. Under what circumstances could an insurance company refuse to pay the insurance claim even though the fire was accidental?
10. What is the role of NIST in the field of fire investigations?

DISCUSSION QUESTIONS

1. Both police and fire are conducting their own investigations and not necessarily cooperating with each other. The AHJ has to decide which department will be the lead in fire investigations for both accidental and criminal cases. To do so, they have asked you to make a list of the pros and cons for each department taking on those responsibilities. Make such a list, and then develop a position for one department or the other to take over investigative responsibilities.
2. Considering question 1, your assignment has changed. Based on the position you adopted to support one department over the other, the AHJ has now asked you to sell them on the department you did not choose to have investigative authority. Develop selling points to achieve your assignment.

ACTIVITIES

1. Do an Internet search for fire investigator associations to identify how many training resources are available for investigators today.
2. Contact local fire investigative authorities to ascertain whether they have training events that are open to the fire service, law enforcement, or fire science students.

NOTES

1. *Blacks Law Dictionary*, 5th ed. (St. Paul, MN: West Publishing Company, 1979), 1279.

LEGAL ISSUES AND THE RIGHT TO BE THERE

Learning Objectives

Upon completion of this chapter, you should be able to:

- Describe legal issues as they pertain to the right of the investigator to be on and stay at the fire or explosion scene.
- Describe the difference between legislative laws and case law.
- Describe and understand Supreme Court decisions that have affected fire investigators and how they carry out their duties.

CASE STUDY

Michigan v. Clifford

The fire occurred on October 18, 1980, about 5:40 A.M. at the home of Raymond and Emma Jean Clifford. At the time, they were away on a camping trip. The Detroit Fire Department responded and extinguished the fire, clearing the scene at 7:04 A.M. Everyone left, fire and police. The fire department had reported the suspicious nature of the fire, but the investigator, Lieutenant Beyer, was not notified until 8:00 A.M. Because he had other appointments, he and another investigator did not arrive until 1:00 P.M. that afternoon, 5 hours after being notified.

In the meantime, a neighbor had notified the Cliffords about the fire, and they had asked the neighbor to contact their insurance company to have them secure the building. When the investigators arrived, they found a work crew at the scene. They were pumping out the basement and putting plywood on the windows and doors. The neighbor told the investigator about notifying the owners and their instructions about getting the house secured by their insurance company.

The Detroit Fire Department policy stated that as long as the owner was not present and the premises were open to trespass, they could search the scene as long as it was a reasonable amount of time since the fire. The investigators waited for the basement to be pumped out, which was completed at 1:30 P.M. In the meantime, they discovered a Coleman lantern in the driveway and they marked it as evidence.

The investigators entered the Cliffords' house without consent or a warrant. They determined the fire started under the basement stairs from a crockpot plugged into a timer that was set to go off at 3:45 A.M. and stop at 4:30 A.M. With this evidence and a strong smell of a flammable liquid in the basement the investigators determined the fire was incendiary in nature. Evidence was secured and properly marked.

They did not stop there; both investigators continued searching the remainder of the house. They called in a photographer to document the contents of the structure. They found the drawers full of old clothing and nails on the wall where pictures used to hang. There were even wires where a cassette player would have set, but the player was missing.

The Cliffords were charged with arson. Their attorney moved for suppression of the evidence based on it being a warrantless search. An interlocutory (occurring between start and end of trial) appeal from the Michigan Court of Appeals held that there was no exigent circumstance and reversed the findings of the lower court, which made the evidence obtained by the investigators during their interior search of the premises inadmissible.

INTRODUCTION

No one text can educate someone about all legal aspects of fire investigation. This chapter addresses the differences between legislated law and case laws. In particular, it concentrates on two United States Supreme Court cases that define the rights the fire investigator has to be on the scene to collect evidence and make a determination on the fire origin and cause.

The first step to ensuring compliance with all legal issues surrounding an investigation is to have access to legal counsel. For most localities, the prosecuting attorneys for that jurisdiction, sometimes referred to as district attorneys, provide this service. In larger jurisdictions, one attorney from that office may handle all fire and arson cases. For the private sector, most insurance companies employ an attorney who gets involved from the very beginning of a case to give direction and recommendations on legal issues.

First responders responsible for conducting the determination of origin and cause must be aware of legal issues that enable them to carry out their task. At the very least, the officer in charge (OIC) of the fire scene must know his or her legal right to be on the scene. The more detailed the investigation, the more potential for legal issues. To understand legal requirements the investigator must know where laws come from, how they are created, and how laws can change.

As a rule, laws, regulations, codes, and ordinances are not created because we believe something might happen but usually because something did happen. In other words, most laws are there for a reason. Knowing why something is legal or illegal can ensure better compliance.

LEGISLATED LAWS AND CASE LAWS

The United States Constitution establishes a balance of power where most laws are developed by the legislative branch of government with approval from the executive branch, the president of the United States. This same process is mirrored in the states and even in most localities. The judiciary branch resolves any conflicts that may arise between those affected by the law.

Eventually as the result of criminal or civil action, a point of law becomes an issue in a court of law. Not all laws are written so that they are 100 percent understood by everyone. Furthermore, not all circumstances involving laws are the same each and every time. For this reason, a judge interprets the law in a court proceeding. When a decision is made, that interpretation relevant to that case under those circumstances becomes a precedent. When similar circumstances come up in other courts on the same point of law, judges should apply that precedent in making a decision. This is case law. Knowing case law can prevent future problems in court.

As cases are adjudicated, parties aggrieved with the decisions can sometimes take their cases to higher courts. When that happens, the entire case is not tried again, but points of law are discussed to make sure it was a fair hearing and all rules were followed.

United States Supreme Court
the highest court of the nation

In the federal court system, including the **United States Supreme Court,** the Constitution allows only certain types of cases to be heard, including cases dealing with the following topics:

- Constitutional law
- Issues between parties from different states
- Conflicts between states
- Issues between U.S. citizens and foreigners
- Cases involving both state and federal laws
- Cases involving maritime law or admiralty
- Cases in which the United States is a party[1]

Note
Supreme Court decisions exist that affect fire investigators and their right to be on the scene.

Several Supreme Court decisions exist that affect fire investigators and their right to be on the scene collecting evidence as well as other factors dealing with criminal law such as Miranda Warnings.

SUPREME COURT DECISIONS

The freedoms and rights of U.S. citizens are spelled out in the Constitution of the United States. The Supreme Court has the responsibility to interpret the Constitution and its amendments through the proceedings from the lower courts. These decisions are not always unanimous but are ruled by a majority.

Note
Fire investigators with full police powers may seize property, conduct searches, detain suspects, and testify in court.

Fire investigators with full police powers may seize property, conduct searches, detain suspects, and testify in court. Each of these activities entails abridging the rights and freedoms of some citizens. Investigators should do this only when necessary to carry out their assigned duties. To do this properly, investigators rely on the findings of the Supreme Court for guidance. Although a multitude of court decisions address investigator duties and responsibilities, in this chapter we address only those dealing with the right to be on the fire scene and when investigators can collect evidence properly.

The first case is *Michigan v. Tyler.* (See **Figure 5-1.**) On January 21, 1970, shortly before midnight, a fire occurred at Mr. Loren Tyler's Auction, a furniture store in Oakland County, Michigan. Around 2 A.M., as the final flames were being extinguished, firefighters discovered plastic containers with what appeared to be a flammable liquid. Chief See, thinking this may be an arson fire, notified the local police. Detective Webb arrived and took photographs but had to leave the building because of smoke. By 4 A.M., the fire was out and the fire department left the scene. Because there was still steam and it was dark, Chief See and Detective Webb also left, taking the containers with them.

Around 8 A.M., Chief See and Assistant Chief Somerville returned to check the building; an hour after that Detective Webb and Assistant Chief Somerville did another examination of the scene, collecting even more evidence: a piece of carpet and taped sections of the stairway.

Michigan v. Tyler

436 U.S. 499 (1978)

Supreme Court of the United States

MR. JUSTICE STEWART delivered the opinion of the Court. . . .

Shortly before midnight on January 21, 1970, a fire broke out at Tyler's Auction, a furniture store in Oakland County, Mich. The building was leased to respondent Loren Tyler, who conducted the business in association with respondent Robert Tompkins. According to the trial testimony of various witnesses, the fire department responded to the fire and was "just watering down smoldering embers" when Fire Chief See arrived on the scene around 2 A.M. It was Chief See's responsibility "to determine the cause and make out all reports." Chief See was met by Lt. Lawson, who informed him that two plastic containers of flammable liquid had been found in the building. Using portable lights, they entered the gutted store, which was filled with smoke and steam, to examine the containers. Concluding that the fire "could possibly have been an arson," Chief See called Police Detective Webb, who arrived around 3:30 A.M. Detective Webb took several pictures of the containers and of the interior of the store, but finally abandoned his efforts because of the smoke and steam. Chief See briefly "[l]ooked throughout the rest of the building to see if there was any further evidence, to determine what the cause of the fire was." By 4 A.M., the fire had been extinguished and the firefighters departed. See and Webb took the two containers to the fire station, where they were turned over to Webb for safekeeping. There was neither consent nor a warrant for any of these entries into the building, nor for the removal of the containers. . . . .

Four hours after he had left Tyler's Auction, Chief See returned with Assistant Chief Somerville, whose job was to determine the "origin of all fires that occur within the Township." The fire had been extinguished and the building was empty. After a cursory examination, they left, and Somerville returned with Detective Webb around 9 A.M. In Webb's words, they discovered suspicious "burn marks in the carpet, which [Webb] could not see earlier that morning, because of the heat, steam, and the darkness." They also found "pieces of tape, with burn marks, on the stairway." After leaving the building to obtain tools, they returned and removed pieces of the carpet and sections of the stairs to preserve these bits of evidence suggestive of a fuse trail. Somerville also searched through the rubble "looking for any other signs or evidence that showed how this fire was caused." Again, there was neither consent nor a warrant for these entries and seizures. Both at trial and on appeal, the respondents objected to the introduction of evidence thereby obtained.

On February 16, Sergeant Hoffman of the Michigan State Police Arson Section returned to Tyler's Auction to take photographs. During this visit or during another at about the same time, he checked the circuit breakers, had someone inspect the

(Continued)

Figure 5-1 Michigan v. Tyler, *436 U.S. 499, full text.*

(Continued)

furnace, and had a television repairman examine the remains of several television sets found in the ashes. He also found a piece of fuse. Over the course of his several visits, Hoffman secured physical evidence and formed opinions that played a substantial role at trial in establishing arson as the cause of the fire and in refuting the respondents' testimony about what furniture had been lost. His entries into the building were without warrants or Tyler's consent, and were for the sole purpose "of making an investigation and seizing evidence." At the trial, respondents' attorney objected to the admission of physical evidence obtained during these visits, and also moved to strike all of Hoffman's testimony "because it was got in an illegal manner." . . .

The decisions of this Court firmly establish that the Fourth Amendment extends beyond the paradigmatic entry into a private dwelling by a law enforcement officer in search of the fruits or instrumentalities of crime. As this Court stated in Camara . . . the "basic purpose of this Amendment . . . is to safeguard the privacy and security of individuals against arbitrary invasions by governmental officials." . . .

[T]here is no diminution in a person's reasonable expectation of privacy nor in the protection of the Fourth Amendment simply because the official conducting the search wears the uniform of a firefighter, rather than a policeman, or because his purpose is to ascertain the cause of a fire, rather than to look for evidence of a crime, or because the fire might have been started deliberately. Searches for administrative purposes, like searches for evidence of crime, are encompassed by the Fourth Amendment. . . .

The petitioner argues that no purpose would be served by requiring warrants to investigate the cause of a fire. This argument is grounded on the premise that the only fact that need be shown to justify an investigatory search is that a fire of undetermined origin has occurred on those premises. . . . In short, where the justification for the search is as simple and as obvious to everyone as the fact of a recent fire, a magistrate's review would be a time-consuming formality of negligible protection to the occupant.

The petitioner's argument fails primarily because it is built on a faulty premise. To secure a warrant to investigate the cause of a fire, an official must show more than the bare fact that a fire has occurred. The magistrate's duty is to assure that the proposed search will be reasonable, a determination that requires inquiry into the need for the intrusion on the one hand, and the threat of disruption to the occupant on the other. . . . Thus, a major function of the warrant is to provide the property owner with sufficient information to reassure him of the entry's legality. . . .

In short, the warrant requirement provides significant protection for fire victims in this context, just as it does for property owners faced with routine building inspections. As a general matter, then, official entries to investigate the cause of a fire must adhere to the warrant procedures of the Fourth Amendment. . . . Since all the entries in this case were "without proper consent" and were not "authorized by a valid search warrant," each one is illegal unless it falls within

(Continued)

Figure 5-1 *Continued*

(Continued)

one of the "certain carefully defined classes of cases" for which warrants are not mandatory. . . .

Our decisions have recognized that a warrantless entry by criminal law enforcement officials may be legal when there is compelling need for official action and no time to secure a warrant. . . . A burning building clearly presents an exigency of sufficient proportions to render a warrantless entry "reasonable." Indeed, it would defy reason to suppose that firemen must secure a warrant or consent before entering a burning structure to put out the blaze. And once in a building for this purpose, firefighters may seize evidence of arson that is in plain view. . . . Thus, the Fourth and Fourteenth Amendments were not violated by the entry of the firemen to extinguish the fire at Tyler's Auction, nor by Chief See's removal of the two plastic containers of flammable liquid found on the floor of one of the showrooms.

Although the Michigan Supreme Court appears to have accepted this principle, its opinion may be read as holding that the exigency justifying a warrantless entry to fight a fire ends, and the need to get a warrant begins, with the dousing of the last flame. . . . We think this view of the firefighting function is unrealistically narrow, however. Fire officials are charged not only with extinguishing fires, but with finding their causes. Prompt determination of the fire's origin may be necessary to prevent its recurrence, as through the detection of continuing dangers such as faulty wiring or a defective furnace. Immediate investigation may also be necessary to preserve evidence from intentional or accidental destruction. And, of course, the sooner the officials complete their duties, the less will be their subsequent interference with the privacy and the recovery efforts of the victims. For these reasons, officials need no warrant to remain in a building for a reasonable time to investigate the cause of a blaze after it has been extinguished. And if the warrantless entry to put out the fire and determine its cause is constitutional, the warrantless seizure of evidence while inspecting the premises for these purposes also is constitutional.

The respondents argue, however, that the Michigan Supreme Court was correct in holding that the departure by the fire officials from Tyler's Auction at 4 A.M. ended any license they might have had to conduct a warrantless search. Hence, they say that even if the firemen might have been entitled to remain in the building without a warrant to investigate the cause of the fire, their reentry four hours after their departure required a warrant.

On the facts of this case, we do not believe that a warrant was necessary for the early morning reentries on January 22. As the fire was being extinguished, Chief See and his assistants began their investigation, but visibility was severely hindered by darkness, steam, and smoke. Thus they departed at 4 A.M. and returned shortly after daylight to continue their investigation. Little purpose would have been served by their remaining in the building, except to remove any doubt about the legality of the warrantless search and seizure later that same morning. Under

(Continued)

Figure 5-1 *Continued*

(Continued)

these circumstances, we find that the morning entries were no more than an actual continuation of the first, and the lack of a warrant thus did not invalidate the resulting seizure of evidence.

The entries occurring after January 22, however, were clearly detached from the initial exigency and warrantless entry. Since all of these searches were conducted without valid warrants and without consent, they were invalid under the Fourth and Fourteenth Amendments, and any evidence obtained as a result of those entries must, therefore, be excluded at the respondents' retrial.

In summation, we hold that an entry to fight a fire requires no warrant, and that, once in the building, officials may remain there for a reasonable time to investigate the cause of the blaze. Thereafter, additional entries to investigate the cause of the fire must be made pursuant to the warrant procedures governing administrative searches. . . . Evidence of arson discovered in the course of such investigations is admissible at trial, but if the investigating officials find probable cause to believe that arson has occurred and require further access to gather evidence for a possible prosecution, they may obtain a warrant only upon a traditional showing of probable cause applicable to searches for evidence of crime. . . .

These principles require that we affirm the judgment of the Michigan Supreme Court ordering a new trial. Affirmed.

Figure 5-1 *Continued*

■ Note
If investigators want to return, they must obtain an administrative search warrant if the fire cause has yet to be determined.

administrative search warrant
warrant issued by a magistrate or judge allowing the investigator to be on the scene to determine the cause of the fire for the good of the people

On February 16, 1970 (26 days later), the state police arson investigator arrived, took photographs, and examined the scene. He returned several more times, taking evidence and collecting information. Mr. Tyler and his partner Robert Tompkins were charged with conspiracy to burn real property as well as other offenses. They were convicted, but the case was appealed based on the evidence being obtained without a proper search warrant. The Michigan Supreme Court agreed and reversed the convictions.

The U.S. Supreme Court agreed to hear arguments on admissibility of evidence in this case. As such, they held that the two plastic containers with flammable liquid were admissible as evidence; because of the exigencies of the situation, the fire chief had the right to secure them as evidence. Because the fire scene was still hazardous with smoke and with problems of seeing resulting from steam and it being nighttime, the Supreme Court agreed that the evidence taken the next morning was reasonable, as an extension of the previous evening's search. The Supreme Court ruled that the subsequent visits were improper and all evidence collected then were inadmissible in court.

This decision makes it clear that investigators can do a search for the origin and cause, and under the exigencies of the circumstances can seize evidence in plain sight. However, if investigators want to return, they must obtain an **administrative search warrant** if the fire cause has yet to be determined. If there is any

Figure 5-2 *The fire investigator should have the building owner sign a consent agreement to search before proceeding with the scene search.*

criminal search warrant a warrant issued by a magistrate or judge based on the sworn (and written affidavit) testimony of an investigator that probable cause exists that a crime has been committed and the person or place to search and the items being sought

evidence to indicate that the fire is incendiary in nature or evidence of any other criminal nature, a **criminal search warrant** must be obtained before continuing the search.

Of course, no search warrant need be obtained if the building owner gives **consent** to search or the renter gives consent to search the leased area. (See **Figure 5-2.**) If the permission is rescinded, then the scene must be secured, preferably by someone guarding the property, and the investigator should go immediately to the magistrate to secure a warrant, administrative or criminal, depending on the circumstances.

■ Note

No search warrant need be obtained if the building owner gives consent.

In *Michigan v. Clifford,* the Supreme Court had a chance to further define when a fire investigator can enter a building under **exigent circumstances** versus when the owners have a reasonable right to privacy. (See **Figure 5-3.**)

consent permission given by the person responsible for or controlling the property, allowing the investigator to search the property

RIGHT TO BE THERE

Private investigators generally do not have an entry problem. They can be on the scene based on the consent of the owner or implied consent based on the provisions of the insurance policy. Should the policy owner refuse entry to the

exigent circumstances for the good of the people, without permission, public safety personnel can enter property on fire to save lives and control the fire

■ Note

Private investigators generally do not have an entry problem. They can be on the scene based on the consent of the owner or implied consent based on the provisions of the insurance policy.

Michigan v. Clifford

464 U.S. 287 (1984)

United States Supreme Court

. . . In the early morning hours of October 18, 1980, a fire erupted at the Clifford home. The Cliffords were out of town on a camping trip at the time. The fire was reported to the Detroit Fire Department, and fire units arrived on the scene about 5:40 A.M. The fire was extinguished and all fire officials and police left the premises at 7:04 A.M.

At 8 o'clock on the morning of the fire, Lieutenant Beyer, a fire investigator with the arson section of the Detroit Fire Department, received instructions to investigate the Clifford fire. He was informed that the Fire Department suspected arson. Because he had other assignments, Lieutenant Beyer did not proceed immediately to the Clifford residence. He and his partner finally arrived at the scene of the fire about 1 P.M. on October 18.

When they arrived, they found a work crew on the scene. The crew was boarding up the house and pumping some six inches of water out of the basement. A neighbor told the investigators that he had called Mr. Clifford and had been instructed to request the Cliffords' insurance agent to send a boarding crew out to secure the house. The neighbor also advised that the Cliffords did not plan to return that day. While the investigators waited for the water to be pumped out, they found a Coleman fuel can in the driveway that was seized and marked as evidence.

By 1:30 P.M., the water had been pumped out of the basement and Lieutenant Beyer and his partner, without obtaining consent or an administrative warrant, entered the Clifford residence and began their investigation into the cause of the fire. Their search began in the basement, and they quickly confirmed that the fire had originated there beneath the basement stairway. They detected a strong odor of fuel throughout the basement, and found two more Coleman fuel cans beneath the stairway. As they dug through the debris, the investigators also found a crock pot with attached wires leading to an electrical timer that was plugged into an outlet a few feet away. The timer was set to turn on at approximately 3:45 A.M. and to turn back off at approximately 9 A.M. It had stopped somewhere between 4 and 4:30 A.M. All of this evidence was seized and marked.

After determining that the fire had originated in the basement, Lieutenant Beyer and his partner searched the remainder of the house. The warrantless search that followed was extensive and thorough. The investigators called in a photographer to take pictures throughout the house. They searched through drawers and closets and found them full of old clothes. They inspected the rooms and noted that there were nails on the walls, but no pictures. They found wiring and cassettes for a videotape machine but no machine.

Respondents [Cliffords] moved to exclude all exhibits and testimony based on the basement and upstairs searches on the ground that they were searches to gather

(Continued)

Figure 5-3 Michigan v. Clifford, *464 U.S. 287, full text.*

(Continued)

evidence of arson, that they were conducted without a warrant, consent, or exigent circumstances, and that they therefore were per se unreasonable under the Fourth and Fourteenth Amendments. Petitioner [State of Michigan], on the other hand, argues that the entire search was reasonable and should be exempt from the warrant requirement.

[T]he State does not challenge the state court's finding that there were no exigent circumstances justifying the search of the Clifford home. Instead, it asks us to exempt from the warrant requirement all administrative investigations into the cause and origin of a fire. We decline to do so.

In Tyler, we restated the Court's position that administrative searches generally require warrants. . . . We reaffirm that view again today. Except in certain carefully defined classes of cases, the nonconsensual entry and search of property are governed by the warrant requirement of the Fourth and Fourteenth Amendments. The constitutionality of warrantless and nonconsensual entries onto fire-damaged premises, therefore, normally turns on several factors: whether there are legitimate privacy interests in the fire-damaged property that are protected by the Fourth Amendment; whether exigent circumstances justify the government intrusion regardless of any reasonable expectations of privacy; and, whether the object of the search is to determine the cause of fire or to gather evidence of criminal activity.

We observed in *Tyler* that reasonable privacy expectations may remain in fire-damaged premises. People may go on living in their homes or working in their offices after a fire. Even when that is impossible, private effects often remain on the fire-damaged premises. . . . Privacy expectations will vary with the type of property, the amount of fire damage, the prior and continued use of the premises, and in some cases the owner's efforts to secure it against intruders. Some fires may be so devastating that no reasonable privacy interests remain in the ash and ruins, regardless of the owner's subjective expectations. The test essentially is an objective one: whether "the expectation [is] one that society is prepared to recognize as 'reasonable.'" . . . If reasonable privacy interests remain in the fire-damaged property, the warrant requirement applies, and any official entry must be made pursuant to a warrant in the absence of consent or exigent circumstances.

A burning building of course creates an exigency that justifies a warrantless entry by fire officials to fight the blaze. Moreover, in *Tyler,* we held that, once in the building, officials need no warrant to remain for "a reasonable time to investigate the cause of a blaze after it has been extinguished." . . . Where, however, reasonable expectations of privacy remain in the fire-damaged property, additional investigations begun after the fire has been extinguished and fire and police officials have left the scene generally must be made pursuant to a warrant or the identification of some new exigency.

The aftermath of a fire often presents exigencies that will not tolerate the delay necessary to obtain a warrant or to secure the owner's consent to inspect

(Continued)

Figure 5-3 *Continued*

(Continued)

firedamaged premises. Because determining the cause and origin of a fire serves a compelling public interest, the warrant requirement does not apply in such cases.

If a warrant is necessary, the object of the search determines the type of warrant required. If the primary object is to determine the cause and origin of a recent fire, an administrative warrant will suffice. To obtain such a warrant, fire officials need show only that a fire of undetermined origin has occurred on the premises, that the scope of the proposed search is reasonable and will not intrude unnecessarily on the fire victim's privacy, and that the search will be executed at a reasonable and convenient time.

If the primary object of the search is to gather evidence of criminal activity, a criminal search warrant may be obtained only on a showing of probable cause to believe that relevant evidence will be found in the place to be searched. If evidence of criminal activity is discovered during the course of a valid administrative search, it may be seized under the "plain view" doctrine. . . . This evidence then may be used to establish probable cause to obtain a criminal search warrant. Fire officials may not, however, rely on this evidence to expand the scope of their administrative search without first making a successful showing of probable cause to an independent judicial officer.

The object of the search is important even if exigent circumstances exist. Circumstances that justify a warrantless search for the cause of a fire may not justify a search to gather evidence of criminal activity once that cause has been determined. If, for example, the administrative search is justified by the immediate need to ensure against rekindling, the scope of the search may be no broader than reasonably necessary to achieve its end. A search to gather evidence of criminal activity not in plain view must be made pursuant to a criminal warrant upon a traditional showing of probable cause.

The searches of the Clifford home, at least arguably, can be viewed as two separate ones: the delayed search of the basement area, followed by the extensive search of the residential portion of the house. We now apply the principles outlined above to each of these searches.

The Clifford home was a two-and-one-half story brick and frame residence. Although there was extensive damage to the lower interior structure, the exterior of the house and some of the upstairs rooms were largely undamaged by the fire, although there was some smoke damage. The firemen had broken out one of the doors and most of the windows in fighting the blaze. At the time Lieutenant Beyer and his partner arrived, the home was uninhabitable. But personal belongings remained, and the Cliffords had arranged to have the house secured against intrusion in their absence. Under these circumstances, and in light of the strong expectations of privacy associated with a home, we hold that the Cliffords retained reasonable privacy interests in their fire-damaged residence, and that the post-fire investigations were subject to the warrant requirement. Thus, the warrantless and

(Continued)

Figure 5-3 *Continued*

(Continued)

nonconsensual searches of both the basement and the upstairs areas of the house would have been valid only if exigent circumstances had justified the object and the scope of each.

As noted, the State does not claim that exigent circumstances justified its post-fire searches. It argues that we either should exempt post-fire searches from the warrant requirement or modify *Tyler* to justify the warrantless searches in this case. We have rejected the State's first argument, and turn now to its second.

In *Tyler,* we upheld a warrantless post-fire search of a furniture store, despite the absence of exigent circumstances, on the ground that it was a continuation of a valid search begun immediately after the fire. The investigation was begun as the last flames were being doused, but could not be completed because of smoke and darkness. The search was resumed promptly after the smoke cleared and daylight dawned. Because the post-fire search was interrupted for reasons that were evident, we held that the early morning search was "no more than an actual continuation of the first, and the lack of a warrant thus did not invalidate the resulting seizure of evidence.". . .

As the State conceded at oral argument, this case is distinguishable for several reasons. First, the challenged search was not a continuation of an earlier search. Between the time the firefighters had extinguished the blaze and left the scene and the arson investigators first arrived about 1 P.M. to begin their investigation, the Cliffords had taken steps to secure the privacy interests that remained in their residence against further intrusion. These efforts separate the entry made to extinguish the blaze from that made later by different officers to investigate its origin. Second, the privacy interests in the residence—particularly after the Cliffords had acted—were significantly greater than those in the fire-damaged furniture store, making the delay between the fire and the midday search unreasonable absent a warrant, consent, or exigent circumstances. We frequently have noted that privacy interests are especially strong in a private residence. These facts—the interim efforts to secure the burned-out premises and the heightened privacy interests in the home—distinguish this case from *Tyler.* At least where a homeowner has made a reasonable effort to secure his fire-damaged home after the blaze has been extinguished and the fire and police units have left the scene, we hold that a subsequent post-fire search must be conducted pursuant to a warrant, consent, or the identification of some new exigency. So long as the primary purpose is to ascertain the cause of the fire, an administrative warrant will suffice.

Because the cause of the fire was then known, the search of the upper portions of the house, described above, could only have been a search to gather evidence of the crime of arson. Absent exigent circumstances, such a search requires a criminal warrant.

Even if the midday basement search had been a valid administrative search, it would not have justified the upstairs search. The scope of such a search is limited

(Continued)

Figure 5-3 *Continued*

(Continued)

to that reasonably necessary to determine the cause and origin of a fire, and to ensure against rekindling. As soon as the investigators determined that the fire had originated in the basement and had been caused by the crock pot and timer found beneath the basement stairs, the scope of their search was limited to the basement area. Although the investigators could have used whatever evidence they discovered in the basement to establish probable cause to search the remainder of the house, they could not lawfully undertake that search without a prior judicial determination that a successful showing of probable cause had been made. Because there were no exigent circumstances justifying the upstairs search, and it was undertaken without a prior showing of probable cause before an independent judicial officer, we hold that this search of a home was unreasonable under the Fourth and Fourteenth Amendments, regardless of the validity of the basement search.

The warrantless intrusion into the upstairs regions of the Clifford house presents a telling illustration of the importance of prior judicial review of proposed administrative searches. If an administrative warrant had been obtained in this case, it presumably would have limited the scope of the proposed investigation, and would have prevented the warrantless intrusion into the upper rooms of the Clifford home. An administrative search into the cause of a recent fire does not give fire officials license to roam freely through the fire victim's private residence.

The only pieces of physical evidence that have been challenged on this . . . appeal are the three empty fuel cans, the electric crock pot, and the timer and attached cord. Respondents also have challenged the testimony of the investigators concerning the warrantless search of both the basement and the upstairs portions of the Clifford home. The discovery of two of the fuel cans, the crock[pot], the timer and cord—as well as the investigators' related testimony—were the product of the unconstitutional post-fire search of the Cliffords' residence. Thus, we affirm that portion of the judgment of the Michigan Court of Appeals that excluded that evidence. One of the fuel cans was discovered in plain view in the Cliffords' driveway. This can was seen in plain view during the initial investigation by the firefighters. It would have been admissible whether it had been seized in the basement by the firefighters or in the driveway by the arson investigators. Exclusion of this evidence should be reversed.

It is so ordered.

Figure 5-3 *Continued*

private investigator assigned by the insurance company, the policyholder risks the chance that the insurance company will deny the claim due to the language in the insurance contract requiring the policyholder to cooperate with the insurance company. Essentially, no scene inspection, no money.

Public investigators have learned from reviews of Supreme Court cases that once the exigencies of the incident are over, public investigators must have

■ Note There are four ways to be on the property legally: exigent circumstances, consent, administrative search warrant, and criminal search warrant.

permission to be on the scene. They must have the right to be there. There are four ways to be on the property legally: exigent circumstances, consent, administrative search warrant, and criminal search warrant.

Exigent Circumstances

■ Note For the good of the public, fire departments are allowed to enter property under the exigencies of the circumstances.

The fire department must have the right to enter private properties without the necessity of obtaining consent or a search warrant. To delay entry may put lives and property at risk. Any delay may also allow the fire to extend, endangering other properties and other lives. For the good of the public, fire departments are allowed to enter property under the exigencies of the circumstances. This is extended to making a determination of the area of origin and cause of the fire. This, too, is for the good of the people in hopes of preventing future fires. This is not an unlimited amount of time; it is not open ended. If there is any delay in conducting the scene examination, it must be justified. Waiting for daylight is acceptable as shown in *Michigan v. Tyler.* However, beyond that, using exigent circumstances to justify a search for the cause of the fire beyond a certain time frame may be questionable.

Consent

The person who owns or has lawful control of the property can give consent to search. Control is the issue: The owner of an apartment building can give consent to a search of the common areas but not the leased apartments. The tenant can give consent to search the apartment or leased area that person controls. Case law shows that 16-year-olds (or other minors) can give consent to the general areas of the apartment where they live with their parents but not to their parents' bedroom.

■ Note It is important that the consent be documented with each individual.

It is important that the consent be documented with each individual. The jurisdiction should create written consent forms for use in the field. These forms should be reviewed and approved by the prosecuting attorney for that jurisdiction. Even when documented, consent can be rescinded. When this happens, the investigator must seek a search warrant to stay on the scene.

■ Note Even when documented, consent can be rescinded.

Administrative Search Warrants

All search warrants must be justified. Unlike a criminal search warrant, the only probable cause necessary is that there is a government interest in the fire investigator completing the investigation of the origin and cause of the fire. An administrative warrant is issued only in the following circumstances:

- There is proof that the investigator has the authority to conduct fire investigations in that jurisdiction.
- There is proof that a fire has occurred.

- The investigator cannot lawfully be on the property either because consent has been denied or the exigencies of the incident are over as identified in the *Michigan v. Tyler* Supreme Court decision.[2]

Once there is any indication that the fire is criminal in nature, the administrative search is over and the investigator must obtain a criminal search warrant.

Criminal Search Warrants

■ Note **Magistrates and judges are considered to be neutral and detached when deciding the merits of the information provided on the search warrant application.**

■ Note **The best rule to follow is to consult in advance the prosecuting attorney to establish a policy and procedure.**

The provisions of the Constitution of the United States as well as the provisions of all states require the issuance of a search warrant by a magistrate or judge based on an independent determination that probable cause exists. Magistrates and judges are considered to be neutral and detached when deciding the merits of the information provided on the search warrant application.[3]

The investigator must swear to the facts written in the application, which is commonly referred to as an affidavit. The request must be specific on what is sought and who or what is to be searched. The application must also state when the property is to be searched as well if the search will be made without giving notice to the property owner. In the course of carrying out the search, the investigator must stay within the boundaries identified in the search warrant. In the course of the investigation, if it becomes apparent that other areas of the structure must be searched, the investigator must seek another search warrant with the probable cause justifying the extended search. To do otherwise may violate someone's civil rights as well as lead to any evidence obtained as a result of that illegal search being inadmissible in court.

The best rule to follow is to consult in advance the prosecuting attorney to establish a policy and procedure to ensure that there is full compliance with all applicable laws.

SUMMARY

Understanding federal and state laws is essential to knowing your rights and the rights of others. Through case law, such as *Michigan v. Tyler* and *Michigan v. Clifford,* fire investigators have been given direction as to what is acceptable and what is not. However, there is no clear, definitive line to show you everything you need to know under all circumstances. Also, courts rule differently from one state to another. To ensure that as an investigator you are a seeker of truth and never deny anyone their constitutionally guaranteed rights it is best to seek recommendations from the prosecuting attorney, who will assist you, in advance, to establish rules, guidelines, approved forms, and policies on how to handle legal situations.

KEY TERMS

Administrative search warrant Warrant issued by a magistrate or judge allowing the investigator to be on the scene to determine the cause of the fire for the good of the people.

Consent Permission given by the person responsible for or controlling the property, allowing the investigator to search the property.

Criminal search warrant A warrant issued by a magistrate or judge based on the sworn (and written affidavit) testimony of an investigator that probable cause exists that a crime has been committed and the person or place to search and the items being sought.

Exigent circumstances For the good of the people, without permission, public safety personnel can enter property on fire to save lives and control the fire.

United States Supreme Court The highest court of the nation.

REVIEW QUESTIONS

1. In regard to creating laws, how does the Constitution ensure a balance of power?
2. The Constitution of the United States establishes the types of cases to be heard by the U.S. Supreme Court. List five types of cases the Supreme Court can hear.
3. Are the U.S. Supreme Court decisions required to be unanimous?
4. What evidence was allowed in the *Michigan v. Tyler* case as per the U.S. Supreme Court?
5. Why was the search deemed unconstitutional in the *Michigan v. Clifford* Supreme Court decision?
6. What right does the private investigator have to be on the scene?
7. Who can give consent to search a house?
8. What is an exigent circumstance?
9. What is the basis for a magistrate to issue an administrative search warrant for a fire investigator to conduct a preliminary fire origin and cause?
10. When an investigator swears before a magistrate for a criminal search warrant, what is allowed to be searched?

DISCUSSION QUESTIONS

1. In the *Michigan v. Tyler* case, there was evidence that implied the defendants were guilty, as was the verdict in the lower court decision. Whose fault is it that their conviction was overturned?

2. On television and in the papers, you see where criminals are set free because evidence was not legally obtained. Potential felons are back on the street. It has been suggested that instead of letting the felons go free we should allow the evidence and also punish the law enforcement agencies. How would this affect our society? How would it affect those law enforcement agencies? How would it affect the actions of the individual officers?

ACTIVITIES

1. Research the decisions of the highest court in your state, sometimes referred to as the state's supreme court. Do any cases affect the fire investigator on search, seizure, or the right to be on the scene?
2. Research the legislated codes of your state to see if there are any statutes that would affect the fire investigator in searching, seizing evidence, or having the right to be on the scene.

NOTES

1. Supreme Court of the United States, "About the Supreme Court," http://www.supremecourtus.gov/about/about.html.
2. J. Curtis Varone, *Legal Considerations for Fire and Emergency Services* (Clifton Park, NY: Delmar, Cengage Learning, 2007), 159–160.
3. Varone, *Legal Considerations,* 149–153.

SPOLIATION

Learning Objectives

Upon completion of this chapter, you should be able to:

- Describe and understand spoliation issues.
- Describe how spoliation issues are affected by first responders as well as the assigned investigator.
- Describe and understand the remedies courts have ascribed to those who have created spoliation issues.

CASE STUDY

When working a fire scene, it is not uncommon for investigators to have a feeling of loss for the family who have just experienced an absolute disaster in their lives. It is even more disturbing when one family member blames another for the incident, creating more pain on top of the fire damage.

On one particular Saturday evening, Engine 15 responded to a reported kitchen fire. They did a quick knockdown of the flames but not before the kitchen was a total loss and smoke spread throughout the entire house. Thankfully, everyone was safe.

The fire scene clearly indicated that the fire had started on top of the stove. In the process of documenting the scene, the investigator noticed a frying pan with an electric burner imprint on the bottom of the pan that indicated that the burner had been on for some time and had broken down the frying pan metal, creating the imprint. The frying pan had black carbon residue that indicates that there might have been some contents in the pan prior to the fire. Burn patterns on the wall pointed right back down to the frying pan. All the rangetop control knobs had been consumed in the fire. The question: Was the burner left on or did the stove fail, activating the burner?

During a brief interview, the investigator discovered that the oldest child, a 15-year-old girl, was babysitting her two younger brothers while her parents were out for the evening. She indicated that she had fried some potatoes in a frying pan on the front right burner earlier in the evening but had not cleaned the kitchen yet. The leftover potatoes were still in the pan.

It would have been so simple to mark this down as an accidental fire in which the teenager had left the burner on under the fried potatoes and left the kitchen for the next two hours. But this was not the case. The girl insisted that she turned the burner off. She even said she had gone into the kitchen earlier to get a snack and because the potatoes were cold, she got something else instead. She also indicated that the burner on the stove had gotten warm on its own without turning the knob. She said her youngest brother was blamed for this, but she knew that he did not turn the burner on, but that the stove was malfunctioning.

The parents arrived home while the fire department was still on the scene, and finding the devastation was quite a shock for both the mom and dad. They were clearly relieved that everyone was safe. But when the investigator advised the father of his findings, the mood changed. The father began chastising the daughter to the point of screams and threats. Getting the father calmed down was no easy task and did not happen until the investigator explained that they could test the switches on the stove to see if the switch was in the on position when the fire occurred. That seemed to mollify the situation for the moment.

Another set of photos was taken of the range, each control, each burner, and of the overall area. In this process, the investigator noticed that the soot around the knobs appeared disturbed. When the investigator checked with the engine company officer and his crew, one of the younger firefighters admitted that he turned the knob posts to make sure the range was off. Of course, the firefighter could not remember the position of the knobs before touching them. After some silence, the investigator

(Continued)

(Continued)

explained to the crew that any chance of discovering the original position of the knobs was now nonexistent. Worse yet, he had to explain this situation to the parents and the young babysitter.

Granted, this was a minor incident in the scope of things. Nevertheless, that family will never have closure; the father will have trouble trusting his daughter again in the future, and the daughter will always feel she was unjustly blamed for something that was not her fault. Had this been a faulty switch, the fire service has lost an opportunity to work toward preventing future events. On the upside, there is no doubt that one engine crew member has a new understanding of spoliation.

INTRODUCTION

spoliation
the destruction of evidence; the destruction or the significant and meaningful alteration of a document or instrument; it constitutes an obstruction of justice

This is one of the most important chapters for the protection of potential evidence and the protection of those handling and collecting the evidence. **Spoliation,** or the spoiling of evidence, is not a new term, but one that has seldom been heard in the fire service in the past. It is essential to know how to pronounce spoliation: (spoh-lee-ay-shun).[1] Regardless of the reason, spoliation can be seen as the destruction of evidence that can constitute an obstruction of justice. The implication of destruction can simply mean the alteration of any evidence or potential evidence that prevents other experts from seeing the value of that evidence.[2] The insurance industry and the civil courts have been dealing with spoliation issues for years. The fire service is not immune to these issues, and neither is it immune to the repercussions should spoliation occur out of careless disregard or ignorance.

■ Note
Every first responder, fire officer, and fire investigator must be well aware of spoliation issues and take steps to prevent spoliation events.

Every first responder, fire officer, and fire investigator must be well aware of spoliation issues and take steps to prevent spoliation events. Each and every department should establish a written agency policy and a training program on the steps to take to prevent the spoiling of evidence at any fire or explosion scene. Part of that training should include other agencies such as dispatchers and other emergency responders such as emergency medical personnel and police patrol officers because they all may handle evidence of one type or another.

■ Note
Failure to maintain and protect evidence can result in court-ordered fines, sanctions, or dismissal of the court case.

Failure to maintain and protect evidence can result in court-ordered fines, sanctions, or dismissal of the court case. Intentional or negligent acts that result in spoliation could result in criminal charges against the first responder or investigator for obstruction of justice. However, it must be remembered that the primary reason for protecting evidence is because it is the right thing to do. Fire investigators protect and serve the people; there is no better way to do this than to ensure we protect the evidence that may lead us to the discovery of how a fire started so that we can prevent future incidents.

WHAT IS SPOLIATION?

The National Fire Protection Association *Guide for Fire and Explosion Investigations* defines spoliation as "the loss, destruction, or material alteration of an object or document that is evidence or potential evidence in a legal proceeding by one who has the responsibility for its preservation."[3] The most import term in this definition is **potential evidence.**

potential evidence
something that may, or could, possibly be used to make something else evident; evidence that is not yet used or that has yet to be used to prove a point or issue or support a hypothesis

Almost any item within and around the structure could potentially become evidence in the determination of the area of origin, fire cause (cause of the explosion), or identification of who or what may be responsible for the event. Thus, every effort must be made to minimize the alteration of the scene before the investigator or investigative team has an opportunity to examine the scene. Depending on the size and scope of the investigation, the investigative team can be suppression personnel collecting data to complete their fire report or a more complex scene where a full-time investigator has been called by the officer in charge to investigate.

■ **Note**
Any object on the scene could have evidentiary value. Oftentimes, the placement or position of that object is just as important as the object itself.

Any object on the scene could have evidentiary value. Oftentimes, the placement or position of that object is just as important as the object itself. So many things can be considered spoliation it is impossible to list them all here. To show how easy it is to commit spoliation, consider two examples. The simple act of flipping circuit breakers eliminates their evidentiary value in the investigation. Once the breaker switch is moved, it is impossible to see which position it was in at the time of the fire. However, if the breaker is taken to the laboratory and X-rayed, you can easily see its position. The next set could be to drill out the pins holding the plastic housing together, as shown in **Figure 6-1**. If due care is taken in the process, there is a good chance the interior will provide evidence of the position of the interior workings of the breaker (on, off, tripped) at the time of or during the incident. The same can be said of the position of control knobs, as shown in **Figure 6-2** on appliances. Once moved by suppression forces, they lose their evidentiary value, possibly preventing the accurate final determination of the fire cause and responsibility.

Spoliation can have a dramatic impact on everyone associated with the fire or explosion scene. Destruction of evidence can affect the ability of both the public and private investigators to determine accurately the area or origin and cause. The home or business owner is affected by not truly knowing what happened, and spoliation can prevent the owner from seeking remedy should the fire have been caused by a third party such as an appliance manufacturer. The insurance companies representing all involved are affected financially because they might not be able to recover funds they paid out for the claim, when in fact they could have taken civil actions against the responsible party for the incident. The spoliation of evidence could also lead to key evidence not being able to be submitted, allowing an arsonist to go free to commit other crimes.

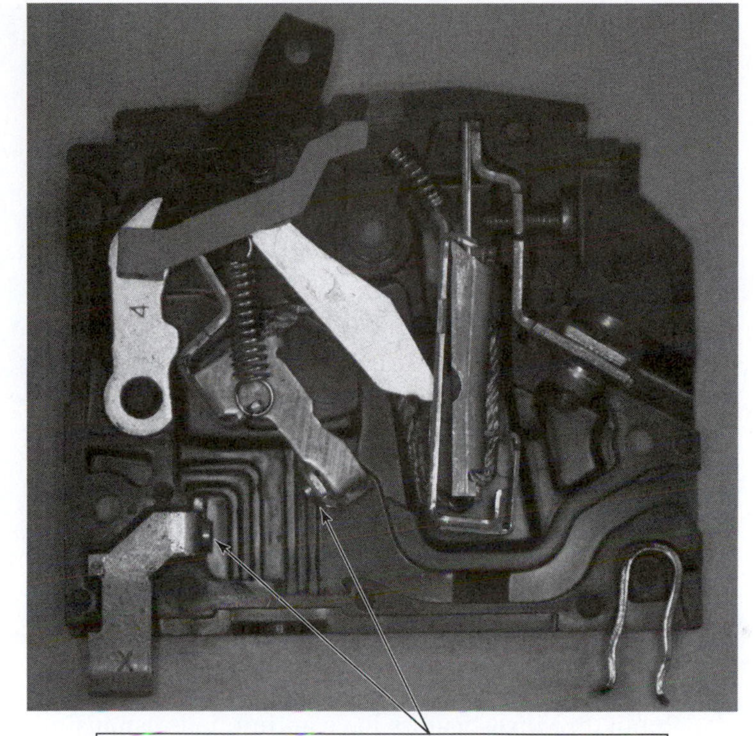

Figure 6-1 *A circuit breaker interior as seen after drilling out the pins holding the two pieces of the plastic housing. If done with due care and according to spoliation recommendations of NFPA 921,* Guide for Fire and Explosion Investigations, *it can provide valuable evidence.*

WHAT SPOLIATION IS NOT

overhaul
the act of searching for hidden fire, ensuring that the fire is completely extinguished

As a general rule, suppression personnel in the normal course of their duty while searching for victims, extinguishing the fire, and searching for fire extension are not committing spoliation. During the **overhaul** stage, suppression personnel must be cognizant of spoliation and limit the moving or alteration of debris to only what is necessary to accomplish their task of searching for pockets of hidden fire and ensuring that the fire is completely extinguished and that there will be no rekindle. Carrying out these duties does not constitute spoliation.

Once the fire is placed under control and the incident switches from suppression to investigation mode, a different set of rules applies about what can or cannot be done to the debris. NFPA 921, *Guide for Fire and Explosion Investigations,* states that the responsibility of investigator "varies according to such factors as the investigator's jurisdiction, whether he or she is a public official or private investigator, whether criminal conduct is indicated, and applicable laws and regulations."[4]

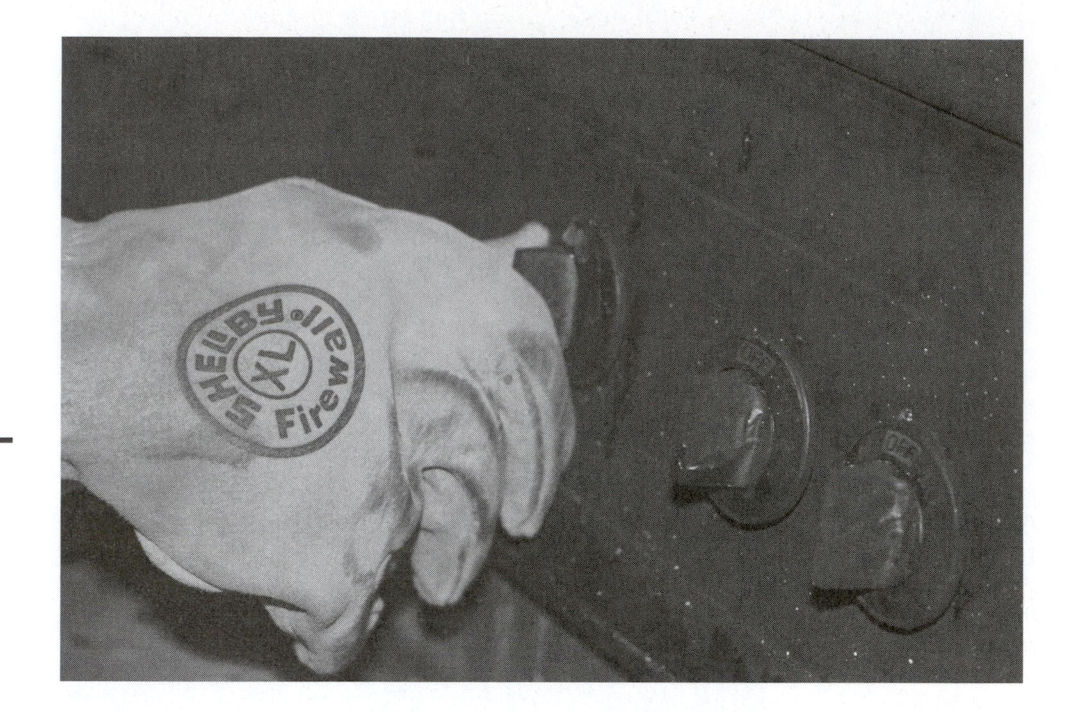

Figure 6-2 *The simple act of turning a knob on an appliance can destroy the knob's use as potential evidence, as such creating spoliation.*

> **■ Note**
> **Public fire investigators have an implied responsibility that for the good of the general public they must endeavor to make an accurate determination of the cause and origin of fires occurring within their jurisdiction.**

Public fire investigators have an implied responsibility that for the good of the general public they must endeavor to make an accurate determination of the cause and origin of fires occurring within their jurisdiction. Legal considerations may also require that they do so expediently, which may put them at the fire scene in the evening and early morning hours. Under these conditions, the public investigator is the only investigator on the scene in most cases. When a first responder is the investigator, the investigation is usually completed before this person leaves the scene of the incident.

It is understood that debris must be disturbed by the public investigator to locate the area of origin as well as to identify the potential cause of the fire. Should the investigator need to alter the debris significantly in the search, the scene should be adequately documented before the debris is moved. Documentation can consist of copious photos from every angle and direction. Videotaping can be beneficial, but investigators should never rely on video alone; it must be used in conjunction with photos. Documentation enables others to see the scene before it was disturbed.

Sometimes during the investigative phase, it is necessary to remove all the debris and room contents and then bring them back into the scene for reconstruction purposes. In this process, the item responsible for the ignition sequence may be removed and brought back into the scene. It may also be necessary to do some disassembly of an item to identify whether it was involved in the ignition sequence. When done by a qualified public investigator in a situation where

Figure 6-3 *The act of shoveling off the scene while searching for evidence in itself does not constitute spoliation.*

■ **Note**
The investigator should endeavor to photograph and document the entire scene.

the scene needs to be processed without delay, none of these actions should be considered to be spoliation. (See **Figure 6-3.**)

It cannot be stressed enough that the investigator should endeavor to photograph and document the entire scene as much as practical. In some circumstances, it is advisable for the investigator to photograph and document those items that were not involved in the ignition sequence, including all utility equipment and appliances capable of producing heat, to show that they were not involved in the ignition sequence.

In addition, the investigator should maintain photos and documentation that support alternate hypotheses that are discarded when the final determination of the fire area of origin and cause is made. This enables others to have the same opportunity to see all documentation and come up with their own determination of the fire origin and cause.

First responders should realize that private sector investigators also have an interest in the fire scene, although they may not arrive at the scene until hours or even days after the incident occurs. The private sector follows different procedures. For example, should private fire investigators identify potential evidence that may require destructive testing, they must notify all known interested parties before such testing is conducted. This enables entities who have an interest in the outcome of testing to be present if they so desire so that they can share the same evidentiary value of the item being tested.

DISCOVERY SANCTIONS AND PENALTIES

The repercussions of not following accepted procedures with evidence can be varied and far-reaching. During a trial, all parties have an opportunity to have access to evidence that will be used by all sides. This process is called **discovery** and is one of the foundations of a fair trial. The evidence on the fire scene may have a dramatic impact on the outcome of the trial. Should someone, such as a first responder or an inexperienced investigator, fail to handle, secure, collect, or store that evidence properly, the courts and all interested parties lose the opportunity to use that evidence in the trial.

discovery
a pretrial device used by both (all) sides in a case to obtain all the facts about the case to prepare for the trial

discovery sanctions
penalty for failing to comply with discovery rules

Note
To ensure that evidence is not mishandled, the courts may impose discovery sanctions, which are punishment for failing to follow discovery rules.

To ensure that evidence is not mishandled, the courts may impose **discovery sanctions,** which is punishment for failing to follow discovery rules. In the case of fire scenes, the sanctions can be quite harsh in the form of monetary fines, the exclusion of the expert witness to testify, or even dismissal of the case. Monetary fines imposed by the court are usually against the individual or the individual's employer. Depending on the circumstances, the employer may stand behind its staff and pay the fines. However, in cases of gross misconduct, the employer may decide not to support the individual, leaving the fine to be paid by the one responsible for the spoliation. Not only can this be costly financially, but it also jeopardizes the individual's credibility in future court cases.

When the testimony of the origin and cause expert is disallowed, the perpetrator of the crime may be allowed to go free, in essence allowing this person to commit other crimes. The far-reaching implications may be that others who were thinking of committing arson might now have the encouragement to carry out the act. Then there is the victim. The dismissal of the case not only leaves victims with a sense of injustice but also may reduce or eliminate their possibility of any financial recovery.

Should the destruction of the evidence be intentional or negligent, tort (civil lawsuits) action can be taken against the firefighter, investigator, or their department. In these situations, the plaintiff will most likely file the suit against the individual, the individual's immediate superior, the fire chief, the senior government official such as the county administrator or city manager, and even the political entity governing the jurisdiction. In doing so, the plaintiff is seeking the largest remedy to its favor.

The lawsuit itself can be damaging to the credibility of the individual even if the individual is innocent. The case is usually played out in the press long before it is ever heard in court. In most cases, the locality represents the first responder or investigator in court. Nevertheless, their primary responsibility is to look out for the interest of the jurisdiction and not just the individual. As a result, some first responders or investigators in this situation have retained their own legal counsel at their own cost. As you can imagine, there is also a psychological cost for the individual and his or her family as well.

■ Note
Criminal charges could also be levied against an individual who negligently or intentionally conducts spoliation.

Criminal charges could also be levied against an individual who negligently or intentionally conducts spoliation. A charge of obstruction of justice has broad implications. It can be during the investigation as well as during the trial. There need not be actual proof of the obstruction, but the evidence that the individual endeavored to obstruct is sufficient for a conviction. Penalties vary from jurisdiction to jurisdiction; however, this is considered a felony and may include both a financial penalty as well as prison time. It may be harder for investigators to face such charges because they are in a position of trust and as such may face stronger scrutiny.

DESIGNING TRAINING PROGRAMS

As mentioned at the beginning of this chapter, a training program is essential to ensure that the agency prevents spoliation incidents. Too many variables make it impossible for one program to fit each and every department. The following subsections describe some of the considerations that must be made in designing a training program.

■ Note
An investigator must believe in the Constitution and court system, that everyone is entitled to a fair and impartial trial. A fair trial constitutes one in which both sides get to see and have the benefit of all the evidence.

Proper Investigative Attitude

The first step is to ensure the investigator has a proper attitude. An investigator must believe in the Constitution and court system, that everyone is entitled to a fair and impartial trial. A fair trial constitutes one in which both sides get to see and have the benefit of all the evidence. When it comes to fire scenes, that includes all of the evidence from the scene—not just what the investigator thinks is important. Investigators must remember, regardless of how certain they are about their findings, that they are neither the judge nor the jury. They must remember that they are *seekers of truth and not case makers,* as stated in the International Association of Arson Investigators (IAAI) Code of Ethics.[5]

In the past, some public and private sector fire investigators had scenes bulldozed and destroyed at the completion of their scene examinations. This is spoliation at its worst. Their reasoning became evident when they would admit that no one was going to second-guess their findings. This brings up the second attitude that every investigator must possess. Every investigator must be prepared to have a second expert come onto the scene and make a different determination of the fire origin and cause. Investigators do not have to like it, but they have to realize it will happen and that they should not personalize it at all. Very capable and talented fire investigators with equal credibility and expertise can look at the same scene and come up with the same or different causes of the fire. With a proper attitude comes an open mind to ensure that investigators really are seekers of truth.

Training Program Design Committee

A design committee must ensure that all aspects of necessary training are included in the program. At a minimum, the committee should consist of a representative from the local prosecuting attorney's office, a member of the department's training division, and a representative from the local fire and arson investigating team. Consideration should also be made of including someone from the private sector such as a claims manager from an insurance company as well as an investigator or engineer from the private sector who specializes in fire investigations. These individuals have a different outlook on the process and can provide recommendations and insight on how to work together for the good of the public.

Creation of Agency Policies

Before a training program can be established, there has to be a review of or creation of an agency procedures policy on actions to be taken at the scene by both suppression and investigative personnel. For example, a policy must be established on how the electrical power is to be disconnected to a structure should the utility company not be immediately available. Spoliation concerns should not get in the way of safety.

The agency's policy should outline when the investigation is to take place and what requirements are necessary to be on the scene legally. The policy should also address when the investigation can be conducted following the dictates of the U.S. Supreme Court rulings on fire scene investigations, such as *Michigan v. Tyler* or *Michigan v. Clifford*. This relates to NFPA 921, *Guide for Fire and Explosion Investigations*, text description of responsibility being different for public and private sector investigators. When the public investigator starts the examination while the suppression units are still on the scene, it cannot be assumed that this is appropriate but should be written out in the investigator's policy manual that this is a common practice and acceptable. However, it is not appropriate for any person outside the department to come on the scene unless approved by the assigned investigator; this too needs to be addressed in the policy manual.

The agency's policy should also address how evidence is to be handled. Is it maintained by the locality, and if so, what section of the law allows the locality to collect evidence of a noncriminal nature? Of course, the issues of liability and spoliation should be addressed as well. Once decisions on these issues have been made, the policies must be included in the training program.

Policy definition is not complete unless it addresses how other entities such as insurance companies and the private sector will address spoliation. They follow a set of rules and policies different from the public sector, and all fire investigators must know how to work together. For this reason, it is a good idea for the training program design committee to include representatives from the private sector.

SUMMARY

The importance of preventing spoliation of evidence cannot be stressed enough. It must be understood that even first responders can be guilty of spoliation of evidence when they do not follow accepted practices and procedures. Ignorance is not a defense for the first responder or investigator should evidence be damaged.

The first way to ensure that spoliation is avoided is to be sure that all who conduct investigations have a good ethical attitude. Next, a good training program for all parties involved, including firefighting personnel, must be in place to prevent spoliation. Before training is delivered, the agency must create a policy and procedures manual documenting proper agency policy on how to avoid spoliation issues. Once policy is designed, tested, and approved, it should be the primary emphasis of the training program.

The primary purpose of spoliation rules and punishments is simple. Everyone involved in criminal cases or civil litigation must ensure that there is a fair and impartial hearing of the facts. When evidence is destroyed, the ability to find all facts is diminished. Destruction of evidence can jeopardize the opportunity to seek the truth in the investigation, so every effort must be made to protect evidence and resist the temptation to touch and move objects that may be potential evidence.

KEY TERMS

Discovery A pretrial device used by both (all) sides in a case to obtain all the facts about the case to prepare for the trial.

Discovery sanctions Penalty for failing to comply with discovery rules.

Overhaul The act of searching for hidden fire, ensuring that the fire is completely extinguished.

Potential evidence Something that may, or could, possibly be used to make something else evident; evidence that is not yet used or that has yet to be used to prove a point or issue or support a hypothesis.

Spoliation The destruction of evidence; the destruction or the significant and meaningful alteration of a document or instrument.[6] It constitutes an obstruction of justice.

REVIEW QUESTIONS

1. Define *spoliation*.
2. List and describe all parties that could be affected by spoliation.
3. What protections do public investigators have during their examination of the fire scene?
4. Whom should private sector investigators contact if they are going to conduct destructive testing on a piece of evidence?
5. Why should the private sector be involved in a training program on spoliation for the fire department?

DISCUSSION QUESTIONS

1. You are on a fire scene and you hear an investigator arrange for the debris to be bulldozed, as he says that he has no intention of letting another fire investigator second-guess him. How do you feel about his actions? What should you do about his actions?
2. If you are involved in firefighting, discuss what you have seen in the past that could have been a spoliation issue.
3. A fire has claimed the life of a firefighter. A private investigator has been hired by the defense counsel of the alleged arsonist. Why should the investigator willingly provide any and all evidence to the other investigator when ordered by the courts? Should the investigator intentionally hold anything back?

ACTIVITIES

1. Research recent spoliation issues by doing an Internet search for "spoliation" and "fire."
2. Research cases involving spoliation and list any and all sanctions the courts have imposed.

NOTES

1. *Webster's New World College Dictionary*, 2nd ed. (New York: Simon & Schuster, 1984), 1376.
2. *Black's Law Dictionary*, 5th ed. (St. Paul, MN: West Publishing Company, 1979), 1257.
3. National Fire Protection Association, *NFPA 921, Guide for Fire and Explosion Investigations* (Quincy, MA: NFPA, 2008), 921-15.
4. National Fire Protection Association, *NFPA 921, Guide for Fire and Explosion Investigations*, 921-104.
5. International Association of Arson Investigators, "I.A.A.I. Code of Ethics," http://www.firearson.com/insideiaai/codeofethics/index.asp.
6. *Black's Law Dictionary*, 1376.

SCIENCE, METHODOLOGY, AND FIRE BEHAVIOR

Learning Objectives

Upon completion of this chapter, you should be able to:

- Describe the concept of scientific methodology.
- Describe the aspects of fire behavior as they relate to the fire investigator.
- Describe and understand the concept of heating at the molecular level.
- Describe and understand the concept of heat transfer as it relates to the fire investigator.
- Describe and understand the concept of flame spread and complications associated with flashover in an investigation into the fire's area of origin and cause.

CASE STUDY

Engine 33 is out in the district returning from a false alarm at the local grammar school when the tones go off reporting a fire that just happens to be less than a mile away. As the engine nears the scene, black smoke can be seen just above the horizon. As they pull in front of the house, the garage door is open and flames can be seen from floor to ceiling inside the garage with more flames lapping out from the top edge of the door. The homeowner is in the yard telling the crew that he was the only one home and that he has some burns on his hands and arms. One firefighter starts treating the homeowner for his burns while the rest of the crew pulls an attack line, accomplishing a quick knockdown of the fire, saving the rest of the home from fire damage, although smoke has permeated the entire structure.

Fortunately, an ambulance is an automatic assignment for any structure fire and it arrives on the scene shortly after the engine company, taking over the treatment of the homeowner. The temperature is just above freezing, so they take the patient into the ambulance to treat his burns. Soon thereafter, they leave the scene with the patient, taking him to the local hospital for treatment of first-degree and some minor second-degree burns on his hands. The fire officer was busy with the suppression, so did not get a chance to talk further with the victim.

The engine company officer calls dispatch on the radio, asking for a duty fire marshal. Doing the proper thing, they conduct overhaul, making holes in the wall and ceiling to ensure they have extinguished all extensions of the fire. In doing so, they take every precaution not to disturb the fire scene any more than absolutely necessary.

The fire marshal arrives on the scene, conducts his investigation by working from the least damaged area to the most, photographs the area, and takes notes along the way. He notices that burn patterns indicate the fire was at floor level across a large percentage of the garage. Burn patterns indicate that an ignitable liquid may be present. Although this is not unusual to have containers of ignitable liquids in any garage, there are no containers that would indicate a leak or a spill of any kind. Further examination shows no presence of an ignition source; the garage had neither heat nor appliances of any kind. There was electricity, but it consisted of light switches, light fixtures, and outlets mounted at waist height. None of the wiring or devices showed any involvement in the ignition.

Evidence is collected from the floor and is placed in a 1-gallon evidence can for processing by the state forensic lab. Having completed his scene examination, the fire marshal tells the engine company officer to fill out the fire report as the fire still being under investigation. He then delivers the evidence to the laboratory and goes to his office to write the report. Although the investigator did not intentionally sniff the debris, he did detect an odor similar to diesel fuel while cleaning his tools. This is not scientific and neither is it admissible in court. Only the lab can confirm the presence of an ignitable liquid in the sample provided. Several other characteristics of this case indicate the fire was incendiary in nature. However, having completed his fire scene examination, the fire marshal's job is done. In this particular jurisdiction, the fire department investigates the scene only and the local sheriff's department then

(Continued)

(Continued)

takes over the investigation if there are any indications that the fire may have been incendiary in nature.

This is not necessarily a bad thing, provided the sheriff's detective is a trained fire investigator with full knowledge of fire and fire behavior. Regretfully, in this scenario, this was not the case. At the fire scene, a sheriff's detective caught up with the homeowner, who had been treated and released from the hospital and was standing in the yard when the detective arrived.

The two went into the garage where the homeowner explained that he was using diesel fuel to clean some engine parts; they were soaking in a 5-gallon container. He further explained that when he went to move the bucket, he bent over and the cigarette in his mouth fell into the bucket causing an explosion that knocked him off his feet. In his haste, he accidentally tipped over the bucket, spilling its contents across the entire floor. When asked where the bucket was now, the homeowner explained that in his haste to pick up the bucket and put out the fire he received his burns. He said the bucket was now in the back of his pickup where he rescued it from the fire.

Sure enough, an empty bucket was in the bed of the truck, and indeed, it did reek of diesel fuel. The detective thanked the homeowner for his time and wished him well in the healing of his wounds. He returned to his office where he closed his investigation, noting that the fire as described by the homeowner was plausible and the fire was accidental in nature.

The only saving grace was a thorough investigation by an insurance investigator, who uncovered ample evidence to allow the insurance company to deny the claim based on the fire being intentionally set by the insured to collect insurance proceeds. Although the individual was never charged with arson, at least he did not profit from his actions.

So, what was the real issue? Diesel fuel is combustible, but it is difficult at best to get it to ignite from a cigarette, especially with the volume described. Under most circumstances, the cigarette would have dropped in the liquid and extinguished. At the very least, the detective should have looked further into the case rather than accept an unlikely scenario. As a fire investigator, you must know the properties of the fuels you are dealing with in a fire situation. Recognizing a plausible explanation from an implausible one can make or break your case.

■ Note
To conduct detailed fire investigations effectively, you will also need additional training in topics such as college chemistry and physics.

INTRODUCTION

To fully appreciate and understand the science of fire as it pertains to fire investigations you will need to obtain a basic understanding of the behavior of fire. To conduct detailed fire investigations effectively, you will also need additional training in topics such as college chemistry and physics. You will learn about heat and molecular movement as well as heat transfer and other basic scientific concepts that relate to fire. Basic concepts in electricity and mechanics, including friction, are topics you will also need to master.

■ Note
Good science is that which is based on proven and reproducible scientific principles, whereas junk science is that which is unproven, founded on speculation, conjecture, and outdated concepts and principles.

However, this is just the beginning; books, research papers, and professional articles can add to your knowledge as an investigator. Continual studies of such documents are a necessity for every investigator, young and old, new and experienced. In this education endeavor, you need to ensure you are studying good science that can be used in the fire scene evaluation process. Good science is that which is based on proven and reproducible scientific principles, whereas junk science is that which is unproven, founded on speculation, conjecture, and outdated concepts and principles.

To ensure a more accurate determination of the area of origin and the cause of the fire it is best always to use a process called the scientific method. This is a means of providing a systematic framework to assist the investigator. The scientific method is a technique to acquire new knowledge by gathering evidence and evaluating that evidence using reasoning.

SCIENTIFIC METHODOLOGY

Walking onto a fire scene where there has been massive destruction can be daunting. Even a single-family dwelling fire can give you pause, as you stand in the front yard and ponder what happened.

scientific method
the systematic pursuit of knowledge involving the recognition and formulation of a problem, the collection of data through observation and experiment, and the formulation and testing of a hypothesis

One aid in the process is to use a systematic approach to examine the fire scene, which is referred to as the **scientific method.** This method was not invented by the fire service but takes a fundamental principle from the scientific community and adapts it to the process of conducting fire investigations. There are variations to the steps in the process; however, the NFPA 921, *Guide for Fire and Explosion Investigations,* outlines the scientific method to include the following steps: Recognize the Need, Define the Problem, Collect Data, Analyze the Data, Develop a Hypothesis, Test the Hypothesis, and Select a Final Hypothesis.[1] (See **Figure 7-1.**)

The first step is *Recognize the Need;* in other words, identify the problem. The problem is that a fire or explosion has occurred, thus the need for an investigation. More definitive is to state that there is a need to investigate the fire's area of origin and the cause of the fire. With these findings, the investigator may be able to identify the need for an education program to prevent future incidents. The culmination of all data as to a fire's origin and cause provides statistical analysis for the prevention of future events.

■ Note
The culmination of all data as to a fire's origin and cause provides statistical analysis for the prevention of future events.

For a municipality, the need can be established well in advance. The very reason a fire investigator position may exist is that the jurisdiction recognizes that fires will occur and they will need to be investigated. One additional fact: Recognizing the need can take on a legal tone as well, for example, did the fire occur within the investigator's jurisdiction or area of responsibility? Even if there is a suppression mutual aid agreement with a neighboring jurisdiction and the fire apparatus respond, this may not give the fire investigator the legal

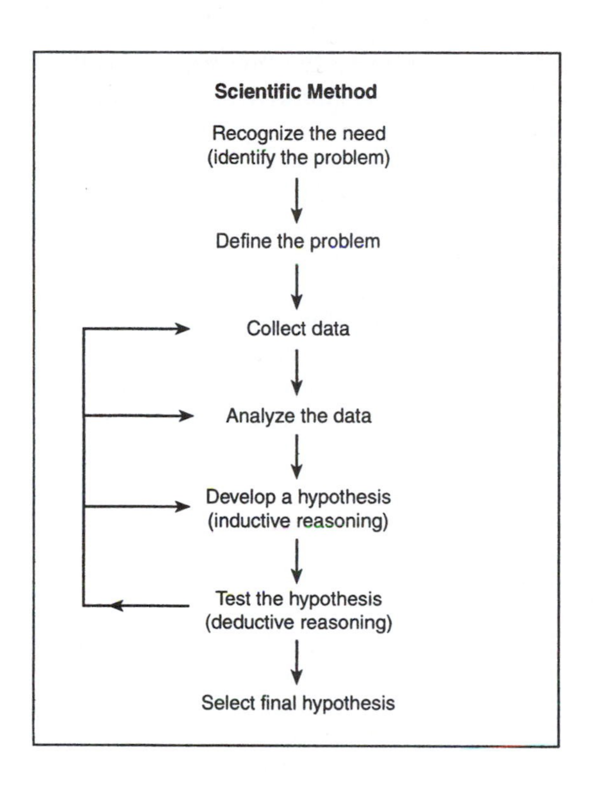

Figure 7-1 *The scientific method chart shows the steps to follow and to retrace when a proposed hypothesis cannot withstand the test.*

Source: Reprinted with permission from NFPA 921-2008: *Guide for Fire and Explosion Investigations*, Copyright © 2008, National Fire Protection Association, Quincy, MA. This reprinted material is not the complete and official position of the NFPA on the referenced subject, which is represented only by the Standard in its entirety.

right to cross that jurisdictional line and investigate the fire in the neighboring jurisdiction.

Next is to *Define the Problem*. In the first step, you identified that there is a problem. In this step, you understand the problem and must determine how to come up with a solution. The answer seems quite simple: Conduct a detailed investigation into the fire's origin and cause. Most likely, the locality has already created a policy that will dictate how fires are to be investigated. Most fire scenes receive an initial investigation by the fire officer in charge or a person assigned by the fire officer. This investigation is usually limited to a brief scene examination along with possible brief interviews of witnesses or the property owners or tenants. Depending on the fire officer's findings, a duty investigator may be called to the scene to do a more detailed investigation, perhaps because the fire officer cannot find the origin or cause or the circumstances discovered are unusual or suspicious. In some instances, when the fire is a multiple alarm or when circumstances are unusual, the fire investigator may go to the scene immediately and conduct the investigation in lieu of the fire officer.

■ Note
The data to be collected include everything and anything that can, or may, be used in making a decision as to the fire origin and the fire cause.

Next, the investigator must *Collect the Data*. Every aspect of the incident must be recorded in one form or another; from photographing the fire scene to recording interviews, the investigator must collect all evidence about the scene. The data to be collected include everything and anything that can, or may, be used in making a decision as to the fire origin and the fire cause. What is collected is considered **empirical data,** which is based on observations, experiments, or experience.

empirical data
data that are based on observations, experiments, or experience

After the data is collected, the investigator must *Analyze the Data* using logic and reason. The only data that can be analyzed are those data that are obtained from observation or experiments. In doing this analysis, the investigator uses his or her training, education, knowledge, and experiences. An investigator's level of expertise is based on knowledge and experiences. Knowledge is a culmination of the individual's formal education, such as obtaining a degree, along with additional education from seminars, conferences, and other specialty schools. Just as important are the investigator's experiences, both those directly and indirectly connected to the investigation of fires. Someone with a strong background as an electrician may have a better understanding of an electrical fire cause. Likewise, someone with experience in military ordinance may understand explosive devices better than other investigators can. Regardless, we are the sum of our experiences. All experiences add to an investigator's expertise.

Note
Regardless, we are the sum of our experiences. All experiences add to an investigator's expertise.

Based on the data collected and analyzed, the investigator can *Develop a Hypothesis* as to the fire origin and the fire cause. A hypothesis is an assumption made based on the empirical data collected by the investigator. This is not the final determination until all other steps in the scientific method are completed.

Before establishing a final determination of the fire origin and cause, the investigator must *Test the Hypothesis*. This is done using **deductive reasoning,** taking the facts of the case and doing a thorough and meticulous challenge to all facts known. Unless the hypothesis can stand up to all reasonable challenges, it is an unacceptable theory and must be discarded. When this happens, the investigator must then go back and reexamine and re-analyze the data to develop a new hypothesis; the investigator must subject this new hypothesis to a thorough and detailed test to see whether it can withstand the test of deductive reasoning.

deductive reasoning
taking the facts of the case and doing a thorough and meticulous challenge to all facts known, using logic

Note
Unless the hypothesis can stand up to all reasonable challenges, it is an unacceptable theory and must be discarded.

If no hypothesis withstands the tests, the investigator must go back and collect new data to enable a proper determination of the fire origin and cause. New data may be additional interviews, following new leads from those interviews, or conducting additional scientific testing. This process continues until a hypothesis can withstand all tests, showing itself to be the final determination as to the fire origin and cause. The testing of the hypothesis is dynamic, and it must be retested whenever new data are presented.

However, this is not to say that all fires have a proven hypothesis. In fact, many fires may very well have to be labeled as having an undetermined cause and origin. This is hard to accept sometimes, especially when there are injuries or a loss of life. Nevertheless, the facts are that the investigator cannot solve every fire, sometimes because there is just not enough data to come up with a plausible explanation as to what happened, sometimes because there are inadequate resources in that jurisdiction to allow adequate time to do an in-depth investigation. Regardless, the investigator must resist the temptation to attach a cause

■ **Note**
In these situations, where an investigator has no success in identifying the fire's origin or cause, the fire must be labeled as undetermined.

based on limited data. In these situations, where an investigator has no success in identifying the fire's origin or cause, the fire must be labeled as undetermined.

PRESUMPTION OF THE FIRE CAUSE

Just as dangerous as labeling a fire based on limited knowledge after an extensive investigation is presuming the fire cause before you even start a fire investigation. All too often individuals make presumptions on the fire cause based on false impressions, frequency of certain types of fires, or a piece of loose information heard when arriving on the scene.

The fundamental problem is that an individual may look only for evidence that supports the presumption rather than keeping an open mind and collecting all data. This is why many fire investigators want to look at the fire scene before conducting their interviews. This way, the investigator will not be tainted by something someone said in the interview, such as the homeowner indicating she had continual problems with the furnace. Even though the investigator is more than capable of keeping an open mind following interviews, others, such as a jury, may believe the fire investigator could have been influenced during the interviews.

A case in point is when the homeowner indicates that the fire had to be arson because of an argument he had with the neighbor only hours before the fire. Physical evidence does indicate the neighbor was involved, an arrest is made, and a trial date is set. During testimony, the defense counsel asks whether the investigator had any indication that the neighbor was involved before the fire scene investigation was conducted. Counsel could then paint the picture that the investigator looked only for incendiary indicators based on the information gathered in the interview and not on the scene. Although this may sound like a desperate maneuver for the defense, it is one that can put the investigator in a bad light with the jury.

■ **Note**
The investigator should do a systematic search of the fire scene, working from the least damaged area to the most damaged area.

Had the investigator been able to say, "No, Counselor, I looked at the fire scene and found evidence of an incendiary fire before conducting my interview with the homeowner." The bottom line is for investigators to keep themselves beyond reproach as much as possible.

SYSTEMATIC SEARCH

systematic search
a search based on a system that is used for each fire, each and every time

Doing a thorough search of the fire scene is essential to a successful conclusion of an investigation. As with using the systematic approach of scientific methodology, the investigator should do a **systematic search** of the fire scene, working from the least damaged area to the most damaged area. This method may allow the investigator to follow the path of the fire back to its origin. This process can work extremely well for most.

However, this is not the only method for conducting a search. Working from the highest point of the structure to the lowest can work as well, as long as that is what is done consistently in each and every investigation by that investigator. This process is critical to aid in a thorough and complete investigation. A secondary benefit relates to the investigator's courtroom testimony. If you systematically conduct each and every fire the same way, day after day, month after month, or even year after year, you will have no problem answering in the affirmative if and when challenged in court as to your procedures for conducting an investigation.

You have to have somewhere to start to conduct a systematic approach. With a fire in a typical single-family residence, the starting point could be on the street in front of the house. Starting from this point, you can work in a clockwise direction, observing the house and property from all sides, not only looking at burn patterns but also looking for evidence. Working toward the structure, you look at the damage, observing the least damaged area and eventually working into the most damaged area. This process can enable investigators to follow the path of fire travel right back to where the fire started.

One note of caution regarding recommended procedures: No one procedure or process can fit each and every situation. For this reason, you should always be prepared with alternative, viable plans. When conducting a systematic search, you may have a logical reason to work from the bottom up or the top down in a particular fire. Investigators need only be ready to provide a logical explanation as to why they varied from their usual procedure should it be challenged in court.

CHEMISTRY OF FIRE FOR THE FIRE INVESTIGATOR

This text assumes that readers have a fluent knowledge of basic fire behavior and fire chemistry. A good refresher in this area is available in the chapter titled "Fire Behavior" in *Firefighters Handbook: Firefighting and Emergency Response,* 3rd edition (Delmar, Cengage Learning, 2008). For example, one of the basic principles that firefighters learn is that to extinguish a fire they must remove one of the sides of the fire triangle: Remove the oxygen, and the fire smothers; remove the fuel, and there is nothing to burn; remove the heat, and the fire will go out. For the full-time investigator, it is good to review the *Firefighters Handbook* because it discusses the basic principle of atoms, elements, and compounds, knowledge of which is key to a fire investigator being successful in his or her career.

Fire investigators also consider heat, fuel, and oxygen important. Investigators are interested in learning what action or event brought these items together to start the fire. Generally speaking, oxygen is always available. That leaves heat and the fuel. To determine how the fire started, you have to know how the first

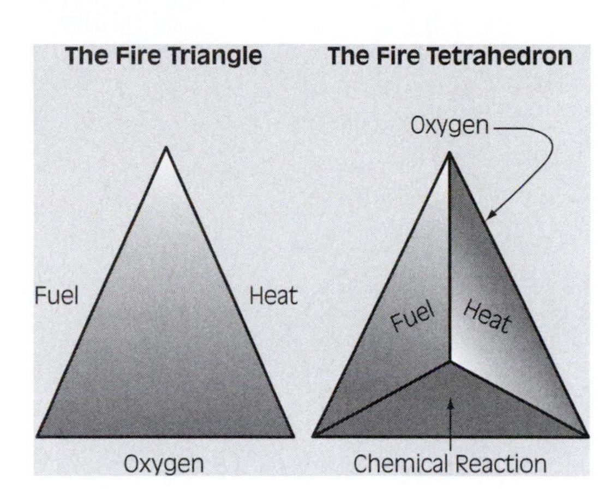

Figure 7-2 *The fire triangle and the fire tetrahedron are the basic models that explain how to extinguish a fire as well as how a fire started.*

fire tetrahedron
tetrahedron is a solid figure with four triangular faces; when each face represents the four items necessary for a fire (heat, fuel, oxygen, and uninhibited chemical reaction), the figure then becomes the fire tetrahedron

item ignited (fuel) came in contact with a capable ignition source (heat). If you cannot explain this, you do not have a true hypothesis on how the fire started.

Fire Tetrahedron

The fire service also uses a model called the **fire tetrahedron,** which is a four-sided solid object with four triangular faces, a pyramid. The first three sides are the same as the fire triangle: heat, fuel, and oxygen. The fourth side is the self-sustained chain reaction, sometimes referred to as the *uninhibited chain reaction.* This one side explains why the fire continues to grow rather than stay in one place. (See **Figure 7-2.**)

To truly understand fire the assigned investigator must be able to bring everything down to the molecular level. Sharing electrons or the excitement of a molecule that causes rapid movement can explain how a fire starts, reacts, grows, and is extinguished. Any investigator without this knowledge is limited in understanding the full aspect of fire science and will be handicapped in defending a proper hypothesis.

■ Note
To truly understand fire the assigned investigator must be able to bring everything down to the molecular level.

Self-Sustained Chain Reaction Combustion is a complex process that results in the rapid oxidation of the fuel, producing heat and light. When this fuel continues to burn on its own in a self-sustained manner through an **exothermic reaction**[2], the fire continues to grow. An exothermic reaction is a chemical reaction created when the bonds of molecules are broken, which releases heat and light. Another term for this chain reaction is "uninhibited chain reaction," which is one side of the fire tetrahedron.

exothermic reaction
the release of heat from a chemical reaction when certain substances are combined

For example, an exothermic reaction occurs when you hold a piece of paper horizontally and expose the underside to the heat from a match, as shown in **Figure 7-3.** Even without the flame of the match touching the paper, the heat is sufficient to start the ignition process. The heat begins to break down the paper,

Figure 7-3 *A piece of paper shows the chain reaction of fire.*

■ Note
An exothermic reaction is a chemical reaction created when the bonds of molecules are broken, which releases heat and light.

the paper starts to turn brown, and vapors from the decomposing paper are given off. These vapors are combustible, and when they are heated to their autoignition temperature or come in contact with the flame, they ignite. Now the paper is burning. The leading edge of the flames heats up more of the paper surface area, releasing more vapors that in turn ignite, contributing to the growth of the fire. This self-sustained chain reaction continues across the surface of the paper until the fire is extinguished. If disturbed by no other outside factors, the fire will eventually go out as a result of the lack of fuel because the paper is completely consumed, which leaves behind carbon and other compounds in the form of ash. If the paper is next to other combustibles, the fire could continue to grow as the nearby materials are exposed to the heat and those items too begin to give off vapors and eventually ignite, creating a larger fire.[3]

Investigators sometimes might have to explain fire in the simplest terms to a judge and jury when testifying how such a small fire can eventually burn down a large structure. Explaining how fire behaves at the molecular level can help others to understand fire and its behavior.

■ **Note**
Fires cannot occur without an oxidizer.

oxidizer
any material that forms with a fuel to support combustion

Oxygen Fires cannot occur without an **oxidizer.** In most situations, the atmosphere, which contains approximately 21 percent oxygen, is the primary oxidizer to complete the combustion process. However, fires can occur in the absence of atmospheric oxygen. Many compounds, when mixed with other chemicals or when heated, can give off oxygen in sufficient amounts to allow combustion to occur or to continue. One such chemical, ammonium nitrate, was the fuel for one of the United States' largest accidental disasters and was also used to make explosive devices on multiple occasions.

Ammonium nitrate fertilizer (NH_4NO_3) was the chemical stored in the hold of the *S.S. Grandcamp,* a French liberty ship that was tied up at the Monsanto plant in Texas City, Texas, on April 16, 1947. Smoke was seen coming from the hold at approximately 0800 hours. The ship already held 2,300 tons of ammonium nitrate fertilizer in 100-pound bags. The master of the ship decided to have the hatches battened down and steam pumped into the hold to extinguish the fire.

■ **Note**
One of the ship's anchors, weighing approximately 3,200 pounds, was thrown 1.62 miles from the ship and buried itself about 10 feet into the ground where it landed.

Around 0830 hours, the hatches blew off from pressure building in the hold. A thick column of orange smoke was seen coming from the hold. At 0900, flames were coming from the hold, and then at 0912 hours, the ship disintegrated in an unbelievable explosion that was heard 150 miles away. One of the ship's anchors, weighing approximately 3,200 pounds, was thrown 1.62 miles from the ship and buried itself about 10 feet into the ground where it landed. The end result of that explosion and the other explosions and fires that resulted from the initial blast resulted in 568 dead or missing and more than 3,500 injured. (See **Figure 7-4** and **Figure 7-5.**) The key issue is that ammonium nitrate, when broken down, is its own oxidizer.[4]

Figure 7-4 *The remains of one of the three engines from the Texas City Volunteer Fire Department responding to the initial alarm. (Photo courtesy of Moore Memorial Public Library, Texas City, Texas.)*

Figure 7-5 *Smoke from the fires created from the explosion of the* S.S. Grandcamp, *as seen from Galveston about 8 miles away. (Photo courtesy of* Moore Memorial Public Library, *Texas City, Texas.)*

■ Note
Many chemicals are capable of being oxidizers; many are in the form of nitrates, nitrites, peroxides, and chlorates.

■ Note
Investigators should also be aware that an atmosphere of chlorine could also support combustion.

Many years later, the same chemical, ammonium nitrate fertilizer, was used in a different manner but also with a devastating, horrific effect. It was the key component of a bomb that was loaded into the back of a rental truck that was parked just outside the Murrah Federal Building in Oklahoma City. When detonated, the two and a half tons of fertilizer mixed with fuel oil tore off the face of the building and resulted in 168 deaths and hundreds of injuries. (See **Figure 7-6.**) The explosion tore off the entire front of the Federal Building, as shown in Figure 7-6.

Many chemicals are capable of being oxidizers; many are in the form of nitrates, nitrites, peroxides, and chlorates. The key component of their makeup is oxygen, which under certain circumstances can be released, enabling combustion. Investigators should also be aware that an atmosphere of chlorine could also support combustion.

Another factor the investigator must be aware of is the effect of the amount of oxygen present. Chemically speaking, the oxidation process occurs when oxygen bonds to other elements at the molecular level. This process can be slow or fast—slow in the form of rusting or fast during combustion. The rate of combustion is based on amount of oxygen available. Not all compounds containing oxygen in their molecular makeup are capable of providing enough oxygen for combustion. Typical combustion (if there is such a thing) can occur in the ambient atmosphere, which usually contains 21 percent oxygen. But in cases where there is a higher concentration of oxygen the fire can be more intense. This is evident when you observe a blacksmith's forge where blowing air across the coals

Figure 7-6 *At 0902 hours on April 19, 1995, an explosion rocked downtown Oklahoma City, destroying the Alfred P. Murrah Federal Building and damaging more than 300 other buildings in the vicinity. Fatalities occurred in 14 separate buildings, taking 168 lives and injuring 759 other victims.*

creates more intense heat. Today, high concentrations of oxygen can be mixed with combustible gases to produce an intense flame that is capable of cutting through iron.

In an investigation if there are reports of white, intense heat, far from ordinary combustion, the investigator should consider the proximity of an oxidizer to create such a situation.

Fuels Simply stated, a fuel is anything that can burn. Fuels can be in any physical state: solids, liquids, or gases. (See **Figure 7-7.**) The physical state of the fuel affects the combustion process. Most fuels are required to be in the gaseous state before combustion can occur, but this is not always the case.

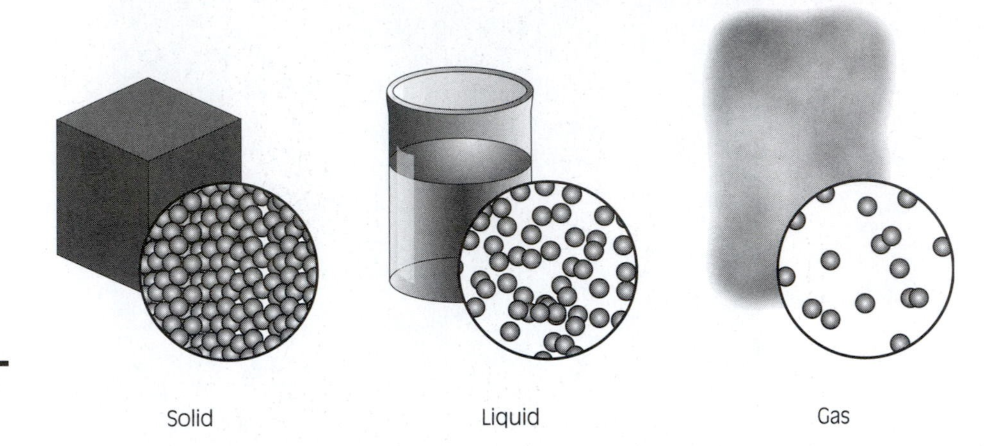

Figure 7-7 *Status of matter.*

Solids Solids have a physical size and shape. To ignite, solids must be heated to the point of decomposing and, thus, producing vapors. Sometimes you can see this, such as in a fireplace where a log is burning and you can see a space between the surface of the log and the visible flame. The space is the layer where the solid log is being vaporized and the vapors above the log are burning. This can be seen even more clearly on a burning candle. It is not so much that the wick is burning but that the solid, wax, in this case, is heated and turned into a liquid. Capillary action allows the liquid to be drawn up the wick where the liquid is vaporized. Once these vapors reach the correct proportion of vapor/air mixture, there is combustion.

■ Note
The only way to truly understand combustion is to break the process down into its molecular components.

The only way to truly understand combustion is to break the process down into its molecular components. With solids, the molecules are packed closely together by bonds that help give the material its shape and form. Some molecules hold a tighter bond than others do. This is evident in how easy it is to break or pull the material apart. All atoms and molecules are in motion. When heated, this motion increases. As the rate of movement increases so does the number of times when the molecules collide. The more heat, the faster the molecular movement and the greater the impact of these collisions. Eventually, these collisions result in the breaking of some of the molecular bonds. When these bonds are broken apart, energy is released in the form of heat and light.

As the heat continues to rise and the collisions become more intense and more frequent, resulting in more bonds being broken, the heat from this process intensifies as well. When these molecules break apart, some combine with oxygen; this is the oxidation process. As a result of the oxygen bonding with these other molecules, heat is given off; this is an exothermic reaction. In time, this process becomes self-sustaining in a form known as combustion.

The heat being applied to the solid matter must be sufficient to overcome the amount of heat that solid object is capable of absorbing. A good example is an

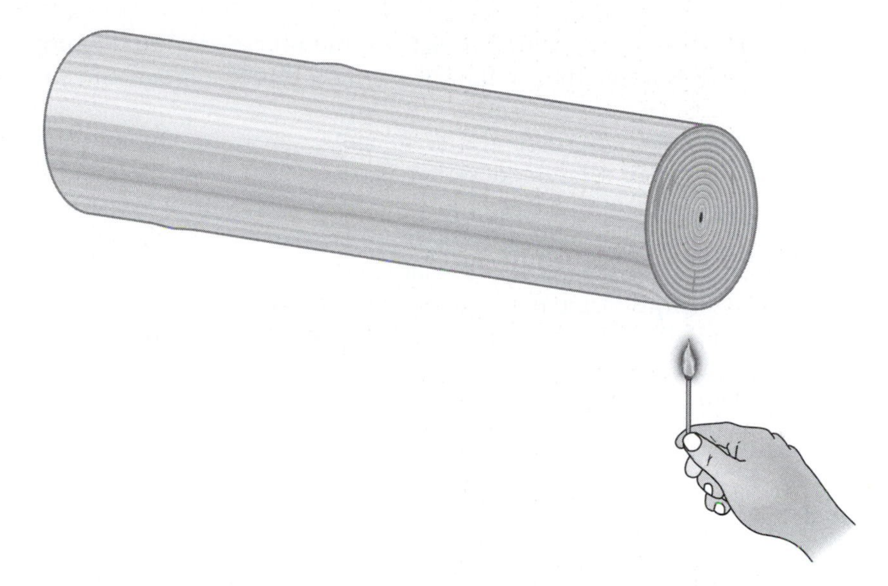

Figure 7-8 *Attempts to ignite a log with a match fail because the mass of the log is able to absorb the heat from the match, distributing it but not raising the log to its ignition temperature.*

8-inch-diameter log placed in a fireplace by itself. It is impossible to just place a match to the bottom of the log, as shown in **Figure 7-8,** and expect it to ignite. What takes place is the heat from the flame of the match is absorbed by the mass of the log. There is localized charring, but the log will not ignite.

However, if kindling in the form of wood shavings or balled-up paper is placed under the log, a fire should readily ignite. When the flame of a match is applied, the kindling cannot absorb the energy because of its small mass and the heat overcomes the fuel, allowing the breaking of bonds, release of molecules that combine with oxygen, and eventual self-sustained ignition.

The surface-to-mass ratio explains what happens in the fireplace. A large mass with little surface area, such as the log, absorbs the heat from the match, transferring in into its mass. But the paper has more surface than mass, and cannot absorb the energy of the match; the molecules are rapidly heated to the point where their bonds are broken down; they bond with oxygen molecules in the oxidation process, and give off heat and light: fire.

Smoldering fires are another situation. Smoldering can be seen, for example, when upholstered furniture burns, where air can readily permeate the material and allow the material to char without a visible flame. The same can happen in sawdust piles and some combustible insulation products. Smoldering is even designed to happen in some products such as cigarettes and other tobacco products. Smoldering is also a phase of the combustion process, which is discussed later in this chapter.

Note
It is important to explain what did or did not happen in the fire being investigated.

It is important to explain what did or did not happen in the fire being investigated. Was the heat source sufficient to ignite a particular item (match on a log)? Was the mixture of vapors with oxygen sufficient to create an ignitable vapor? Was the heat source sufficient to ignite the vapor? The investigator may very well have to account for all parts of this process in the development of a hypothesis and may eventually have to explain this to a judge and jury.

■ Note
Liquids will not ignite in the liquid state; the fuel must be vaporized and those vapors will burn. As with solids, investigators should consider the fuel in its molecular state.

flash point
the temperature of a liquid at which point it gives off sufficient vapors that, when mixed with air in proper proportions, ignites from an exterior ignition source; because of limited vapors only a flash of fire across the surface occurs

fire point
the lowest temperature at which a fuel, in an open container, gives off sufficient vapors to support combustion once ignited from an external source

autoignition temperature
the minimum temperature at which a properly proportioned mixture of vapor and air will ignite with no external ignition source

Liquids A liquid's ability to ignite and burn depends on its form. When diesel fuel in a large container is exposed to heat, to some extent that heat can be absorbed, preventing ignition. However, if you were to take that liquid, spray it into the air or upon a piece of paper (increasing its surface area); ignition requires perhaps only the energy from a match.

Liquids have two other important characteristics. The *density* of a liquid fuel dictates whether it will sit on top of water or sink and rest below the water's surface. Fuels such as gasoline and kerosene are lighter than water and will rest on the water's surface and allow the fuel to continue its vaporization process and to burn if ignited. The *solubility* of a liquid fuel dictates whether it will readily mix with water. Polar solvents such as alcohol will mix with water. This could be an important factor when suppression forces apply their water streams in and around the area of origin. Alcohols are soluble in water. If an alcohol was used as the first fuel ignited, the residue and unburned fuel may very well mix with the suppression water, essentially being washed away and leaving little trace.

Liquids will not ignite in the liquid state; the fuel must be vaporized and those vapors will burn. As with solids, investigators should consider the fuel in its molecular state. In basic terms: When heat is applied to the liquid, the molecules become agitated and move faster and faster. Atmospheric pressure and surface tension keep many of these molecules from flying from the liquid. But, as collisions between the molecules occur and the temperature rises, some molecules will have sufficient force to break free of the surface tension and enter the atmosphere where they will be suspended in the air as vapor. This process—the collision of the molecules and the breaking of bonds—creates heat in the form of an exothermic reaction, the same as in solids.

These processes continue until sufficient vapors are given off to support a flaming fire across the surface of the liquid, but not sufficient to allow combustion to continue. This is considered to be the **flash point** of the liquid.[5] As the temperature of the liquid rises, usually just a few more degrees above the flash point, the liquid generates sufficient vapors to allow the flame to continue to burn. This is considered the liquid's **fire point.**[6] In both circumstances, there is an external source of ignition. If the liquid was allowed to continue to heat without an external source of ignition and the liquid heats to the point that it ignites on its own (reaching its ignition temperature), this point is called its **autoignition temperature.**[7]

As with the vapors from a solid, the vapors from liquids must be in the right proportion with oxygen to allow ignition. Likewise, the ignition source must be sufficient to ignite the mixture. All of this must be explained by the investigator, should the ignition of the vapors be involved in the cause of the fire or even in the extension of the fire from one point to another.

Gases Relatively few gases are ignitable at room temperature. However, with increased temperature other solids and liquids begin to break down and produce vapors. Vapors are indeed in a gaseous state, but a true gas is one that is in a gaseous state at normal temperatures and pressure. Vapors and gases can mix with air and ignite under the proper conditions. However, not all gases burn; in fact, some such as carbon dioxide are used to extinguish fires. Dealing with gases requires considerable scrutiny to make sure that all the facts match the hypothesis. Many variables must be taken into account along with a lot more science, such as vapor density.

Note
Dealing with gases requires considerable scrutiny to make sure that all the facts match the hypothesis.

vapor density
density of a gas or vapor in relation to air, with air having a designation of 1

Vapor density is the weight of a gas when compared to air. This measurement is made at sea level where air is considered to be 14.7 pounds per square inch (psi) and is given a designation of 1. Should a gas be heavier than air, it has a designation greater than 1; should a gas be lighter than air, it has a designation less than 1, such as 0.66.

Note
As investigators examine the scene, they discover available fuel sources along with burn patterns or explosion patterns.

As investigators examine the scene, they discover available fuel sources along with burn patterns or explosion patterns. In the process of developing the hypothesis, the investigator must determine whether the fuel gas was capable of producing the damage seen at the scene. Should there be evidence of an explosion at the upper levels of the structure, it may be an indication of a lighter-than-air gas, such as natural gas (vapor density of 0.55) that may be from a kitchen range. In a case where there is low-level explosive damage, it may be an indication of a heavier-than-air gas such as propane, which has an vapor density of 1.6.[8] Caution must be exercised, however, because the type and method of building construction may result in damage contrary to the expected properties of the gases or vapors.

Figure 7-9 shows that both gases and vapors can ignite only when mixed with an appropriate amount of oxygen. Each gas has an explosive (flammable) range; if the concentration of the gas or vapor is below that range, it is considered to be too *lean*. If the concentration is above the flammable range, the mixture is considered to be too *rich*. In either case, there is little likelihood of an explosive ignition. The flammable ranges for some gases are narrow, for example, propane, which has a lower explosive limit (LEL) of 2.15 percent and an upper explosive limit (UEL) of 9.6 percent. Other gases can ignite in almost any atmosphere, such as acetylene, which has a LEL of 2.5 percent and a UEL of 81 percent. These figures can change dramatically as the temperature increases.[9]

Heat and Temperature

As you already know, molecules are always in motion. This motion, and the collision of molecules, is always producing energy in the form of heat. When you touch an object with your hand, it may feel cold to the touch, but that is only relative to the fact that the object has a lower surface temperature than your hands do. However, that object still contains heat. When there is a release of energy—an exothermic reaction—it could simply make the object warmer. This could eventually lead to *thermal runaway*, a phenomenon where more and

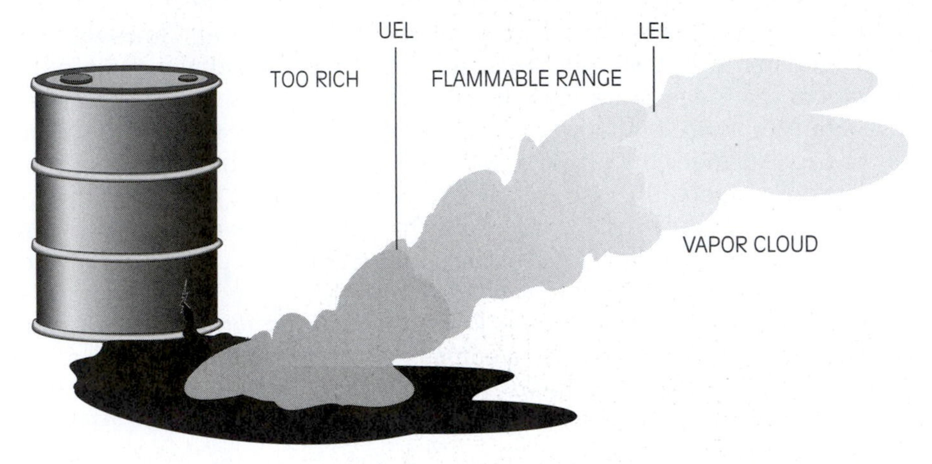

Figure 7-9 *Flammable/explosive range: The vapors closest to the spilled liquid are too rich to ignite. As the vapors mix with air, they may achieve the correct proportion of vapors to air to enable ignition to occur; this is the flammable range, which is also known as the explosive range. As the vapors spread farther from the spill, they dissipate (become too lean), diluting sufficiently so that the vapor-to-air ratio does not allow ignition. The threshold where the vapor concentration can be ignited but just before the vapor is too rich is known as the upper explosive limit (UEL). The threshold just before the fuel becomes too lean but can still ignite is known as the lower explosive limit (LEL).*

■ Note
Heat is kinetic energy, and temperature is the measurement of that energy.

■ Note
The ignition temperatures for materials listed in texts and publications are the result of laboratory testing of the product. As such, in the real world, the actual temperature may be higher than what was experienced in the lab.

more energy is created that eventually results in the release of heat and light, which is fire. This chain reaction was discussed earlier in the section titled "Fire Tetrahedron."

Temperature is a measurement of the amount of heat. Heat is kinetic energy, and temperature is the measurement of that energy that can be obtained by using an instrument such as a thermometer. Heat and temperature are related but are not the same—you measure the amount of heat using a device that gives you the temperature of the object.

Ignition Temperature and Ignition Energy

The ignition temperature is the minimum temperature a substance must attain before ignition can occur. Many variables must be taken into account for a wide range of circumstances. With the exception of a few solid materials that can burn at the surface, such as coal or magnesium, all other solid fuels must be heated sufficiently to release vapors; the vapors ignite and burn. Liquids are the same: mostly, they need to be heated to release vapors that in turn can be ignited at the proper temperature for that vapor. Some liquids already produce sufficient vapors at their ignition temperature below ambient air temperature. These fuels, such as gasoline, are capable of ignition if in the right mixture with air and heat.

The ignition temperatures for materials listed in texts and publications are the result of laboratory testing of the product. As such, in the real world, the actual temperature may be higher than what was experienced in the lab. Most important, it is imperative that the investigator research all aspects of any fuel to be listed as the first material ignited. Unusual variables can affect the ignition temperature, making it higher or, in rare circumstances, lower. The investigator must also take into account the size, shape, and form of the first material ignited, and all must be explained in the investigator's final hypothesis.

Just as important as the ignition temperature is the amount of energy necessary to make the ignition occur. For ignition to occur, it needs to be in only a tiny space. Take the example of someone walking across a carpeted floor. Under the right circumstances, the individual may pick up a sufficient static charge so that when he touches a grounded object such as a door knob or another person, he experiences a discharge of energy. The individual being touched will feel the arc. This represents the fact that there was a high temperature in the discharge but in such a minute space that no harm occurred.

However, that same discharge in an atmosphere of ignitable vapors, such as when fuel is handled at a service station, can cause an explosive ignition. The fuel vapors may be within the explosive range (LEL and UEL) at the source of the energy, and there is enough energy for a sufficient duration of time to ignite that vapor in and around the discharge area. The resulting fire or explosion could be devastating, and it came from the energy of a static discharge. The amount of energy necessary in such a situation is only 25 millijoules.[10]

Sources of Heat

Heat is a factor in providing sufficient energy to release vapors from solids and liquids. It is also a factor in providing the energy to cause the ignition of a fire. Heat also promotes flame spread and subsequent fire growth. There are four sources of heat: mechanical, chemical, electrical, and nuclear. Some you experience every day and take for granted, whereas others are more unusual and infrequently a source of ignition.

■ Note
The investigator must be familiar with all potential heat sources.

The investigator must be familiar with all potential heat sources. Before identifying the actual heat source involved in a fire, all other heat sources must be eliminated. The investigator with a weak level of knowledge about any of these heat sources is limited in the ability to form a viable hypothesis about the source of ignition.

A multitude of resources and opportunities to learn more about sources of heat exist. College classes in chemistry and physics can help you understand chemical and mechanical heat sources. Investigators can find that developing a professional relationship with college professors in these subjects can be invaluable. If you need to learn more about electricity, college physics classes can give basic knowledge of electricity, electron movement, electrical potentials, and so on. You can learn practical applications from a local electrical contractor. Spending a little time on a construction site can be very rewarding and help you understand installation components and proper installation. Many seminars at the national level as well as local training from various chapters of the International Association of Arson Investigators or the National Association of Fire Investigators also are available.

Mechanical Mechanical heat is the heat of friction. Two or more objects rubbing together create friction. Friction by itself does not mean there will be a fire. Factors such as the roughness of the surface, the types of materials rubbing together, dryness of the materials in question, and the speed at which they come in contact are variables that must be taken into account by the investigator. For example, two smooth pieces of plastic coated with a silicone lubricant slowly rubbed together with little pressure will produce heat, but it is not likely that they will produce enough energy to start a fire.

Truck brakes are another example of two materials coming in contact that create mechanical heat. Take into account the surfaces of the materials and the fact that they are designed to slow down the rate of movement, combined with pressure of the brake cylinder they will do as they are designed on a regular basis. However, with a mechanical failure of misalignment or when the cylinder creates pressure (gets stuck) when not activated during a high rate of speed, pushing the brake pads together, sufficient energy may be created to ignite nearby combustibles in the form of grease, plastic, and rubber components or even debris caught in the workings. Brake fires by themselves may cause only local damage. However, in some circumstances they can result in the extension of the fire into the vehicle, causing massive destruction of the vehicle and its contents.

Another form of mechanical heat is the heat of compression. This process is used in diesel engines where the fuel is compressed to the point where ignition can be accomplished without an outside source such as a sparkplug. Firefighters can experience the phenomenon of compression heat when they fill self-contained breathing apparatus (SCBA) air bottles. The bottles heat up as they are being filled. This is a form of mechanical heat.

Chemical A mixture of two or more chemicals can create heat (exothermic reaction) and in some circumstances may heat sufficiently to cause ignition. Although not an everyday circumstance, chemical heat can be found in industrial applications, and even some household chemicals are capable of creating such a reaction. Chemical heat can also be an arsonist's tool and can even be used as a time delay. Regardless of the circumstances, investigators need to collect evidence in the form of debris for analysis along with possible examples of chemicals for further analysis.

■ Note
Spontaneous heating can occur from biological action as well as chemical.

spontaneous heating
a process in which a material increases in temperature without drawing heat from the surrounding area

One form of heating can occur without an external source of energy. **Spontaneous heating** can occur from biological action as well as chemical. Biological items such as damp hay or manure or sawdust give off heat as they break down and decay. If these products are well insulated so that the dissipation of the heat is prevented, the heat can increase to the point of reaching the ignition temperature of the material. When this happens in the presence of sufficient oxygen,

spontaneous ignition initiation of combustion from within a chemical or biological reaction that produces enough heat to ignite the material

spontaneous ignition can occur, where the fire starts with no external source of heat or energy.[11]

Electrical Electricity is all around. You need only to lose electricity to see how much you depend upon it. Electricity and electrical devices are heat producers either as a by-product of their workings or intentionally, as in electric baseboard heaters. Fires caused by electrical devices can result from malfunction because of poor design, improper construction, or poor installation. Just as likely, a failure can occur from misuse of electricity and associated appliances. Investigators must be aware that just because electricity is present does not mean that it is responsible for the fire.

Note
Electrical sources can be as small as a static electrical arc or as massive as a lightning bolt.

Electrical sources can be as small as a static electrical arc or as massive as a lightning bolt. Investigators most often come in contact with the other forms of electrical heating, which are generally man-made. Resistance heating, which occurs in two forms, is the result of an appliance or fixture design. This can be a light bulb or a heater. Heaters can heat rooms or can cook food. Then, there is resistance heating in the form of an appliance or fixture failure when it does not act as designed, manufactured, or used. This unintended resistance heating is more than capable of igniting nearby combustibles.

The electrical wiring designed to deliver energy throughout a structure can fail in the form of an overcurrent or an overload. Although this too can be a form of resistance, it can also create arcs or sparks that are capable of being an ignition source.

Electricity?

Engine 21 responded on a fire in a small single-family residence. The two-bedroom house was vacant with very little contents in the kitchen. Fire was limited to paint supplies in the corner of the kitchen, and there was a considerable amount of paper spread around the floor as if someone was getting ready to paint.

In addition to the material in the corner burning, there were scorch marks on each of the electrical outlets above the kitchen countertop, six outlets in all. The engine company called for an investigator. The investigator looked at the scorch marks, and then found an outlet in the wall exactly where the paint supplies were burning in the corner. That outlet showed heavy scorch marks as well, more prominent than what would have been caused by the extension of the fire.

The investigator told the engine company to list the fire as accidental on the fire report and put down the source of the heat as electricity from a faulty electrical system. When the fire report was completed, it was sent to administration as policy dictated. The local newspaper had called the fire marshal's office and asked for particulars on the fire. The next morning a short article was run in the local newspaper about the electrical fire.

(Continued)

(Continued)

By midmorning the local volunteer fire chief, who was not on the call the previous day, saw the article in the paper. He called the fire marshal's office and requested that the investigator meet him on the scene, which he did within the hour. While standing in the front yard, the chief asked if the investigator noticed anything unusual. Much to the embarrassment of the investigator, there on the power pole at the street was a coiled wire that had once gone to the house in question. This feeder line was neatly coiled and there in the bottom loop was a bird's nest. The power to the property had been disconnected for several weeks.

Walking around the house, they discovered there was no alternate power supply. Going inside the house, they took apart the charred outlets. The inside of the outlet was undamaged, not even showing sooting. The same was true for all the outlets, including the one at the area of origin.

The investigator had the knowledge to find this out for himself but did not take the time to do so. Instead, he chose to take the easy path based on limited knowledge. Had he followed the scientific methodology, he would have known this prior to leaving the scene on the day of the incident.

A complete and thorough investigation was then done and the laboratory analysis of the debris showed gasoline. The homeowner confessed to burning the outlets with a plumber's torch and setting the fire. The motive was to collect insurance monies to pay for a kitchen remodel.

If an investigator is going to look at any fire scene, he or she must be prepared to do a complete and thorough investigation and look at all the evidence. Using a systematic approach and scientific methodology can prevent similar events in the future.

More about electricity as an ignition source is covered in Chapter 10, "Electricity and Fire."

Nuclear The first generation to deal with the term *nuclear* thought of it only in the aspect of an unbelievably devastating bomb. Today's populace is still aware of the military use of radioactive materials, but nuclear energy is used in the making of electricity as well as in industrial applications and medical treatments.

By its very nature, radioactive materials are unstable. As such, they are constantly releasing atomic particles and in this process heat is a by-product. When used in a reactor, the controlled release of these atomic particles heat water, turning it to steam that powers the turbines and creating electricity. There are various types of radioactive isotopes with a wide variety of industrial applications for testing steel, quality of welding points, and the thickness of materials. Agricultural uses include improving nutritional value of some crops, making pest-resistant plants, and for research. Medical uses include medical testing, treatments, and medical research.

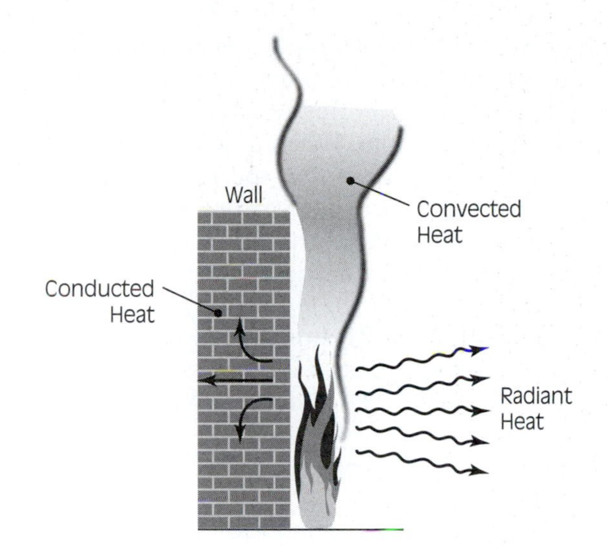

Figure 7-10 *Examples of heat transfer in fire.*

■ Note
Just because something is capable of creating heat does not mean that it is capable of creating enough heat to cause or sustain ignition.

Just because something is capable of creating heat does not mean that it is capable of creating enough heat to cause or sustain ignition. The amount of radioactive material you might find in a medical lab or in an industrial site most likely is not sufficient to create enough energy to be the heat source that starts a fire. Remember, it is just as important to be able to eliminate a heat source.

Heat Transfer

Heat transfer is a common topic in most fire suppression manuals. The intent is to teach firefighter students how heat travels so that they can understand how to extinguish or where to look for fire extension. (See **Figure 7-10.**) Likewise, the investigator, whether for the fire report or for conducting a detailed investigation, needs to understand heat transfer to interpret the direction of fire travel. As the fire extends into one area or another, investigators will need to explain how this happened or how it could not have happened through normal heat transfer.

Heat can be transferred through conduction, convection, radiation, and direct flame contact. At any common fire scene, the investigation may find examples of each mode of heat transfer. While developing a final hypothesis, investigators may find themselves explaining how the heat from the heat source came in contact with the first material ignited. Thus, a thorough understanding of heat transfer is essential for any fire investigator.

Conduction The transfer of heat through a solid object is considered to be conduction. Some materials are good conductors and others are not. The key for the preliminary investigation is to decide whether that material at the fire scene was capable of conducting the heat from one point to another or whether that material acted as an insulator, preventing or retarding the flow of heat.

Metals are good conductors. When a flame is applied to one end of a metal rod, the heat is conducted along that rod to the other end. Not many rods go through walls, but pipes do travel from one point to another, acting as potential sources of heat transfer. Metal trusses or I-beams are also good conductors. Because of their mass, it takes a while longer for the heat to build up before transferring across the beam.

Note A detailed investigation requires the investigator to understand all aspects of the thermal properties of materials; more important, investigators need to understand what is happening at the molecular level.

It is important for the investigator to know the material they are considering as a potential source of heat transfer, explaining how the fire traveled. A detailed investigation requires the investigator to understand all aspects of the thermal properties of materials; more important, investigators need to understand what is happening at the molecular level.

Convection The transfer of heat through the movement of liquid or gases is considered to be heat transfer by convection. For the most part, heated gases are in fact smoke and fire gases. As the heated gases flow throughout the structure, they come in contact with solid objects, transferring the heat, eventually spreading the fire. Because hot gases tend to rise, the heat builds up in the higher levels of rooms or compartments. Without resistance, the heated gases continue to flow, moving into cooler areas and heating up the contents of the structure. As the fire increases, the gases become more buoyant and have a greater propensity to rise. This is especially evident in vertical openings where convection enables heated gases to rise into other areas of the structure.

Investigators can look at the burn patterns to discern the direction of fire travel. The principle of heat travel through convection can help explain the extension of the fire. However, the investigator needs to be aware of the impact of ventilation. This can occur as a result of the fire breaking windows or the fire suppression forces implementing forced ventilation. The change in the movement of hot gases around the structure must be explained to test and complete the final hypothesis.

Radiation Radiation is the transmission of energy through electromagnetic waves. Radiation travels only in straight lines and travels at the speed of light. You have felt radiated heat, whether from the sun's rays or from the heat of a hot woodstove or fireplace. Radiant heat is capable of igniting nearby combustibles. If the fire is of sufficient volume, it may be capable of igniting combustible products on adjoining properties.

Firefighters see the damage from radiant heat when the vinyl siding of an adjacent home deforms or melts. They feel the heat as they approach the fire. Radiant heat plays a role in every fire. In a compartment fire, the heat moves upward and outward. As this happens, it collects in higher elevations and is restricted from escaping by the ceiling of an enclosed space. This heat, hot air, and gases collect in these areas, traveling by convection, but they also radiate downward from the heated area to the less heated area. This radiant heat can then play a role in extending the fire to combustibles below the heated layer.

As with all aspects of the transfer of heat the investigator needs to first understand the principle, be able to defend it scientifically, and then be able to put that principle in simple language for a jury or judge to understand.

■ Note
As with all aspects of the transfer of heat the investigator needs to first understand the principle, be able to defend it scientifically, and then be able to put that principle in simple language for a jury or judge to understand.

Thermal Layering

The products of combustion in the form of heated gases and smoke rise from the fire because they are more buoyant than the surrounding air. If in a compartment, they rise until they meet resistance, usually in the form of a ceiling. These gases rise in the form of a **plume,** hit the ceiling, and spread horizontally. This horizontal spread continues until the plume comes in contact with a wall. As this buoyant mass of hot gases and smoke hit the wall, there is a slight downward motion that is dictated by the amount of energy in the hot gas layer. Another phenomenon that occurs at this time is the creation of a **ceiling jet** that consists of a thin layer of buoyant gases moving rapidly just under the ceiling in all directions away from the plume. The ceiling jet is what is responsible for the activation of smoke detectors, sprinkler heads, and so forth.

plume
the column of smoke, hot gases, and flames that rises above a fire

ceiling jet
a thin layer of buoyant gases that moves rapidly just under the ceiling in all directions away from the plume

■ Note
The ceiling jet is what is responsible for the activation of smoke detectors, sprinkler heads, and so forth.

As these gases form in the upper levels in the compartment, they start to form thermal layers of varying temperatures. Sometimes these layers are visible until disturbed by ventilation, application of a suppression stream, or the introduction of a larger volume of burning fuel. These temperatures can vary to the extreme between the temperature in the ceiling and the floor.

The upper thermal layer continues to bank down all the way to the floor unless it finds an escape route through a window or any other opening that leads out and away from the compartment. The damage created, the marks on the structural components and contents, may be indicators of value to the fire investigator studying the burn patterns.[12]

Heat Release Rate

As fuels burn they lose mass. This process results in the release of energy in the form of heat and light. The amount of energy produced over a period of time is considered to be the **energy release rate (ERR)** more commonly referred to as the **heat release rate (HRR).** The heat released can be measured in British thermal units (BTUs) or in kilowatts (kW). The rate at which heat is released is important to the fire investigator. Just as important is the total amount of energy that is released for a certain size of fuel package. The fuel package can consist of a sofa or a pile of wood shavings. Knowing both the rate at which heat is released and the total energy that will be released can help the investigator in analyzing the fire scene and discerning whether the known contents could have caused as

■ Note
The amount of energy produced over a period of time is considered to be the energy release rate (ERR) more commonly referred to as the heat release rate (HRR).

energy release rate (ERR) amount of energy produced in a fire over a given period of time

heat release rate (HRR) the rate at which heat is generated from a burning fuel

much damage as is seen. These factors can also explain what should or should not have happened. A match at only 50 kW ERR cannot provide enough energy to ignite a solid 12-cubic-inch block of wood. However, it can readily ignite a piece of paper.

A multitude of resources can assist the investigator in learning about the HRR of various objects and other scientific data. The Internet is a common searching ground to obtain up-to-date information. However, use reputable sites to ensure accuracy of the information you seek. Most government sites such as the National Institute of Standards and Technology and private Web sites like Underwriters Laboratories have great reference materials. Nothing, however, can take the place of maintaining a good up-to-date library. Publishers such as Delmar, Cengage Learning and the National Fire Protection Association publish many valuable texts. The NFPA *Fire Protection Handbook* is a good example of a text that is of immense value to everyone in the fire service.[13]

Compartment Fires

■ Note
The configuration, construction, and contents of a compartment affect the growth and development of a fire.

The configuration, construction, and contents of a compartment affect the growth and development of a fire. Every fire differs according to a multitude of variables, but there are some basic concepts and factors that define the progression of all fires. Both NFPA 921, *Guide for Fire and Explosion Investigations,* and NFPA *Fire Protection Handbook,* 20th edition, describe in detail how fire progresses from the first flickering in a room to the destruction of the room and contents, and then through the decay of the fire when it might self-extinguish. Of course, there is always the possibility that the fire may be extinguished by sprinkler systems or the actions of the fire department.

In the beginning stage of the fire, heat and smoke become buoyant and rise to the ceiling. This creates a plume, a column of smoke rising up to the ceiling. Hitting the ceiling, the heat spreads horizontally to create the thin layer of high heat as the ceiling jet. It continues to travel until it hits resistance such as a wall. If there is a natural vent at or near the ceiling, and the fire does not increase the heat release rate or extends to other combustibles, this is the limit of the damage and actions until the fuel has been consumed.

■ Note
Occasionally, a flameover, sometimes called a rollover, occurs. This is where the gases in the upper layer ignite, sending flames rolling across the buoyant layer.

As shown in **Figure 7-11** the fire progresses, with smoke and heat escaping from openings in the room of origin. As the fire grows in size, more heat and smoke are produced than can escape, and both begin to bank downward, as shown in **Figure 7-12,** affecting other items in the room and heating them up to their potential ignition temperature.

As the fire continues to grow, the heat continues to rise. This process creates currents of hot air moving upward that are then replaced by air at the lower level entering the room, filling the void. The process of new air being introduced is called *air entrainment.* (See **Figure 7-13.**) The fresh air aids in the combustion process, keeping the fire growing.

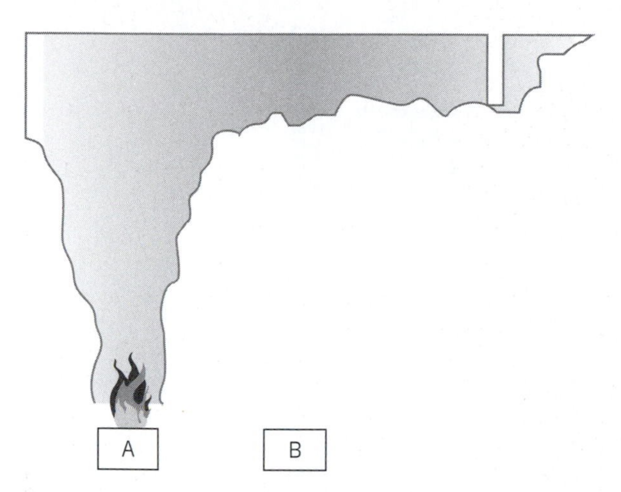

Figure 7-11 *Compartment fire in the initial stage where the heat and smoke rise to the ceiling, and then travel horizontally, filling the area.*

Source: Reprinted with permission from the *Fire Protection Handbook®*, 2008 Edition, Copyright © 2008, National Fire Protection Association, Quincy, MA. This reprinted material is not the complete and official position of NFPA on the referenced subject, which is represented only by the Standard in its entirety. *Fire Protection Handbook®* is a registered trademark of the National Fire Protection Association, Quincy, MA.

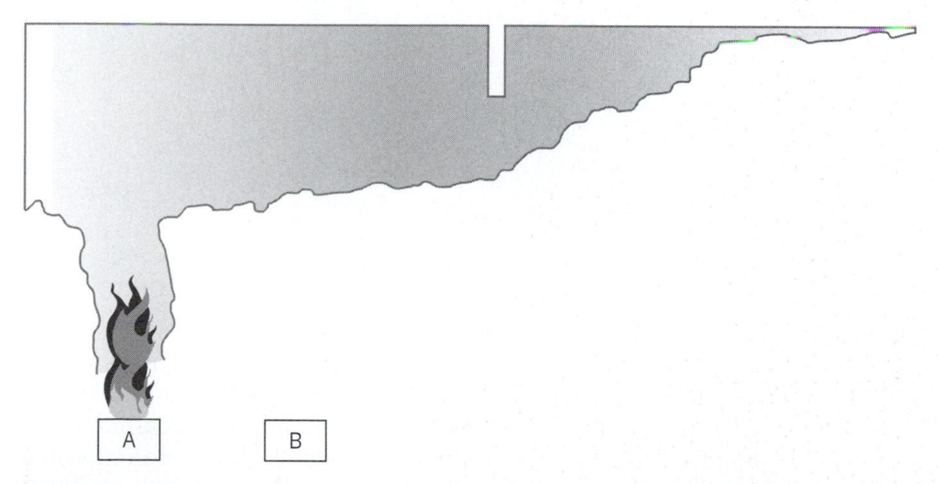

Figure 7-12 *Compartment fire in the stage when the fire progresses, releasing more heat and smoke, both of which bank down and affect more items in the room and possibly heat them up to their ignition point.*

Source: Reprinted with permission from the *Fire Protection Handbook®*, 2008 Edition, Copyright © 2008, National Fire Protection Association, Quincy, MA. This reprinted material is not the complete and official position of NFPA on the referenced subject, which is represented only by the Standard in its entirety. *Fire Protection Handbook®* is a registered trademark of the National Fire Protection Association, Quincy, MA.

Figure 7-13 *Through an open door the hot thermal layer flows outward at the ceiling level. Below that cooler replacement air flows inward that contains oxygen that aids in the combustion process.*

flameover
the point at which a flame propagates across the undersurface of a thermal layer; same as rollover

rollover
the point at which a flame will propagate across the undersurface of a thermal layer; same as flashover

Occasionally, a **flameover,** sometimes called a **rollover,** occurs. This is where the gases in the upper layer ignite, sending flames rolling across the buoyant layer. This may be but is not always a precursor to the phenomenon called **flashover,** as shown in **Figure 7-14.** This is a transition where all the combustible materials in the room are at their ignition temperature and they ignite. A flashover does not always occur. Many fires start to decay because of a lack of fuels before a flashover occurs. The air (oxygen) supply also may be consumed before the room reaches a sufficient temperature for flashover to occur. If a fire in its early stage has natural ventilation, it never builds up sufficient heat to create the environment necessary for flashover to occur.

If the compartment is in the post-flashover stage and ventilation occurs, either through a natural event or from suppression forces, a volume of smoke

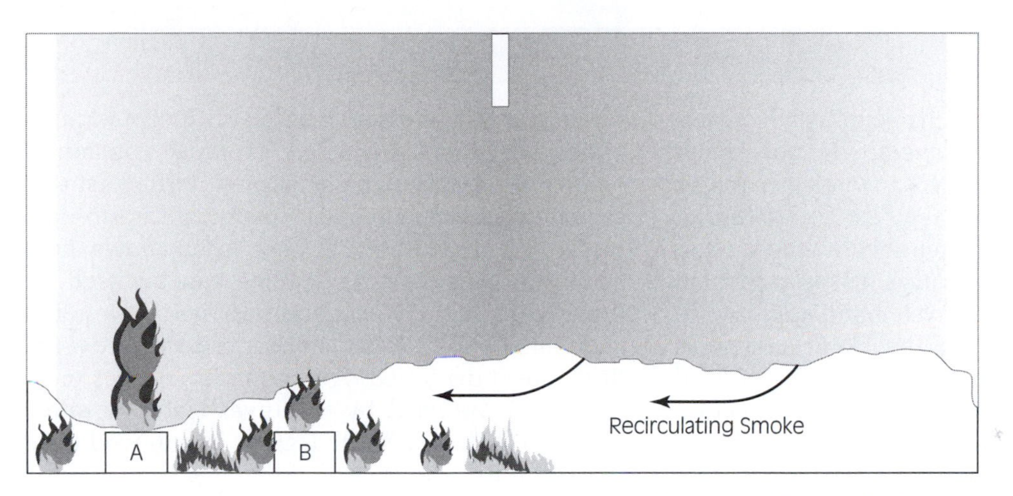

Figure 7-14 *Compartment fire flashover where all exposed combustibles reach their ignition temperature.*

Source: Reprinted with permission from the *Fire Protection Handbook®*, 2008 Edition, Copyright © 2008, National Fire Protection Association, Quincy, MA. This reprinted material is not the complete and official position of NFPA on the referenced subject, which is represented only by the Standard in its entirety. *Fire Protection Handbook®* is a registered trademark of the National Fire Protection Association, Quincy, MA.

flashover
a transition phase of fire where the exposed surface of all combustibles within a compartment reach autoignition temperature and ignite nearly simultaneously

backdraft
an explosion resulting from the sudden introduction of air (oxygen) into a confined space containing oxygen-deficient superheated products of incomplete combustion

escapes the compartment. This smoke, which is incomplete particles of combustion, is a fuel that is at its self-ignition temperature. Reaching the outside air provides the final side of the fire triangle—sufficient oxygen. The smoke may ignite. Witnesses sometimes report seeing flames within the smoke, or this phenomenon may manifest as a fireball above the structure.

In a situation where there is adequate ventilation, the fire will eventually decay as the fuel is consumed, and there will be fewer and fewer flames. In situations where there is little or no ventilation, the compartment retains the heat, and oxygen content falls well below 16 percent, and there will be fewer flames. This can create a dangerous situation—a firefighter's nightmare. If for any reason the compartment has an influx of air, the vapors in the compartment will be at the self-ignition point and will mix with the newly introduced oxygen and will ignite. Firefighters call this term **backdraft,** where air rushes into the compartment, mixes with the vapors, and then ignites, sending a wall of fire out through the ventilation opening such as a door or window. Depending on the source, this is sometimes called a *smoke explosion* or *flashback*. This is a rare condition, but nonetheless it does occur.

The likelihood that all variables of one compartment fire ever being exactly like another is so remote it is unimaginable. The description of the compartment fire in this text is only one scenario of many, but it should give you an understanding of the development and growth of fire in a structure.[14]

SUMMARY

Fire investigators need not be scientists to investigate fires. However, those who do not understand or have good scientific knowledge endanger themselves and the public they serve. The use of scientific methodology is essential to a proper determination of fire origin and cause. Just as important, investigators should use a systematic approach to the examination of the fire scene. Doing so ensures that the entire scene is properly searched for all evidence necessary to formulate a proper hypothesis as to the fire's area of origin and cause.

In the process of making the final determination of the fire cause, investigators must use good science, which is science based on accepted principles and facts. Investigators must constantly challenge what they have learned in the past and seek new knowledge based on current up-to-date research and studies. This chapter shows how some basic firefighting science can be used by fire investigators. Items such as the fire tetrahedron model and process of heat transfer show the firefighter how to extinguish the fire. Other principles can be used by the investigator to show others how the fire started, grew, and moved from one point to another.

KEY TERMS

Autoignition temperature The minimum temperature at which a properly proportioned mixture of vapor and air will ignite with no external ignition source.

Backdraft An explosion resulting from the sudden introduction of air (oxygen) into a confined space containing oxygen-deficient superheated products of incomplete combustion.[15]

Ceiling jet A thin layer of buoyant gases that moves rapidly just under the ceiling in all directions away from the plume.

Deductive reasoning Taking the facts of the case and doing a thorough and meticulous challenge to all facts known, using logic.

Empirical data Data that are based on observations, experiments, or experience.

Energy release rate (ERR) Amount of energy produced in a fire over a given period of time.

Exothermic reaction The release of heat from a chemical reaction when certain substances are combined.

Fire point The lowest temperature at which a fuel, in an open container, gives off sufficient vapors to support combustion once ignited from an external source.

Fire tetrahedron Tetrahedron is a solid figure with four triangular faces. When each face represents the four items necessary for a fire (heat, fuel, oxygen, and uninhibited chemical reaction), the figure then becomes the fire tetrahedron.

Flameover The point at which a flame propagates across the undersurface of a thermal layer. Same as rollover.

Flash point The temperature of a liquid at which point it gives off sufficient vapors that, when mixed with air in proper proportions, ignites from an exterior ignition source; because of limited vapors only a flash of fire across the surface occurs.

Flashover A transition phase of fire where the exposed surface of all combustibles within a compartment reach autoignition temperature and ignite nearly simultaneously.

Heat release rate (HRR) The rate at which heat is generated from a burning fuel.

Oxidizer Any material that forms with a fuel to support combustion.

Plume The column of smoke, hot gases, and flames that rises above a fire.

Rollover The point at which a flame will propagate across the undersurface of a thermal layer. Same as flashover.

Scientific method The systematic pursuit of knowledge involving the recognition and formulation of a problem, the collection of data through observation and experiment, and the formulation and testing of a hypothesis.[16]

Spontaneous heating A process in which a material increases in temperature without drawing heat from the surrounding area.

Spontaneous ignition Initiation of combustion from within a chemical or biological reaction that produces enough heat to ignite the material.

Systematic search A search based on a system that is used for each fire, each and every time.

Vapor density Density of a gas or vapor in relation to air, with air having a designation of 1.

REVIEW QUESTIONS

1. Describe heating at the molecular level.
2. Give the methods of heat transfer and give examples of each.
3. What are the differences between heat and temperature?
4. Describe in detail flaming combustion and smoldering combustion.
5. Define *spontaneous heating*.
6. Why is heat release rate so important to the fire investigator?
7. What is flashover and how can it affect the fire scene?
8. What is meant by a systematic approach for the fire investigator?
9. What are the four sources of heat? Define each.
10. What is thermal layering?

DISCUSSION QUESTIONS

1. Discuss scientific methodology. What is a hypothesis? Is it nothing more than an educated guess?
2. Discuss the necessity of following the systematic approach as defined in NFPA 921 or using a different approach.

ACTIVITIES

1. Create a list of all materials found to be capable of spontaneous combustion.
2. Research your own community and identify where you would find a good resource to answer basic and advanced questions in chemistry.

NOTES

1. National Fire Protection Association, NFPA 921, *Guide for Fire and Explosion Investigations* (Quincy, MA: National Fire Protection Association, 2008), 921–916.
2. National Fire Protection Association, *Fire Protection Handbook*, 20th ed. (Quincy, MA: National Fire Protection Association, 2008), section 1, chapter 2.
3. Dennis P. Nolan, *Encyclopedia of Fire Protection*, 2nd ed. (Clifton Park, NY: Thomson Delmar Learning, 2006).
4. Linda Scher, *The Texas City Disaster.* Code Red Series (New York: Bearport Publishing, 2007).
5. National Fire Protection Association, *Fire Protection Handbook*, section 1, chapter 2.
6. National Fire Protection Association, *Fire Protection Handbook*, section 1, chapter 2.
7. National Fire Protection Association, *Fire Protection Handbook*, section 1, chapter 2.
8. National Fire Protection Association, *Fire Protection Handbook*, section 1, chapter 2.
9. National Fire Protection Association, *Fire Protection Handbook*, section 1, chapter 2.
10. National Fire Protection Association, *Fire Protection Handbook*, section 1, chapter 2.
11. National Fire Protection Association, *Fire Protection Handbook*, section 1, chapter 2.
12. National Fire Protection Association, *Fire Protection Handbook*, section 1, chapter 2.
13. National Fire Protection Association, *Fire Protection Handbook*, section 1, chapter 2.
14. National Fire Protection Association, *Fire Protection Handbook*, section 1, chapter 2; and National Fire Protection Association, NFPA 921, *Guide for Fire and Explosion Investigations*, chapter 5.
15. Dennis P. Nolan, *Encyclopedia of Fire Protection*, 2nd ed. (Clifton Park, NY: Thomson Delmar Learning, 2006).
16. NFPA, NFPA 921.

PHYSICAL EVIDENCE COLLECTION AND PRESERVATION

Learning Objectives

Upon completion of this chapter, you should be able to:

- Describe steps (including preliminary assessment) to take to protect the scene and preserve evidence.
- Describe the process of identifying evidence.
- Describe the proper process for collecting and preserving evidence.
- Describe the processing of the evidence to include the tagging, analysis, reports, and documentation of the chain of evidence.

CASE STUDY

It was a senseless accidental fire that ended up destroying 58 apartments, putting more than one hundred individuals on the street. The investigation took 3 days, which included seeking not only the cause of the fire but also the reason for the extension of the fire throughout the sprinklered building.

On scene were the local fire marshals as well as investigators contracted by the insurance industry. The primary investigation was done by the local authorities prior to letting the private sector come onto the scene. Once the local fire marshals completed their investigation, they shared the findings with the building owners, tenants, and insurance representatives.

The cause of the fire was attributed to an exterior light fixture overheating as the result of a tenant changing its configuration and then insulating the fixture with a shirt to block the light from entering his bedroom window. Flames extended up the exterior of the building and into the building's large attic space. This fire extension circumvented and eventually overran the sprinkler system.

This should have been a simple case, but it was not. What was once thought of as a minor lapse in protocol for evidence retention ended up with a fire investigator, the chief fire marshal, the fire chief, and the jurisdiction defending themselves in a civil suit.

During the investigation, the investigator wanted to compare the light bulb from the scene with a sample, an exemplar bulb, from the spare supply of the complex. The bulb at the area of origin was broken, but the fixture, element, and base remained. The investigator broke the new bulb to see, in a nontechnical visual manner, if it was truly a 60-watt bulb. The outer glass of the new bulb was broken away and a comparison of the filament as well as the bulb stem proved to be inconclusive. This left the investigator with a bulb with jagged glass edges on the base, so to be safe, he set the bulb in a small Styrofoam cup and placed it in a metal evidence can for safety.

The fire marshals completed their work and cleared the scene. The light fixture and bulb were taken from the scene by the fire marshals and placed in the department's evidence locker but not logged in, as required. All other materials examined at the scene were left in a covered area on the property.

Sometime later, a lawyer and a private investigator representing the renters from the apartment complex showed up at the fire marshal's office wanting to see the evidence from the fire. Various tenants did not have fire insurance and were seeking recourse to recover from their losses. The light fixture and a bulb were shown as requested. It was later realized by the investigator from the fire marshal's office that the bulb was actually the representative bulb and not the actual bulb from the light fixture. At that point, the lawyer was contacted and told of the situation. Based on this, the lawyer felt that the investigator was fabricating evidence.

Civil charges were filed against the investigator and his supervisor, the chief fire marshal. Immediately the press, both newspaper and TV, had put the investigator and the department before the public with the allegations. This type of situation can be a career-changing event, a major stress on families, and induce a loss of confidence. Creditability is certainly at stake.

(Continued)

(Continued)

On the day of the trial, cameras were rolling outside the court as the defendants approached. Everyone was seated, and the plaintiff, the tenants' lawyer, brought forward their complaint. All evidence was presented and then the plaintiff rested their case. The defense lawyer representing the fire marshal and the fire investigator asked for dismissal based on the lack of evidence. Fortunately, the judge dismissed the case, giving summary judgment on behalf of the fire officials. Truth prevailed and all parties were deemed innocent of any wrongdoing.

All this from what may seem as an innocent action but which was in fact a major breach in protocol for proper evidence handling, storage, and documentation.

INTRODUCTION

■ Note
First and foremost, it is critical that each department or investigative division establish a physical evidence policy.

Evidence issues are discussed in other chapters as specific needs arise. In this chapter, we cover the basic concept of evidence recognition, collection, documentation, and preservation. First and foremost, it is critical that each department or investigative division establish a **physical evidence policy.** This policy must cover all aspects of handling and storing evidence for each incident investigated. It should also cover what type of evidence is to be collected by the first responder investigator and under what circumstances the first responder should collect such evidence.

physical evidence policy
a locality's policy that guides and directs the proper collection, storage, handling, use, and disposal of all evidence

The policy is only as good as those who have input in its creation. In addition to the resources within the department or division, collect input from the local laboratory that will be processing the evidence. Lab personnel can specify the type of container, the documentation, and retention of the evidence to match their own policy.

Everyone within the department must be trained on the policy to the level that they will be involved with handling evidence. Everyone within an investigation unit must be thoroughly trained on all aspects of the policy. The engine company officer should be trained on evidence recognition and how to protect evidence from being damaged. As the first person to conduct an investigation, he or she needs to be well aware of identifying evidence and under which circumstances to call for an assigned investigator. Engine company officers should also know how to collect and hold the evidence under extreme circumstances.

■ Note
The engine company officer should be trained on evidence recognition and how to protect evidence from being damaged.

Even the line firefighter should be trained to recognize evidence and to protect it from harm. Firefighters need to know what not to do so as to avoid spoliation (see Chapter 6). They must be aware of contamination issues, especially because there are several firefighting tools that are fuel driven and should not be taken onto a scene. The National Fire Academy has a handoff course titled "Arson Detection for the First Responder." This course highlights what to look for with an incendiary fire and what actions need to be taken by the firefighter. There may be a similar course provided by your jurisdiction, state, or province.

■ Note Even the line firefighter should be trained to recognize evidence and to protect it from harm.

AUTHORITY TO COLLECT EVIDENCE

Who should or should not collect evidence varies. The first variable is the role of the investigator on the scene. The fire officer, first responder investigator, doing the preliminary investigation should not take any evidence except as a last resort to protect and preserve the evidence. The public assigned fire investigator should collect any and all evidence that would be used as part of a hypothesis that may be incendiary in nature. Neither the fire officer nor the assigned fire investigator should take evidence from a fire scene that has been determined to be accidental in nature, except under extreme circumstances.

■ Note The fire officer, first responder investigator, doing the preliminary investigation should not take any evidence except as a last resort to protect and preserve the evidence.

If a fire is accidental in nature, there is no crime. If there is no crime, there is a question as to the legality of the investigator taking, seizing, evidence of any kind for any reason. This caution is based on the fact that the Constitution of the United States is very specific on unreasonable search and seizure. The localities should research with both local legal counsel for their jurisdiction and their local prosecuting attorney as to what they can or cannot take from a fire scene that has clear indication of being accidental in nature. Once this determination has been established, it should be documented and placed in the department's physical evidence policy. There is one given that is common across the land: Investigators can search a fire scene to make a determination of the fire cause based on it being good for the general public.

■ Note If there is no crime, there is a question as to the legality of the investigator taking, seizing, evidence of any kind for any reason.

Arguments can be made for the fire investigator taking a faulty appliance and having it tested. To do so without the permission of the owner of the appliance opens the locality to liability. Also, when the government takes any evidence for an accidental fire and wants to test it, an entirely different procedure must be followed as recommended in the section on spoliation in NFPA 921, *Guide for Fire and Explosion Investigations.*[1] As with so many other things, there are exceptions. In joint operations between the public and private investigators, it has sometimes been decided that the public investigator could hold the evidence until testing. The locality's physical evidence policy should cover these topics as well as give examples when it would be appropriate to take evidence from an accidental fire scene.

Many laws, case law in particular, dictate that fires should be classified as accidental until proven otherwise. Technically, there should be no such thing as a suspicious fire. The investigator can be suspicious of circumstances, prompting further investigation, but fires are going to be classified as natural accidental, incendiary, or unknown.[2] The unknown category simply means there is insufficient evidence to prove the actual cause. But the fire is still considered accidental. Put in another light: Laws in the United States dictate that everyone is

■ Note Many laws, case law in particular, dictate that fires should be classified as accidental until proven otherwise.

innocent until proven otherwise. Thus, using this same principle, all fires are accidental until proven otherwise.

The actual cause of the fire may not be known until the fire scene examination has been completed. Regardless of the eventual outcome of the fire investigation, every effort must be taken to preserve the fire scene as if it were a crime scene. This is not a conflict with the previous paragraph but a logical process, recognized by the courts, as an appropriate step to take to preserve any evidence that can identify the fire as being accidental or incendiary in nature.

When there is an indication that the fire many not have an accidental cause, the public investigator should collect any and all evidence for processing, taking all appropriate steps to protect, collect, preserve, and submit the evidence for the required testing.

■ Note Private investigators and insurance investigators have an entirely different set of rules in such situations.

Private investigators and insurance investigators have an entirely different set of rules in such situations. This is because they also have a different focus on the fire scene as well. When a property owner has a fire and the property is insured, the insurance company may send one of its own investigators from their Special Investigations Unit or it may hire a private insurance investigator. Private sector investigators can collect and keep evidence from a fire that is either accidental or incendiary in nature. Should the property owner refuse to allow this to happen, the insurance company, based on the insurance policy, which is a legal contract, can simply refuse to pay the claim. With this type of incentive, property owners seldom refuse to allow the investigator to take the evidence needed. In fact, in any situation other than incendiary fires, the insurance company acts as an agent of the homeowner; both have a common interest in the outcome of the findings. The process of collection and preservation of evidence of an incendiary nature is no different for the private investigator than it is for the public investigator. The procedures and processes are very similar if not exactly the same.

■ Note Private sector investigators can collect and keep evidence from a fire that is either accidental or incendiary in nature. Should the property owner refuse to allow this to happen, the insurance company, based on the insurance policy, which is a legal contract, can simply refuse to pay the claim.

Evidence of an accidental nature takes on a different light in that the insurance company may be seeking evidence to show that the fire was the result of a responsible third party. This may be in the form of a faulty appliance or recently contracted repair work. Should either of these two examples be the cause of the fire, the insurance company will pay the property owner the appropriate value of the loss up to the limits of the insurance policy. The insurance company can then recover its loss through subrogation against the responsible party. (See Chapter 4 for more information on the interests of insurance companies in fire investigations.)

Situations in which the fire starts on one property and spreads to another can create some interesting scenarios on the right to be on the scene to collect evidence. If the fire started in a pizza shop in a strip mall, for example, and extended into the clothing store next door, the public investigator can look at both scenes under the auspices of determining the fire area of origin, cause, and the reason for the fire extending into the neighboring tenant space. On the other hand, the private investigator from the insurance company may represent the clothing store and has no legal right to go onto the pizza shop premises without permission from the tenant or owner. Even if permission is given, it can be rescinded at any time, and the private investigator will have to leave the pizza shop. If the insurance company investigator represents the strip mall owner, however, under most circumstances that investigator can look at the entire property and all tenant locations.

PROTECTING EVIDENCE

■ Note
It is important to preserve all evidence, regardless of whether the fire is later determined to be accidental or incendiary.

First responders are the first to encounter any evidence on the scene. It is important to preserve all evidence, regardless of whether the fire is later determined to be accidental or incendiary. That evidence can help to prevent future fires through education, identifying needed recalls on appliances, or putting an arsonist behind bars. The latter not only stops that one individual from setting fires, but may act as a deterrence, causing others to think twice before committing arson.

With the proper training, suppression forces can recognize evidence and take the steps necessary to protect it. Taking extra precautions when advancing attack lines, looking at doors before forcing to see if they already have tool mark impressions, or limiting salvage and overhaul are all steps that can go a long way in aiding the investigation.

Overhaul is the process of searching for hidden extension of the fire. Overhaul relies on the opening of walls, ceilings, and even going through the debris on the floor or ground to find cinders and to ensure they are extinguished to prevent a rekindle of the fire. In some instances, flaming combustion can still be within the walls. Thus, suppression forces must continue the overhaul process to stop the fire and to prevent rekindle.

salvage
a suppression activity used to protect the contents of a property from smoke or water damage by removing objects from the structure, covering materials within the structure with tarps, and/or removing water with squeegees, pumps, water vacuums, and so forth

Salvage is a process of taking items that have not been damaged or severely damaged and protecting them by covering them with salvage covers or removing them from the structure. There is a tough balance to strike between the overhaul and salvage process and preservation of evidence. Both suppression functions must continue but be limited if possible to prevent loss or contamination of evidence.

The first responder who does a preliminary investigation to fulfill the requirements of the Fire Incident Reporting System may discover that the fire is not accidental but incendiary in nature. When this happens, two things must take place. First, the officer in charge must follow that jurisdiction's procedure for contacting an investigator to come to the scene. Second, the scene must be protected to prevent the destruction of evidence. This means limiting access to the scene to anyone who does not need to be on the scene. Suppression forces will still need to seek hidden spots of fire and carry out extinguishment, so they will still need to make the scene safe by providing for water removal or cordoning off areas that are unsafe.

Securing a Fire Scene

■ Note
Steps to preserve evidence include first establishing a barrier using barricade tape or rope to cordon off the area to prevent unwarranted entry.

Steps to preserve evidence include first establishing a barrier using barricade tape or rope to cordon off the area to prevent unwarranted entry. The size of the scene must be established so that the entire scene is within the barrier perimeter. Rarely is the perimeter barricade perfectly square because firefighters or police personnel will tie the barricade tape or rope to existing objects, such as trees, fence posts, or even parked vehicles. The barricade must make a statement that it should not be crossed. Thus, the use of commercial barricade tape preprinted with the words "FIRE LINE—DO NOT CROSS" or "CRIME SCENE" is often used. Either one sends a clear message to not cross that line. (See **Figure 8-1.**)

However, barricade tape alone will not completely deter individuals from entry. Personnel must be assigned to watch the perimeter to make sure that the scene stays secure. This can be accomplished with a minimum of personnel, depending on the site. As shown in **Figure 8-2,** a simple small structure can be watched by two or three people. As long as all sides are under the watchful eye of an assigned person, the scene should be secure.

The department must be aware of the legal issues dealing with setting up barricade lines. For example, can the department prevent the owner of the property from entering his or her own property? In most jurisdictions, this can be done under the provision of protecting the person from harm during exigent circumstances such as the building still burning or being unsafe. However, the

Figure 8-1
Commercial barricade tape can send a clear message to not cross the line.

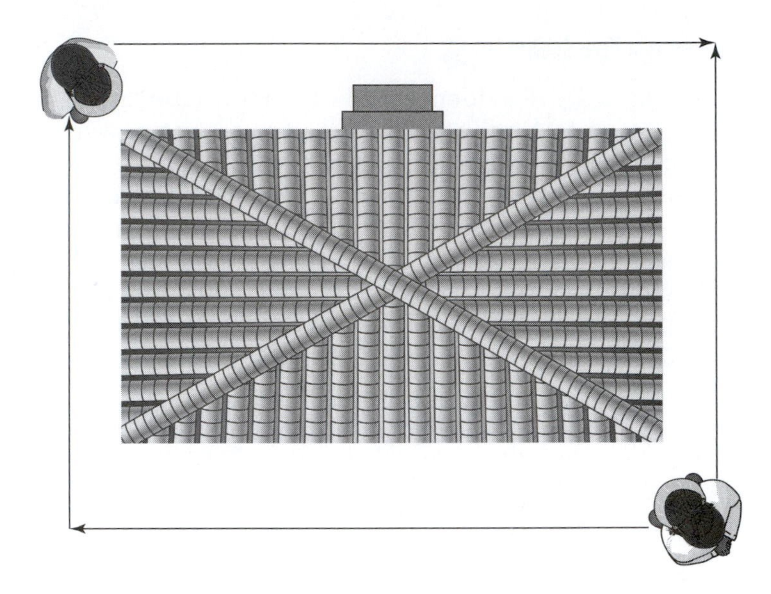

Figure 8-2 *A simple property can be observed by as few as two or three assigned personnel. Their primary function is to make sure that no unauthorized personnel enter the scene.*

department may not be able to keep out the press when they show up on the scene. Specific legal provisions might allow the press to cross fire lines provided they do not hinder the emergency operations. Researching and inserting a section in the physical evidence policy that covers the securing of the scene should prevent any future problems or potential conflicts.

Note If the scene is secure when the investigator arrives, the investigator must ensure that it is in that same condition when he or she leaves.

As a general rule, the private sector does not need to barricade a scene. Sometimes the barricade tape is still in place when the private sector investigator arrives and sometimes the structure may even be boarded up and secure. The bottom line is that there generally is no reason to establish a barricade line because the crowd has gone and the exigencies of the situation have long gone. The private sector investigator still has a responsibility for public safety. If the scene is secure when the investigator arrives, for example, the building is boarded up and locked, the investigator must ensure that it is in that same condition when he or she leaves. This will keep out neighborhood children and the curious and limit the liability of the investigator.

Note Additional safety concerns include actions such as cutting samples from walls, ensuring that it will not change the structural integrity of the building.

IDENTIFYING AND COLLECTING EVIDENCE

Before even starting the process of identifying and collecting evidence, the investigator should be fully cognizant of all safety issues as outlined in Chapter 2. Additional safety concerns include actions such as cutting samples from walls, ensuring that it will not change the structural integrity of the building.

Evidence can take many forms—not just the obvious sorts such as the gas can in the kitchen or the burnt road flare on the seat of the car. Burn patterns on a wall or the structure itself may be the evidence necessary to come up with an accurate and plausible hypothesis. Once identified, all evidence must be preserved. Again, this can easily be done by picking up the evidence, properly packaging it, logging it, and placing it in a secure area, such as the trunk of the investigator's vehicle. But not all evidence can be preserved so easily. Especially if that pattern pointing back to the area of origin is on an unstable wall. It must be documented through photography and notes about observations in the investigator's field notes.

Almost all patterns in a structure fire help to tell the story of what happened. The interpretation of these patterns is something that is learned over time and through experience as well as from classroom learning and texts. The *burn pattern* is the physical effects from the fire on structural components and contents. (See **Figure 8-3.**) These patterns can come from char of combustible products such as wood in the structure framing and contents. It can come from the oxidation of metal from structural components, appliances, and contents. Metal may also discolor or distort from the heat. The total consumption of structural components is part of the pattern to be observed. In addition, soot or discoloration of wall or ceiling surfaces is a type of pattern as well. These patterns are discussed in detail in Chapter 11, but observation of all patterns should be properly photographed and documented in the investigator's notes and investigative report.

Figure 8-3 *A burn pattern can tell the story of the fire's origin and direction of travel.*

Evidence can include devices and appliances too. Smaller items could be circuit breakers, switches, cords, or electrical multistrips. Larger evidence can consist of an entire electrical panel, clothes dryer, or even a refrigerator. As the investigation continues, the investigator may collect other evidence relevant to the investigation, such as papers, documents, or even signed statements.

Each type of evidence has a unique collection process. However, the typical type of evidence from the fire scene will include liquids, solids, and gases. Each has its own unique collection process. All are essential to the final accurate determination of the fire origin and cause.

Residue Gases and Vapors

Knowing the potential gas that may have been involved in the incident dictates where to look for a sample. The vapor density of the gas dictates whether the vapor settles into low areas or rises to collect in the upper portions of a structure. Most lighter-than-air gases dissipate before a sample can be taken. This is not to say that sampling cannot be done, but the gas is less likely to be present when the investigator arrives. Gases that are heavier than air settle in low areas. In particular, they may settle in floor drains or sink drains.

■ Note

Gases that are heavier than air settle in low areas. In particular, they may settle in floor drains or sink drains.

When gaseous samples might still be at the scene, commercial sampling kits can provide vacuum containers. These containers have had the air removed and then were sealed to create a vacuum. Using the attached tube and valve, you can place the container in an atmosphere or insert the tube into an atmosphere, such as a drain pipe, and then open the valve. The pressure within the container is less than the atmospheric pressure and the gas present at the tip of the tube will flow into the container and be trapped by the control valve. This gas can then be sent to the laboratory for examination. The obvious tests are to see whether the gas is a combustible vapor and exactly what type of vapor was present.

Other means of collecting gas and vapor samples include the use of mechanical devices that pull the air sample through a charcoal trap or other absorbent material to collect the residue for testing. Other pumps pull the gas into a sample chamber where it can later be collected and processed by a laboratory.

■ Note

Liquid samples are the most common type of evidence taken at a potential incendiary fire scene.

Liquid Samples

Liquid samples are the most common type of evidence taken at a potential incendiary fire scene. Evidence is usually in the form of debris or flooring solids that might have absorbed a liquid **accelerant.** More often than not, the debris sent in for analysis is solid items cut up or scooped up from the floor and placed into an evidence container. This type of evidence is collected and sent to a lab to be examined for the presence of an ignitable liquid.

accelerant
an item or substance to ignite, spread, or increase the rate of fire growth

When taking debris samples such as carpet, wood flooring, or concrete, the investigator is generally looking for the liquid residue trapped within the material. These types of samples are better suited to be preserved and transported in

Figure 8-4 *A 1-gallon evidence can with debris should be filled no more than two-thirds to three-quarters full.*

metal cans. Most laboratories prefer the 1-gallon cans, encouraging you to collect as large a sample as possible. However, only fill the can two-thirds to about three-quarters full to allow head space at the top for sampling by the laboratory. (See **Figure 8-4.**)

■ **Note**

Tools must be thoroughly cleaned and tested with a hydrocarbon detector before being used to ensure that there is no oil residue on the blades or cutting edges that could contaminate the samples.

Each type of flooring requires specific tools to cut up or break it up so that samples can fit inside an evidence can. These tools must be thoroughly cleaned and tested with a **hydrocarbon detector** before being used to ensure that there is no oil residue on the blades or cutting edges that could contaminate the samples. The forensic laboratory can assist in making this determination in advance of any tool's use.

New saw blades must be scrubbed because most manufacturers use a thin coat of oil to keep blades from rusting. At no time should oil be used on these devices because the oil on the tool could contaminate the sample being taken. Thus, many of the saw blades may very well rust. Under most circumstances, they are still usable. Common utility knives can be used to cut up carpet. Wood flooring can be cut up and removed using hand tools or battery-operated saws. Concrete floors can be broken down with masonry chisels, a hammer, and a strong arm. Some masonry power saw blades can also make the job easier; before using them be sure they will not contaminate the scene.

Before using any tools, check again with a hydrocarbon detector to be sure that they are clean and free of any hydrocarbon contaminates. If they are used to

hydrocarbon detector an electronic device intended to be used in the field to identify the location of ignitable liquids; used to improve the probability of obtaining the best sample to submit to the laboratory for testing

get a sample from more than one location, the tools must be cleaned between the collection of samples from the different areas.

The petroleum sheen seen on standing water is not necessarily an indication of the presence of a ignitable liquid. Many structural contents and components are petroleum based and, in burning, leave a slight residue that may be seen as a sheen on standing water. However, if large quantities of a liquid accelerant were used, there *might* be residue left in the debris.

■ Note **If during the investigation there is any reason to believe that the containers were involved in the incident, they should be marked as evidence.**

If ignitable liquid containers are found at the scene, they should be protected. If during the investigation there is any reason to believe that the containers were involved in the incident, they should be marked as evidence. If there is any residue within the container and the container cannot be sealed, every effort must be taken to recover the residue so that it can be examined by the laboratory. If an eyedropper is to be used, remember it must be clean and checked before use. If there is not enough of the liquid to pour into an evidence container, other items, such as sterile cotton balls or gauze pads, may be used to absorb the liquid. Then, these items must be placed into the evidence container.[3]

If there is reason to believe that the liquid sample is in the concrete, you can use certain methods to extract the liquid without having to break up the concrete. You can use a wet broom to spread out any liquid and allow it to soak into the top layer of the concrete. Hydrocarbon fuels are lighter than water and will sit on the surface. Then, put down a layer of an absorbent material to pick up the liquid residue. Absorbent materials such as kitty litter or even dry flour should sit on the surface for 30 minutes. Then, you can pick it up using a clean tool and place it into an evidence container. Whichever material you use—cotton balls, kitty litter, or something else—an unexposed, clean sample of the absorbent material must be packaged in a sealed container and delivered to the laboratory as a comparison sample to show that there was no contamination of the original sample submitted to the laboratory.

■ Note **If there appears to be more than one type of solid residue, special precautions must be taken not to accidentally scoop them up together, mixing them in the process.**

Solids

Solid evidence is sometimes collected for verification of the material's identity. Both accidental and incendiary fires can be started when solid objects in granular form mix with other products and result in a chemical reaction. These can sometimes be oxidizers that when mixed with an oily substance can react, resulting in ignition. Whether the mixing of these two components is accidental or intentional can be determined by a thorough investigation of all the facts, following the scientific methodology. (See Chapter 7 for more details.)

Collecting solids requires precautions. If there appears to be more than one type of solid residue, special precautions must be taken not to accidentally scoop them up together, mixing them in the process. The results could be devastating should the two chemicals react and cause an exothermic reaction that could result in a secondary fire. Collect both samples and ensure that they will

K-9 accelerant dog
a dog specifically trained by a reputable organization to identify the presence of an accelerant and give the appropriate signal of its location

■ Note
The Bureau of Alcohol, Tobacco, Firearms, and Explosives has taken a lead role in the training and use of accelerant detection dogs.

not come in contact. Also be aware that the material you are collecting may be corrosive. If you have any suspicion that this may be the case, wear double or even triple layers of latex gloves and consider obtaining corrosion-resistant gloves to handle this type of material. For some corrosive materials, you can use glass evidence containers to preserve them.

Finding the Best Sample

Depending on the scene, the investigator may be able to identify where it is best to take a sample. The leading edge of a floor burn or behind the baseboard that may have protected the accelerant product are both good locations. But is the accelerant in those locations? Many investigators resort to the use of a hydrocarbon accelerant detector. There are many on the market with quite different price tags. There are single-sensor units and dual-sensor units. There are multiuse units that can find the trace evidence as well as measure the lower explosive range. Each investigator must decide which product fits his or her needs and budget.

The best "tool" to use to look for evidence is a properly trained **K-9 accelerant dog.** (See **Figure 8-5.**) The Bureau of Alcohol, Tobacco, Firearms, and

Figure 8-5 *K-9 accelerant dogs are a great help in finding the best sample to submit for laboratory analysis. (Photo courtesy of Karl Mercer.)*

Explosives has taken a lead role in the training and use of accelerant detection dogs. The insurance industry, in particular State Farm Insurance, has provided many of these animals to the investigations community across the country. The dog's ability to detect the location of an accelerant can enable the fire investigator to find the best sample to submit to the laboratory for analysis.

However, using a K-9 does not guarantee a positive sample. Although a dog's ability to detect accelerants is far more sensitive than any technology presently available, investigators should not rely totally on the K-9's ability but should use all knowledge, skills, and resources to do a complete and thorough investigation.

Evidence Containers

What type of evidence container to use depends on the situation, the instructions from the laboratory, the preferences of the investigator, and fiscal resources. Metal and glass evidence containers have pros and cons. Regardless of the type of container, all evidence containers must be new and unused. Glass does not corrode or rust, can last indefinitely, and you can see the contents. Glass can be sealed adequately if done properly and with care. Do not depend on the rubber seal that comes with glass jars because these seals can react with vapors, contaminating the sample. They can also fail, resulting in the loss of vapors from the sample. A laboratory can recommend the proper sealing method. As a general rule, a layer of aluminum foil between the glass edge and the rubber seal can create a proper seal without having vapors reacting with the rubber seal. Drawbacks to glass containers are its fragile nature and the need for precautions to make sure containers are not damaged.

There are some alternate products on the market such as nonpermeable heat-sealing bags made of nylon. These bags will keep the evidence indefinitely, and you can see what is inside the container. However, if the lab is not set up to process these bags, it makes no sense for the investigator to stock such items.

Note
Metal cans are easy to handle and seal properly and they are durable.

Metal cans are easy to handle and seal properly and they are durable. (See **Figure 8-6.**) However, they will rust—even lined cans may eventually rust—and you cannot see the contents. Labs tend to ask for the use of lined cans so as to prevent the degradation of the can from oxidation with the liquid it contains. Sometimes it can take a year or more to get a case to court. In that time, unlined cans may rust through, losing their integrity. Even though a can is new, investigators should always check each and every can with the hydrocarbon detector just before use on the fire scene. This can prevent any contamination of the evidence placed in the container.

Note
Along with looking for evidence of the fire origin and cause, there are times when investigators need to collect more traditional evidence.

Traditional Crime Scene Forensics

Along with looking for evidence of the fire origin and cause, there are times when investigators need to collect more traditional evidence such as blood or

Figure 8-6
One-gallon metal paint cans are the containers of choice for most investigators and laboratories.

other body fluids. Weapons such as knives, guns, or blunt objects such as a baseball bat or heavy stick may also need to be recovered and protected. Fingerprints are a concern should the scene prove to be a crime scene. New research in fingerprinting can enable the investigator to recover fingerprints even though the area is heavily sooted.

tool mark impressions
potential identifiable marks from tools used by the perpetrator of the crime

Tool mark impressions need to be documented, and then carefully collected. In some cases, fire investigators have been known to take the entire door for analysis and eventual presentation in court. Along with the impressions, if tools that may have been used to make these impressions are found around the scene, they too must be collected and preserved. The laboratory can do a comparison to see if a specific tool was used; metal flakes or other characteristics could match the tool with the damage.

Investigators must keep an open mind and look at all evidence. For example, a restaurant was having fiscal difficulties and the owner decided to burn it down to collect the insurance money. In the process of setting up the fire scene to look like an intruder broke in and lit the fire, the owner made tool mark impressions on the kitchen door. The door was open when the fire department arrived, but the tool marks were forcing the door into the frame rather than opening the door. (See **Figure 8-7.**) This along with other evidence was enough for the insurance company to deny the claim, and the individual eventually was arrested for the arson.

Figure 8-7 *Tool mark impressions on doors can indicate a forced entry or an attempt by someone to make it look like there was forced entry.*

If the scene is criminal in nature, and the investigator finds documents at the scene, the documents may be collected under certain circumstances. One reason to collect is to protect them from being damaged by the elements. Should these paper artifacts already be wet, it may be necessary to collect them to be dried to prevent mold and mildew that could damage the documents. Documents could show motive, and if left behind by the investigator, they could be destroyed by the perpetrator of the crime.

On the other hand, if a fire started in the kitchen of a house, for example, even if the fire is incendiary in nature and looks like insurance fraud, depending on the circumstances, the investigator may have no right to search the file cabinet in the home office, which has no fire damage at all. If there are indications that evidence could be in those files, a guard should be left on the premises and the investigator should go to the proper legal authority, such as a magistrate, to obtain a search warrant. This ensures the rights of all concerned are protected and protects the developing case and allows any evidence found to be admitted into court.

Not-So-Traditional Evidence

Whenever a first responder encounters a victim within a burning structure, every effort must be made to save the individual by getting the person out of the fire scene atmosphere. Once removed and immediate life-saving actions have been taken, the first responder should start thinking about evidence issues. As clothing is removed, it should be kept separate to prevent contamination. The patterns on clothing in some circumstances can provide valuable evidence. There may be trace evidence as well, such as an accelerant. If the fire was set to cover up other crimes, the clothing may contain evidence of the other crimes, such as assault among others.

Burn injuries on victims also need to be investigated and documented. However, this should be done only after the victim has received proper medical care. Documentation should include photographing the burns; these patterns are evidence.

When it is obvious that the first responder has encountered a body that is well beyond saving, the body should be left in place, undisturbed. Burn patterns on the body and the associated clothing can also tell a story if properly documented and preserved. The location of the body and relationship to the surrounding area are important, which is why it is vital not to move a body if not necessary. It is critical to remember that the first and most important task is to save victims in fire situations; this is a far greater priority than anything else, including collecting evidence, at the fire scene.

■ Note
Accidental fires have been associated with large appliances since their inception.

Accidental fires have been associated with large appliances since their inception. (See **Figure 8-8.**) The cause has varied from bad design, poor manufacturing, to improper use. Sometimes misuse can be obvious, such as storing open containers of paint thinner adjacent to the gas furnace. There have been many recalls for various appliances over the years when these appliances were found to malfunction and create a heat source sufficient to ignite parts of the appliance or nearby combustibles.

■ Note
The placement of cans or jugs of gasoline around a furnace could indicate a totally ignorant act or the intentional design to start a fire and make it look accidental.

Some have taken advantage of the situation and have used large appliances to set a fire intentionally. The placement of cans or jugs of gasoline around a furnace could indicate a totally ignorant act or the intentional design to start a fire and make it look accidental. In some fires, an appliance was intentionally altered by the intended arsonist so that it malfunctioned and caused the fire.

When large appliances are involved in the ignition sequence, they pose a problem of keeping them as evidence. Most evidence lockers are not large

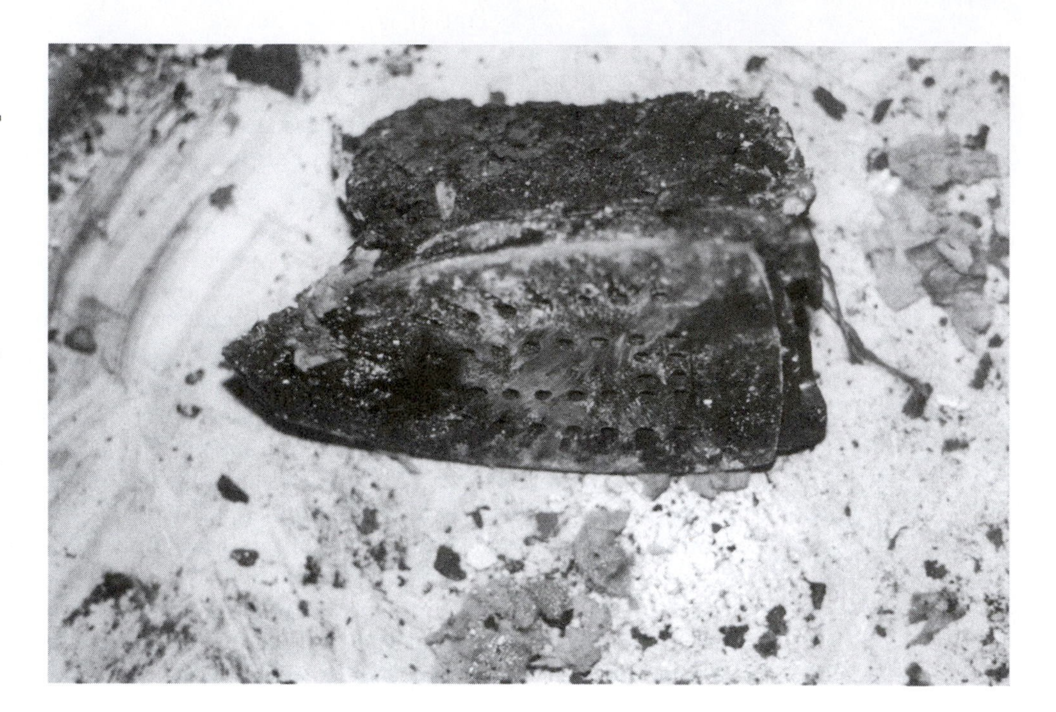

Figure 8-8
Appliances can be involved in the ignition sequence and as such may need to be collected as evidence. If the fire is determined to be accidental, the appliance can then be turned over to a responsible party. If the appliance was used as part of an incendiary device, the appliance should be kept and processed as criminal evidence.

enough to store a large appliance. Wherever the evidence is to be placed, it must be in a space that is controlled, where the investigator has complete control and there is no chance that the appliance will be tampered with or altered in any way. This control of the appliance is important for both accidental cases when the private sector will be seeking subrogation and when the public investigator has determined that the appliance was used as an incendiary device.

Sometimes the entire appliance need not be collected, but the switch, control, or circuitry must be taken for examination. Put the item in an evidence envelope and pack it so that it is protected from movement to keep it safe until examination can be conducted. Smaller appliances such as microwave ovens, coffee makers, and toasters might be part of the ignition sequence. The smaller items are easy to collect and protect. Be sure to place these items in a container that will protect them from damage.

Comparison Samples

When there are indications of the presence of an accelerant, samples are taken to the laboratory for analysis. These samples can consist of materials from the structure that have absorbed the liquid residue of an accelerant. The materials could be flooring such as carpet and padding, wood planking, or concrete.

comparison samples
an uncontaminated sample of what is being tested that helps examiners to identify any contaminates

If possible, it is also a good practice to find another portion of the unburned flooring that does not contain the accelerant. Samples taken from unburned, uncontaminated areas are called **comparison samples**. Unburned areas may have been protected because they were under furniture or on the opposite side of the room. When collecting comparison samples, investigators should follow the same process used for other evidence, ensuring there is no contamination and that it is properly documented. This enables the laboratory to compare the two samples. Using comparison samples provides a more reliable analysis, showing the accelerant if it is present.

■ Note
When collecting comparison samples, investigators should follow the same process used for other evidence.

When dealing with a liquid samples, sometimes investigators use absorbing materials such as cotton balls or gauze. The investigator should supply a clean sample of the absorbents, sealed separately, as a comparison sample. The comparison sample should show whether the absorbent did or did not contaminate the accelerant sample submitted.

Comparison samples with other types of evidence such as electrical or mechanical devices are commonly referred to as **exemplars.** Electrical switches, electrical circuit breakers, and even electrical panels can all benefit from the collection of comparison samples for analysis. The investigator should be able to see how the item appeared before the fire and all components to make it work properly.

exemplar
a comparison sample of an electrical or mechanical device that enables the investigator to compare the damaged item to an undamaged version of the product

Improper Install

The rear of the house was fully involved in the fire when the fire department arrived. The enclosed porch was gone, the rear area of the basement was destroyed, and the rooms above were complete consumed. The fire was knocked down pretty quickly, and there was not much overhaul to be conducted because most everything was gone but metal and concrete.

The investigator looked at the back of the structure. What stood out was a tee-connector for a stovepipe, still hanging on the outside wall where the pipe came through the basement wall. On the inside a wood stove still contained hot coals. On the ground all around the tee-connector were sections of double-walled pipe. The tee-connector (shown in **Figure 8-9**) was designed to have the smoke enter and flow upward. The bottom of the tee-connector was designed to be a clean out. A cover secured with three screws was to be in placed on the bottom of the tee. However, as shown in **Figure 8-10,** there was no cover. It was on the ground under the tee-connector. The screws were in place on the cap but had pushed in at the base of the tee instead of penetrating and securing the cover. (See **Figure 8-11.**) The only thing holding the cover in place was the pressure of the three screw points against the tee-connector. The chimney was enclosed by a wooden chimney chase; proof was shown by the remains of a 2 × 4 on the wall adjacent to the tee-connector.

(Continued)

(Continued)

Because the investigator was not sure whether this was a design flaw or an installation error, he ordered and tested a new tee-connector (see **Figure 8-12** and **Figure 8-13**). The screws that came with the new plate were self-tapping screws. The screws on the failed plate were ordinary sheet metal screws. During an interview with the installer, the installer admitted that he lost the original screws and just used three old screws he had in his tool box. The insurance company paid the homeowner the limits of the policy and sought subrogation from the installer's insurance company.

Figure 8-9 *Tee-connector as seen at the fire scene. The 2 × 4 on wall adjacent to the connector is the only remaining part of the chimney chase. Further evidence at the scene indicated that the homeowner used the chimney chase for storage. Poles for a badminton set are sitting on the tee support posts.*

Figure 8-10 *The bottom of the tee-connector does not have a cover plate to keep smoke, fire, and soot from escaping into the chimney chase. Cover found directly below under the debris indicates that it fell off before the fire. Notice indentures on the connector where screws did not penetrate the steel but bent it in. Base plate held in position by tension of the points of the three screws.*

Figure 8-11 *Base plate and the tee-connector.*

Figure 8-12 *A new tee-connector showing the bottom plate.*

Figure 8-13 *Base plate (shiny surface reflecting wood decking) in position.*

contamination
anything introduced into the fire scene or into the evidence that makes test results unreliable or places doubt on the results of the laboratory testing

CONTAMINATION

Of concern at almost every fire scene is **contamination.** Some of this occurs as the suppression activities occur. Fire streams tend to move objects, materials, and residue from one point to another. This cannot always be avoided. Some claim that positive samples for accelerants can be caused by firefighters walking around the fire scene tracking in contaminates. However, a 2004 report published by ASTM International showed that footwear does not transport (track) identifiable ignitable liquid residue throughout a fire scene during an investigation.

■ **Note**
The use of any suppression appliance such as a fan or saw should be avoided in and around the area of origin, especially those run with a gasoline engine.

The use of any suppression appliance such as a fan or saw should be avoided in and around the area of origin, especially those run with a gasoline engine. Any presence of such an appliance, even if it did not contaminate the scene, may cast doubts on the validity of any hydrocarbon evidence recovered.

Under most circumstances, first responders need to concentrate on protecting what they find. An unknown footprint in the mud can be covered with a bucket, road cones can surround a tire impression in the ground, or a road cone can cover a bottle, can, or cigarette butt found in the backyard. The most important thing suppression forces can do is to limit overhaul until the investigation is completed either by the officer in charge or an assigned investigator.

Use of Gloves and Safety Gear When Collecting Evidence

It is essential that all safety equipment be worn all the time. If your helmet falls off or repositions when you bend over, readjust the helmet and make sure it is secure on your head. Gloves protect your safety and also help you avoid contaminating any evidence handled. Heavy gloves are critical to the safety of the investigator. Many sharp objects in fire debris can easily penetrate lightweight gloves. Because leather gloves may contain contaminates from other areas or other scenes, do not wear them when handling evidence that will be subjected to laboratory analysis. Handle evidence wearing latex gloves, and even then, discard the gloves after collecting a sample and use new gloves for the next sample to ensure that no cross contamination occurs. Obviously, investigators must have an adequate supply of latex gloves before starting the process of collecting evidence.[4]

■ **Note**
All boots and gear must be thoroughly cleaned between each scene to prevent cross contamination. After cleaning, a calibrated hydrocarbon detector should be run over the gear to make sure there is no contamination.

Never place the gloves inside the evidence container. If the lab needs the gloves for some reason, tape them to the outside of the container or place them in a plastic bag and attach that bag to the outside of the container.

The fire investigator must consider safety gear as well as tools before entering the scene. All boots and gear must be thoroughly cleaned between each scene to prevent cross contamination. After cleaning, a calibrated hydrocarbon detector should be run over the gear to make sure there is no contamination. If you use disposable coveralls, the operative word is *disposable*. Use a new disposable coverall at each scene.

A report in the *Journal of Forensic Sciences* titled "The Evaluation of the Extent of Transporting or 'Tracking' an Identifiable Ignitable Liquid (Gasoline) throughout Fire Scenes During the Investigative Process" clearly shows that evidence will not be contaminated from having fire boots, work boots, or sneakers walk through a fire scene.[5] Even though the chance of contamination is slim to none, it is never a bad idea to go in with clean gear and footwear.

Tools take special care. Thoroughly scrub shovels, trowels, and other hand tools with soap and water, and then test them with a hydrocarbon detector before entering each scene. Document this task in your field notes to ensure that there is no doubt that this was accomplished before the investigation.

FIRST RESPONDER CONSIDERATIONS

■ Note **The Supreme Court decision in the *Michigan v. Tyler* case cited that the volunteer fire chief was correct in taking the evidence to protect it from being damaged or from disappearing.**

The Supreme Court decision in the *Michigan v. Tyler* case cited that the volunteer fire chief was correct in taking the evidence to protect it from being damaged or from disappearing the night of the fire at the furniture store. If the evidence were to be lost, not properly protected, or out of the custody of the person taking the evidence, the evidence may not be admissible in court. This could potentially hurt any future criminal proceedings. Proper training of suppression staff on evidence procedures goes a long way in ensuring the preservation of evidence.

In areas where there is no readily available assigned investigator, a first responder can be trained in evidence collection in conjunction with the training received in fire origin and cause. A small evidence kit can be established to collect debris samples or containers, enabling the first responder to tag and document the evidence properly. Of course, there must be a procedure for the first responder to lock up the evidence, ensuring that no one else has access and that the evidence is in the possession of the first responder the entire time before being turned over to an investigator for processing. This maintains the **chain of custody.** If an arrest is eventually made in the case, the first responder can testify in court on the evidence collection and the chain of custody. In the case of an accidental fire, the first responder might testify in civil court should a civil suit result in a product failure or if the insurance company chooses to subrogate.

chain of custody a means of documenting who had control of the evidence from the time of collection through the trial and up to its release or destruction

Of course other measures can be taken. Local law enforcement, even if not fire investigators, can provide assistance in evidence collection, documentation, and storage. The primary course of action is to have a plan in place as part of the department's policy, well in advance of the incident.

DOCUMENTING, TRANSPORTING, AND STORING EVIDENCE

■ Note **Every piece of evidence is only as good as the way it is handled.**

Every piece of evidence is only as good as the way it is handled. The evidence must be properly and fully documented so that it will maintain its value for future trial. Each piece of evidence must be labeled or tagged. Then, the evidence must be properly stored from the time it is collected until the investigator receives a court order to destroy the evidence.

Documenting Evidence

■ Note
Every piece of evidence collected by the investigator must be tagged or labeled.

Every piece of evidence collected by the investigator must be tagged or labeled. The information on the evidence label should include the date and time the sample was taken, the location it was taken from, and the person who took the evidence. In addition, the label must have the case number of the fire location along with a description of the evidence itself.

The next item to be attached to every piece of evidence is the chain of custody. When the evidence transfers from one person to another, that name of the person receiving the evidence must be placed on the chain of custody label, including the date and time that the transfer occurred.

The information about the evidence and the chain of custody must also be included in the investigator's file. This evidence document should be used to track each and every piece of evidence for that case.

Generally, public investigators use commercial evidence tape to seal the container, and then place the investigator's initials on the tape. In the private sector, evidence tape is rarely used; the evidence cans are sealed and then boxed for shipment via a common carrier. The sealing of the box is proof enough that the evidence was not tampered with en route to the lab.

evidence transmittal
a form designed to identify the evidence to be submitted to the laboratory and the tests to be conducted on that evidence; can also serve as the chain of custody of the evidence

Shipment of Evidence

If the investigator is fortunate enough to work close to the laboratory, it is beneficial to deliver the evidence. Upon arrival at the laboratory, the investigator hands over the evidence and the **evidence transmittal** in the receiving process. The evidence transmittal form describes the evidence, the testing desired, and all the contact information of the investigator. It also includes a section on chain of custody. Once turned over, the investigator receives a copy of the transmittal for the investigation file.

■ Note
Evidence needs to be packed so as not to be damaged in shipment; even metal cans can be damaged and lose their integrity if not packaged properly.

Because not everyone lives near a laboratory, sometimes investigators have to rely on a common carrier to transport the evidence. Evidence needs to be packed so as not to be damaged in shipment; even metal cans can be damaged and lose their integrity if not packaged properly. Once the evidence is packaged, the investigator must remember to insert the evidence transmittal. With the package properly sealed, it is sent registered, requiring a receiving signature, via a common carrier such as the United States Postal Service, Federal Express, or United Parcel Service.

Once the laboratory receives the evidence, the investigator receives the signed receipt from the carrier. Most laboratories also sign the evidence transmittal, which constitutes the chain of custody. A copy is then sent to the investigator.

■ Note
Evidence rooms must be climate controlled to keep the environment dry and cool.

Evidence Storage

The storage of evidence should be in a secure location where the investigator or lab personnel are the only persons having access. Evidence rooms must be climate controlled to keep the environment dry and cool. There should be ample space for racks and shelves to hold the evidence. Some evidence lockers contain freezers to hold evidence scheduled for the detection of accelerants. However, care must be taken to ensure that the moisture within a container is not of such a volume that it could freeze, expand, and rupture the container.

If the evidence is not volatile and does not need laboratory testing of accelerants, other care issues should be observed. Those items that are damp or moist may eventually mildew and mold if not properly dried out. Do not keep that type of evidence in sealed containers until it is completely dried out.

If there is any chance of the evidence breaking down, oxidizing, or corroding, then special precautions should be taken. Either place the evidence in plastic boxes or other containers so it does not contaminate other evidence stored nearby.

LABORATORY TESTING

gas chromatography (GC) a laboratory test method that separates the recovered sample into its individual components; provides a graphic representation of each component along with the amounts of each component

mass spectrometry (MS) one of two tests used to identify the presence of an ignitable liquid in a submitted sample, the other test being the gas chromatography (GC); see end of chapter for detailed definition

The most frequent test requested by fire investigators is the testing of debris to determine whether there is the presence of an accelerant such as hydrocarbons. The evidence transmittal usually asks the lab to test for the presence of ignitable liquid residues. It is then up to the laboratory to decide which test to run on the evidence provided to garner the best results. Generally, the laboratory runs two tests, gas chromatography and mass spectrometry. Although this has been the standard for many years, some research institutions are always searching for better ways to improve science, even in the field of fire investigations.

In fire investigations, it is common to run the two tests together in a method called gas chromatography–mass spectrometry (GC/MS). The **gas chromatography (GC)** separates the sample into individual components. The instrument provides a graphic representation of each component, quantifying the amount within the sample. This along with the further analysis of **mass spectrometry (MS)** can enable, if there is a large enough sample, the lab personnel to identify any ignitable liquids present in the contents of the evidence.

Of course, a forensic laboratory can provide many other services. In addition to those commonly expected such as recovering fingerprints or doing DNA testing, a lab can accomplish more obscure jobs such as putting glass fragments back together from bottles or glass panes. They can identify tire mark or shoe impressions and even the source of small paint chips. Thanks to the more recent movies and television shows on crime, the public is more aware of the potential of forensic labs. It is critical that investigators, early in their careers, visit the local forensic lab that will process their evidence. It is always good to meet the members of your team and to discover the true potential of each lab along with its recommendations on evidence collection and preservation.[6]

■ **Note**
The best tool for success with testing at the laboratory is an open line of communications between lab personnel and fire investigators.

The best tool for success with testing at the laboratory is an open line of communications between lab personnel and fire investigators. Forensic scientists can identify the correct amount of debris to collect for obtaining the best test results. They can also make recommendations on most aspects of evidence collection.

The laboratory personnel conducting the tests are responsible for writing the report and for testifying in court as to the testing process and the results of their testing. The investigator may have to testify as to how the evidence was discovered, where it was discovered, and the method used to collect the evidence. Opposing counsel will be meticulous in examining this area to ensure that the investigator followed proper procedures.

RELEASE OF EVIDENCE

Evidence for either a criminal case or a civil suit may have to be kept for years, awaiting trial. Then, it will need to be kept even longer, to a date beyond any potential for an appeal. In criminal cases, much of the evidence submitted in the trial is kept by the courts. However, lab samples, debris, or volatile liquids are usually returned to the investigator. When the evidence is no longer needed, the first step should be to contact the owner of the evidence to see if that person may need it for any reason. Even a can of debris may be of value. Following the criminal trial, the property owner may want to file civil suit, where the evidence may be of use in such litigation. If the owner wants the evidence, it should not be given back until the investigator petitions the courts to release the evidence to the owner.

If the property owner has no need for the evidence and does not want to take the evidence, it is advisable for the investigator to get a signed release stating such. The next step in most jurisdictions is to petition the courts to have the evidence destroyed. After the evidence is destroyed, policy may require the investigator to return the destruction order to the courts stating the date of the destruction. Common sense dictates that copies of all documents should be kept in the investigation file.

■ **Note**
Common sense dictates that copies of all documents should be kept in the investigation file.

Not all evidence collected is used in trial. Usually, the prosecuting attorney decides what will be admitted in court during trial. The investigator must maintain any remaining evidence not used in the trial just as if it had been used in court. When petitioning the court to release or destroy the evidence, all of the evidence should be listed in the request and on the judge's order.

subpoena
the command of the courts to have a person appear at a certain place at a specific time to give testimony

litigant
someone who is involved in a lawsuit

It is similar for civil cases. If the investigator for some reason is the holder of the evidence and has been **subpoenaed** for the civil trial, all evidence must be kept and maintained as if it were a criminal case. However, in some civil trials there may be multiple **litigants.** If this is the case, the evidence must be made available to all parties. In all such situations, advice from the jurisdiction's attorney or the local prosecutor should provide direction for how the evidence is to be viewed and who can examine such evidence. The one thing to remember is that no destructive testing of the evidence should be conducted without all known parties having a chance to be involved or at least to observe. This is clearly a situation in which spoliation of the evidence is of prime concern to all parties. See Chapter 6 for further information on spoliation.

SUMMARY

Evidence consists of a multitude of items from objects to patterns. The investigator must have the training and expertise to identify the evidence and to know the potential value of that evidence to know what is to be collected.

It is critical that investigators know the legal parameters regarding taking and securing evidence. With indicators that the fire was incendiary in nature, there is no doubt that the investigator can, and more important should, collect any and all evidence as to the fire's area of origin and cause. Any evidence that may indicate who might have caused the fire should be collected and preserved as well.

With a fire that is clearly declared accidental at the time of the fire scene examination, where no evidence needs to be collected to prove an hypothesis, most likely there is no legal authority to collect and secure evidence from the scene. However, if there is evidence that may be of value to the property owner who is not present, and leaving it behind may lead to its damage or disappearance, the investigator may want to consider taking and securing the evidence. The physical evidence policy should address this action and it should be cleared by the local prosecuting attorney or jurisdiction before taking on such a liability.

A better solution is to find a relative or neighbor and turn the entire structure over to them until the property owner arrives. If no one is available to take on the responsibility, it may be necessary to post security on the structure until a responsible party can be located.

Once evidence has been collected by the fire officer or assigned fire investigator, it should be locked up and secured. These actions must be documented and a list of who the evidence is turned over to and when should be retained to maintain the chain of custody.

Once collected, the evidence should be processed for testing by a forensic laboratory as soon as practical. This improves the potential of getting the best results from the test process. Testing can range from identifying whether a liquid is ignitable or determining the presence of an ignitable liquid in a debris sample. There may also be the normal forensic testing of fingerprints, tool mark impressions, and so forth.

After the evidence has been used in court and the potential for all appeals have expired, investigators must arrange to dispose of the evidence. This can be done by returning it to its owner or by petitioning the courts for permission to destroy the evidence. The evidence may still have the potential to be used in civil cases, so this must be explored before the evidence is released or destroyed.

KEY TERMS

Accelerant An item or substance to ignite, spread, or increase the rate of fire growth.

Chain of custody A means of documenting who had control of the evidence from the time of collection through the trial and up to its release or destruction.

Comparison samples An uncontaminated sample of what is being tested that helps examiners to identify any contaminates.

Contamination Anything introduced into the fire scene or into the evidence that makes test results unreliable or places doubt on the results of the laboratory testing.

Evidence transmittal A form designed to identify the evidence to be submitted to the laboratory and the tests to be conducted on that evidence; can also serve as the chain of custody of the evidence.

Exemplar A comparison sample of an electrical or mechanical device that enables the investigator to compare the damaged item to an undamaged version of the product.

Gas chromatography (GC) A laboratory test method that separates the recovered sample into its individual components; provides a graphic representation of each component along with the amounts of each component.

Hydrocarbon detector An electronic device intended to be used in the field to identify the location of ignitable liquids; used to improve the probability of obtaining the best sample to submit to the laboratory for testing.

K-9 accelerant dog A dog specifically trained by a reputable organization to identify the presence of an accelerant and give the appropriate signal of its location.

Litigant Someone who is involved in a lawsuit.

Mass spectrometry (MS) One of two tests used to identify the presence of an ignitable liquid in a submitted sample, the other test being the gas chromatography (GC). MS is used to find the composition of a physical sample by generating a mass spectrum representing the masses of sample components; this test is sometimes referred to as mass-spec or MS.

Physical evidence policy A locality's policy that guides and directs the proper collection, storage, handling, use, and disposal of all evidence.

Salvage A suppression activity used to protect the contents of a property from smoke or water damage by removing objects from the structure, covering materials within the structure with tarps, and/or removing water with squeegees, pumps, water vacuums, and so forth.

Subpoena The command of the courts to have a person appear at a certain place at a specific time to give testimony.

Tool mark impressions Potential identifiable marks from tools used by the perpetrator of the crime.

REVIEW QUESTIONS

1. Why should the fire investigator not seize evidence from any accidental fire?
2. Under what circumstances might the fire investigator want to collect evidence at an accidental fire?
3. What incentive is there for the insured to allow the insurance company investigator to conduct the investigation and take any evidence?
4. Why does the investigator test tools with a hydrocarbon detector before using them to collect evidence?
5. What are the benefits of using 1-gallon metal cans for debris samples?
6. What is a comparison sample, and why is it so important?
7. Can an investigator continue using the same tools such as shovels at different scenes? If so, what needs to be done (if anything) to the tool to prevent contamination?
8. What basic steps can be taken by investigators to ensure that they do not personally contaminate the fire scene?
9. What is an evidence transmittal form?
10. What conditions must be met to destroy evidence?

DISCUSSION QUESTIONS

1. What various items might be found around a house that could be used to mark or preserve a shoe print impression, a 2-foot-long tire impression, or a 1-gallon gasoline can found in the side yard? For those already in the fire service, what might be found on the apparatus on the scene that could be used to perform the same tasks?
2. Under what circumstances might a volunteer fire officer decide to take evidence found on an incendiary fire scene, especially if the only full investigator is days away from visiting the scene? What if it was an accidental scene with no one to take control of the scene?

ACTIVITIES

1. On the Internet, research recent case law that involves a fire scene and the collection of evidence.
2. Contact the clerk of courts and ask whether there are any trials pending that involve a charge of arson. Visit the courthouse on the date of the trial and observe the trial, especially the introduction of evidence and the testimony of the investigator and the forensic personnel from the laboratory.

NOTES

1. National Fire Protection Association, NFPA 921, *Guide for Fire and Explosion Investigations* (Quincy, MA: National Fire Protection Association, 2008), 921-103.
2. National Fire Protection Association, NFPA 921, *Guide for Fire and Explosion Investigations,* 921-158.
3. National Fire Protection Association, NFPA 921, *Guide for Fire and Explosion Investigations,* 921-137.
4. National Fire Protection Association, NFPA 921, *Guide for Fire and Explosion Investigations,* 921-136.
5. A. Armstrong, et al., "The Evaluation of the Extent of Transporting or 'Tracking' an Identifiable Ignitable Liquid (Gasoline) Throughout Fire Scenes During the Investigative Process," *Journal of Forensic Science* 49, no. 4 (2004). Paper ID: JFS2003155.
6. Niamh Nic Daeid, *Fire Investigation* (Boca Raton, FL: CRC Press, 2004), 160–181.

SOURCES OF IGNITION

Learning Objectives

Upon completion of this chapter, you should be able to:

- Describe various heat sources capable of being a source of ignition.
- Describe how the forces of nature in the form of lightning and radiant heat from the sun can provide sufficient energy to be potential heat sources.
- Describe how smoking materials can be a potential heat source.
- Describe how chimneys and associated appliances can contribute to a fire as sources of ignition.
- Describe and understand the concept of spontaneous ignition and where it can and cannot occur.
- Describe the role of appliances, large and small, as sources of ignition.

CASE STUDY

We want to believe that all people are smart and that common sense will prevail when a homeowner takes on a task. Alas, this is not always the case. The homeowner was faced with another year of high fuel oil costs. He had a floor furnace and the chimney for the furnace was cement block, attached to the outside wall of the structure.

Faced with the cost of a new chimney, he decided to purchase a woodstove and stove pipe to connect the furnace to the existing chimney with a chimney thimble. This thimble is a terra-cotta ring that accepts the pipe from the woodstove, directing vapors, smoke, and gases up into the chimney.

The plan was simple. Cut a hole in the wall adjacent to the chimney, chisel a hole into the cement block chimney, attach the thimble, place the woodstove, and connect the pipe between the thimble and woodstove.

The thimble was placed against the wood paneling on the wall; the hole was cut to that exact dimension of the exterior of the thimble. The sheetrock was then cut, as was the sheathing and exterior siding, eventually exposing the face of the cinder block chimney. Next, the homeowner took a chisel and carved his way through to the interior of the chimney. The opening he created was only approximately 6 inches in diameter. The opening in the flue, interior of the chimney, is only 9 square inches, 3 inches across. Thus, the opening created tapered from 6 inches to 3 inches (see **Figure 9-1**).

Figure 9-1 *The interior of the flue is shown; when measured, it was only 9 square inches: 3 inches by 3 inches.*

(Continued)

(Continued)

Figure 9-2 *The thimble is 8 inches across with an interior of 6 inches in diameter.*

Because the exterior diameter of the thimble was 8 inches and the opening only 6 inches (see **Figure 9-2**) the homeowner put mortar on the face of the cinder block, pushing the thimble up against the block. The mortar could not have held the thimble in place, so the siding, wall sheathing, sheetrock, and wood paneling—all of which are quite combustible—served to hold the thimble in position. (See **Figure 9-3**.)

With the woodstove in place, the thimble in the wall, and the mortar just barely dry, the homeowner put kindling and wood into the stove, lit the wadded paper to start the fire, and warmth started spreading through the room. The burning kindling had a good draft, pulling the smoke from the woodstove burn chamber into the pipe, through the thimble and into the narrow chimney flue. But as the larger pieces of wood caught fire and started giving off smoke, it was too much for the chimney and the smoke, gases, and heat started coming out the front of the woodstove. Closing the doors helped little because the smoke started coming out the damper vents, from around the pipe as it came into the stove, and from the pipe as it entered the thimble. The family rushed to open windows and doors. Soon they noticed more smoke from the area of the thimble, and then the wall paneling surrounding the thimble burst into flames. Finally, they called 9-1-1 while they filled pitchers, pots, and other vessels to throw water onto the wall.

Unbelievably, they stayed in the house until the fire department arrived. Without a doubt, they kept the fire from spreading on the paneling, but at an extreme risk of losing

(Continued)

(Continued)

Figure 9-3 *Thimble was intended to be held in place with the mortar shown where it is being positioned. The hole cut in the paneling, sheetrock, sheathing, and siding (all combustible) held the thimble in place until they ignited.*

their lives staying in the smoke and gases. More important is the fact that starting the fire in the woodstove that night most likely saved their lives. Had they not used the woodstove, the floor furnace would have been used that night. There is no doubt that carbon monoxide from the floor furnace would have seeped into the thimble, piping, and woodstove, escaping into the home. They very well could have died in their sleep.

There are two schools of thought on filling out this fire report. Was the first material ignited the paper used to start the woodstove or the paneling holding up the thimble? With explanation, both answers could be correct. When the arsonist flicks his thumb across the wheel of a lighter, the first material ignited could be considered the fuel from the lighter itself, but the fire report will most likely show the first material ignited as being the material the lighter ignited. With that in mind for the woodstove case, many would report that the first material ignited was the paneling. The cause of the fire was an improperly installed woodstove, but some may say the cause was a total lack of knowledge by the installer.

INTRODUCTION

Sources of ignition are all around us—we use them in everyday life. At home, in the office, in the factory, and even at play. Through history, primary igniters used by people have changed. When the Europeans started to inhabit the New World in the 1600s, it was the open fire used for cooking and heating that not only kept them alive and, when not controlled, took their lives as well.

Case in point is Jamestown, Virginia, one of the first colonies. More than 400 years ago, English settlers came to North America, built their stockades and homes. They built their structures with thatch roofs, and their chimneys were of sticks coated with mud. Almost the entire settlement burned not once but twice. The source of ignition for these fires was most likely the fireplaces. The mud coverings might have failed or embers escaped from chimneys and landed on the thatch.

In 1827, John Walker invented a match that could be ignited by striking on a rough surface. Several changes in matches had occurred over the year. Since the match was invented, it has been a primary igniter of hostile fires. It is unknown how many accidental fires matches may have caused, especially in the early years because they could be struck on almost anything to ignite. There was concern enough about the danger of the strike-anywhere match that the safety match was invented in 1855, in Sweden, by Johan Edvard Lundstrom. This match could ignite only if struck on a specific surface, rendering the match safe in most circumstances.

Even today, matches or lighters can be a primary igniter, along with the associated smoking materials. However, many small appliances, chemicals, and other heat-producing items can be and are a source of heat for uncontrolled fires. The fundamental issue is heat. Heat is a source of ignition that has caused many fires throughout the years. The devices that create the heat have changed over time.

Note
Heat is a source of ignition that has caused many fires throughout the years. The devices that create the heat have changed over time.

In every fire investigation, the first thing that must be identified is the area of origin. In the area of origin the investigator should find the first item ignited. That ignition was caused by a heat source, which may also be in the area of origin or in close proximity. Sometimes the heat source is missing from the scene because it was destroyed as a result of the fire or was moved, possibly by suppression activities. With a clearly defined area of origin and other supporting evidence, a heat source can be inferred, if it cannot be found. But a heat source must be established to make a final determination as to the fire's origin and cause.

In this process, it is vital that the investigator use scientific methodology, nothing but good science to develop and challenge all hypotheses. This is accomplished when the investigator's abilities have developed through a good education along with the sum of their experiences.

COMPETENT IGNITION SOURCES

A heat source must be able to deliver enough heat to cause ignition. For this to happen, there must be sufficient heat, at or in excess of the ignition temperature of the first material ignited. The heat must also be in contact with the first material ignited for a sufficient duration to allow the heat to be transferred to heat the fuel to its ignition temperature.

LIGHTNING

It seems fitting to first address the one heat source we can do little to control. Lightning is a static discharge between two differing potentials, negative and positive. Lightning can discharge from one cloud to another cloud or from ground to a cloud or cloud to ground.[1] The discharge can even occur within the same cloud when the lower surface of the cloud has a potential different from the upper surface of the cloud.

■ Note **Worldwide there are more than 8 million cloud-to-ground discharges of lightning every day.**

Worldwide there are more than 8 million cloud-to-ground discharges of lightning every day.[2] In a study done in the northern Rockies, it was discovered that 1 in 25 lightning strikes are capable of starting a fire.[3] The 18th edition of the NFPA *Handbook* states that the average strike is 24,000 amperes, but strikes have been measured as high as 270,000 amperes.[4] However, the National Weather Service lists lightning peaking at an average of 30,000 amperes and as high as 300,000 amperes.[5] The point is that lightning contains a lot of energy, more than enough to start a fire if conditions are right.

■ Note **National Weather Service lists lightning peaking at an average of 30,000 amperes and as high as 300,000 amperes.**

Most everyone has seen the result of a lightning strike on a tree. The destruction and splitting of the trunk are the results of the lightning energy being so massive that the moisture in the tree, water, is vaporized violently, literally exploding and ripping open the trunk of the tree. Even with this much energy and heat, it does not mean there will be a fire. As discussed in Chapter 7, all conditions have to be just right for a fire to start. Lightning bolts contain unimaginable amounts of energy, but it only lasts for a very brief amount of time. Was the mass large enough to absorb the heat? The answer in most circumstances is yes. What most studies find is that if there is an appreciable amount of moisture present in both live and dead trees, there will not be a fire.

■ Note **There were 17,400 fires annually started by lightning.**

When lightning strikes structures, it can cause both damage as well as fires. The United States Fire Administration released a report in 2001 that there were 17,400 fires annually started by lightning. Of these fires, 41 percent are in structures where roofs, sidewalls, framing, and electrical wires are the most common materials ignited.[6]

Depending on the amount of damage from the fire, there may or may not be obvious indications of a lightning strike on the structure. There are some clues to look for, items that would usually not melt in a fire, such as copper, may have vaporized as the result of lightning energy. The clues in the physical scene along with witness accounts may provide enough evidence to prove that lightning was the heat source for the fire.

SEISMIC EVENTS AND ESCAPING GASES

■ Note
In many earthquakes when available heat sources are brought in contact with fuels as the result of the movement and destruction of buildings.

A seismic event is not a source of ignition. However, the event itself changes the dynamics of the environment, where what typically would not have been a source of ignition can become responsible for starting many fires. Earthquakes are perfect examples.

The most famous fire associated with a seismic event is the Great 1906 Fire of San Francisco. The ensuing fire was the result of a multitude of events, but primarily resulted from open flames, building destructions, and a multitude of fuel. This is universal in many earthquakes when available heat sources are brought in contact with fuels as the result of the movement and destruction of buildings.

methane gas
colorless, odorless, flammable gas created naturally by decomposing vegetation; it can also be created artificially

Smaller seismic events can create fractures in rock that eventually enable the release of **methane gas** that typically would not have made it to the surface. Although this is not a source of ignition, it is fuel and can show how a simple light fixture can provide the energy for ignition when in contact with fuel.

Deep in the ground, in various areas, are pockets of natural gas. This gas is colorless and odorless as well as deadly and ignitable. This gas can escape to the surface from time to time, but seismic activity can also move the layers of rock, sometimes creating a path for the escape of the gas. Natural gas is lighter than air and will migrate upward.

As this gas migrates to the surface, it may also find a man-made path, allowing a more rapid escape to the atmosphere. Trenches around large pipes that are backfilled with stone to allow water drainage and provide stability also create a path for the gas to migrate along. Wells, both deep and shallow, can create a conduit for escaping gases. Because most wells are capped, the gases may escape along the water pipe trench leading to the structure and terminating in a basement or crawlspace.

■ Note
Once inside a structure, because the gas is lighter than air, it collects in higher locations.

Once inside a structure, because the gas is lighter than air, it collects in higher locations. If a source of ignition is present, the gas may detonate, creating an explosion and potentially a subsequent fire. Such events can come as a surprise as described in the following text box.

RADIANT HEAT

■ Note
When the rays of the sun are concentrated, the focal point can produce enough energy to be a competent ignition source.

Sunlight is the ultimate radiant heat, but by itself is not intense enough to ignite common combustibles. However, when the rays of the sun are concentrated, the focal point can produce enough energy to be a competent ignition source.

Methane Gas Surprise

The insurance investigator received a strange assignment to go to the western, mountainous part of the state to do a preliminary investigation. It was about a 6-hour trip, one-way. The strange part was that the investigator was asked to deliver a check made out to the insured. Usually, the claims manager handed out the checks, not the investigator. The investigator's instructions were to look at the structure, listen to the insured, and if the facts matched the scene, hand over the insurance check; if not, return the check to the main office.

At the insured's home, there was debris all around where the home used to sit. It was a small square home and it appeared that each of the four exterior walls were just laid out like they were pushed over. The roof was almost intact, but it lay behind the house. There were signs of small fires here and there where items such as curtains had burned and been consumed, but the fire did not go further. The only other structure on the small property was a small well house that was built as a replica of the original home with white siding and a peaked roof.

The investigator made a cursory investigation by walking around the structure and noticed that there was a full basement. The homeowner arrived, and the two talked in the side yard about 20 feet from the well house. The homeowner explained that he got up early in the morning and went to the basement to get some tools he was going to use on a project. The only light in the basement was a ceiling-mounted ceramic base light fixture with a single bulb and a pull string to turn it on and off.

The homeowner reached for the string and that was the last thing he remembered. When he recovered, he looked around, and the house was gone. The string from the light was still in his hand, and he knew that he had burns on his head and face. Neighbors had called emergency services, and the homeowner was transported to the hospital, where he was treated and released. He did experience first-degree burns on his face, but his hair, including his eyebrows, were gone.

The homeowner had lived in this community all his life, and he worked in the nearby coal mine. He said he knew exactly what had happened, that it was methane gas from an abandoned mine that seeped up through his well and then into his house. The investigator, not being from this part of the country, knew little about mines or methane problems. The investigator felt some skepticism as he listened to the story.

About that time there was a loud explosion. Ducking down more from reflex, the investigator saw the roof of the pump house hit the ground about 4 feet away. All four walls of the well house were laid out, exposing the above-ground pump and pressure tank. There was a slight whiff of smoke but no fire.

The investigator gave the check to the homeowner, had him sign a receipt, shook his hand, and headed back to the main office, better educated, adding more to the sum of his experiences.

Bouncing the sun's rays off a shiny concave surface, such as the bottom of a metal can, can focus a beam of light sufficient to create heat to ignite a fuel source.

■ Note Focal length of a focused beam of light tends to be short, so the first material ignited must be in close proximity.

Focusing the sun's rays can be done with glass such as a magnifying glass, a glass sphere, or pieces of broken glass. Cut glass, broken glass, bottles, or decorative glass pieces must be able to refract light, creating a focused beam that can then be applied to an ignitable fuel. The focal length of a focused beam of light tends to be short, so the first material ignited must be in close proximity.

The focused rays of the sun are capable of starting a fire, but there is no indication that this is a frequent event. In years past, this type of heat source was usually associated with wild land fires. Today more and more products sold as decorative items for homes are capable of this phenomenon, and it is occurring more frequently in residential structures. As with all fires, if glass is suspected of being a potential heat source, a systematic and thorough investigation should be able to prove a reliable hypothesis.

SMOKING MATERIALS

■ Note Cigars generally go out if unattended.

■ Note Pipes can be involved in the ignition sequence when the bowl of the pipe is emptied into or onto potential fuel.

■ Note The common cigarette is the culprit that has been blamed for so many fires across the United States.

■ Note smoking-related fires ranked fifth behind cooking fires, heating equipment fires, electrical fires, and arson. Smoking fires resulted in 35,100 fires, killing 890 civilians and injuring 2,130.

Cigarettes, cigars, and pipes are common smoking materials. Statistics show fewer people smoke today than did 10 years ago. However, a sufficient amount of tobacco is still used to consider smoking materials a source of heat for an uncontrolled fire.

Cigars are less frequently considered a fire hazard. Cigars generally go out if unattended. To keep a cigar burning, the smoker needs to draw on the cigar, pulling air across the coals at the end of the cigar to keep it burning. When placed in an ashtray, cigars eventually self-extinguish.

Pipes are the same—pulling air across the burning leaf in the bowl supplies more oxygen, causing the tobacco to glow more intensely and burn. But left unattended, pipes tend to go out. This is why you will see smokers relight both cigars and pipes from time to time. However, pipes can be involved in the ignition sequence when the bowl of the pipe is emptied into or onto potential fuel. With a breeze, the coals may heat up the same way they do when someone draws on the pipe. The air passing the coals can increase the burning rate, release more heat, and potentially ignite the combustible fuel. But, again, like the cigar this is not a common occurrence.

As far as smoking materials are concerned, the common cigarette is the culprit that has been blamed for so many fires across the United States. There is conjecture that in the past, cigarettes have been blamed for many fires where they were not the appropriate heat source. If no other heat source can be found, it can be easy to blame smoking materials. This is not to say that smoking materials do not cause fires. Statistics compiled between 1999 and 2001 show that smoking-related fires ranked fifth behind cooking fires, heating equipment fires, electrical fires, and arson. Smoking fires resulted in 35,100 fires, killing 890 civilians and injuring 2,130.[7]

The trend is decreasing, however, with the lower number of smokers in the United States. Hopefully, this downward trend and the possibility of new legislation requiring self-extinguishing cigarettes will decrease the number of smoking-related fires and associated deaths and injuries even more.

A cigarette will continue to burn when left unattended or when carelessly discarded. A cigarette by itself may not always have sufficient heat to cause ignition. Studies have shown that a cigarette placed on a sheet will char the sheet but will not ignite it. When a cigarette is placed between the folds of a sheet or between the cushions of a couch, it is insulated. The heat can then build up because it is not dissipating. Then, there may be enough heat to lead to ignition of combustibles.

In an investigation, cigarette butts or cigar stubs may be found on the scene. If there is any reason to believe that they may have come from the perpetrator of a crime, they can be collected and submitted to the laboratory for forensic testing. Even if there is an insufficient sample to show DNA, the evidence may reveal the blood type of the user. Science is improving every day. New research indicates that the cigarette butt tossed into a hydrocarbon fuel for ignition is capable of providing DNA data that could lead to the identification of the smoker.[8]

■ Note
The safety match is the most prevalent and, as its name implies, is the safest type to store, carry, and use because it relies on the match striking the chemical surface on the match book or box to ignite.

MATCHES

There are two basic types of commercial matches in use today. The **safety match** is the most prevalent and, as its name implies, is the safest type to store, carry, and use because it relies on the match striking the chemical surface on the match book or box to ignite. The **strike-anywhere match,** which may be found in camping or survival stores, will ignite by striking on almost any surface. Both matches have a common construction that consists of a head containing a chemical to initiate the fire; tinder to keep the fire burning, which is the wood or cardboard that the chemical is attached to at one end; and the handle, which in this case is the same as the tinder.

safety match
a match designed to ignite only when scraped across a chemically impregnated strip

strike-anywhere match
a match where the head is chemically impregnated to ignite when scraped across any rough surface

The safety match became a necessity as more and more fires were started accidentally by the strike-anywhere match and its predecessors. Earlier matches contained chemicals such as white phosphorus and yellow phosphorus. One type of match could ignite when a small glass ball of sulfuric acid attached to the head of the match was broken with the teeth so that the acid mixed with chemicals coated on the head of the match and a chemical reaction occurred that resulted in ignition. These devices were just as hazardous to the user as they were a fire hazard.

A third type of match on the market today is the wind and waterproof match. This match has a different look, as shown in **Figure 9-4;** to overcome the potential of wind blowing out the initial ignition the match contains a higher chemical content. The match has also been treated to resist moisture.

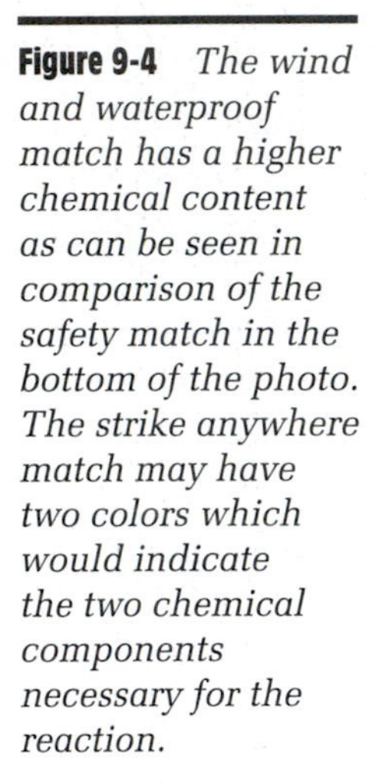

Figure 9-4 *The wind and waterproof match has a higher chemical content as can be seen in comparison of the safety match in the bottom of the photo. The strike anywhere match may have two colors which would indicate the two chemical components necessary for the reaction.*

Matches as Evidence

Because of its size, the residue match is not always easy to locate. This is not to say you should not look. On the contrary, always consider the match as well as any other heat source once the area of origin has been located. With the residue match found on the scene, the first consideration is whether suppression activities placed it there or whether it was there long before the fire actually started.

The very characteristic of the match allows it to continue to burn. Some cardboard matches are impregnated with paraffin to allow burning and to provide moisture proofing. However, as a safety feature, cardboard and wood matches are also chemically treated to retard the afterglow when the match is fanned or blown out. Thus, when the match is blown out and dropped it should not have a glowing ember that can reignite and ignite other combustibles.

■ **Note**

Cardboard and wood matches are also chemically treated to retard the afterglow when the match is fanned or blown out.

One feature of the cardboard match is that it is torn from the base of the matchbook. This tearing might sometimes leave a characteristic pattern, allowing the laboratory to prove that a specific match was torn from a specific matchbook. Today most matchbooks have manufactured perforations—small cuts at the base of the match where it attaches to the matchbook. This limits the amount of area that must be torn so that the match can be more easily removed from the matchbook. This also reduces the surface area that actually tears, in turn limiting the potential, but not preventing, making a comparison.

The laboratory can also compare the cardboard match against the matchbook, looking at width, color, thickness, and composition of the cardboard. The limited comparison by itself may not sound like much, but with any case, it is the culmination of all the facts that leads to knowing the truth. Remember, it is just as important to eliminate suspects and innocent parties as it is to find the culprit.

Matching the Match

The investigator was conducting the investigation of the scene of a vehicle fire. The late-model car had been parked beside the owner's house. The events, according to the owner's statement to the fire officer in charge, was that he had gone out first thing in the morning and tried to start the vehicle. He stated that he could smell gasoline, so he assumed that he had flooded the engine. Knowing a flooded engine would not start, he went back inside to get a cup of coffee, thinking he would try to start it later. He looked outside about 10 minutes later and found the vehicle burning. Fortunately, the fire department's engine company was nearby and was able to extinguish the fire in short order.

Working from the least damaged area to the most damaged area, the fire investigator looked into the trunk and found no problems. There was little fire damage and no indication that the fire started in the trunk area. The next area to be examined was the engine compartment. The exterior of both the trunk and engine compartment showed little damage. There was little damage to the engine as well. The only real damage was to the insulation under the hood, which was damaged from the outside where the flames had escaped from the windshield, reflecting heat downward on the top surface of the hood.

In the engine compartment, there was an obvious odor of gasoline. This was an older vehicle with a V-8 engine and a carburetor. This configuration created a small depressed area under the carburetor that was capable of holding a small quantity of liquid. Upon close examination, the investigator found there about a quarter inch of a liquid, later identified as gasoline. Within this liquid were a dozen cardboard matches with the heads of the matches showing signs of carbon as if they had already been burned. The liquid and the matches were recovered, placed in appropriate containers, and labeled for identification and tracking of the chain of custody.

Looking in the passenger compartment, which was severely damaged, the investigator found the front seat consumed of all its combustible materials, leaving only the metal frame and wire springs. In the center of the seat on the driver's side was a white residue. It was about a half inch in diameter and about a foot long, somewhat intact. This was collected as well and properly sealed and tagged.

In the interview with the vehicle owner, the investigator heard the same story given to the fire officer. Evidence was clear that the fire started in the passenger compartment and most likely by a road flare that was dropped onto the seat because the

(Continued)

(Continued)

white reside was on top of the seat springs. The owner, when questioned, advised that he was not smoking in or around the vehicle, that the vehicle was locked when he went out to start it first thing that morning, and that he had no road flares in the passenger compartment.

The last question the investigator asked was if the owner had any matches in his possession. He reached into his pocket and came out with a matchbook. Thinking that the investigator wanted to light a cigarette, the owner apologized that there were no matches left. The investigator then asked if he could have the matchbook; producing a small evidence bag, he asked the homeowner to drop the matchbook into the bag, which he did. The investigator then asked the homeowner to sign a release for the matchbook, which he did.

The laboratory was able to prove that the matches in the engine compartment came from the matchbook surrendered by the vehicle's owner. The white material in the passenger compartment was consistent with the residue from a road flare. Faced with these facts in an interview, the homeowner confessed to setting the fire. He further stated that the vehicle just came back from the shop with an estimate that it would take over $1,000 to repair. The insurance company would not pay for the repairs but would pay if the vehicle was burned. He stated that he poured gasoline onto the engine, stepped back, lit each match from his matchbook and tossed it toward the engine compartment. Each went out and he ran out of matches. He then confessed that he went to the trunk, took out a road flare, ignited it, and tossed it onto the driver's seat, and then closed the door.

In addition to demonstrating a good scene examination, proper collection of evidence, and good interview skills, this fire resulted in an arrest and successful conviction. The one lesson to be learned is that the question not asked is a missed opportunity. In asking for the matchbook, it was given freely. It is surprising how often a perpetrator will give up evidence willingly when asked so as not to appear guilty.

LIGHTERS

■ Note
Lighter fluid is usually naphtha and sometimes benzene.

Two liquid fuel lighters are on the market today. The older style is usually metal and has a fuel cavity that can continually be refilled with liquid lighter fluid. Lighter fluid is usually naphtha and sometimes benzene. The fuel cavity is filled with a fiber material that absorbs the liquid, which limits leakage. The lighter uses a wicking process to bring the liquid to the top of the lighter where it can vaporize near the ignition device. The cap of the lighter keeps the vapors from totally escaping, allowing enough vapor to be present when the lighter is opened. Should the lighter be unused for any appreciable time, the lighter fluid may vaporize and escape, requiring it to be refilled on a regular basis. The leakage dissipates in small quantities over time, rarely constituting a hazard.

■ Note
The igniter is a sparking device that consists of a piece of flint and a rotating steel (a wheel with a rough surface).

The igniter is a sparking device that consists of a piece of flint and a rotating steel (a wheel with a rough surface). The flint is held against the wheel with a spring. The friction of the steel across the flint creates a series of sparks that fly into the area where the vapors are present and ignition occurs. To extinguish, the user simply closes the lid, smothering the fire.

The predominant lighter in use today is the disposable butane lighter. The body is plastic and serves as a pressure vessel for the fuel (butane). Under pressure, butane becomes a liquid; when a valve is opened on the top of the lighter, the vapors escape and become the fuel for the lighter's operation. Because the gas continues to escape when the lighter is activated, it is essential that there is immediate ignition or the collection of the vapors would grow to an unsafe volume. Although some units on the market still use the flint and steel, science has provided a better and more reliable igniter. This technology is in the form of **piezoelectric ignition.**

piezoelectric ignition
certain crystals that generate voltage when subjected to pressure (or impact)

■ Note
This crystal is of a variety that releases an electrical discharge when it receives an impact or pressure. The electrical discharge is in the form of an arc, which provides sufficient energy to ignite the escaping gas.

The pressure on the activator that opens the fuel valve also strikes a quartz crystal. This crystal is of a variety that releases an electrical discharge when it receives an impact or pressure. The electrical discharge is in the form of an arc, which provides sufficient energy to ignite the escaping gas. One other feature of interest is that many of these lighters enable the operator to adjust the amount of fuel escaping, which in turn can increase or decrease the flame height.

Although most of today's butane lighters are disposable, there are some on the market that are the refillable type. These units can be refilled from a larger cylinder of butane with a simple pressure filling device. Because these units are not disposable, they tend to be sturdier and are usually metal instead of plastic.

Because most, but not all, butane lighters are disposable, they may be left at the scene for a variety of reasons. If not consumed by the fire, there may be latent prints on the surface of the lighter. If the fire consumes the lighter, the plastic pieces will be destroyed, but the metal parts may survive. Searching for these small items may require a more detailed search. If found, they may identify a particular brand that can be traced to locations where they are sold in the area. It is important to remember that a culmination of all the evidence and facts of the case may lead to a successful investigation.

Hobby Micro Torch

micro torch
a miniature torch that uses butane as a fuel. The mechanism can inject air into the chamber, allowing for a high heat output; usually used by hobbyists

A device similar to the operation of a lighter is a hobby **micro torch.** This device is usually sold to hobbyists to braze and solder small particles in hobby type projects. These torches, as shown in **Figure 9-5,** are capable of reaching much higher temperatures, in the range of 2,400°F, which is much higher than the average lighter. Both lighters and mini torch use the same fuel and sometimes have the same type of igniter. The difference that enables the higher temperature in the hobby torch is the head assembly, which allows the introduction of air into the burn chamber to achieve a more efficient flame. Larger models may be

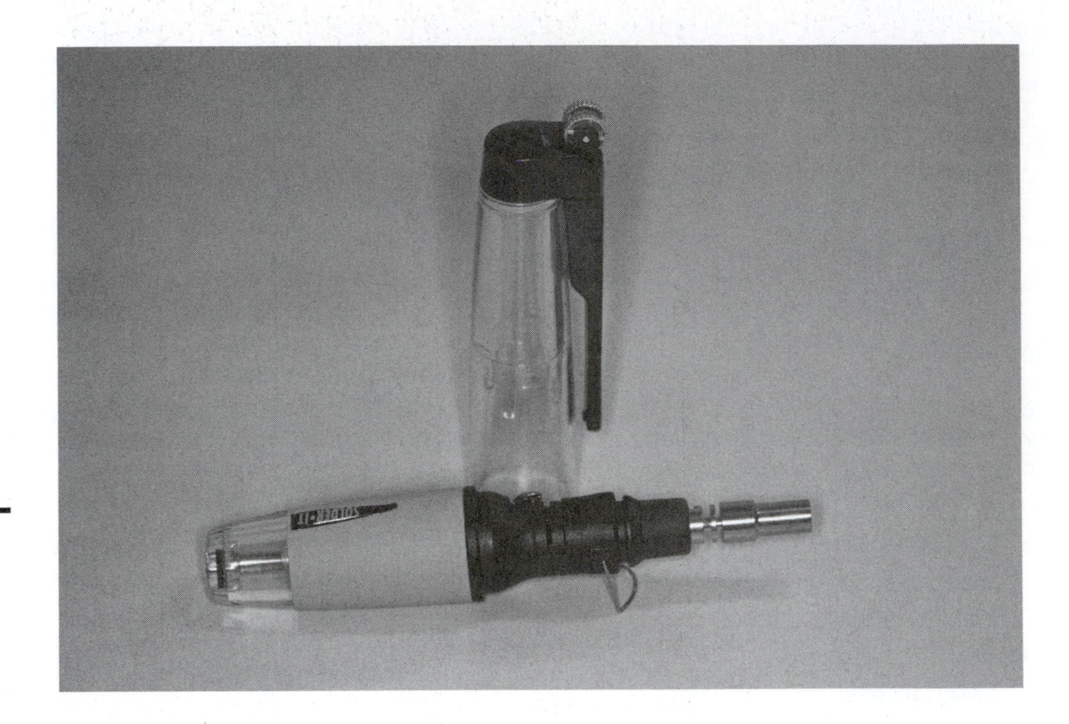

Figure 9-5 *This micro torch, which uses butane, is only 4 inches long and can easily be carried in a pocket.*

■ Note
These torches, as shown in Figure 9-5, are capable of reaching much higher temperatures, in the range of 2,400°F.

as much as 6 inches long, whereas the smaller units may only be 3 inches tall. This size device can easily be concealed.

The units are not very expensive, but it is unlikely it would be left at the scene. Of interest to the investigator would be finding a suspect at the scene and finding such a unit on that person or in his or her vehicle. By itself a micro torch may mean little, but it can be just one more piece of the puzzle that requires further investigation.

Other Hobby Ignition Sources

rocket motors
hobby propulsion kits that provide thrust through the burning of an ignitable mixture

electric matches
electrical devices that can act as an igniter from a remote distance

Every investigator should take a trip to the local hobby store just to see the type of materials available to the curious juvenile, careless hobbyist, or the clever arsonist. In addition to micro torches there are **rocket motors** (see **Figure 9-6**) for small model rockets. They come in a variety of styles for different models, with varying burning durations for the rocket's load or anticipated flying height. Hobby shops even carry **electric matches.** These are remote igniters of low voltage but carry enough energy to ignite common materials. Some even look like a matchbook with wires attached. There are other electric matches used for firework displays and explosives; they all work using the same principle of delivering an electric charge with sufficient heat to ignite an explosive charge.

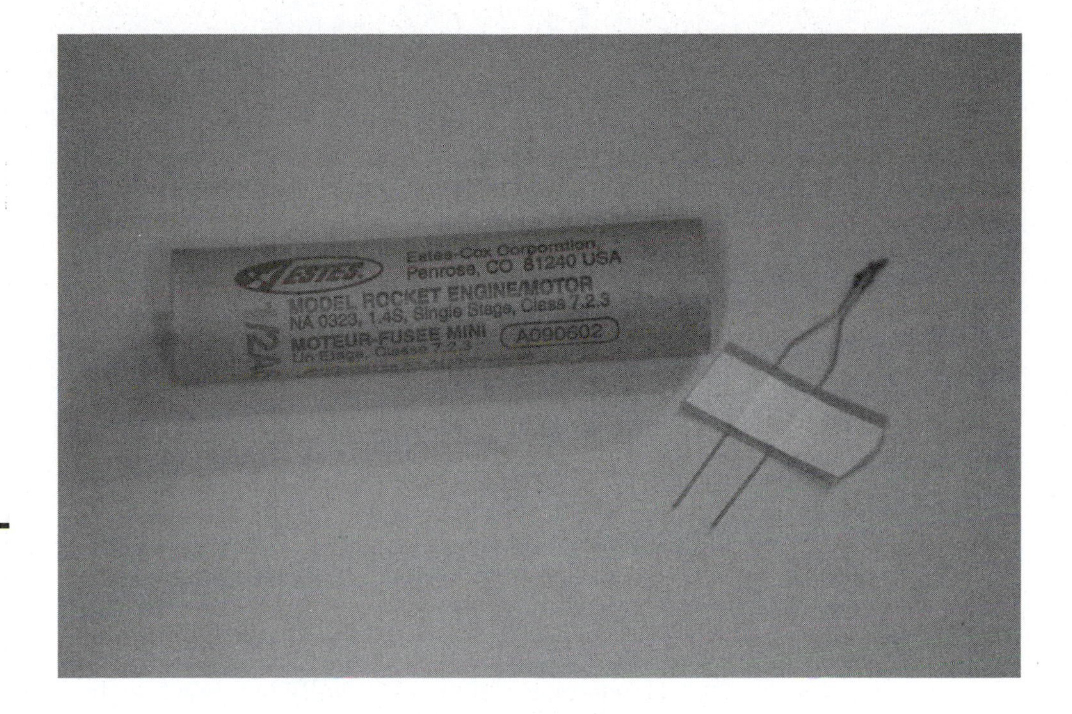

Figure 9-6 *Model rocket motors and their igniters can be found in hobby stores.*

FRICTION HEATING

■ **Note**
Friction heat is created when two items rub together sufficiently to produce heat.

kinetic energy
energy as the result of a body in motion

rotor
in every electrical motor, there are two basic parts, the rotor and the stator; the rotor rotates in an electric motor

Friction heat is created when two items rub together sufficiently to produce heat. Not all friction results in fire or in measurable heat. The slow rubbing of branches may eventually rub off bark, but there is no expectation of combustion. The movement (kinetic energy) of an object with sufficient friction (resistance from that movement) can translate **kinetic energy** into thermal energy. The primary variable is the roughness of the surface. Items forced together with pressure create more friction and in turn more heat. The opposite is true when liquid is involved. The friction of objects wet from water or oil is decreased and less heat is produced.

Rotors

There are many definitions of a **rotor.** They can range from the mathematical rotor, which is a convex figure that can rotate inside a polygon and which touches every side or inside face of the polygon, to the blades of a helicopter. When speaking of rotors in the context of fire cause, a common heat source is the rotor of an electric motor or alternator. The rotor is the rotating portion of the electric motor; the stationary portion is called the **stator.**

stator
in every electrical motor, there are two basic parts, the rotor and the stator; the stator is the part of the motor that stays stationary, housing the rotor

By their very nature these electric motors and alternators get warm when working properly. However, when the bearings that hold the rotor in place fail, a massive amount of friction occurs that can heat up to the point of becoming a competent heat source. The availability of a nearby fuel source dictates whether a fire occurs. With the advent of plastic parts in and around the motor, such as fan blades, the fuel could be part of the assembly. Another reason a motor may produce sufficient heat to become a heat source is oxidation as the result of moisture. Rust on the surface can create more friction, which in turn creates more heat.

■ Note
When the bearings that hold the rotor in place fail, a massive amount of friction occurs that can heat up to the point of becoming a competent heat source.

When a motor is found in or near the area of origin and there are indications it may have been involved in the ignition sequence, every effort must be made to secure the motor without spoiling it, being careful to ensure that there is no spoliation of the evidence. Completely document the evidence and take several photographs to document the condition of the motor, and then collect it as evidence. An engineering laboratory may be able to examine the unit to tell whether the heat was internal or external, adding to the data being collected to make a final hypothesis.

Brakes

Vehicles have rotors associated with the braking system. The source of heat in most braking system failures is usually involving the brakes themselves and not the rotor. When brakes fail, the pads sometimes lock in place. As the vehicle continues to roll, it creates a massive amount of friction, which in turn creates massive amounts of heat. This energy can then ignite grease, plastic, rubber, and other materials. The brakes themselves can start to fly apart, sending hot fragments flying, igniting nearby combustibles such as grass and brush. This can go on for miles before the driver realizes what has happened. In some incidents, the brakes on tractor trailers have overheated so much that the heat was transferred upward into the trailer, igniting the contents of the trailer.

■ Note
The brakes themselves can start to fly apart, sending hot fragments flying, igniting nearby combustibles such as grass and brush.

Muffler Systems

Mufflers of motor vehicles or other metal parts hanging down sufficiently to touch the ground can become a source of sparks. This can include hanging metal bumpers or safety chains used when pulling trailers. As the vehicle increases in speed, the amount of sparks increases as well. Each of these sparks is a glowing metal ember that is more than capable of igniting any fuel found along the road such as grass, leaves, or paper and plastic litter. The road surface need not be asphalt. There are enough rocks on dirt roads to also create sparks, and this source has been the cause of many wild land fires.

Muffler problems are not limited to passenger vehicles, recreational vehicles, or freight trucks. Multitudes of vehicles that are used off road, ranging from bulldozers, logging vehicles, and farm machinery, each has its own characteristics that could provide a heat source.

Logging vehicles and those vehicles designed to be used in wild lands have screens around engine compartments to keep out leaves, pine needles, and other combustibles. All too often these screens get clogged up and instead of crews cleaning them on a regular basis, they are removed. When this happens, leaves and other items can collect in the engine compartment, resting up against the exhaust manifold. The heat from the engine is sufficient to eventually ignite these materials. This probability is increased by the air flow created by the vehicle's fan, which helps any smoldering items to break into open flame.

All too often when dealing with sparks from a motor vehicle the investigator will have to depend on eye witness accounts. Even if sparks were seen coming from a vehicle but the witness did not see the fire actually start, it may be hard to propose this as an hypothesis; it will be conjecture, not fact.

SPARKS FROM CATALYTIC CONVERTERS

catalytic converter
a device installed as part of the exhaust system of an automobile; designed to reduce emissions by burning off pollutants

A **catalytic converter** is a device designed to reduce the toxicity of emissions from the exhaust of an internal combustion engine. Catalytic converters can be found on cars, buses, and trucks. Some construction equipment and forklifts also use catalytic converters. The basic principle is to allow exhaust gases to flow through a ceramic honeycomb (some models use ceramic beads) that is coated with metals that first remove the nitrogen oxides and then separate out the oxygen molecule, creating oxygen and nitrogen. The second stage then burns off any residue hydrocarbons or carbon monoxide, combining these molecules with the oxygen from the first stage. With an internal operating temperature of 1300°F, the ignition is perpetual.

■ **Note**
The average exterior temperature of 600°F is usually not a problem under most circumstances.

Most manufacturers provide adequate shielding around the catalytic converter to dissipate as much heat as possible. The average exterior temperature of 600°F is usually not a problem under most circumstances. The average road provides sufficient clearance of combustibles to prevent any problems. However, taking a vehicle and parking it on the side of the road where there is vegetation that can come in close proximity or contact with the catalytic convert creates a fire hazard.

New Insurance Policy

The local farmer had asked his insurance agent to reevaluate his farm property to make sure he had enough insurance coverage should something go wrong. The agent checked all the structures with the farmer. They needed to check one more remote barn on the property to get the square footage of that building. The farmer rode out to the building with the insurance agent because the agent had air conditioning in his vehicle.

(Continued)

(Continued)

The barn was rarely used because of its condition and remote location. As such, the farmer had plowed all the land around the barn, planting hay for his livestock. He left no path to the barn. It had been exceptionally dry, and the weather was hot. Rather than walk the distance, the hay was only about 18 inches tall, so they drove the car to the barn. Parking just outside the door to the barn the insurance agent left the vehicle running to keep the car cool for their return trip to the farmer's house.

By the time the rural fire department arrived, the vehicle was fully involved. About half the barn was engulfed in flames and at least 4 acres of young hay was burning. An investigator was requested and, upon arrival, examined the scene, working from the least damaged area to the most. He then interviewed both the insurance agent and the farmer separately. They both explained that while taking measurements inside the barn, they both remarked that they smelled smoke. Outside smoke was coming from under the vehicle, and through the smoke they could see the glow of flames. The glow was exactly in the area of the catalytic converter.

With no cell towers in the area, they started to run for help, but because of the long distance back to the farmhouse and the hot weather, it turned into a brisk walk. Before they could get to a phone to call for help, in the meantime a neighbor saw smoke in the distance and called the fire department.

Months later, the new barn was almost complete. The investigator had seen the farmer at lunch and asked about his insurance claim, curious as to the final adjustment on the property. The farmer just smiled and said all was taken care of by the insurance company.

Certain conditions can cause a catalytic converter to be hazardous. Impact injuries to the converter could break up the ceramic interior. As the converter heats up and exhaust passes through, small pieces of the ceramic, heated to incandescence, could be blown through the exhaust system. As these ceramic pieces come out of the exhaust pipe, they are more than hot enough to serve as heat source for leaves, grass, and other vegetation.

An improperly running engine can sometimes create problems with the operation of a catalytic converter. Anytime that the vaporized fuel injected into the cylinders does not ignite, it is then forced into the exhaust system. When this occurs, that fuel enters the catalytic converter where, at 1300°F, it is ignited. Evidence of this action is a rough-running engine and a glowing red (orange) catalytic converter. Eventually, there may even be a burning smell within the interior of the vehicle. This is caused by the transfer of the heat from the converter to the carpet padding just above the converter. This can eventually lead to ignition of the carpet material along with any combustibles sitting on the carpet.

Signs of this occurring are usually first reported by witnesses such as the vehicle's driver and occupants, who can verify that the motor was running rough and that there was a smell of smoke. The driver may even indicate that the floor area was warm or hot. A witness exterior to the vehicle may have noticed the glow from under the vehicle in the area of the catalytic converter. Burn patterns will show low burning above and adjacent to the converter. The converter itself may show degradation and damage from the presence of the vehicle's fuel.

CHIMNEYS AND FIREPLACES

The chimney has been a staple of the American home since the time of the early settlers. The common chimney today is constructed of brick with a liner or steel, and hopefully it is at least double walled. The steel chimney may be placed within a chimney chase that is usually wood framed with siding that matches the structure.

Properly used and maintained, today's chimneys provide few problems and can be an asset to any home or business. It is only when construction is improper or chimneys are not used properly that they can be involved in an unwanted fire situation.

Metal Chimneys for Fireplaces

Metal chimneys are designed for specific applications. They can range from hot water heaters to crematories. Historically, those dedicated to appliances have provided few problems. Metal chimneys for fireplaces and woodstoves have been involved in residential fires, but usually because the user failed to follow proper precautions and burned the wrong fuel or installed the metal pipes improperly.

■ Note
When the examination of the fire scene points back to the metal chimney in the area of origin, it is beneficial to obtain the instructions for assembly from the manufacturer.

When the examination of the fire scene points back to the metal chimney in the area of origin, it is beneficial to obtain the instructions for assembly from the manufacturer. Under most circumstances, the appliance will come with the proper chimney. The instructions will indicate the amount of heat that can be handled by the device and the chimney. Many of the current metal fireboxes (fireplaces) with metal chimneys can easily handle a wood fire from logs and kindling. They may not be able to handle the heat from a pressed log consisting of sawdust and paraffin.

Some firebox assemblies are advertised and installed as zero-clearance units. This implies that wood assemblies can be installed in close proximity or against the exterior of identified metal portions of the firebox. Under most circumstances, this design feature holds true as long as the unit is used as per the instructions provided.

On rare occasions, investigators might pull onto the scene and be in awe of the pattern greeting them as they step from their vehicle. The burn pattern

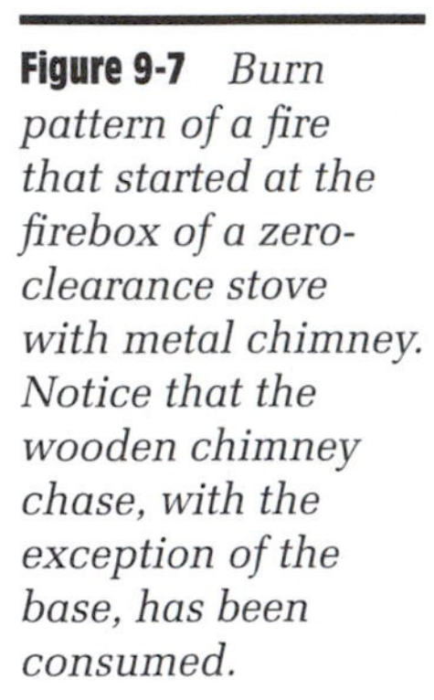

Figure 9-7 *Burn pattern of a fire that started at the firebox of a zero-clearance stove with metal chimney. Notice that the wooden chimney chase, with the exception of the base, has been consumed.*

shown in **Figure 9-7** is too obvious to be true. In this case, it was true. Chapter 11, "Patterns," discusses finding a V pattern. Follow the sides of the V downward and it points to the area of origin in many circumstances. In this case, the area of origin is the firebox of the metal fireplace. (See **Figure 9-8.**)

Notice that the metal fireplace as seen from the interior (see **Figure 9-9**) is heavily damaged. This was the result of the first fire in the unit. What is difficult to see is that the metal is heavily warped and pitted—not what would be expected from usual use of the unit. In **Figure 9-10,** you can see the base of the fireplace unit. The clearances and installation all meet manufacturer requirements.

chimney liner
the covering inside the chimney that enables the flow of hot gases and smoke from the chimney and keeps them from seeping through the mortar into the structure

Brick Chimneys for Fireplaces or Woodstoves

The brick chimney appears to be mortar and brick from the outside. When looking at the top of the chimney, under most circumstances, you will see a portion of the **chimney liner.** Codes require this liner to be installed through the entire length of the chimney from the firebox to the top of the chimney. The liner provides a conduit for the hot fire gases and smoke to continue their upward rise to the top of the chimney to be released into the atmosphere. The liner is essential to the proper operation of a chimney.

■ Note
The liner provides a conduit for the hot fire gases and smoke to continue their upward rise to the top of the chimney to be released into the atmosphere.

A chimney without a liner is a hazard. The mortar that holds the bricks in place is not designed or capable of containing the hot gases or smoke from the fire. In time, both permeate the mortar, leading to a potential uncontrolled fire.

Figure 9-8 *Close-up of the rear outside surface of the firebox. This is the area of origin.*

Figure 9-9 *Interior of a new firebox after being used just once to burn paraffin logs, which is against the recommendations of the manufacturer. Instructions against using pressed logs were found on the coffee table in the same room.*

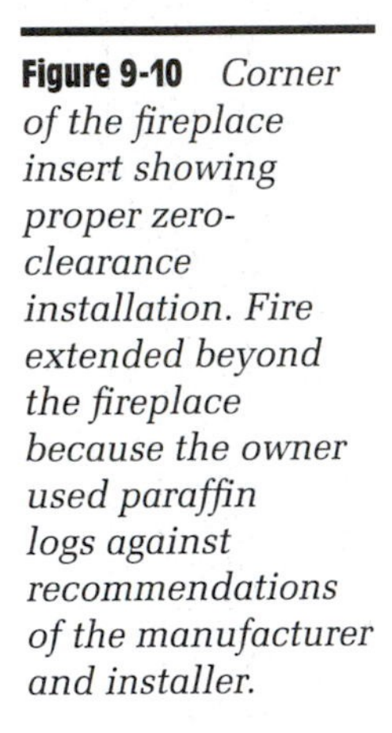

Figure 9-10 *Corner of the fireplace insert showing proper zero-clearance installation. Fire extended beyond the fireplace because the owner used paraffin logs against recommendations of the manufacturer and installer.*

■ Note
A chimney without a liner is a hazard.

■ Note
The first question to ask is whether there was a fire in the fireplace.

A common residential fireplace chimney liner is clay or terra-cotta. However, many other materials are used such as steel, ceramic, concrete, and other combinations of specialized materials. All of these liners, when properly installed, do as they are designed to do: safely remove hot gases and smoke.

When the examination of the fire scene indicates that the fire started in the close vicinity of the chimney, examine the chimney closely. The first question to ask is whether there was a fire in the fireplace. Then, if so, the occupants may be able to indicate the maintenance history of the fireplace, such as when the last time the chimney was cleaned.

Conduct an exterior detailed examination to look for cracks and missing mortar; take care not to damage the chimney in this examination—do not pry or prod. **Figure 9-11** shows a chimney in relation to the adjacent burn patterns. The presence of a flue liner coming out of the top of the chimney indicates that a liner was in place, implying the chimney would be safer than one without a liner. The use of a high-resolution digital camera can aid in the investigation. From the fireplace damper or through a thimble, insert the camera and use auto focus to take a series of photos, changing the angle slightly for each photo. **Figure 9-12** shows the result of such a process; no flue liner was installed below the top portion of the chimney.

The metal strap was clearly installed by the brick mason at the time of construction. This cost-saving measure resulted in an extensive fire in this home. If it is safe, use the same photographing process from the roof, looking down

Figure 9-11 *The liner extending above the brick indicates the chimney is lined.*

Figure 9-12 *Photo of the interior showing that the liner was in fact just one piece of tile held in place with metal hangers. No other liner pieces were installed in the chimney.*

Safety
It is critical to use extreme caution when approaching any chimney. If the structure surrounding the chimney has sustained any damage, it is wise to presume that the chimney is damaged as well.

through the chimney. However, use extreme caution in this process. If looking down the chimney is important, see whether a truck company can provide an opportunity to take these photos safely from the platform of an aerial apparatus. It is critical to use extreme caution when approaching any chimney. If the structure surrounding the chimney has sustained any damage, it is wise to presume that the chimney is damaged as well. Chimneys experience degradation from a hostile fire and lose their structural integrity. Leaning a ladder against the chimney may be all that is needed to cause it to collapse, risking your life and the lives of others. If there is any question, the use of an aerial ladder may be the only platform safe enough from which to continue your investigation.

Local resources may be available that can assist in your examination of the chimney. Many chimney sweeps have camera systems used to inspect chimney conditions. Some of these systems even have recorders for future reference. This could prove to be a beneficial tool in your examination.

Note
The first thing to look for is proper construction, check to make sure there is a flue liner in place.

The first thing to look for is proper construction, check to make sure there is a flue liner in place. Check the condition of the liner; see whether it is broken up or pieces are missing. Is there creosote in the chimney? How much creosote is adhered to the chimney? How much has it restricted the size of the chimney opening?

When conducting interviews, find out whether anyone saw fire coming from the chimney top. This may indicate a chimney fire, which could have been the cause of the fire in the structure. Sometimes occupants might have heard a roaring sound, which could indicate a chimney fire. Such events are caused by the ignition of a buildup of creosote in the chimney and can also be the result of other items in the chimney burning, such as bird's nests, large quantities of cobwebs, sticks, and leaves.

Also, check the construction of the chimney in both the living spaces and concealed spaces such as attics. On rare occasions, the brick mason may choose to brick around wood beams, leaving them inside the brick assembly, exposing them to heat, which breaks down the wood and eventually leads to ignition of the beam. A proper installation has the beams headed off with a wood faceplate; essentially a box is made around the opening for the fireplace to go through the roof. To state the obvious, if the area has been consumed and the chimney is still intact, a telltale opening in the chimney where it was built around the wood beam will exist.

Chimney chase
decorative hollow covering around a cement block or metal chimney; designed to be more aesthetic; usually framed in wood and covered with the same material as the siding of the structure

Another type of masonry chimney is one made from concrete blocks that is then boxed in and covered with a wood frame **chimney chase.** The chase is covered with the same siding material as used for the rest of the house. The top cap is usually metal. On occasion the top surface of the chimney is concrete, and when this is so, it requires a wood temporary framing. Fires have resulted from the construction crew not going back in and removing this framing (see **Figure 9-13**), which rests against the flue liner. As much as a year or two later, depending on the frequency of use, the wood eventually breaks down and can possibly ignite.

Figure 9-13 *A fire starting in the chimney chase in a subdivision consumed all material in that area. The examination of a neighboring house built by the same contractor and crews revealed that they failed to remove the boards for a temporary casting to create the concrete top.*

WIND-BLOWN SPARKS

Wind can cause a burning object to become more incandescent, raising its temperature, and if the burning particles are small enough, the wind can pick them up and deposit them elsewhere. The movement may place the burning ember (spark) in a place where it comes into contact with combustible material. Under the right conditions, the mass of the fuel and its ignition temperature might start an unwanted fire.

■ Note
Another phenomenon is the Bernoulli effect where wind blowing across the top of the opening of the chimney creates a strong updraft.

Chimneys and Wind

The wind can have an amazing affect on both chimneys and sparks. Occasionally, downward-blowing winds can create a down draft in a chimney, possibly stirring the sparks in the fireplace and even blowing them into the room. Sparks landing on a hardwood floor most likely do not have the energy to ignite the wood into flaming combustion but will char it. Most modern carpets char as well, but with enough energy they may ignite eventually. Combustible items such as newspapers, dried flower arrangements, and kindling can be ignited under the right circumstances by a spark from a fireplace.

Bernoulli effect
as it applies to chimneys, the wind blowing across the opening of a chimney decreases the pressure in the flue, increasing the updraft

Another phenomenon is the **Bernoulli effect** where wind blowing across the top of the opening of the chimney creates a strong updraft. The negative pressure created by this effect pulls the air out of the chimney faster than the usually upward draft. The void of the air leaving the chimney is replaced by air from the room. As this air passes over the burning fire and ashes at a faster-than-normal rate, it fans the fire, increasing the burning rate and possibly picking up sparks and embers that would usually not rise up the chimney. Because the sparks are larger, they have more mass and they burn longer. Sometimes they burn long enough to travel up through the chimney and out into the atmosphere. Sparks, if still incandescent, can land on the roof and could have sufficient energy to ignite leaves or vegetation debris on the roof. If the roof is constructed of combustible materials, such as cedar shingles, sparks could even ignite them over time. The fact that the fireplace was in use, the location of the area of origin, the history of high winds may support this theory of the heat source.

Placing Christmas or birthday wrapping paper in the fireplace may seem like a good way to get additional heat and dispose of the paper at the same time. The balled-up paper is easily ignited. It already has little weight and as it is consumed the weight becomes even less. There is the potential that the draft created by the fire, along with the additional draft created when the paper ignites and rapidly burns, is sufficient to pull some of the charred paper into the chimney with the draft. It then might be capable of igniting the creosote on the chimney walls or it may leave the chimney and land on the roof or in the yard where it can ignite dead vegetation, creating an unwanted and hostile fire.

Evidence to indicate this type of ignition is the residue or wrapping paper at or near the area of origin. Presence of other pieces of carbonized wrapping paper at the scene may support the theory. As with all situations, a thorough interview and witness statements may help make the determination as well.

Burning Leaves

When a leaf-burning event gets out of control, it is usually because of human error. Mistakes such as starting a fire with gasoline can result in a fire instantly being too large to handle. Pouring the gasoline and delaying the ignition can result in the vapors migrating away from the leaf pile, possibly igniting other combustibles as well as injuring the person igniting the vapors.

Not staying with a pile of burning leaves can allow the fire to get out of control, endangering structures and wild lands. It only takes a small wind to stir the embers, pick up burning particles, and transport them to other vegetation. The burning leaves themselves can blow away in a heavier wind. When the burning pile has been reduced to ash, it is still burning inside the pile. What may look like an extinguished fire can flare up with a fresh breeze, spreading embers into unburned areas.

■ Note
What may look like an extinguished fire can flare up with a fresh breeze, spreading embers into unburned areas.

When conducting the investigation, the pile of debris and ash may serve as an indication of a potential heat source. The wind at the time of the fire may indicate a direction. However, the wind direction at the time of the investigation may be different from at the time of the fire. An eyewitness account of the smoke travel direction may be a better indicator.

Credible Proof of Wind Direction

Two fire investigators and an engineer, all hired by different insurance companies, were debating the burn patterns of an accidental fire. There were some conflicting patterns. The conflicts all came down to the wind direction at the time of the event. One of the investigators had interviewed the night guard at a neighboring car dealership, and the watchman said the wind was to the northwest. Yet the engineer had called the local television station whose weather department said the wind was consistently to the southeast. The debate would have continued except that the investigator revealed that the guard was an off-duty fire investigator who discovered the fire because the smoke filled the car lot, which was situated just northwest of the fire location. The group went with the smoke to the northwest.

Trash Burning

Piling up trash and burning it is hazardous because the toxic gases released from burning plastics and other chemically impregnated items. As such, trash burning is usually outlawed, but regardless it still takes place in rural communities. The different materials burned can lead to the fire flaring up from time to time as well as the fire releasing sparks that can fall in unburned vegetation, resulting in the extension of the fire.

When trash is placed in an open burn barrel, the sparks can rise in the hot air currents. They travel upward until the hot air currents stop providing enough updraft; then, the spark falls back down. Should sparks have sufficient energy to remain burning when they fall, they could ignite nearby vegetation, spreading the fire away from the barrel. The placement of a small-weave screen over the barrel usually prevents the sparks from leaving. But not everyone takes such a precaution.

Note
Burn piles and burn barrels, like leaf piles, show a path of char leading right back to the source.

In fire investigations, both burn piles and burn barrels, like leaf piles, show a path of char leading right back to the source. Again, it is good to document whether there was a fire in the barrel in the first place.

CHEMICAL REACTIONS

Several chemicals when combined with others react, releasing heat at sufficient levels to ignite nearby combustibles. Chemicals such as organic peroxides and metals may ignite when in contact with moisture or air—some with explosive force. Although the average investigator may not run into too many igniting metals, he or she may encounter other chemical processes that produce heat. Something as simple as fiberglass repair operations require a catalyst to be mixed with the resins. This chemical reaction under the right conditions can also rise to surprising temperatures.

Essentially, many chemicals can react and produce heat that can ignite nearby combustibles. These chemicals are in use in almost every community. It is up to you to be diligent and resourceful in identifying these hazards as heat sources that could have started a fire.

Spontaneous Combustion

spontaneous combustion
a combustion process in which a chemical reaction takes place internally, creating an exothermic reaction that builds until it reaches the ignition temperature of the material involved

Several materials are susceptible to spontaneous heating, which in turn may lead to **spontaneous combustion.** These materials produce heat under various circumstances. If the heat is not allowed to dissipate, it can continue to increase in temperature until the heat is sufficient to spontaneously ignite or ignite nearby combustibles. Throughout the NFPA *Fire Protection Handbook* there are examples of different materials and their capability to spontaneously combust.

For spontaneous reaction to occur there must be an exothermic reaction taking place at normal temperature and in the presence of oxygen. The reaction must accelerate rapidly as the temperature increases. The arrangement of the material must be such that it does not allow, or drastically limits, the dissipation of the building heat. Last, the material undergoing this process must be capable of smoldering.[9]

■ Note
For spontaneous reaction to occur there must be an exothermic reaction taking place at normal temperature and in the presence of oxygen.

Drying Oils

drying oils
organic oil used in paints and varnishes that when dried leave a hard finish

Drying oils such as linseed oil, tung nut, or fish oil are all susceptible to spontaneous heating. The reaction is dependent on the percentage of the oil in contact with oxygen along with the ability to dissipate heat. The crumpled rag soaked with linseed oil left on the work bench will build up heat within the rag, possibly heating to ignition. That same rag hung on a clothes line can dissipate the heat as the reaction occurs, not allowing heat to build up, until it has safely dried.

Once you have located and examined the area of origin, you may find traces of such a rag contaminated with a drying oil. Even after being fully involved in fire, the rag residue may still hold its shape enough to be recognized. The lack of any other potential heat source may also be an indicator—but not proof—that the rag was involved in the fire ignition. Conducting an interview can identify whether and when the occupants were using the rag and what substance was on the rag. The laboratory analysis may be able to confirm the presence of an oil. Finding the container and the project that the oil was being used for can help to support or refute an hypothesis.

Hay and Straw

The self-heating of hay or straw is one of the more well known spontaneous ignition events. Although this does not happen frequently, it is important for the investigator to know and understand how this occurs.

Hay is usually cut and left on the ground to cure, or to dry out. When the hay is dried, it is raked into rows, and then picked up and pressed into bales, which are tied up and ready for storage. (There are other methods of collection and storage, but we concentrate on this one process for now.) The theory of self-ignition is the same for all types of storage materials should the conditions be similar. When placed in the barn, the bales are usually stacked to maximize the use of space. The usual pattern of stacking alternating layers of rectangular bales the long way and then the short way makes the stack stable and creates a tightly packed area, preventing any movement of air within the core of the pile. (Because some farmers are aware of the danger of self-ignition, they might leave open areas in the stacks to allow dissipation of heat. Pipes can also be driven into the stacks so that a thermometer can be inserted to measure the heat being given off within the stack.)

The conditions must be just right for self-heating to begin. Moisture must be present. Different studies have produced varied moisture percentages in different types of hay and straw. The process of evaporation creates heat, but this by itself usually is not enough to cause ignition. However, the presence of microorganisms in the moist bales can start the fermentation process, which creates even more heat release. With no means for that heat to dissipate, it continues to build and eventually reaches the ignition temperature of the baled material. If the reaction within the mass slows down, the farmer may find charred hay in the pile. If the heating process reaches its peak and extends to the outer layer of the hay mass, enough oxygen is present to enable the process to proceed to flame production.

■ Note
The presence of microorganisms in the moist bales can start the fermentation process, which creates even more heat release.

Because of the nature of this event and the ready supply of fuel, all too often the fire results in major destruction, limiting the opportunity for the investigator to find any evidence. However, in a structure that was secure and in the absence of other heat sources, the investigation may turn up evidence to prove what happened. Should there be sufficient remaining debris, it may be obvious whether the fire started inside the stack or outside. Spontaneous ignition starts on the inside, moving upward. A fire that started on the outside is the result of other heat sources.

Time is a factor as well. Spontaneous self-heating is not something that happens overnight. It takes several weeks for this process to begin, and even longer for the temperature to rise sufficiently to cause ignition.[10] As the temperature rises, the reaction increases, creating even greater heat, increasing the rate of heat output, in turn speeding up the process.

■ Note
Sometimes in the process of spontaneous combustion in hay, a clinker is created.

One piece of evidence may help you. Sometimes in the process of spontaneous combustion in hay, a **clinker** is created. This glassy, irregularly formed mass might be found in the center of the area of origin. It can be gray to green in color and various sizes from small particles to chunks that are 2 and 3 inches long. The size is not an indicator of anything other than the basic components being in the hay at the time of the fire. But it does indicate that in the fire the first material ignited was hay and the heat source was spontaneous combustion.

clinker
solid glass-like object found as residue in a large hay fire

Plywood and Pressboard

Mixing wood particles with a resin compound, placing the mixture in a form, and pressing it into shape using high heat and pressure is the usual process for making plywood and particleboard. These products, if allowed sufficient ventilation and placed in smaller stacks with the ability to cross ventilate, are usually no problem at all. However, if sheets are stacked in such a way as to prevent the dissipation of heat, self-heating will take place that could result in spontaneous ignition. Like hay and straw, the location of the area of origin at the center of the plywood stack is the first indicator that this was a case of spontaneous ignition.

■ Note **Like hay and straw, the location of the area of origin at the center of the plywood stack is the first indicator that this was a case of spontaneous ignition.**

APPLIANCES

Appliances come in various sizes, small and large. The larger appliances can be categorized as being electric or gas. In most homes, the most common small appliances run off electricity. Some require simple motors or servos for movement. More frequently, the fire service encounters the small appliance that uses heat for its intended purpose, including toasters, toaster ovens, electric skillets, and coffee makers, to name a few.

What investigators must first determine is whether the appliance was involved in the ignition sequence. If so, what was the reason—was it a malfunction of the unit itself, was it misuse or abuse by the user, could the appliance have been improperly installed, or was there a design or manufacturing defect? Then, you must ask whether setting the fire was an intentional act intended to look accidental.

Small Appliances

■ Note **Misuse and abuse of small appliances cause failures that result in fire.**

Misuse and abuse of small appliances cause failures that result in fire. The bagel that was too thick for the toaster may jam the appliance in the cooking position, continuing to cook, eventually igniting the bagel. If the toaster is under the cabinets, flames could impinge on the overhead wood, spreading the fire.

Toaster ovens have similar problems when food is left in them too long or when paper and plastic materials are left on top of the units, which creates a potential fire hazard. Cooking grease–laden items also pose problems. Foods such as bacon that is cooked in a small flat pan too close to the burner is a potential source of fire. As the grease heats up and spatters, it comes in contact with the heating elements with the potential of ignition.

Coffee makers had a rough start with failure after failure in both design and manufacturing. Today there still may be recalls from time to time, but the safety problems are nowhere near as significant as previously. Many units have automatic shutoffs that limit the potential fire hazard.

■ **Note**
Portable heaters have been a serious problem, especially in the workplace.

Portable heaters have been a serious problem, especially in the workplace. Workers wanting to keep warm often place the heater under the desk, which may also be the location of the plastic trashcan or a cardboard box of papers. In the investigation, the burn pattern may help narrow this down to portable heaters as a potential fire cause.

When you find a small appliance at the area of origin, it is imperative that you do not alter or experiment with the device at the scene. The movement of a knob could constitute spoliation because it would be impossible afterward to determine whether the device was on or off at the time of the fire. If there is any indication that the device was involved in an incendiary act, the device must be protected and saved for a full-time investigator to collect to send it to a laboratory for testing. If there are no indications of any incendiary nature, because the appliance actually still belongs to the homeowner it cannot be seized. You might obtain the owner's permission to take the appliance. However, a vested interest in the device may lay with the insurance industry, not the homeowner (see more about spoliation in Chapter 6). Most insurance companies will share their findings with the local investigator upon request, so this may be the more logical path to follow to examine the appliance.

■ **Note**
If there are no indications of any incendiary nature, because the appliance actually still belongs to the homeowner it cannot be seized.

Small appliances can also be used to set fire intentionally. Timers can turn on appliances when the perpetrator is well away from the property with a solid alibi. Heaters and combustibles can be left in close proximity with no denial from the occupant that it happened. Was it a foolish action or an intentional act that created this situation? These are answers that can be uncovered only with a complete analytical investigation.

Larger Appliances

As a general rule, a properly operating refrigerator or washing machine is infrequently involved in the ignition sequence of a fire. There are a few gas-operated refrigerators on the market, but there have been few problems with them. Appliances involved with heating do have a history of fire-related incidents. Like smaller appliances, the majority of fire-related problems with large appliances involve misuse or abuse.

The primary energy source coming into the appliance is the investigator's first consideration. Gas lines on stoves usually include a loop or enough line to allow the occasional movement of the unit, such as for pulling out the stove for cleaning. However, constant movement of the appliance could create a leak resulting from loosened fittings or can eventually create a small stress hole in the pipe. Electric appliances are not immune to damage from this type of movement. Most larger electrical appliances have an electrical pig tail, which is a 2- to 4-foot electrical connection made with flexible (stranded) wire that can withstand movement. Constant movement can create problems, but there have also been instances in which the appliance has abraded or nicked the insulation on the wire, creating a hazard. There have also been instances where the appliance has ended up sitting on the cord, but this is usually obvious because the appliance no longer sits level.

■ Note
When a larger appliance is in the area of origin, all aspects of the appliance use must be explored.

When a larger appliance is in the area of origin, all aspects of the appliance use must be explored. A careful removal of debris may indicate what was happening at the time of the fire, whether there is cloth debris near or on the burners of the stove. Were there clothes in the dryer? Document in writing and in photographs the position of all knobs on the appliances along with proof that the appliance was or was not hooked up to power or to a fuel line.

A fire resulting from food left on the stove or in the oven too long is something almost every firefighter has encountered. The electric burner element can leave an impression on a steel pan if the pan is left on the burner too long. Although you might tend to think that is the source of the fire—and it might be—you should also consider that the damage on the pan may have been from a prior incident and has nothing to do with this fire.

Both the fittings for the gas appliance and the electrical cord insulation may be consumed in the fire. This may limit the investigation, but there are other things to examine. Signs of electrical shorting on the wire may be evident. However, exercise caution in assuming which occurred first; it may initially be unknown if the short caused the fire or the fire caused the electrical short. A flattened out area of the electrical feed may indicate damage. A kink or hole in the gas line may be an indication as well. During the interview phase, ask questions about appliance movement and whether any problems with the appliance were experienced prior to the fire.

Gas problems may exist in other areas in addition to fittings and piping. Gas coming into an appliance must be regulated down to a usable pressure. The gas control houses the regulator and sends the fuel when needed to heat the water or dry the clothes. Should the regulator fail, it could allow too much gas to flow, creating a larger than wanted flame that can result in an unwanted fire. However, most appliances have thermal-limiting switches that detect an unwanted high-heat situation and cut off the regulator, shutting off the gas; in electrical appliances, the limiting switch shuts off the electricity.

■ Note
Most appliances have thermal-limiting switches that detect an unwanted high-heat situation and cut off the regulator, shutting off the gas; in electrical appliances, the limiting switch shuts off the electricity.

These appliances are insulated to prevent the unwanted escape of heat. This is an energy-saving effort as well as a safety feature. In older appliances, the insulation may settle, exposing upper areas and allowing them to radiate more heat than desired. Sometimes the insulation is taken out, such as in the access areas for hot water heaters. The insulation behind the panel allowing access to the heating elements is intended to be removed for replacing the element, and the insulation is replaced when the job is done. Not putting this insulation back in place generally will not cause a fire problem except when the water heater is improperly used and combustible materials are crowded against uninsulated areas to the point that the residual heat could not dissipate. **Figure 9-14** shows just such a fire where the tenant used the space around the hot water heater for storing dirty clothing until laundry day.

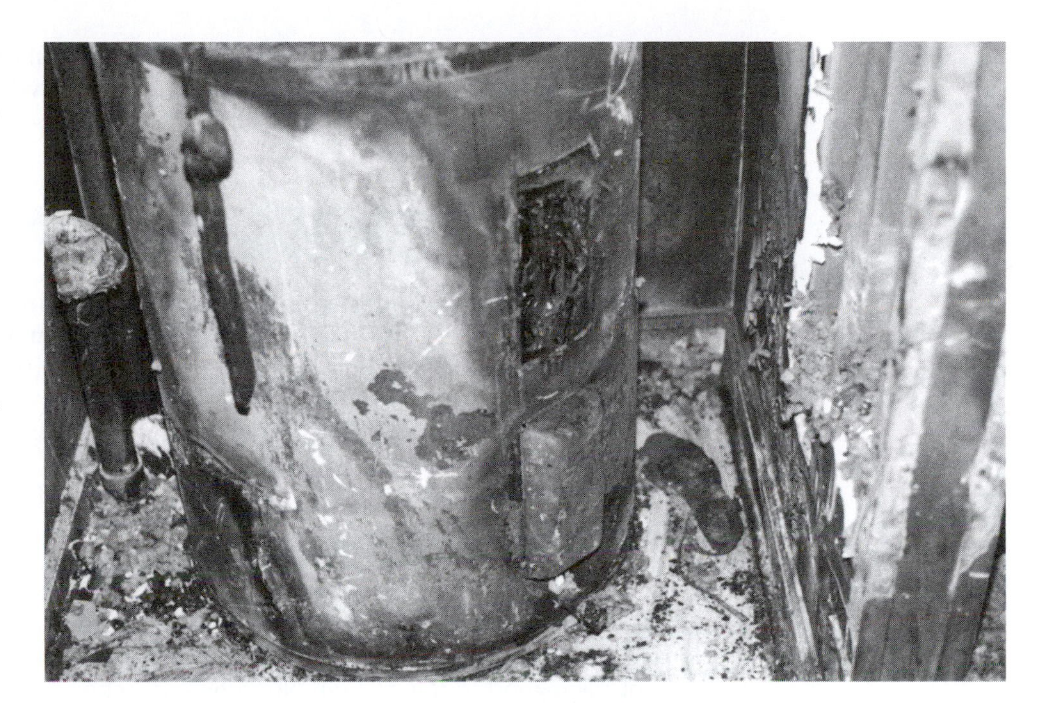

Figure 9-14 *Small heater in the closet area became a hazard when the occupant stuffed dirty clothes in and around the water heater, forcing combustible materials up against the unit. The area of origin was the access area for the heating element where insulation had been removed; this allowed heat to transfer to the clothing.*

Gas appliances also rely on proper ventilation with air entering the unit to allow for proper combustion and for the removal of the gas residue following combustion. Blocking the air intake of a gas furnace prevents proper combustion and may cause flame to roll out or allow radiant heat to heat up nearby combustibles. Materials such as sticks and leaves that fall into the gas furnace or hot water heater flue can block the escape of gases, endangering the occupants and also forcing heat back into the appliance area, which creates a potential problem.

LIGHT BULBS

■ Note

A 60-watt light bulb can produce temperatures as high as 255°F, and a 100-watt bulb can produce 300°F.

Typically, the proper wattage light bulb in a light fixture does not cause a problem. When bulbs of higher wattage than recommended are used, especially in enclosed fixtures, heat can build up to the point of potentially igniting nearby combustibles. A 60-watt light bulb can produce temperatures as high as 255°F, and a 100-watt bulb can produce 300°F. If these bulbs are insulated and ventilation is restricted, they can reach higher temperatures.

Also on the market are high-intensity bulbs that are capable of reaching even higher temperatures. Some of the smaller units can easily reach 500°F.

Quartz bulbs can reach in excess of 1,000°F. Close examination of the area will reveal the residue from the bulb as well as the fixture, which will aid in the investigation.

ANIMALS AND INSECTS

Animals and insects are not frequently involved in an ignition sequence. But they make for such great stories that it is best to mention them here, if only to clarify a few issues. Insects come into play infrequently, but it can happen.

The light fixture that is not used for some time and then turned on can sometimes be involved in a fire situation. Especially if it turns out to be the base of a hornet nest. Hornet nests are the consistency of paper, and the insulated bulb, in time, could ignite the material. There are reports of slugs climbing into outside timing devices and shorting out the circuitry, which results in a fire.

Animals, too, can be involved in an ignition sequence, usually as the result of eating through insulation on electrical wires. Some in the scientific field have explained that there is no reason for this to happen because the wire insulation would not taste good to the animal. Nonetheless, when you find a rodent at the exact area of origin, its body charred and stretched out, with its teeth still in contact with the wires, you might never know its motive, but the animal did chew through the wires.

Not discussed in this chapter are chemicals and incendiary devices. Discussions on such items should be limited to organized training events where careful attention is paid to who attends and how the information is disseminated. That said, many people, including teens, have an abundance of information on how to make chemical incendiary devices from their Internet research. Sometimes even Hollywood gives this potentially dangerous information away, prompting the impressionable to experiment with chemicals and improvised devices.

SUMMARY

We are surrounded by heat sources every day. Even on the warmest day of the year, people may use electric blow dryers or may want a morning cup of hot coffee or tea. Smoking materials still abound and add to the number of accidental fires each year. This chapter covers only a few of the available heat sources fire officers and investigators may encounter. It would be impossible to identify every heat source in all possible situations.

As products change, so do the dynamics of heat sources being involved in unintentional fires. The strike-anywhere match would still be at the top of the list as a leading cause of fire, but new science and technology created the safety match, which has most likely prevented a countless number of fires. It is up to the fire officer/fire investigator to stay current on technology changes like this. With the availability of the Internet, this should not be hard to do.

The most important aspect of identifying a heat source is that it be done scientifically and accurately. This is accomplished through a systematic and thorough fire scene investigation. In addition to keeping an open mind, it is essential that you are able to research and uncover information about all aspects of the potential heat source. In this manner, future fires may be prevented by better understanding and controlling potential sources of heat.

KEY TERMS

Bernoulli effect As it applies to chimneys, the wind blowing across the opening of a chimney decreases the pressure in the flue, increasing the updraft.

Catalytic converter A device installed as part of the exhaust system of an automobile; designed to reduce emissions by burning off pollutants.

Chimney chase Decorative hollow covering around a cement block or metal chimney; designed to be more aesthetic; usually framed in wood and covered with the same material as the siding of the structure.

Chimney liner The covering inside the chimney that enables the flow of hot gases and smoke from the chimney and keeps them from seeping through the mortar into the structure.

Clinker Solid glass-like object found as residue in a large hay fire.

Drying oils Organic oil used in paints and varnishes that when dried leave a hard finish.

Electric matches Electrical devices that can act as an igniter from a remote distance.

Kinetic energy Energy as the result of a body in motion.

Methane gas Colorless, odorless, flammable gas created naturally by decomposing vegetation. It can also be created artificially.

Micro torch A miniature torch that uses butane as a fuel. The mechanism can inject air into the chamber, allowing for a high heat output. Usually used by hobbyists.

Piezoelectric ignition Certain crystals that generate voltage when subjected to pressure (or impact).

Rocket motors Hobby propulsion kits that provide thrust through the burning of an ignitable mixture.

Rotor In every electrical motor, there are two basic parts, the rotor and the stator. The rotor rotates in an electric motor.

Safety match A match designed to ignite only when scraped across a chemically impregnated strip.

Spontaneous combustion A combustion process in which a chemical reaction takes place internally, creating an exothermic reaction that builds until it reaches the ignition temperature of the material involved.

Stator In every electrical motor, there are two basic parts, the rotor and the stator; the stator is the part of the motor that stays stationary, housing the rotor.

Strike-anywhere match A match where the head is chemically impregnated to ignite when scraped across any rough surface.

REVIEW QUESTIONS

1. What is the definition of *competent ignition source*?
2. How much energy can a lightning bolt produce?
3. How could methane gas enter a house? Would it be detectable by the occupants?
4. Why are cigars less of a potential heat source than cigarettes are?
5. What type of evidence can a paper match and matchbook provide?
6. Describe the different types of matches.
7. Name three types of ignition devices that can be found at a local hobby shop.
8. What would cause the sparks that are coming out of the exhaust from the catalytic converter?
9. What keeps hot gases, smoke, and sparks from coming out through the mortar of a brick chimney?
10. Define *spontaneous combustion.*
11. Give three examples of spontaneous heating and how each occurs.
12. How can small appliances be involved as potential heat sources?

DISCUSSION QUESTIONS

1. When you look around your classroom (or home), what potential sources of heat do you see?
2. Fire-safe cigarettes are capable of being manufactured. Some states have already legislated laws allowing only fire-safe cigarettes to be sold in that state. Other states are considering similar legislation. Why won't the tobacco industry switch to manufacturing fire-safe cigarettes on their own instead of waiting for mandates?

ACTIVITIES

1. On the Internet, research whether today's science would allow DNA to be extracted from a cigarette butt found at the scene, even if the butt has been exposed to hydrocarbons.
2. The spontaneous heating of hay requires moisture. Yet several different scientific reports identify different levels of moisture that must be present. On the Internet, find three different documented percentages required to prompt spontaneous combustion in hay.

NOTES

1. National Fire Protection Association, *Fire Protection Handbook*, 18th ed. (Quincy, MA: National Fire Protection Association, 1997), 1–66.
2. National Fire Protection Association, *Fire Protection Handbook,* 10-78.
3. National Fire Protection Association, *Fire Protection Handbook,* 1-69.
4. National Fire Protection Association, *Fire Protection Handbook,* 3-53.
5. National Weather Service, Melbourne, Florida, Forecast Office, Lightning Center, "What Is Lightning?" http://www.srh.noaa.gov/mlb/ltgcenter/whatis.html.
6. U.S. Fire Administration, "Lightning Fires," in *Topical Fire Research Series* 2, no. 6 (August 2001, rev. March 2002).
7. John R. Hall, Jr., *Fire Protection Handbook*, 20th ed. (Quincy, MA: National Fire Protection Association, 2008) , 3-14.
8. M. Pizzamiglio, A. Marino, G. Maugeri, and L. Garofano, "STRs Typing of DNA Extracted from Cigarette Butts Soaked in Flammable Liquids for Several Weeks," International Congress Series, *Progress in Forensic Genetics 11 - Proceedings of the 21st International ISFG Congress* 1288 (April 2006), 660–662.
9. National Fire Protection Association, *Principles of Fire Protection Chemistry and Physics* (Quincy, MA: National Fire Protection Association, 1998), 243–247.
10. National Fire Protection Association, *Principles of Fire Protection,* 243–247.

ELECTRICITY AND FIRE

Learning Objectives

After completing this chapter, you should be able to:

- Describe basic electrical theory.
- Describe the electrical wire size, sheathing, and usage.
- Describe a typical residential electrical system and its components.
- Describe various sources of ignition that can be created by electricity.
- Describe some common electrical failures.
- Describe how the use of improper electrical components can create sufficient heat for ignition.
- Describe the role of static electricity in an ignition sequence.

CASE STUDY

The house was brand new, sitting in an affluent new subdivision. It was a two-story, wood frame, vinyl-sided house. The electricians had finished their final work only a few days prior to the fire. The homeowner was moving in and turned on the air conditioning for the first time as boxes were being carried into the house.

The fire was discovered by neighbors who were out walking their dog. When the fire department arrived, flames were visible in the left rear corner of the home. By the time the fire was knocked down and placed under control, it had consumed a good portion of the back entrance, laundry room, and utility room. The remainder of the home had both smoke and heat damage.

The fire officer in charge called for a fire investigator, but he also started getting statements from the witnesses before they disappeared. The couple who reported the fire stated they first saw the flames in the front right corner of the house, which is the exact opposite side of the house from where the most damage was found. Furthermore, there was little to no damage in the front right corner.

Looking at the scene there was a clear, definitive area of most damage, which was in the back left corner of the house. Exactly in the center of this area, on the underside of where the flooring would have been, there was a charred electrical junction box under the debris. The floor joist where this box would have been secured was completely consumed. However, this did not match where the witnesses said they saw the fire.

Even though it was daylight, the firefighters were using large floodlights inside the structure. The investigator took these lights and moved them to the heavy damage area, and then aimed them down at the hole in the floor, into the crawlspace. This was the area that the fire scene evidence indicated as the area of origin. The investigator then went to the sidewalk and stood in the exact place where the neighbors said they saw the fire. The mystery was solved. The neighbors were looking at the house from across the street in direct line between the right front corner leading back to the left rear corner. Remember, the house was on a slight rise that was approximately 5 feet above the level of the sidewalk. From the witnesses' perspective, they were looking into the open vent for the crawlspace. Directly back from this angle of view, the investigator could see the glow from the fire department lights that were located in the left rear of the structure. The couple had seen the flames under the house, in the crawlspace, that were located in the left rear corner of the house. In the same way, the investigator was observing the glow from the work lights from this same area.

With this confirmation, the investigator went back to his designated area of origin. The crawlspace burn patterns clearly indicated that the fire went upward from this point, in the area of the junction box. The only available heat source was the number 6 aluminum electrical wire and a junction box where the wire fed from the panel down to the box. The wire then fed to the two outside heat pump units located on the right side of the house.

The aluminum wiring in the area of origin had melted. The junction box was sitting on the ground in the carbon debris from the burning structure. It was at the

(Continued)

(Continued)

bottom of this debris. The steel box was heavily damaged. After taking the screws off the cover plate, the investigator discovered that all that remained inside were the metal spring residue from six wire nuts (wire connectors) along with some carbon and melted aluminum.

Tracing the wire back to undamaged areas showed that it was an aluminum number 6 conductor (wire) designated as two sixes and eight. This means that the plastic vinyl sheath contained two insulated number 6 wires with a number 8 bare ground wire, all aluminum.

On the inside of the cover plate there were clear signs of electrical arcing. After taking the box and the wire nuts as evidence a comparison was made on what was required by the code. It was discovered that the box was too shallow for the number of wires and connections being made. A comparison of the wire nuts also indicated that they were too small for the size wire they were trying to connect. The box from the area of origin can be seen in **Figure 10-1,** showing it was just over one and a half inches deep. In contrast the correct size box for the wire size and number of connectors can be seen in **Figure 10-2;** it is more than 2 inches deep.

As stated, the electricians had only finished the job of wiring a short time prior to the fire. Taking the electrical box, the burnt wire nuts, and photographs, the investigative team met the electricians at the scene the day following the fire

Figure 10-1
Electrical box from the area of origin showing the size of the box.

(Continued)

(Continued)

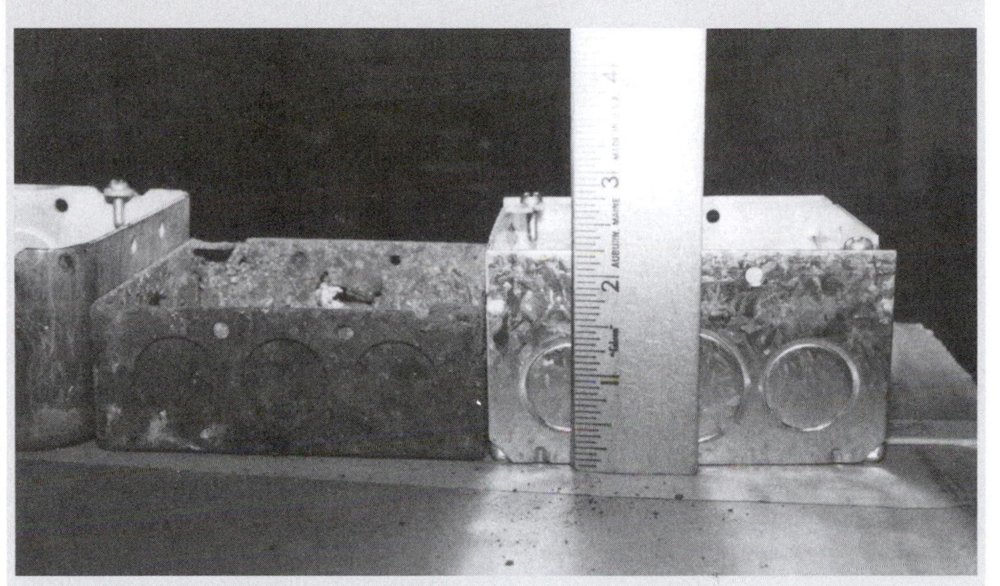

Figure 10-2 *New electrical box showing the size required by the National Electrical Code for the size of the wire and number of connections to be made in the box.*

scene examination. The master electrician was the owner of the electrical company. He was accompanied by his journeyman electrician, who was not quite 20 years old.

The most interesting part of an interview is making sure to ask even the most obvious of questions, because the dynamics of those being interviewed can indicate if they are being honest or deceptive. It was more than commendable that the master electrician stated that he had not personally wired or checked every connection done on the job. He turned to his apprentice and was quite clear that he expected the assistant to answer every question honestly and clearly so that the fire investigation team could find out what really happened.

When the fire investigators showed the electricians the area of origin, explained the process of coming to such a conclusion, as well as showed them the evidence collected at the area of origin, the apprentice hung his head and said that he was worried that that connection might have been a problem.

Stating that the truck still had the supplies used on the last day they worked, the apprentice pulled out new wire nuts and stated that he knew he should have gotten the proper size but that he had none on the truck and wanted to get the job done. He also stated that he found a nick in the wire leading to the heat pumps and discovered there was enough play in the wire to allow him to cut the wire and create a splice.

He knew he needed a larger box, but none was on the truck. Rather than driving the twenty-plus miles back to the shop, he made do with the box he had. He confirmed

(Continued)

(Continued)

that the box was smaller than the one needed and that he had secured the box to the floor joist in the crawlspace in the area where the fire started.

Not once did the owner of the electrical company, the master electrician, stop the younger employee during his story. In fact, he encouraged the young man to continue. Wow, such ethics were certainly commendable.

The last description on what had happened gave an indication as to why the fire occurred. When the wire nut did not fit on the two wires, the apprentice shaved the aluminum wire down to get the two wires to fit in the connector; even then he knew that he did not get as good a connection as he would have liked. After making the connections, he found that no matter how he twisted or turned the wires he could not get the cover plate in position over the connections. Laying on his back in the crawlspace, he found that he could put his knee on the cover, pushing it up in place enough to get the screws started. He then tightened it down and crawled back out.

He turned on the HVAC unit and it powered up; so he shut it down and left the job. After his story, it hit home to him that he knew he was responsible for the fire. The master electrician apologized on behalf of his company and stated he would contact his own insurance company. Before leaving, he let the investigator keep the wire nuts from the truck, which were placed with the burnt wire nuts from the scene and are shown in **Figure 10-3.** (See also **Figure 10-4.**)

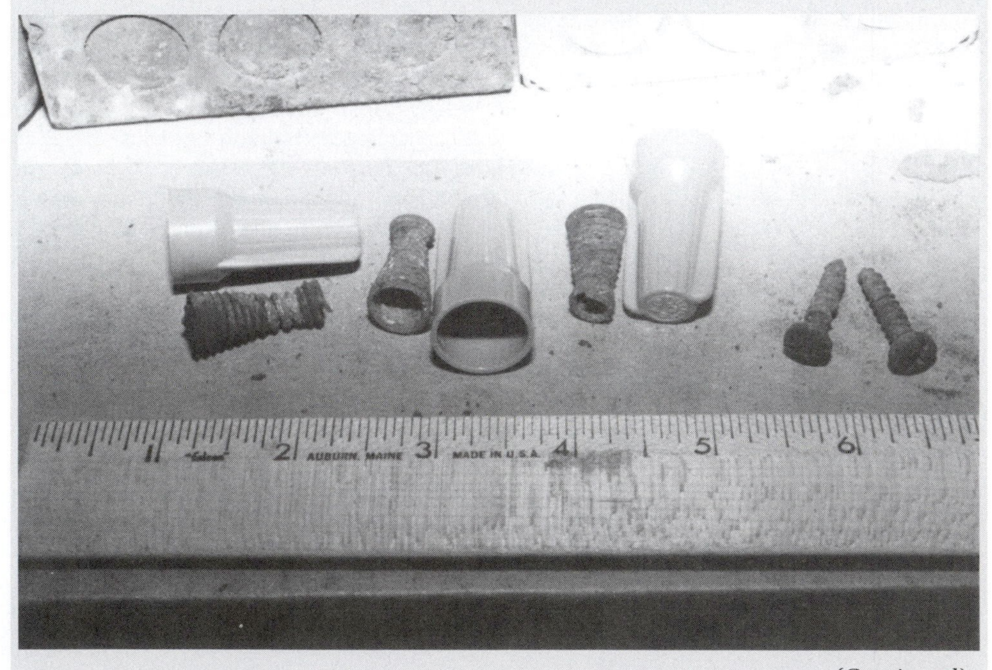

Figure 10-3 *The remains of the wire nuts from within the electrical box in the area of origin alongside wire nuts taken from the same box used by the electrician when the box was wired.*

(Continued)

(Continued)

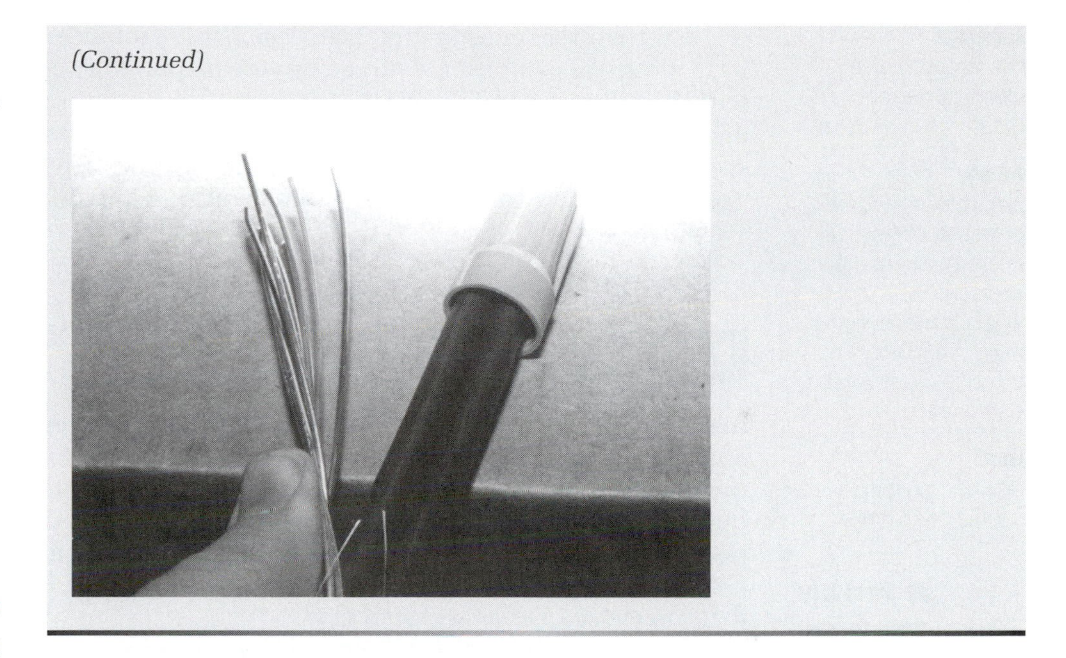

Figure 10-4 *An attempt to get two wires of the same size as were used on the scene to connect using the wire nuts used by the electrician.*

Note
An incendiary fire cannot be affirmed until all other potential heat sources are eliminated.

Note
Technical investigators must understand Ohm's law and how to apply it to understand the relationship between current and resistance.

Ohm's law
mathematical equation that describes the relationship between the voltage (V), current (I) and resistance (R); it can be expressed in three ways (see end of chapter definition)

INTRODUCTION

Electricity plays into many of the other chapters of this text: from *spoliation* to *sources of ignition*. A thorough understanding of electricity is critical to every investigator from the fire officer doing the initial cause and origin for the fire incident report, to the technical investigator looking at an incendiary fire. An incendiary fire cannot be affirmed until all other potential heat sources are eliminated. The investigator cannot eliminate electricity as a cause unless the investigator understands how electrical systems can fail or not fail.

The first step toward understanding electricity is to understand the basic components and how they are installed. Not just the components installed today, but those that met the code in the past and are still in use today. This chapter covers basic electricity and points to electrical failures that lead to a fire cause. Just as important, it points out some items that are not necessarily proof that the fire started from an electrical source.

Neither this chapter nor an entire book could provide the information necessary to supply all the knowledge for an investigator to be competent in this area. Technical investigators must understand **Ohm's law** and how to apply it to understand the relationship between current and resistance. They must be able to work the formulas to obtain power, **amperage, voltage,** and **watts** for any given situation. Every person conducting an investigation should seek additional training in this area.

amperage
the strength of an electrical current measured in amperes

voltage
electromotive force or pressure, difference in electrical potential; measured as a volt, which is the ability to make 1 ampere flow through a resistance of 1 ohm

watts
unit of power equal to 1 joule per second

A further recommendation is to establish a professional relationship with a local electrical contractor, who can be very helpful in explaining different issues with electricity as well as give the investigator an opportunity to get some hands-on training on how components are supposed to be installed and how they are supposed to operate. You should look within your own department—it is surprising how many off-duty firefighters do electrical work on their days off as a second career.

BASIC ELECTRICITY

In Chapter 1, we covered the history of alternating current (AC) and direct current (DC). As a result of that history, today, direct current is restricted to batteries in vehicles and various portable devices. Our homes and industry use alternating current for power needs. At the most basic level, both sources of electricity provide the energy necessary to run common electrical devices. They both provide electricity as the result of the flow of electrons, and they both have hazards that can cause injuries and start fires.

Note **Today, direct current is restricted to batteries in vehicles and various portable devices. Our homes and industry use alternating current for power needs.**

The electrical energy itself is capable of being a source of ignition, but it really is the appliance or wiring that creates the situation. When appliances, electrical components, or wiring has been improperly designed, manufactured, or used, the electricity has the potential to be energy that starts the unwanted fire.

Note **The electrical energy itself is capable of being a source of ignition, but it really is the appliance or wiring that creates the situation.**

Flow of Electrons

Typically, the comparison of water to electricity is a good way to understand how electrical current will flow though wires. To keep it as simple as possible, consider a faucet in your kitchen sink. The water flows through the pipes, up to the faucet, into the sink, where the water is used, and then out through the drain. The pipes are like electrical wires, and the water is under pressure, which we measure in pounds per square inch. In wires, we measure this pressure in volts. The larger the pipe, the larger the amount of water that can be carried, which is measured in gallons per minute. The larger the wire, the more electricity that can be carried, which is measured in amperage.

The water when it reaches a sink is used to wash dishes, clean vegetables; essentially, it is used to do the work. The electricity powers the light, runs the fan, essentially does the work, just like the water in the sink. From this point, both the water and the electricity flow back to the ground, neither under pressure.

Failures can also be compared to water. If we develop a leak in the pipe, we have less pressure and not as much water as we need to do the washing. A failure of the wiring insulation that allows the electrical energy to leak to ground will do the same: It will provide less power to do the work. Lights may dim and electrical motors may slow down. Likewise, if the pipe was over-pressurized, it could burst or maybe an appliance could fail and start leaking. An overload of electrical current on a wire could cause it to fail, melting the insulation. The overload could go into an electrical appliance causing the appliance to fail as well.

That energy flow can be taken down to the molecular level. Consider a pipe with an interior diameter of 1 inch. Fill it with marbles that are 1 inch in diameter. The only reason for using such large marbles is to keep them from falling out in this experiment. With the pipe full, push just one more marble into one end of the pipe. When this happens, one marble will be pushed out of the pipe on the other end. The energy to make this happen was the impact of the marble you pushed in, affecting the first marble. That marble pushed on the second marble and the second marble pushed or impacted the third marble, on and on down the line, pushing out the last marble.

electrons
negatively charged particles that are a part of all atoms

conductors
anything that is capable of allowing the flow of electricity; commonly used to describe the wires used to provide electricity within a structure

insulators
something that is a nonconductor of electricity; commonly thought of as glass or porcelain, but today includes plastic and rubber as in the coverings of electrical wire

Electrons do not go in a straight line like the marbles in the pipe, but as one more electron is introduced into the wire, the impact is felt all the way down the line, pushing out electrons on the other end, back into the ground. Granted the process is not exactly the same as marbles in a pipe, but the effect is the same. The electrons flowing through the wire are the energy potential to make things such as light bulbs and appliances work. We want to keep the electrons under control. However, electrons love ground. They would like nothing better than to escape to ground and will do so through anything that will allow them to do so. Those items that are capable of allowing the flow of electrons are considered to be **conductors.** Those things that block the flow of electrons are **insulators.** Keep in mind that this is an overly simplistic way to look at electron flow and other aspects of electricity. Electrons do not feel any real emotion of *like* or *dislike* and neither is there any thought whatsoever; the effects of electricity and the actions that make things happen are based on science and the laws of physics. It is imperative for you to seek additional training in this area.

TODAY'S ELECTRICITY

service drop
overhead wiring from the electrical company that is attached to the weather-head to deliver electricity to the structure

service lateral
underground wiring that comes from the electrical company that comes up to the meter base

In North America, the primary delivery of electrical power for distribution and for larger occupancies, such as malls and factories, is referred to as three-phase power. Three-phase power is ideal for large electrical motors because the delivery system provides for consistent power, allowing a smoother operation and reducing vibration and motor movement.

In this chapter, we discuss single-phase power, which is the most common for residences and small businesses. It is delivered with two wires and a ground either to the weather head, sometimes called a service head, for above-ground delivery or up through the ground to the meter for underground service. The wire coming from overhead is sometimes called a **service drop,** and the wire coming from underground is usually referred to as a **service lateral.** The meter is

! Safety Never rely on the power company meter being gone as a basis of deciding that the electricity is cut off.

meter base
the receptacle for the electrical meter installed by the power company

service panel
electrical wire travels from the meter base to the service panel; the panel provides a means to provide overcurrent protection and the ability to distribute electricity throughout the structure through branch circuits

the next part of the delivery system where the electricity is measured by the utility company. Removal of the meter from the meter base is also a way of cutting off the electricity to the structure.

Safety Note: Logic would indicate that the absence of the meter means that the electricity to the structure has been cut. This may not always be the case. Either intentionally or by a failure in the system, some current can leak through, energizing the circuits of the structure. Never rely on the power company meter being gone as the basis of deciding that the electricity is completely cut off. Always use a multimeter to test the circuitry at the panel and in other locations where working around electrical lines.

SERVICE PANEL

From the **meter base,** the electricity enters the structure to the main electrical service panel. The power comes in on two wires with a bare neutral wire. Each power wire provides 120 volts, together providing 240 volts. Once inside the panel, the current is broken down into branch circuits. (See **Figure 10-5.**) For larger appliances, 240 volts are supplied. For lights, outlets, and smaller appliances, 120 volts are supplied.

Inside the **service panel** is a ground bar that allows the connection of all neutral and ground wires. The sides of the panel box have partial precut knockouts that with a slight force applied will allow openings for the insertion of bushings

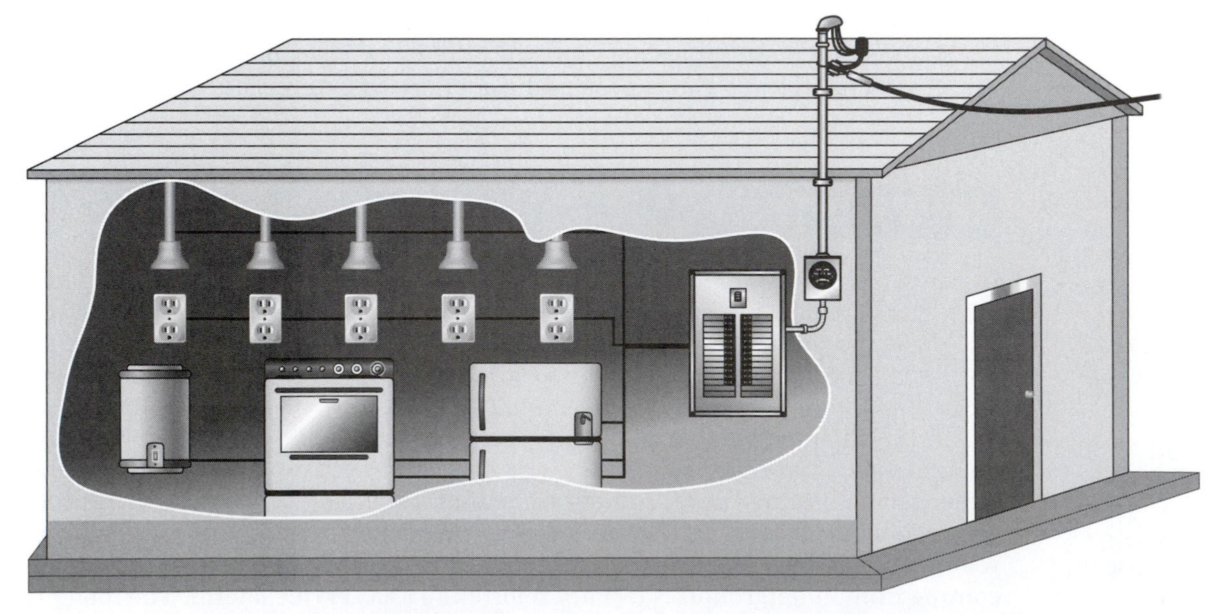

Figure 10-5 *Diagram of an electrical circuit breaker service and sample branch circuits.*

that protect the wire as it enters the panel. There is an assortment of these connectors on the market, from metal screw types to plastic units that snap in place. Blank covers for these holes are available in case a hole is no longer necessary or was knocked out by mistake.

GROUNDING

■ Note
It is essential that ground exists for the entire electrical system.

grounding block
a block of metal within an electrical panel, affixed with holes and screws to allow for the insertion of wires where the tip of the screw will bind and hold the wire in place

It is essential that ground exists for the entire electrical system. The **grounding block** in the service panel should be connected via an adequate size wire to a grounding source. The ground wire should be attached to metal plumbing pipes if they extend at least 8 feet into the ground or to a ground rod, which must be driven into the ground to the depth where the ground stays moist. Usually, this requires an 8-foot ground rod of copper or anodized steel. Every section of the electrical service must be grounded. The continuous use of the bare conductor will lead to the system maintaining ground. If metal conduit is used, this too must be grounded to ensure a safe atmosphere.

Floating Neutral

■ Note
When the neutral wire separates, it is called a floating neutral.

floating neutral
the electricity which is usually 120 volts from two lines, creating 240 volts, fluctuates; this can create 60 volts on one line and 180 volts on the other, or any combination that equals the 240 volts

The service connection coming into the structure consists of two hot wires and one neutral. With a properly working system, the neutral helps balance the system and each hot wire provides 120 volts. This is because the neutral provides a fixed point of zero. When the neutral wire separates, it is called a **floating neutral.** Without this reference, the voltage can vary between the two hot lines. It could be 180 volts on one line and only 60 volts on the other. This will vary, and when testing with a voltmeter, you might notice that the voltage is only 80 volts on one branch, and when you test that same branch again in a few minutes, it could be 140 volts. It will become obvious to the occupants because incandescent lights will get bright or may go dim. The problem would be too much current, which could cause fires, or too little current, which could cause damage to motors. A floating neutral is not a grounding issue, but one strictly related to the parting or breaking of the neutral wire that comes in from the power company.

OVERCURRENT PROTECTION DEVICES

For the common residential structure, the protection devices come in the form of **fuses** or **circuit breakers.** These devices are designed to protect from a short circuit or dead short, which could cause overheating of wires, equipment, or devices and result in damage or fires. However, a 20-ampere (amp) fuse does not blow, and neither does a 20-amp circuit breaker trip, when current instantly reaches 21 amps or even 30 amps. These protection systems must be more complex to handle the various components in a residence. Motors when starting up will take considerably more energy than when they are operating. The system must be designed to handle the surge of a motor starting, yet protect as rapid as possible to an arcing wire to prevent a fire.

There is a delay built into the overcurrent protection devices to allow for a steady current experienced in a starting motor. This short delay could, however, damage delicate electronic equipment such as computers. Thus, the necessity to use surge protectors on these devices.

Overcurrent protection devices may activate from an extreme draw of electricity or from heating. As such if a fire near an electrical panel was to occur, it could trip the breakers from the external heat. This must be kept in mind when conducting a fire investigation and examining the electrical panel.

Note
A floating neutral is not a grounding issue, but one strictly related to the parting or breaking of the neutral wire that comes in from the power company.

Fuses

The fuse service panel, depending on the age of the panel, may be anywhere from 30 amps to 400 amps. The older fuse panels could have as few as two plug fuses, screw type. More frequently, on older panels you find two primary cartridge fuses that act as a disconnect, and another set of cartridge fuses that would feed one appliance. There are branch circuits where one fuse would search each branch. The code will dictate how many devices (light fixtures or outlets) can be fed off one branch, and this will be dictated by the size of the wire among other things.

fuses
a device used as a screw-in plug or a cartridge that contains a wire or thin strip of metal designed to melt at a specific temperature that relates to certain overcurrent situations; the melting of the wire or metal strip stops the flow of electricity as designed

circuit breakers
a protection device as part of an electrical system that will automatically cut off the electricity should there be an excessive overload

The two primary types of fuses are plug fuses and cartridges. The plug fuse has a brass screw base; a safety version has a smaller diameter ceramic screw base and screws into a brass holder base that is the same size as the standard plug fuse base. The top of the plug will give information as to the amperage and type of fuse such as a delay fuse, and so forth. The clear view center window allows you to look inside to see the fusible link to check its status whether it is intact or blown. If the fuse blew as the result of a dead short, the lead from the link will have splattered on the inside surface of the fuse window, as shown in **Figure 10-6.** If the fuse blew because of other reasons, you will see that the link has just melted and separated, stopping the flow of electricity.

The window of the fuse will also be either round or hexagonal (six sided). The hexagonal shape indicates a 15-amp fuse, the round indicates a 20-amp or higher rating. Many manufacturers also place the amperage on the button, which is on the bottom of the fuse. (See **Figure 10-7.**)

Note
These devices are designed to protect from a short circuit or dead short, which could cause overheating of wires, equipment, or devices and result in damage or fires.

Cartridge fuses also have a fusible link on the inside connecting the two ends of the fuse. If too much electricity flows through the link, it will melt, breaking the connection. Some cartridge fuses have replaceable links, but most are disposable. The ratings of the fuses are printed on the outside cylinder. Some also have the amperage stamped on one of the copper ends. To check whether a cartridge fuse has blown, you have to use a multimeter using the ohms settings to see if the circuit can be completed. If not, then the fuse is blown.

Figure 10-6 *Plug fuse blown as the result of a dead short; splatter of ink can be seen on the underside of the fuse window.*

■ Note
There is a delay built into the overcurrent protection devices to allow for a steady current experienced in a starting motor.

■ Note
Overcurrent protection devices may activate from an extreme draw of electricity or from heating.

Circuit Breakers

Circuit breakers perform the same function as fuses, but more efficiently. It is obvious that when a fuse blows a new one must be purchased and installed. Breakers trip and can be reset by just flipping the switch back on. It is not a cost thing so much as a matter of convenience. The breaker also provides better protection than the fuse does because it is more sensitive and will trip earlier in the case of an electrical failure. A circuit breaker can also trip from heat as well as from an increase in the flow of electricity.

Many manufacturers of breakers provide a way for the breaker to indicate its status. Once tripped the breaker either goes to the center position, as shown in **Figure 10-8,** or it will have a small window in which green will show when the breaker is operating normally and red will show if the breaker has tripped. To reset, the homeowner simply turns off the breaker, and then turns it back on.

The downside of breakers for the investigator is that once a breaker has been flipped back to the on position, it can essentially ruin any chance of knowing what happened to that breaker or knowing for sure which branch failed. This constitutes spoliation, as discussed in Chapter 6. (See **Figure 10-9.**) However, left

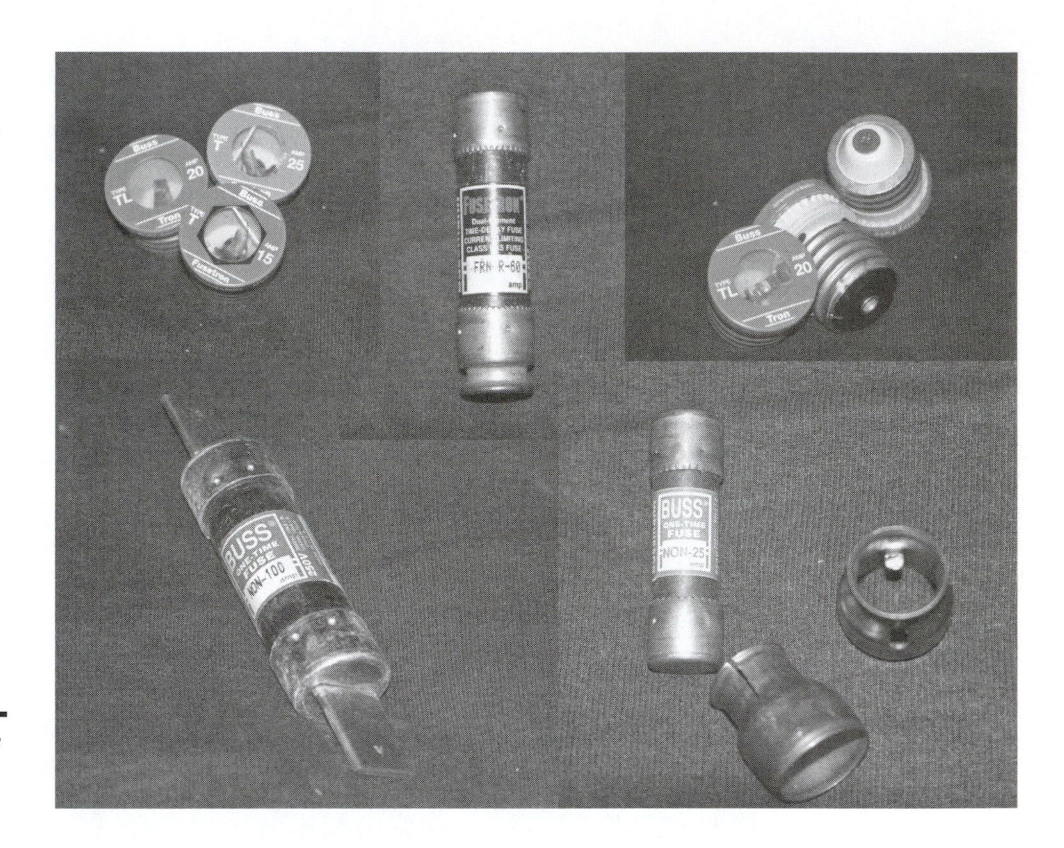

Figure 10-7 *Samples of both plug and cartridge fuses.*

■ Note
The two primary types of fuses are plug fuses and cartridges.

■ Note
The breaker also provides better protection than the fuse does because it is more sensitive and will trip earlier in the case of an electrical failure. A circuit breaker can also trip from heat as well as from an increase in the flow of electricity.

untouched the breakers can be secured and sent to a laboratory for X-raying and examination.

Circuit breakers for 120-volt branches have a single pole, essentially one place to attach a wire. Breakers for 240 volts tap into both lines coming into the panel and are considered double pole. They have places for two power wires to be attached. The double-pole breaker sometimes has one switch or may have two that are attached at the factory. There are also specialized breakers such as **ground fault circuit interrupter (GFCI)** breakers for use outside or interior locations subject to water use, such as kitchens, bathrooms, and laundry areas, to provide greater protection from electrocution, and delay breakers that are used when motors may start up and make a large draw of electricity at that time.

Breakers come in smaller sizes as well. A half size allows for two to fit in the space of one breaker, and even a quad size allows four 120-volt breakers to fit in the space of two breakers. When two half-size breakers are joined together, they are sometimes referred to as tandem breakers.

The best safety issue to evolve through the years was the change of the 2002 National Electric Code where it requires arc fault circuit interrupters (AFCIs) to be installed on all bedroom circuits in new construction. These

Figure 10-8 *Breakers in the center position have tripped. The investigator can then examine those branches to determine whether there was any involvement in the fire's cause.*

■ Note
The downside of breakers for the investigator is that once a breaker has been flipped back to the on position, it can essentially ruin any chance of knowing what happened to that breaker or knowing for sure which branch failed.

breakers not only sense a short and shut down, they also sense an electrical arc from things such as wiring with damaged insulation, frayed cords, or overheated cords. This movement is an effort to prevent fires in residential occupancies during the night that would endanger the occupants while they are sleeping.

Disconnect Box

Additional electrical distribution boxes are fed by the main electrical panel but are designed to service a specific machine or appliance. Even in homes you can find these boxes being used for furnaces or hot water heaters. In the industry, you will see them more frequently used to provide electrical current to specific large machines.

These boxes can use plug fuses, cartridge fuses, or circuit breakers. Those with breakers rely on the breaker itself to be the mechanism to shut off the appliance. Those with fuses usually have a handle assembly of some type to allow it to be shut off in one motion.

Figure 10-9 *Circuit breakers having been removed from the panel and then flipped on and off by an enthusiastic firefighter. This constitutes spoliation and could hamper the investigator's ability to make an accurate determination of fire origin, cause, and responsibility.*

ground fault circuit interrupter (GFCI)
a circuit breaker that is more sensitive than the standard breaker, tripping at a slight ground fault with the intent of preventing electrocution

Fuses Doing Their Jobs

Sometimes what looks like a bargain can in fact be very costly. A small printing company needed to replace their commercial printing machine. With more orders, they needed a larger machine, but because they had not been paid for those new orders, there was a cash flow problem. The solution was to purchase a used larger printing machine, which is exactly what they did.

When installed the larger machine needed more current to operate. The existing subpanel for the older printer could not be upgraded and to replace it with new breakers was in excess of the capital they had on hand. However, the electrician advised them that they could use an older cartridge fuse panel disconnect that he had in his shop.

(Continued)

■ **Note**
The best safety issue to evolve through the years was the change of the 2002 National Electric Code where it requires arc fault circuit interrupters (AFCIs) to be installed on all bedroom circuits in new construction.

(Continued)

Installation went well, but there was an ongoing problem with the machine. It was constantly blowing the cartridge fuses. The operator would pull out the large blown fuses and install new fuses. The blown fuses were just dropped on the floor. After several weeks of a continuous problem of replacing fuses, having to wait for someone to run to the store and get new fuses, the operator made a discovery that he thought was the solution to his problems.

When the installation was done on the machine, it was an industrial application, so the National Electrical Code required that all wires be run in conduit. Small pieces of the conduit that had been trimmed off when doing the installation were on the floor near the burnt fuses. The operator noticed that the two items, fuses and conduit, were almost the same diameter. He took the conduit to his shop and cut off pieces of the conduit to be the same size as the fuses. He then inserted the pipe in the disconnect box in place of the fuses, closed the box, and pulled the handle to the on position. The printing press started working.

Things were great for about an hour; then, employees noticed a slight smell of smoke and in minutes smoke was coming from the printing press. They called 9-1-1, but before the fire department could arrive the entire printing press had gone up in flames. When the Engine Company interviewed the operators, nothing was said about the fuses. It was only during the fire scene examination by the investigator that they found the disconnect as shown in **Figure 10-10.**

Figure 10-10 *Electrical disconnect fuse box with electrical conduit inserted instead of the required fuses.*

COMMON FAILURES ASSOCIATED WITH BREAKERS AND FUSES

Sometimes there may be an obvious indicator of an electrical problem such as the stack of blown fuses sitting near the electrical panel. This would indicate an obvious problem and can create an entire string of questions for the interview with the structure occupant or owner. Attempts have been made to bypass the protection offered by fuses by placing a penny behind the fuse, allowing it to come in contact with both the center button and the outside surface of the brass screw base. Several other techniques to bypass the plug or cartridge fuse would render any protection useless.

■ Note **Circuit breaker design has improved over the years to prevent tampering and bypassing.**

Circuit breaker design has improved over the years to prevent tampering and bypassing. Older breakers could be prevented from tripping if the switch were held in the closed position by tape or even jamming materials into the breaker to keep the switch from moving. Today's breakers will trip regardless of the physical position of the switch, thus preventing this type of tampering.

The investigator should know the basics of the code and know there is a reason for each section of the code. There are requirements that the covers on fuse or breaker panels be in place with doors closed. Besides the obvious of having the cover off and creating a potential for electrocution, there is the potential that hot melted metal fragments might fly from the box, igniting nearby combustibles, should there be a massive system failure.

Another reason for covers being placed on panel boxes, disconnects, and even over electrical boxes is to reduce the amount of debris that enters these devices that could be a potential fuel. Contaminates such as dust could also serve to impair the proper operation of switches and devices. Thus, the code requirement to keep covers on all components of the electrical system.

■ Note **Always check for attempts to bypass the overcurrent protection devices, such as nails through cartridge fuses.**

The first part of an electrical system examination should be to ensure that each branch circuit had the proper level of protection. A common branch with electrical outlets may be protected with a 20-amp breaker or fuse. If the branch is providing energy to just light fixtures, it may have a 15-amp breaker. However, if either are protected by a 30-amp breaker, it would be cause for further examination, not just on that branch but on other branches, disconnect boxes, and fixtures.

Always check for attempts to bypass the overcurrent protection devices, such as nails through cartridge fuses. Just as the homeowner, or even the arsonist, can be unlimited in their imagination to overcome protection devices, the fire investigator must be open minded and unlimited in creative thinking when looking at a system where the overcurrent protection devices have failed.

■ Note **It is not the strength of the metal but its ability to share electrons that will allow current to pass freely.**

WIRE

When considering conductors, metals are the first thing that come to mind. It is not the strength of the metal but its ability to share electrons that will allow current to pass freely. This was measured as relative conductance[1] of which silver is 100 percent. Obviously, silver is too expensive for common structural wiring. Thus, copper at 98 percent has been the metal of choice for wiring homes and businesses.

Many stores sell audio and video cords with gold connectors, advertising that they are the best. This is not because gold is a better conductor; in fact, gold has only 78 percent relative conductance. The reason for the gold connectors is that gold is the least chemically reactive of the metals. In other words, it will not as readily corrode or oxidize as will other metals, making for better audio and video connections.

Again, fiscal concerns were behind not using copper for large transmission lines. Aluminum, at 68 percent relative conductance, was chosen for the task because of the much lower cost. In the 1960s, the cost of copper began to rise. As a cost savings for new construction the industry started providing aluminum wiring for residential and commercial structures. To get the same amount of electricity through the lines as copper, the electrical codes called for aluminum wiring to be used that was one size larger when replacing copper or being used instead of copper.

American Wire Gauge the standard used by the industry to determine the size and designation for electrical wiring

Wire size numbering may not seem to make sense. Small bell wire is size 22 gauge; the wire size to power an electrical outlet is size 12 gauge. That may not seem to make sense, but there is a reason. The **American Wire Gauge (AWG)** is used to standardize the size of electrical conductors. Be cautious: There is an entirely different scale to measure steel wire (the W&M Wire Gauge).

■ Note

The size of the wire comes from the number of times it takes to draw the wire down to a particular diameter.

To make wire, it is drawn (pulled) through draw plates with a preset hole. Usually, the wire is drawn from one size down to the next and down to the next, and so forth. The size of the wire comes from the number of times it takes to draw the wire down to a particular diameter. Thus, it took more draws to create size 22 wire than it did to create size 4 wire. Wire sizes from gauge 5 through 14 have another measurement in this process. The wire is equal to the number of bare solid wires it took to span 1 inch. For example, if you took the same size solid conductors, laid them side by side, and found that it took 10 conductors to span 1 inch, then the size of each conductor is 1/10th of an inch, or size 10.

Solid wire is used for electrical systems that are permanent in nature, not moved around. This is true up to the larger size wires such as size 8 AWG or larger. At this size, the larger solid wire would be hard to handle, especially making bends to get to panels or electrical equipment. Making the wire in strands enables better flexibility for better handling. Also around this size you start to see aluminum being used as a cost-saving measure.

Smaller wires used for plugging in appliances and lamps are also stranded, this time for flexibility and durability. Any wire being bent time after time will stress and eventually break. The smaller strands are more flexible and resilient to a certain point, but even they will fail after a period of time. In most vehicles, the electrical wires are stranded to account for the constant movement and vibrations.

NM
stands for *nonmetallic* and refers to a covering over wire conductors intended for use within a structure to deliver electricity to various points within the structure

■ **Note**
Some ground wires are coated in green vinyl, which designates them as grounds.

UF
the type of insulation on the electrical wiring that is used as an underground feeder. The insulation tends to be thicker and solid up to the insulated conductor

conduit
a tube or trough made from both plastic and metal that is designed to provide protection for electrical conductors

Greenfield
the manufacturer's name for a flexible conduit; the name has become a common term to describe any flexible conduit, both metal and plastic

WIRE INSULATION

Electrical conductors today are coated with a plastic vinyl to insulate the wire to prevent the escape of electrons. These conductors are designated as **NM,** which stands for nonmetallic. The pliability of this plastic enables the wire to be bent at 90-degree angles without damage or failure. Of course, constant bending will cause the insulation to crack or tear. The obvious reason for the insulation is to prevent the escape of electrons, which could cause injuries or heating that could lead to ignition of nearby combustibles.

For structural wiring, each of the conductors carrying and returning the electricity is individually coated. A ground wire is present as well but is usually not coated. All of these wires are then coated with another layer of plastic vinyl for protection. Some manufacturers place paper between the coated wires and the outer vinyl sheath and additional paper around the ground wire as well. Special note: Some ground wires are coated in green vinyl, which designates them as grounds.

There is a different type of sheathing for underground wire. To protect it from the damp conditions of being in the ground, the sheathing is solid plastic vinyl and not just a coating. It encompasses each insulated wire as well as the bare ground wire. This type of wire carries the designation of **UF,** which stands for underground feeder.

The sheathing, as seen in **Figure 10-11,** also provides information about the wire, which is printed on the surface or imprinted in the sheathing itself. This identifies whether the wire is NM or UF, plus identifies the size of the wire along with the number of insulated conductors. The bare wire is not included in the number but is designated as "G" or "Ground." A designation of 12/2 G contains two insulated number 12 wires with a bare ground; 10/3 G is three insulated number 10 wires with a bare ground wire.

Some wires do not need the outer plastic sheathing. Individual insulated wires are available to be used in plastic or metal **conduit.** (See **Figure 10-12.**) This is tubing designed to carry wiring from one point to another. Conduit is usually used in areas where the wiring may be exposed and susceptible to damage. Also available is a flexible metal sheath called **Greenfield** but more aptly described as flexible armored cable.

The color of the insulation on the individual wire usually indicates its use or function. The bare wire or wire with green insulation is the ground wire. The black and red wires are carrying current. The white wire is the neutral and carries current at zero voltage, completing the circuit back to the panel. Caution should be exercised because in some wiring it is necessary to use the white wire to carry a full load. When this is done, the electrician should tape or paint the end of the wire black for this designation; however, this is not always accomplished on the job.

Figure 10-11 *Sheathing of wire with plastic vinyl protects the insulator, ensuring no leakage of current. Size of wire, number of wires, and usage of wire are indicated on the outside sheathing.*

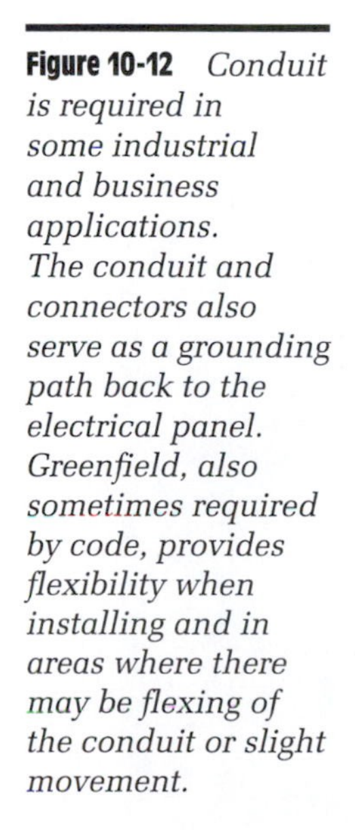

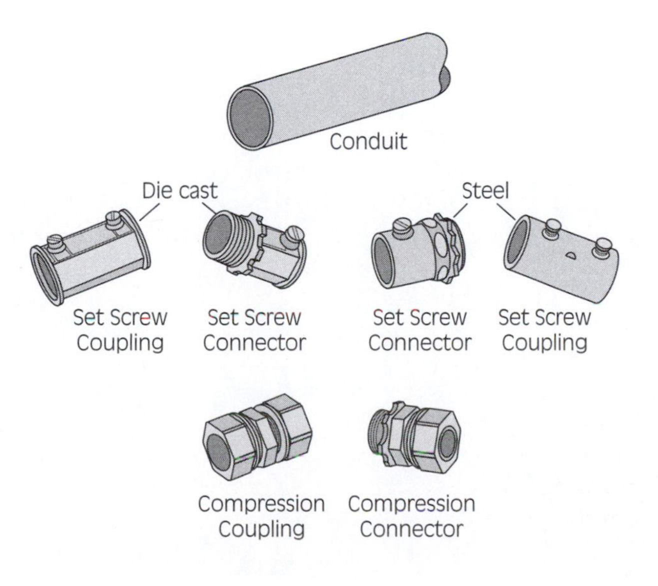

Figure 10-12 *Conduit is required in some industrial and business applications. The conduit and connectors also serve as a grounding path back to the electrical panel. Greenfield, also sometimes required by code, provides flexibility when installing and in areas where there may be flexing of the conduit or slight movement.*

■ **Note** The color of the insulation on the individual wire usually indicates its use or function. The bare wire or wire with green insulation is the ground wire. The black and red wires are carrying current. The white wire is the neutral and carries current at zero voltage, completing the circuit back to the panel.

OLDER WIRING

Just because the wiring is older does not mean it is dangerous. There are complications in some residential homes that have older, ungrounded wiring and outlets (the two-prong-only type). Using adapters can sometimes provide a means for a ground, but this is not always the case. Even in some older homes that are upgraded to newer wire, electricians may not replace all circuits. Those that are replaced usually leave the older disconnected wire in the walls.

In **Figure 10-13,** on the left side of the photograph, is the turn-of-the-century wiring known as knob and tube. Both the knob and tube were insulators made of ceramic that held the wire in place as it is run throughout the structure. The wiring on the right was post–World War II wiring that was insulated with a rubber and cloth covering, eliminating the need for ceramic insulators. But this wiring had only two wires where newer wiring provided two wires and an additional ground wire. In the center is a piece of conduit, a metal tube containing the wires; the conduit itself will act as an additional ground if properly installed.

Figure 10-13 *Examples of older knob and tube wiring, post–World War II wiring, and newer conduit wiring.*

National Electrical Code

■ Note **The National Electrical Code (NEC) is actually a product of the National Fire Protection Association code development process.**

The National Electrical Code (NEC) is actually a product of the National Fire Protection Association code development process. This panel of experts only requires items in the code based on the fact that to do less would be a hazard. Many of the items in the code are there as a result of a known fire; incorporating corrections into the code is an attempt to prevent future similar occurrences.

As such when the NEC requires a certain size box for a certain size wire and number of connections, it is because that is the safest way to do that specific installation. The connectors, wire nuts, are also specifically designed for a certain number of wires of a certain size. To exceed that number of wires results in an unstable connection; one that will fail at a later date when the wires creep out of the connection, resulting in a loose connection that will have arcing that could result in a fire.

This brings up the point that any time there is a poor connection it could result in heating. When a wire is connected to an outlet or light switch there should be sufficient contact to ensure that all current is being passed through without excess heating. When looking at an outlet or other similar connection, observe the condition of the wire and screw to see if there is pitting or damage. Loose connections will result in arcing, which in turn can cause damage.

■ Note **A tight connection is critical to the proper operation of any device and for the safety of the system.**

A tight connection is critical to the proper operation of any device and for the safety of the system. Sometimes a loose connection can occur as a result of the type of wiring being used. But it is not just the fault of the wire so much as other factors in its use and properties of different metals.

RESISTANCE HEATING

Resistance is present in all aspects of electrical energy delivery. There is less resistance with copper than with steel, and thus we use copper for the distribution of electricity. In many appliances, we want the resistance to carry out the task for which the appliance is designed. A portable electric heater is designed to have resistance across the heating element, which in turn causes the element to heat up, turning red, giving off heat, accomplishing the intent of the device.

The problem is when the resistance is not wanted. With a loose connection or a corroded wire, both situations limit the amount of surface area for the flow of the electricity. With the same amount of energy going through a smaller area, there will be resistance. This increased resistance creates heat, sometimes to the point of creating a glowing connection. Should combustible materials be in the vicinity there may be the potential of an unwanted fire. The rest of this chapter looks at other resistance issues that can create problems.

Common Failures Associated with Wire and Insulation

■ Note A failure of an electrical extension cord will be insufficient to trip a 15-amp breaker.

■ Note Caution: There is a considerable safety factor built into most electrical systems. Just using 14-gauge wire instead of the required 12-gauge may cause the wire to warm up, but it is unlikely there will be any energy that could result in ignition of combustibles.

■ Note By no means does finding such a hazard elsewhere support this reason as the cause of the fire in another location.

It would be impossible to cover all types of failures, but here are some that are more common and warrant discussion. Most important, remember that the fuses and circuit breakers cannot prevent all electrical failures. A failure of an electrical extension cord will be insufficient to trip a 15-amp breaker. As such, the cord could continue to heat, starting a fire, and the breaker will not trip.

Common failures of an electrical system can be using a wire too small for the load to be carried, causing the wire to overheat. Caution: There is a considerable safety factor built into most electrical systems. Just using a 14-gauge wire instead of the required 12-gauge may cause the wire to warm up, but it is unlikely there will be any energy that could result in ignition of combustibles.

Damaged insulation, which can happen during installation, would allow continued degradation and may eventually lead to a short circuit. Wires that are nicked or stretched cannot carry the normal load and they too could heat up to the point of ignition of nearby combustibles. Stretching can be caused by the pulling of the wire by electricians during installation. Sometimes others can stretch wires too, such as plumbers or carpenters who find wires in their way and pull and stretch them to get them in another position out of the way of their project. The nicking of wires or the insulation can be caused by the pinching or improper use of staples to hold the wire in place. A properly installed staple will just hold the wire snug.

Wires are run in the walls and through the studs through drilled holes. Without careful planning nails used for attaching the sheetrock or other materials could penetrate the wall and be driven into the wires, causing harm or short circuits. Finding this after the fact is not always easy, but it is possible with a thorough investigation. Sometimes finding such damage in other areas of the structure, such as shown in **Figure 10-14,** can support the findings that this could have happened. By no means does finding such a hazard elsewhere support this reason as the cause of the fire in another location. There must be direct evidence.

Other items that can be observed that may give indication of conditions of the electrical system or use of electricity in the structure will be the excessive use of extension cords along with frayed or patched electrical cords as shown in **Figure 10-15.** Degradation of structural wiring from vibrations causes chaffing of the insulation. This can result in some even slight arcing from the hot wire to ground but may not actually cause the breaker to trip. The wire in **Figure 10-16** ran through a hole in concrete pipe and was attached to an electrical water pump. The constant vibration of the pump also led to vibrations of the wire that chaffed it on the concrete. Notice the small beading on the side of the wire; this is where occasional arcing took place. Fortunately, this was discovered before it caused more damage.

Other causes of damage to the insulation of power cords can come from an exterior source of heat, such as from a portable heater. There can also be heat

Figure 10-14 *Nails put in wall penetrate wiring within the wall. Fire started in another room but from what looked electrical in nature; thus, the search to see if there were other electrical problems in the structure.*

Figure 10-15 *This wiring was not involved in the fire but gives an indication of the potential condition of wiring elsewhere in the structure.*

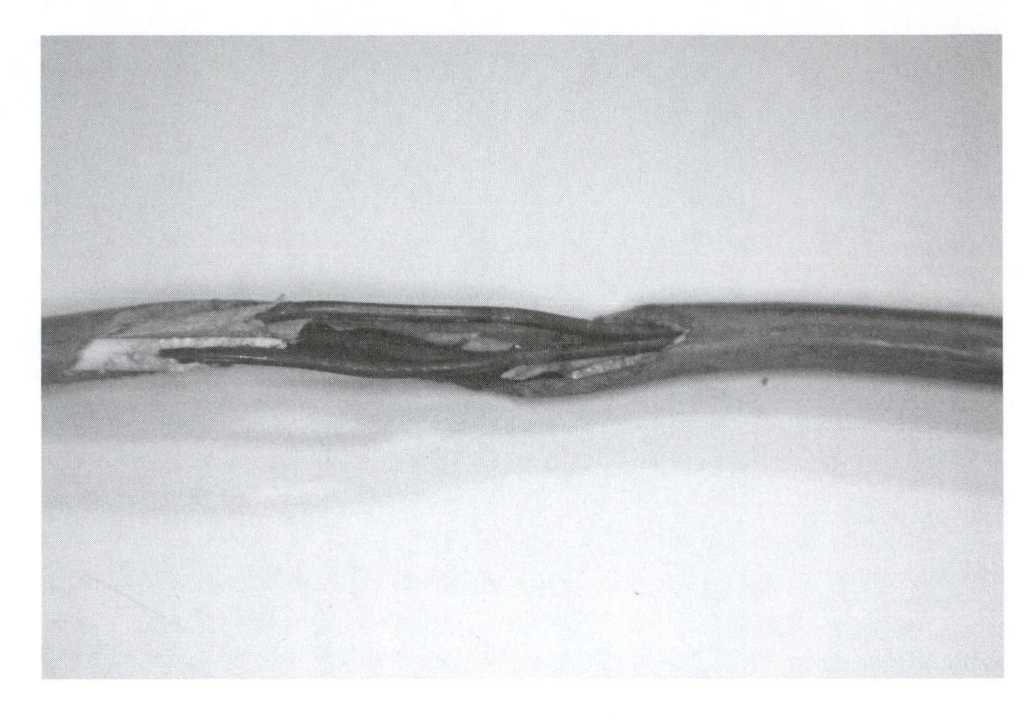

Figure 10-16 *Wire that ran through a hole in concrete chaffed to the point of removing the insulation. Occasional arching occurred on the wire, which caused the beading of the side of the wire.*

Note
Sleeving of the insulation is an indicator of internal heating, which is a sign of overcurrent in that branch. The characteristics of sleeving are when the insulation melts enough to allow it to slide easily back and forth over the wire.

sleeving
the slight separation of the insulation from around the electrical conductor(s), producing an effect that allows the insulation to loosely slide back and forth on the conductors

buildup from cords being insulated by carpets or other items in a living space such as newspapers or magazines. Power cords for most small appliances and lamps and small extension cords are designed to dissipate the heat. When this does not happen, the heat buildup could lead to failure of the insulation, potentially exposing wires, which could lead to an electrical short.

Sleeving of the insulation is an indicator of internal heating, which is a sign of overcurrent in that branch. The characteristics of sleeving are when the insulation melts enough to allow it to slide easily back and forth over the wire. (See **Figure 10-17.**)

Aluminum Wire Problems

As mentioned previously, the cost of copper raised the cost of wiring in the mid-1960s. A good substitute for copper was aluminum. It was considerably less expensive, was an abundant resource, and is flexible for ease of handling in running the wire and making connections. The only drawback to aluminum is that it does not have the same ability to carry current as copper does. An easy fix was to use a slightly larger aluminum wire that would allow it to carry the same load as a smaller copper wire. This proved to be a minor consequence because a number 12 aluminum wire could carry the same current as a number 14 copper wire.

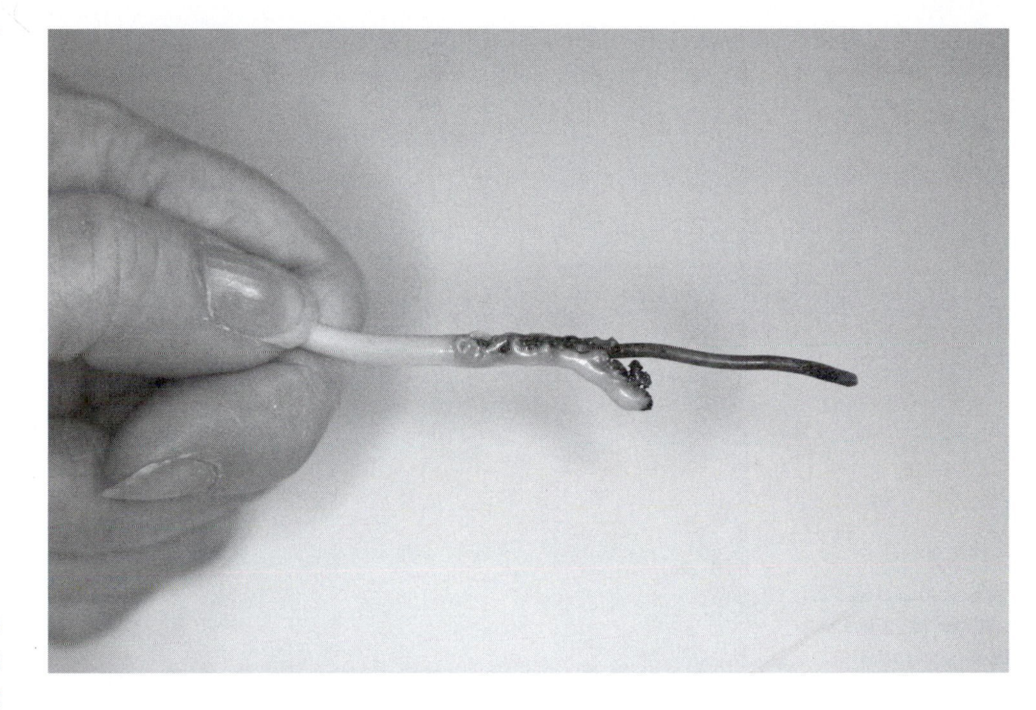

Figure 10-17
Sleeving on a single conductor, the result of the wire heating up the interior, allowing the insulation to loosen. Wire slides freely within the insulation.

■ Note
The only drawback to aluminum is that it does not have the same ability to carry current as copper does. An easy fix was to use a slightly larger aluminum wire that would allow it to carry the same load as a smaller copper wire.

CO/ALR
a designation given to electrical devices that can safely use aluminum wiring; this is only given to those devices that meet a predetermined test accepted by both Underwriters Laboratory and the industry

Aluminum wire does have a characteristic that proved detrimental in the early years of installation. With the distribution of the new wire on the market, electricians just substituted the wire and gave no thought to the potential of using copper connectors with aluminum wiring. Where the aluminum wire wrapped around the screw post on an outlet or other fixture it was found that the aluminum would expand slightly when heated. The heating was caused by the normal flow of electricity to the device. When the aluminum wire expanded, it put pressure on the copper or copper-clad screw post. As such over time, the screw would ever so slightly back off, causing it to loosen on the wire. When this happened, there was less surface area in contact between the wire and the device. This caused the same amount of electricity to flow through a smaller area, creating even more heat. Eventually, the problem created what is referred to as a *glowing arc*, producing excess heat and producing light. These conditions have been known to cause fires.

In 1972, Underwriters Laboratory and the wiring device manufacturers came to an agreement on the testing to be used on devices designed to be used with aluminum wire. If a device could pass this test, it would be designated as **CO/ALR.** Prior to this, the industry was allowed to label their devices for use with aluminum with no testing or documentation. At that time the designation used on the devices was **AL/CU.** With these new advancements, there have been fewer fire-related events with aluminum. Testing done by Wright-Malta Corporation[2] included 1,000 devices with 4,000 wire terminations; half

Figure 10-18 *An AL/CU connector. Newer, safer designs are the CO/ALR connectors.*

AL/CU
a designation given to electrical connectors by the industry to indicate that the device is safe for use with aluminum and copper wires; however, no definitive test by a third party could validate that the devices were designed or safe for aluminum wire

arc
the discharge of electricity from one conductive surface to another, resulting in heat and light

of the devices were the older technology AL/CU and the others were the newer CO/ALR. The samples used were also from the various manufacturers in the market. All devices were wired properly, according to the National Electrical Code. There were many failures and burnouts with the AL/CU devices and only one failure with the newer CO/ALR device. It should be mentioned that the use of aluminum wire, installed properly, is not a fire hazard in itself. (See **Figure 10-18.**)

Arc

When electrical current jumps across a gap between differing electrical potentials, there will be a discharge of energy, an **arc,** producing a high-temperature luminous discharge. The distance between the gaps can be dictated by the amount of energy potential. NFPA 921, *Guide for Fire and Explosion Investigations,* describes arc as a high-temperature luminous electric discharge across a gap or through a medium such as charred insulation.[3]

Usually the arc is of such short duration that it will not be a competent ignition source for most nearby combustibles. However, under the right conditions and with the proper fuel, a fire could occur. In an atmosphere of ignitable dust or vapors in the correct proportion there could be ignition.

■ Note
When electrical current jumps across a gap between differing electrical potentials, there will be a discharge of energy, an arc, producing a high-temperature luminous discharge.

A parting arc occurs when the path of the electricity is disrupted, such as an electrical wire separating from a connection. For the brief moment when the wire starts to leave, the flow of electricity will attempt to continue to flow across the gap, producing an electrical discharge. This can be seen when an electrical plug is pulled from the outlet, especially if there is a load on the cord attached to the plug.

Arc welding is a parting arc. The circuit is completed when the rod is touched to the grounded metal. As the rod is pulled away from the metal, a parting arc is created; this energy is enough to melt metal. Of course, this arc is not sustained with common electrical current. A special arc welder must produce the current specific to make this a continuous process.

A **spark** from an arc is a molten piece of metal that is capable of flying some distance and still having sufficient heat to act as a competent ignition source for ignitable materials. The distance traveled depends on many factors, including the energy being carried by the wire and the size of the molten globule.

■ Note
A parting arc occurs when the path of the electricity is disrupted, such as an electrical wire separating from a connection.

spark
spark; glowing bit of molten metal debris

Arcing Through Char

The insulation material surrounding the wire is combustible. In its designed form, it prevents the escape of electricity. However, when wiring is exposed to fire, the insulation will burn and char, and as the insulation turns to carbon it will allow the flow of electricity, thus arcing through char. This is usually a condition that occurs as a result of the fire and is not necessarily the cause of the fire.

Because the amount of current flowing through the char may be insufficient to trip breakers or fuses, the arcing may continue and will create damage back toward the panel. As the arcing takes place, there will be the splatter of metal. Careful examination may point back to where the arching originally occurred, farthest from the panel, which can help to pinpoint the area of origin.

■ Note
Careful examination may point back to where the arching originally occurred, farthest from the panel, which can help to pinpoint the area of origin.

STATIC ELECTRICITY

Static electricity is just as it implies: It is *static,* not in motion. In the process of movement, different materials may pick up extra electrons. This can be when you walk across a carpet on a dry day, or a conveyer belt or even the flow of liquids may pick up the electrons. As electrons are picked up, a negative charge is created, and as electrons are given off, a positive charge is created.

The electrical potential exists on the surface of a nonconductive body; this body is said to be charged. The electrons are essentially trapped and have no potential of movement until they meet a conductive body. At that time, the excess electrons flow to seek balance, and there is a small discharge. This discharge is sufficient to create light and sometimes it will even make a snapping

static electricity the buildup of a charge, negative or positive, as the result of items coming in contact and then breaking that contact, either taking away or leaving electrons that results in the change of the static charge of the items involved

or popping noise, and sometimes it can be felt such as when your finger touches a door knob. Under most circumstances there is no damage, although some may experience a momentary pain if they are part of the completion of the circuit. The static arc discharged from the tip of a finger will not burn the skin or cause any damage because the discharge is extremely small in amperage and very short in duration.

However, if this discharge were to take place in an atmosphere of ignitable dusts or vapors in the right proportion, there is a potential that the discharge may contain enough energy to be a competent source of energy to cause the ignition.

■ Note When investigating whether static electricity was involved in the ignition sequence, the investigator must determine if there could be a potential for such a discharge.

When investigating whether static electricity was involved in the ignition sequence, the investigator must determine if there could be a potential for such a discharge. There are several safety regulations for the flow of ignitable liquids that require bonding and grounding. For example, a 55-gallon drum of an ignitable liquid sitting on its side in a cradle is required to be bonded to the ground. If liquid is to be flowed from the spigot of the drum to a container, a ground strap attached to the drum must be attached to the container so that they are both on the same potential and are not able to collect additional electrons. Without this bonding and grounding, a discharge could build up and cause a static discharge. Under the right conditions, it could be volatile.

The investigator should be familiar with all the regulations associated with any process being used in and around the fire scene. With this knowledge, a line of questioning for the witnesses and victims will be more insightful and beneficial toward the ultimate discovery of the fire origin and cause.

MELTING OF CONDUCTORS

■ Note Copper will melt at 1,981 degrees Fahrenheit and aluminum will melt at 1,220 degrees Fahrenheit.

Not directly associated with the ignition sequence but something that will aid the investigator in the fire scene examination are issues dealing with the melting of conductors. Copper will melt at 1,981 degrees Fahrenheit[4] and aluminum will melt at 1,220 degrees Fahrenheit.[5] Because of the low melting temperature of aluminum, it will most likely melt in any fire. Any patterns created by the melted aluminum wire will be of little use.

■ Note The energy of the short may cause small globules of wire to fly off and fall back on the wire, resolidifying.

Copper can provide information by its physical condition during the fire scene examination. If there is no impact on the copper and insulation is intact, you know that there was limited heat in that area. When bare copper is found, that is a clear indication that the heat was more intense in this area. Because the insulation burned off, you should see some discoloration of the copper, such as dark red or black, as a result of the oxidation.[6] The copper wire itself can start to show damage as heat increases. Blisters will appear as the copper starts to break down. Then, the wire will start to melt, showing dripping on the underside of the wire. As this continues, more melting will start to thin out the wire, reducing

bead
the melted end of a metal conductor that shows a globule of resolidified material with a sharp line of demarcation to the remainder of the wire

■ **Note**
A laboratory can examine the wire to make a more definitive evaluation as to whether the wire was energized, and whether the damage is the result of an electrical short or melting from flame or hot embers.

it in size. There will be no sharp line of demarcation and the end of the exposed wire may appear pointed.

Tracing the wire damaged from external heat from the undamaged area to the damaged area will aid in determining the direction of fire travel: from more damage to less damage. If the wire is missing, it will be difficult to find evidence of the melted copper. But there will be evidence of wire installation such as drilled holes and staples that may help the investigator trace the wire path.

When there is an electrical short in the energized wire, there will be a different appearance. The energy of the short may cause small globules of wire to fly off and fall back on the wire, resolidifying. The size of the globules may be easier seen with a magnifying glass. The end of the wire will have a **bead** showing a clear line of demarcation.

Some electrical arcing can be less severe, causing a notch in the wire instead of a separation of the wire. The notch may also have small globules attached to the area near the notch, and the notch will appear melted. In contrast, a mechanical notch may have scratch marks or misshaped wire.

A laboratory can examine the wire to make a more definitive evaluation as to whether the wire was energized, and whether the damage is the result of an electrical short or melting from flame or hot embers. The most important aspect about an electrical short and the fire scene is determining which came first: Did the fire cause the electrical short or did the electrical short cause the fire?

SUMMARY

The old adage that "a little bit of knowledge could be dangerous" is never more true than when talking about someone making a determination of a fire that involved electricity. Every investigator, especially the technical investigator, must become very knowledgeable about electricity, the electrical code, the potential for electrical failure, and electrical issues to ensure that he or she can conduct an accurate and credible investigation of the fire origin and cause.

Electrical theory is important to understand, and as an investigator, you must be able to explain it to others to sell them on your theory of the fire's area of origin and electrical cause. A fundamental understanding of the components of an electrical system, the code mandates, and common practices are also a great tool for the investigator.

Knowing what is not a potential heat source can be just as valuable and important as knowing what can be a competent source of heat for a fire. The limitations on overcurrent protection devices go along the same line—know what they cannot protect as well as what situation they should prevent.

The most important tool the investigator has is the same that has been mentioned in every chapter: The fire investigator must approach every scene with a clear and open mind and use a systematic approach to ensure that there is an accurate and plausible finding in the final hypothesis developed.

KEY TERMS

AL/CU A designation given to electrical connectors by the industry to indicate that the device is safe for use with aluminum and copper wires. However, no definitive test by a third party could validate that the devices were designed or safe for aluminum wire.

American Wire Gauge The standard used by the industry to determine the size and designation for electrical wiring.

Amperage The strength of an electrical current measured in amperes.

Arc The discharge of electricity from one conductive surface to another, resulting in heat and light.

Bead The melted end of a metal conductor that shows a globule of resolidified material with a sharp line of demarcation to the remainder of the wire.

Circuit breakers A protection device as part of an electrical system that will automatically cut off the electricity should there be an excessive overload.

CO/ALR A designation given to electrical devices that can safely use aluminum wiring. This is only given to those devices that meet a predetermined test accepted by both Underwriters Laboratory and the industry.

Conductors Anything that is capable of allowing the flow of electricity. Commonly used to describe the wires used to provide electricity within a structure.

Conduit A tube or trough made from both plastic and metal that is designed to provide protection for electrical conductors.

Electrons Negatively charged particles that are a part of all atoms.

Floating neutral When the service entrance neutral wire separates, it creates a floating neutral situation where the electricity which is usually 120 volts from two lines, creating 240 volts, now fluctuates. This can create 60 volts on one line and 180 volts on the other or any combination that equals the 240 volts.

Fuses A device used as a screw-in plug or a cartridge that contains a wire or thin strip of metal designed to melt at a specific temperature that relates to certain overcurrent situations. The melting of the wire or metal strip stops the flow of electricity as designed.

Greenfield The manufacturer's name for a flexible conduit. The name has become a common term to describe any flexible conduit, both metal and plastic.

Ground fault circuit interrupter (GFCI) A circuit breaker that is more sensitive than the standard breaker, tripping at a slight ground fault with the intent of preventing electrocution.

Grounding block A block of metal within an electrical panel, affixed with holes and screws to allow for the insertion of wires where the tip of the screw will bind and hold the wire in place.

Insulators Something that is a nonconductor of electricity; commonly thought of as glass or porcelain, but today includes plastic and rubber as in the coverings of electrical wire.

Meter base The receptacle for the electrical meter installed by the power company.

NM Stands for *nonmetallic* and refers to a covering over wire conductors intended for use within a structure to deliver electricity to various points within the structure.

Ohm's law Mathematical equation that describes the relationship between the voltage (V), current (I) and resistance (R); it can be expressed in three ways, depending on what you need to solve:

$$I = V \div R$$

$$V = I \times R$$

$$R = V \div I$$

Service drop Overhead wiring from the electrical company that is attached to the weather-head to deliver electricity to the structure.

Service lateral Underground wiring that comes from the electrical company that comes up to the meter base.

Service panel Electrical wire travels from the meter base to the service panel. The panel provides a means to provide overcurrent protection and the ability to distribute electricity throughout the structure through branch circuits.

Sleeving The slight separation of the insulation from around the electrical conductor(s), producing an effect that allows the insulation to loosely slide back and forth on the conductors.

Spark Spark; glowing bit of molten metal debris.

Static electricity The buildup of a charge, negative or positive, as the result of items coming in contact and then breaking that contact, either taking away or leaving electrons that results in the change of the static charge of the items involved.

UF The type of insulation on the electrical wiring that is used as an underground feeder. The insulation tends to be thicker and solid up to the insulated conductor.

Voltage Electromotive force or pressure, difference in electrical potential; measured as a volt, which is ability to make 1 ampere flow through a resistance of 1 ohm.

Watts Unit of power equal to 1 joule per second.

REVIEW QUESTIONS

1. What is the difference between a conductor and an insulator?
2. What is a floating neutral and will it be found in the structural wire or on the service wire from the power company?
3. How can a static electrical arc that can hardly be felt be a source of ignition of a flammable gas?
4. What can be seen on the inside of the glass on a 20-amp plug fuse to show that it was a dead short?
5. Circuit breakers will trip from an overload. What other event or condition will cause the breaker to trip?
6. Arc fault circuit interrupters are installed in which room of a house and why?
7. Why does the code require covers to be placed on electrical panels and boxes?
8. If aluminum wire is to be used instead of copper, what size aluminum wire should be used if the code requires size 14 copper wire?
9. Describe a copper wire that has melted as the result of flame.
10. Describe a copper wire that is damaged as the result of an electrical short.

DISCUSSION QUESTIONS

1. Based on the basic concept of electricity, why is it so important for the investigator to have a strong knowledge of electrical components, their operation, and their failures?
2. Knowing the ramifications of spoliation, discuss the potential for firefighters working a fire scene and what they may do that could be labeled spoliation.

ACTIVITIES

1. Using the Internet, research issues dealing with fires that result from the use of aluminum wiring, in particular, the use of AL/CU devices and connectors.
2. Using the Internet, research the science of electricity and the actual path electrons take when flowing down a stretch of wire.

NOTES

1. Research and Education Association, *Basic Electricity* (Piscataway, NJ: Research and Education Association, 2002), Table 7.2, "Relative Conductance," 126.
2. InspectAPedia/ElectricAPedia, "Using COALR or CU-AL Electrical Outlets and Switches with Aluminum Wire," http://www.inspect-ny.com/aluminum/COALR.htm (accessed August 25, 2008).
3. National Fire Protection Association, *NFPA 921, Guide for Fire and Explosion Investigations,* (Quincy, MA: National Fire Protection Association, 2008), Chapter 3, "Definitions," section 3.3.7, 921-11.
4. Richard L. P. Custer, *NFPA Field Guide for Fire Investigators* (Quincy, MA: National Fire Protection Association), Table 4.38, 4-80.
5. Custer, *NFPA Field Guide for Fire Investigators,* Table 4.38, 4-80.
6. National Fire Protection Association, *NFPA 921, Guide for Fire and Explosion Investigations* (Quincy, MA: National Fire Protection Association, 2008), Chapter 8, "Electricity and Fire," 924-77.

PATTERNS: BURN AND SMOKE

Learning Objectives

Upon completion of this chapter, you should be able to:

- Describe and understand the concept of normal fire growth, direction, and the patterns fires create on building components.
- Describe the impact of suppression activities on fire growth and direction and how suppression can affect the resulting burn patterns on the structure and contents.
- Describe unique patterns associated with concrete, glass, and steel springs.
- Describe various burn pattern indicators that point to where the fire originated.

CASE STUDY

The restaurant was located in a typical strip mall where the storefront was mostly glass with a glass door. The floor plan showed the seating area in the left half of the occupancy going from front to back. The right half of the structure starts in the back with the kitchen. Restrooms occupy the front section of the right half with an office in the middle of the right half.

The business owner recently had employee problems and had changed the locks on the day of the fire. He was the only person with keys, all of which were in his possession at the time of the fire. The structure had been checked by the owner as he left the building and he was adamant that there were no persons left in the building when he locked up. He also affirmed that he personally locked both the front and back doors before leaving close to midnight.

He had parked his car near the road, quite some distance from the stores. As he was getting in his car, he heard glass breaking and looked up to see three young men in front of his store where the front glass had been broken out. He then saw one of them throw something else in the store, and he immediately saw flames. He yelled and the three young men took off. Instead of giving chase, he ran to a nearby phone booth and called the fire department.

Firefighters put a quick knockdown on the fire, and police were doing a search for three males. Several young people had been pulled over and questioned based on a weak description from the restaurant owner. However, shortly after the fire investigator arrived, he did a brief scene examination and then did an interview with the owner. He called law enforcement and called off the search for the three young men.

The fire consisted of the burning of a fuel, later identified as gasoline, and little else had a chance to reach ignition temperature. There was a brick just inside the window. However, the only fuel container found on the property was a five-gallon red plastic container located on the floor just outside the restroom door, near the front door. Both doors were still locked; there was no other access to the property except the broken front window.

The ignitable liquid pour pattern on the floor extended from the window, across the seating area to the restroom, and then made a left turn and went toward the back of the building. The pour pattern then took a right-hand turn, through the kitchen door opening and around the food preparation table in the middle of the kitchen and back out the second kitchen opening, looping back to the path leading from the restroom.

The owner, by his own admission, had exclusive opportunity to gain entry into the restaurant and pour the liquid. He had the keys and was in the vicinity when the fire started. Faced with the facts, the owner finally confessed. There were so many scenarios that could have played out in this particular case had the owner been more resourceful in his planning. But the burn patterns clearly refuted the owner's statement and allowed for a quick resolution to the case.

■ **Note**
The experience of seeing the action of ventilation on a burning structure as it changes the direction of fire travel can be invaluable to the investigator.

■ **Note**
Seeing the smoke and fire patterns created on structural surfaces and contents are an education.

■ **Note**
This is not to imply that only an experienced firefighter can be a proficient fire investigator, but it certainly makes sense that firefighters' experiences may help them transition from fighting the fire to exploring the cause of the fire.

■ **Note**
Fires move upward and outward.

INTRODUCTION

Neither text nor any class by itself can educate someone on interpretation of burn patterns and normal fire progression. The sum of one's experiences will help the individual become proficient in this field. These experiences as a firefighter include studying fire behavior from watching fires burn and observing how fires react under various conditions. The experience of seeing the action of ventilation on a burning structure as it changes the direction of fire travel can be invaluable to the investigator. Ventilation can be natural, such as results from high heat causing a window to break, or caused by a firefighter cutting a hole in the roof. Watching these things occur and then seeing the smoke and fire patterns created on structural surfaces and contents are an education all in themselves.

Through these experiences, the investigator will then be able to recognize the patterns left behind by fire and smoke on structural surfaces and contents so that the investigator can interpret the patterns and make a presumption on the fire direction that will lead back to the fire area of origin.

This is not to imply that only an experienced firefighter can be a proficient fire investigator, but it certainly makes sense that firefighters' experiences may help them transition from fighting the fire to exploring the cause of the fire. Others from the police community or engineers who investigate fire also have their own experiences and specialized knowledge that can help in many other aspects of the investigation. To enable them to interpret burn patterns they will need to undergo additional specialized training and take opportunities to see fires burn and to watch burn patterns develop.

Burn and smoke patterns are nothing more than the indicators left behind after the fire has been extinguished. By examining these patterns, the investigator can interpret the direction of fire travel, sudden growth of the fire, and eventually the area where the fire originated.

Burn patterns used in the analysis of the fire scene must also pass a test themselves. Does the scenario chosen by the investigator meet sound scientific principles? If this is not the case, then the analysis may be flawed. Science tells us that fires move upward and outward. It also tells us that with other influences, fires can increase dramatically in volume, can change direction, and can move downward as well. All of these factors must be taken into account and the final conclusions of the interpretation of the burn patterns should be supported by science.

In this chapter, you will learn about these patterns, including basic concepts about plumes, fire travel, and the impact on fire patterns of fire suppression actions. We discuss the simple patterns and the impact caused by a phenomenon called flashover on existing fire burn patterns. You will learn about the fuel load impact on patterns and the patterns on different types of materials.

PATTERNS

Note
Fire patterns are just as the name implies: they are the pattern of damage created on any material affected by the fire.

Fire patterns are just as the name implies: they are the pattern of damage created on any material affected by the fire. These damages can be from the flames, hot gases, or radiant heat. They can be seen on combustible and even on noncombustible surfaces in the form of heat marks or soot. The patterns are what the investigator examines with the hope of determining the direction of fire travel and eventually where the fire started.

Note
The patterns are what the investigator examines with the hope of determining the direction of fire travel and eventually where the fire started.

These patterns can be the charring of combustible materials such as wood or plastics. In some cases, the material may have been consumed and the absences of these materials are part of the burn pattern as well. Metal surfaces may show charring of the paint. Once the paint has burned away, or if the item had no paint, the metal surface may then **oxidize.** This oxidation pattern can also help the investigator. Some metals such as pewter, lead, and some pot metals have a low melting temperature and may melt if in the path of the fire. This too can show the direction of fire travel or can help the investigator by pointing back to the area of origin.

oxidize
the effect that heat has on a metal surface, in which it consumes any covering, resulting in the rusting of the metal surface

In addition to the partial burning, oxidizing, or total consumption of materials, other indicators can help as well. In the lesser damaged areas, materials may simply be discolored from heat, and some items may be warped or distorted as well. Of course, the smoke and sooting can aid in the investigation. (See **Figure 11-1.**) All of these indicators must be taken into account in the creation of the hypothesis of the origin and cause of the fire.

Figure 11-1 *Patterns on furnishings may include sooting, heat changes, and oxidation.*

■ **Note**
The use of these patterns may not be as simple as it may sound.

■ **Note**
Under most circumstances, normal fire growth is upward and outward.

Not to complicate things, but to put this in perspective, the use of these patterns may not be as simple as it may sound. The fire could start in an area of limited combustible materials. As the fire grows and moves from one room to the next it may encounter materials that are capable of great heat release. That same material may also produce high volumes of smoke and soot, creating a more devastating-appearing image than in the room of origin. Then, take into account that the fire breeches a window, which may then allow air currents to enter the structure and change the direction of fire travel, creating deeper char as the fresh air feeds the fire.

The ultimate goal of examining the patterns is to find the least damaged area. The investigator then works toward the most damaged area, which may lead to the area of origin. The investigator must take into account, and eliminate, any patterns created by ventilation of a larger fuel load in the path of the fire.

BASIC FIRE PATTERNS

■ **Note**
Because of direct fire impingement, radiant heat, and other factors a fire will also burn downward but at a much slower rate.

■ **Note**
This brings us to one of the tools every investigator needs: common sense.

Under most circumstances, normal fire growth is upward and outward. In this process, the initial patterns on the wall will have the appearance of V if the fire is near a vertical surface. The V is created by the flames and heat of the fire as it impacts the wall. The width of the V is dependent on the size and width of the fire that is burning, nothing more. As the fire increases in size, so will the pattern that is created. The longer the fire burns in any area, the more damage there will be in that area. Not all fire damage is directed upward and outward, however. Because of direct fire impingement, radiant heat, and other factors a fire will also burn downward but at a much slower rate. This brings us to one of the tools every investigator needs: common sense.

The question that needs to be asked is whether the patterns left behind make sense. Is this what should be expected to be seen under the current known circumstances? If something does not seem as it should, it is the investigator's duty to search until all questions can be answered. This is all part of the analytical process.

If the fuel has a low heat release rate, which creates a fire with shorter flames, the effect on the wall may be an inverted V pattern. As the fire grows in volume and intensity, this pattern may end up being masked by new patterns with the potential of destroying any evidence of the initial inverted V.

Natural progression of the fire may create an hourglass shape. (See **Figure 11-2.**) This occurs when the bottom of the pattern is created by the natural shape of the flames created by air entrainment where the flames lean inward. This air is not always equal from all sides, so the fire may appear skewed to one side or the other depending on the air currents. As the fire intensifies, it releases more heat and larger flames that produce the upward and outward flow. This in turn creates an hourglass shape, which is a natural pattern shape that may be seen by the investigator.

Figure 11-2 *The inverted V can be a small fire that burned itself out or one that continues to burn, possibly creating the hourglass shape, which was partially created as the result of air entrainment. The third type of pattern is the V pattern that points back down to the area of origin.*

This same type of pattern can sometimes be seen in areas of the structure other than the area of origin. These patterns could be caused by burning material that falls on other combustibles, possibly at the floor level. An example would be curtain material that is ignited at ceiling level as a result of the buildup of the heat layer in that area. As the curtains start to burn, they then fall off the curtain rod, dropping to the floor. This creates a second, floor-level pattern on the wall. The location and the fact that the burn pattern is less intense may explain that it was a drop-down fire and not a second fire set.

There may in fact be multiple fire sets in the structure. The investigator must be cautioned not to jump to conclusions that the fire was incendiary in such cases. By following the analytical approach, the investigator will search for a reason for these multiple areas of origin. An example is an industrial occupancy where a large electrical wire shorts out and the overcurrent protection does not activate. The wire will heat evenly across its entire length, igniting similar materials in proximity to the wiring.

■ **Note**
The actual patterns to be interpreted are usually two-dimensional, but the investigator should be visualizing the fire in a three-dimensional form.

The actual patterns to be interpreted are usually two-dimensional, but the investigator should be visualizing the fire in a three-dimensional form. The fire as it starts, with no outside influences, will resemble a cone, starting narrow at the bottom with the wide section at the top. Depending on the distance from objects, the cone may eventually intersect and come in contact with walls or contents. When the cone comes in contact with the wall, as shown in **Figure 11-3,** it creates a two-dimensional pattern on the wall. This pattern may resemble a U shape. By

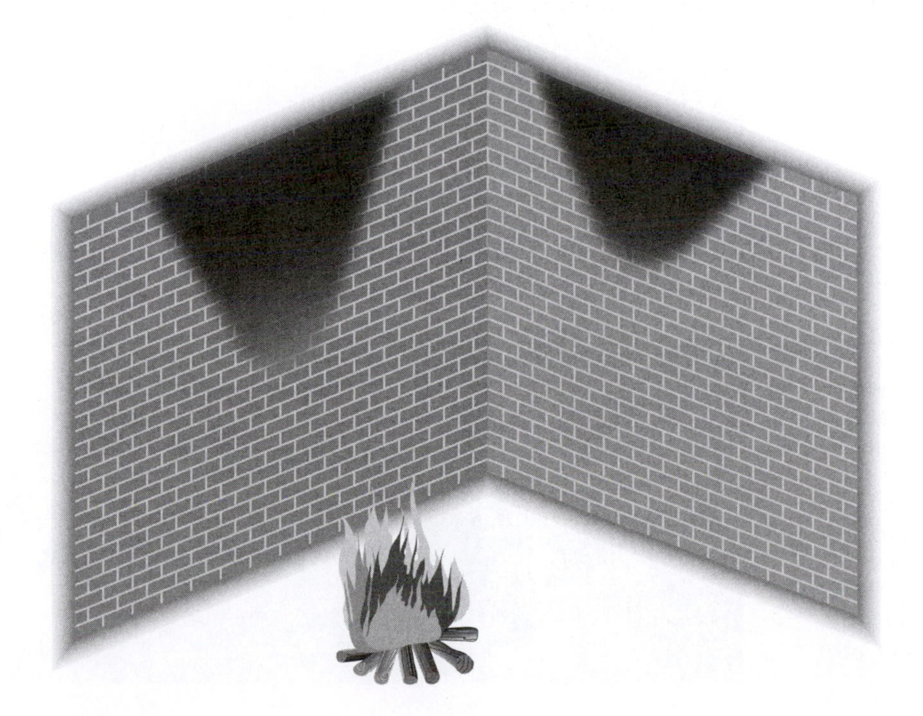

Figure 11-3 *A cone pattern from the fire intersects and leaves marks on the wall and contents.*

using the height of the pattern on the walls and marks on the contents, it may be easy for you to see the entire cone pattern in your mind's eye. This in turn may help the investigator narrow down the search for the area of origin.

The ceiling area directly over the cone, if still intact, will show its own pattern. (See **Figure 11-4.**) It may be circular in nature or slightly skewed if there were any winds or air movement. The amount of fuel can also influence a pattern. The fire could start in a limited-fuel area, but when it reaches an abundance of fuel it can create a large pattern, even sometimes masking the original pattern.

When fuels have a high heat release rate they burn with more vigor, giving off gases and flames that create more of a vertical pattern, straight up or out, creating the V pattern. This type of fuel creates more heat at a faster rate, building up in confined spaces in the upper areas near ceilings, and banking down and leaving distinct marks on walls and furnishings. This upper heated area also affects other items in the room by radiating the heat back downward, creating patterns on the top surface of contents.

■ Note
Generally, the banking down of the heat layer creates a solid line, a clear delineation along the walls and furnishings.

Generally, the banking down of the heat layer creates a solid line, a clear delineation along the walls and furnishings. However, the heat layer can be somewhat fluid, creating occasional wavy lines. This fluidity is evident when the thermal layer of hot gases, flowing across the ceiling, comes in contact with an obstacle such as a wall. At this point, the hot air and gases hit the wall and roll down and then back up. This creates a unique pattern with a slightly lower burn than the level of the rest of the thermal layer. An example is shown in **Figure 11-5,** where the fire started in the room to the right; you can see the

Figure 11-4 *A circular pattern directly over the area of origin. What was initially a sooted area was then burned clean as the fire continued to burn and grow.*

Figure 11-5 *The patterns show the lower burn in the room to the right. The impact of the flowing gases from right to left created a rollover pattern on the wall to the left as the gases and heat hit the wall, and then rolled down and back up, causing a slight dip in the pattern.*

upward movement of the heat on the trim work of the opening to the room. On the left, you can see the result of the rolling effect on the wall. This slightly lower mark is where the thermal layer hit the wall and rolled down and then back up, staying buoyant.

As the fire spreads, it will go through openings. One way to look at the flow of this layer of hot gases is to think of it as a fluid. But unlike liquids collecting downward, this layer collects and flows upward. In your mind's eye, consider the fluidity of the gases as they collect at the ceiling and bank down. Once they reach the door opening to the next room they will roll over the upper frame of the door and spill into the ceiling area of the next room. If the door is closed, the hot gases and smoke would then leak through the cracks and small openings, no different from what a liquid would do at floor level. As these hot gases pass along their way out of the room of origin, they will continue to damage the ceiling area and upper portions of the room, creating patterns on the wall and on the doorframe.

The patterns will lead back to the area of origin. It can be in the basement, living area, or in the attic. That origin can be at floor level, on the wall such as at an electrical outlet, about 3 feet off the floor as on a range or countertop, or at the ceiling such as a failed light fixture. An open mind and an analytical methodology will lead the investigator to a sound conclusion.

What you also must remember is that fuel load in a given area may show the heaviest char and will need to be examined to eliminate whether that is the case. Ventilation plays a role in the char patterns as well. Where there is ventilation, there will be an abundance of air that will enhance the combustion process, creating more heat and in turn creating more char. This too may need to be eliminated if it is not the area of origin. The culmination of all the facts leads to a determination of an area of origin and an eventual cause of the fire. The following sections discuss some areas where patterns can be found and examined.

CEILINGS

Patterns on the ceiling are a result of the mere fact that heat rises and is stopped only by an obstacle such as a ceiling. The ceiling area may or may not be intact. Because it may be the first area to be affected by the buildup of heat, it is also the area that may be the first to fail. Most ceilings are made of drywall, and as the gypsum in the drywall breaks down, it weakens and may fall on its own accord, or, with the addition of suppression water, it may get heavy and drop to the floor, taking most of its pattern with it. However, the debris on the floor from the ceiling may hold clues, as will the structural elements that held the ceiling drywall. First and foremost, the heaviest area of damage may be in the area of origin.

Note Most scenes will have patterns across the entire floor of origin.

Note The investigator needs to observe the patterns on the ceiling and compare them to the patterns on the walls.

It is rare but sometimes the patterns are limited to an oval pattern on the ceiling when the area of origin is directly below that pattern. Most scenes will have patterns across the entire floor of origin. The original direction of fire travel may have been down a south-facing hall only to have the structure ventilate itself and strong winds push the pattern back on itself, taking another direction.

The investigator needs to observe the patterns on the ceiling and compare them to the patterns on the walls. Together they may show heavier damage in the direction of the kitchen. Working toward that area, the patterns on the wall may be lower to the floor. Just outside the kitchen, there is a breakfast nook that has massive damage. There is a lot of debris in this area, which shows that the bench seats were padded, there were lots of newspapers on the table, and evidence indicates that the window failed early, allowing the fire to burn more intensely.

Realizing that this may be a fuel load and ventilation effect, you must observe the patterns in the rest of the room, which show heavy damage in the kitchen cabinets. A "v" pattern goes up the cabinets, the ceiling directly overhead is missing drywall, and the ceiling rafters have fire damage.

By itself, this is not a final determination. Other things must be determined such as potential source of heat and first material ignited. Interviews and a thorough investigation will solidify the final determination. But it was the observation of the ceiling and wall damage that lead the investigator to the room of origin.

WALL SURFACES

■ **Note**
As an investigator, do not rely on what the code mandates, but go into any structure with an open mind—you never know what you may find.

drywall
also called sheetrock; a wallboard of gypsum with coating of paper on each side

There are several types of wall surfaces, both commercially installed and self-designed. In commercial establishments, there may be fire resistance requirements different from those of residential structures. However, occupants of either commercial or residential structures can do some unusual designs that may change the fuel load dynamics. This could range from a concrete wall with no covering to an office in an airplane hanger where the occupant uses old silk parachutes for wall coverings. As an investigator, do not rely on what the code mandates, but go into any structure with an open mind—you never know what you may find. There are some specific issues that may come up depending on certain wall finishes. The next few sections address specific issues that you may encounter with various construction features and finishes on walls.

Drywall

■ **Note**
The gypsum will start to break down as the moisture is driven off in a process referred to as calcination.

calcination
the process of driving off the moisture in the gypsum, discoloring and softening it in the process; the process includes the driving off of moisture that is chemically bonded within the gypsum

The most commonly encountered wall material is **drywall,** which in some areas is referred to as *sheetrock* or *gypsum board*. Drywall consists of gypsum plaster in a core that is wrapped with strong paper. The drywall surface can be painted, covered in stucco or wallpaper, and can even be covered with veneered wood paneling or any other finish. The addition of these coverings changes the way the wall will react in a fire situation. Sometimes wall coverings retard the fire progress, but more often than not they increase the fuel load.

In general, drywall itself is fire retardant. It will resist burning but will eventually fail as the fire progresses. The paper surface will be the first to burn away, and then the gypsum will start to break down as the moisture is driven off in a process referred to as **calcination.** As this takes place, the gypsum allow heat to transfer through itself and burn off the paper on the other side. When the moisture is driven off, the gypsum takes on a whitish appearance. In this state, it may start to lose its cohesiveness and will weaken. All it could take to fail completely is the application of a fire stream from suppression forces. The drywall in the ceiling may fall from its own weight or the added weight of moisture from the fire streams.

■ **Note**
When conducting the examination of the calcination of the drywall, it is critical to make sure that like surfaces are being compared.

The degree of damage to the drywall paper or the gypsum around the room may vary, with more damage in one area than another. Any difference may be from either a localized high fuel load or may indicate a longer burn time and thus indicate the direction of fire travel. (See **Figure 11-6.**) Gauges on the market allow the investigator to measure the depth of the calcination to measure the amount of damage at any given point in the room. When conducting the examination of the calcination of the drywall, it is critical to make sure that like surfaces are being compared.

Figure 11-6
Sheetrock with the paper burned away, leaving the gypsum exposed to the fire and fire gases. Comparing these walls to adjacent walls may help determine the direction of fire travel. The fire, in this case, came from the room to the left and flowed into the hall on the right.

Suppression water applied to areas of drywall calcination may cause it to fail and fall to the floor. These fire streams may have been necessary to stop the fire. Although firefighters are trained not to flow water unless necessary, the less experienced firefighter may not give this a thought. Such firefighters may need to be educated by the fire officer or fire investigator as to how water flow when not necessary can destroy evidence.

Note that manufacturers also produce some drywall products to meet certain fire-resistive ratings. To allow a longer fire protection, additional bonding fibers are added to the gypsum during the manufacturing process. Drywall also comes in different thicknesses and lengths. The length is to make the hanging process easier, creating less joints to tape and seal. The thickness of the drywall is for fire resistance and soundproofing. In some industrial occupancies there may be more than one layer of drywall to give the wall a better fire protection rating.

Wood Walls

Paneling is wood based but may have a plastic surface. The plastic and the wood of similar pieces of paneling can be compared to determine the direction of fire travel or to show where there was secondary burning of a large fuel load in the path of the fire. Be sure to compare like surfaces in the comparison, just as for drywall.

Paneling also has a tendency to curl as it burns, and because of its combustible nature, it may disappear altogether. When the paneling has burned away, it exposes the material underneath. If this should be a concrete wall such as in a basement or in commercial structures, the installer may have used furring strips attached to the concrete to provide spacing and a surface to attach the paneling. Unlike a wood stud wall where the 2 × 4s go from floor to ceiling, the furring strips may not do so. Individual strips may be spaced at the top half and matching that vertical line a short strip may be placed on the bottom part of the wall. This allows scrap wood or less expensive lumber to be used. The point is that the investigator should be cautious to interpret the gap in the strips as being burned away when in fact the were always just two separate pieces of wood. (See **Figure 11-7.**)

■ Note
Oak will not char and blister the same as pine or ash will. This type of differing char blisters is of no significance to the investigator.

Wood boards may be placed either horizontally or vertically on the walls to give a rustic appearance. There could also be different types of wood as well. Oak will not char and blister the same as pine or ash will. This type of differing char blisters is of no significance to the investigator. What is important is whether there is a line of demarcation between where there is char or no char, or in some circumstances deep char with much lighter char. This **demarcation** is a measurement that can be used to show fire or heat path.

demarcation
a distinct line between a fire-damaged area and nondamaged areas; not a gradual charring but an abrupt end

Wall Paper

Wallpaper can be actual paper, cloth, or plastic. It can even be a combination of any of those materials. The charring of the paper needs to be compared with the same type in that room or area. Some paper with plastic coating will bubble or blister, leaving an unusual pattern. Some cloth coverings may have raised areas in the design that will be more prone to ignition. If using the patterns on a wall to indicate a fuel load or the direction of fire travel, the investigator must be sure to compare like surfaces for a more accurate interpretation.

■ Note
Wallpaper can be actual paper, cloth, or plastic. It can even be a combination of any of those materials.

Figure 11-7 *Missing sections of furring strips on concrete walls require close examination; they may not have been burned away so much as shorter pieces were used in construction.*

A kitchen is an area where there may be differing burn patterns on walls. In one area, the wall surface may be covered with wallpaper that has a plastic film. This may have been done to keep grease and spattered oils from staining the walls. The adjacent wall in the dining room may be just drywall with a coating of paint. The plastic in the wall covering is an additional fuel, which may create a visibly different char than any burning covering on the drywall in the dining room, even though the dining room may have been exposed to more heat or for a longer period of time. This char effect from the plastic film on wallpaper is shown in **Figure 11-8.**

Concrete Walls

First, you must ensure that you are dealing with true concrete. New design techniques enable builders to create architectural designs by using Styrofoam, which is then covered with a cloth material and sprayed with a water-resistant material or concrete. This results in a finish that looks like stucco. This has been used predominantly on exteriors but has sometimes been used in interiors as well.

spalling
when the surface of concrete pops off as a result of water in the concrete reaching boiling temperature and turning to steam

Concrete and stucco on walls will pop off as the result of heat. This is called **spalling** and is the result of moisture within the concrete being heated to its boiling point, 212°F, where it converts to steam. In the transformation from a liquid to a gas (steam), it expands 1,700 times its volume. That means 1 cubic inch of water now becomes 1,700 cubic inches of steam. This action can be forceful enough to pop off the surface of the concrete, known as spalling. The ability of

Figure 11-8 *Wallpaper with a plastic film on the surface will burn differently from other types of paper surfaces. The baseboard has been removed, revealing a pour pattern where an ignitable liquid ran down the wall and collected behind the baseboard. Sample was collected and came back from the laboratory as positive for gasoline.*

the moisture in concrete to reach the boiling point can be the result of captured heat, such as under stairwells or in small areas where reflected heat can cause this to occur, and may not indicate the presence of addition fuel or proximity to the area of origin. Depending on the structure, the concrete may not contain any appreciable moisture, so there may not always be spalling.

Some concrete walls are sealed with a finish to limit moisture absorption or for decoration. The sealer in some cases can become a fuel as well. It would be wise to find out as much about the construction and postconstruction changes to the structure should there be an unusual burn pattern on a concrete wall.

Like any other wall surface, the investigator should examine the wall for heat and soot patterns. If there was a finish on the concrete walls, it may have burned off in some areas and not others. All these factors add clues and indicators, allowing the investigator to work toward the area of origin.

DOORS

As the fire grows and the thermal gas layer expands, moving outward away from the area of origin, it will eventually impact at least one door. Doors can exhibit the same patterns as walls; in some cases, there is more damage on doors because the walls are made of fire-resistive sheetrock and the doors may be combustible wood. The patterns may also indicate whether the door was open or closed at the time of the fire.

When a door is closed, the gap between the door and the doorframe may be tight enough to limit escaping gases and heat. Frequently, this is more true on the door edge with the hinges. As such, the surface of the door may show more damage than the edge if the door was closed at the time of the fire. By examining the patterns on the door, the investigator should be able to tell whether the door was open or closed at the time of the fire. The most telling clue will be the lack of char on the inside edge of the door at the hinges, which indicates that the door was closed at the time of the fire.

When the door is closed at the time of the fire, the top of the door will most likely experience heat. As the fire burns, air is pulled in through any space at the bottom of the door. Thus, the top will be burned and the bottom may show a lesser degree of fire damage. However, as with most issues in fire investigations, there is no one answer for everything. Should there be drop-down near or at the base of the door, with the available air you may find fire damage along the bottom of the door. Likewise, should the fire be incendiary in nature and an ignitable liquid was used and poured through the doorway, then the bottom of the door will be burnt from this as well. By looking at the scene indicators, you should be able to determine which scenario may have caused the charring under the bottom edge of the door.

If the door was open at the time of the fire, it should have the same level of damage on the face of the door as on the wall. Depending on the distance of the door from the area of origin, the door may even be part of the V pattern showing the upward and outward flow of the fire, as shown in **Figure 11-9.**

■ Note

Flashover within a room where the doors were open or closed may create so much charring as to obscure and eliminate all previous patterns.

Flashover within a room where the doors were open or closed may create so much charring as to obscure and eliminate all previous patterns. This does not necessitate you discard any patterns found or stop looking for patterns in a room that experienced flashover.

■ Note

It would be of interest if suppression forces had to force any doors or windows in the carrying out of their duties.

Exterior doors should be checked to determine whether they were open or closed at the time of the fire. Close examination should be conducted to see if there were any signs of forced entry. During the follow-up meeting with fire suppression personnel, the investigator should verify the condition of the doors when the firefighters arrived on the scene. It would be of interest if suppression forces had to force any doors or windows in the carrying out of their duties. Also verify whether they tried to see if the door was unlocked prior to forcing. This may seem like a foolish question, but in the heat of the moment there have been incidents where given an order to force the door, the firefighter did just that—forced it before trying the doorknob.

The investigator also needs to check with other emergency services personnel that arrived on the scene prior to the fire suppression personnel. A police officer or rescue squad personnel may have attempted a rescue, forcing the door. Or they may have found the door open or unlocked; both situations are

Figure 11-9 *Door was open at the time of the fire and the V pattern is evident on the surface of the door. The edge of the door and doorjamb where the hinges are located are also charred, indicating that the door was open at the time of the fire.*

of interest. Neighbors and even strangers have been known to attempt a rescue when they see flames. Check with witnesses to see if this may or may not have been the case.

Any marks on doors that would indicate forced entry should be examined closely, documented, and photographed. If you are not experienced in this area yet, check with local law enforcement. Most police agencies have experts in this area and should be able to provide someone who has expertise in taking tool mark impressions for later use in the investigation. During this process, there may be additional evidence discovered in the form of paint chips or metal fragments that may identify the tool used to force the entry.

WINDOWS AND GLASS

Windows have a lot to tell the fire investigator. Patterns on windows close to the area of origin all too often are heavily damaged because of the availability of air and the fire burning through the window. Glass will fail relatively quickly in a fire. If there was any pressure from an explosion of any type, the glass may fail, sending shards into the yard. When working from the least damaged area to the most damaged area, at some point in the process the investigator will look at the yard and surrounding area of the structure. Glass shards found in this area need to be examined closely. If they are sooted, it may indicate that the fire was burning, soot collected on the glass, and then an explosion occurred, sending the broken, sooted shards of glass to the outside of the structure.

■ Note
If there is an indication that the glass found in the yard came from a window of the structure and it is not sooted, it may indicate that the explosion came first, which in turn may have started the fire.

If there is an indication that the glass found in the yard came from a window of the structure and it is not sooted, it may indicate that the explosion came first, which in turn may have started the fire. The investigator should wait before rushing into such an assumption because more investigation will need to be done. Ask yourself whether the fire could have been burning in another area of the structure with a room being protected by a closed door. The fire could have reached a fuel that caused an explosion, causing the door to that room to fail and in turn impacting the unsooted glass and sending it into the yard. Granted, it may seem far-fetched, but now is the time to get these answers to ensure a proper and accurate hypothesis of the fire area of origin and cause can be formulated.

■ Note
If the investigator finds clean glass at the bottom, the window may have been broken prior to the fire and will bring on a whole new range of questions as to what or who may have done this and why.

The glass may not be far out in the yard. In the process of a detailed investigation, the investigator will want to look directly below a window that has failed as the result of the fire. By looking on the ground under the window on the outside, and on the floor under the window on the inside, you can remove the fire debris in layers until you get to the floor or the grass/ground. If the glass in both circumstances is sooted, then the window will have most likely failed as the result of the fire. If the investigator finds clean glass at the bottom, the window may have been broken prior to the fire and will bring on a whole new range of questions as to what or who may have done this and why. Most important, it is critical to keep an open mind and let the evidence solve the puzzle, not speculation.

In any given structure fire, varying types of damage to glass can occur. Sometimes the glass will crack in half-moon-type cracks from heat. When more heat is applied, the glass might melt, falling down on itself in multiple folds, or it can craze.

Crazing of the glass is when multiple small cracks that don't quite go all the way through the glass appear. This same effect occurs in pottery when the pots are pulled from the kiln and sprayed with a mist of water. This crackling or crazing of the glaze on pottery has been used for centuries. Thus, crazing of glass in a structural fire is most likely the result of rapid cooling, possibly from the application of suppression water spray, and nothing more.

Reconstruction of broken glass bottles may have a value in a fire investigation. They can be cracked, broken, or even crazed. If enough of the glass can be collected and if there is good evidentiary reason for the reconstruction, then a laboratory may have the expertise and ability to undertake the task. However, tempered glass, which is found on glass doors, car windows, and television screens, cannot be reconstructed. When it breaks, it shatters into many small fragments rather than sharp shards, making it safer than regular glass. (See **Figure 11-10.**)

■ Note
Burn patterns on windows also give some clues as to the fire's development.

Burn patterns on windows also give some clues as to the fire's development. It is understood that if the window has been mostly consumed, there is limited value in its remains. If it was a wood sash window, the burn pattern may indicate whether the window was open or closed as the flames impinged on the window. An open single-hung window, where only the bottom section moves up and down, will have protected area where the window was sitting at the time of the fire. If there was no damaged from the flames, soot patterns may give the same information.

Other types of windows may offer clues as well. Several models of windows require a crank to open the window. On jalousie windows, slats of horizontal glass all tilt open as the crank handle is turned. On awning and casement windows, one, two, or more sections of glass open independently when the handle is turned.

These windows can have wood or aluminum framing. Aluminum frames melt easily and can possibly leave a pattern on the outside showing that the window was pivoted out at the time of the fire. The handles themselves are made of a metal alloy, such as aluminum, and may melt early in the fire. Sometimes the residue from the handle may be present, but in mass destruction of that area it may be impossible to tell whether the handles were off or on at the time of the fire.

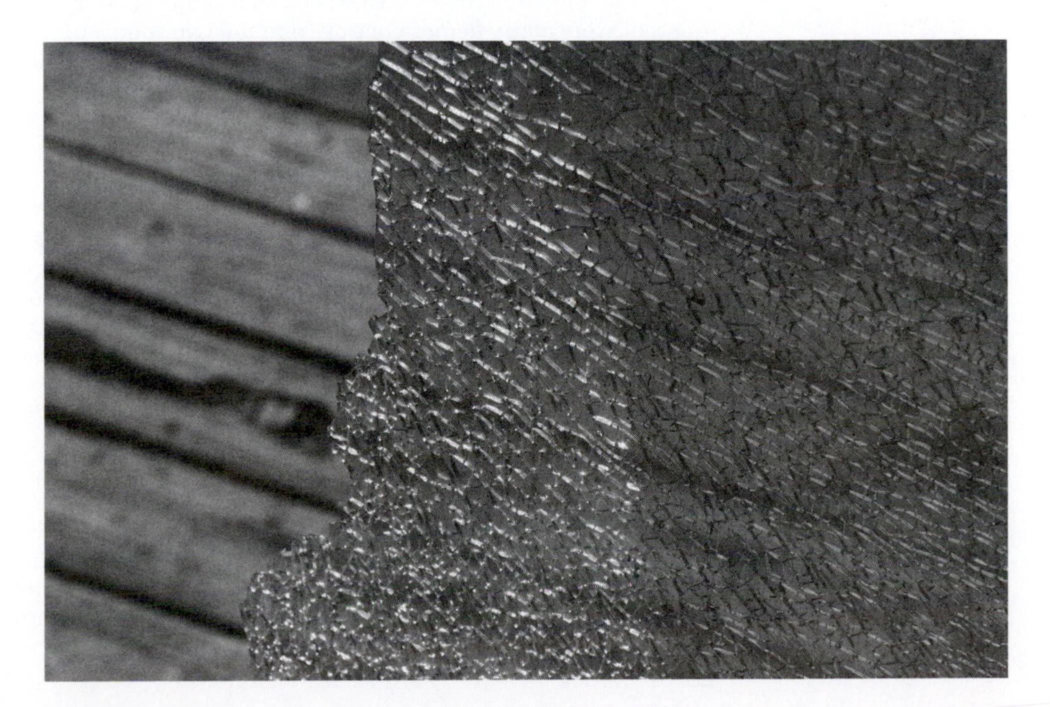

Figure 11-10
Tempered glass broken but still remaining in the door frame. This is not crazing but the result of design to make the glass safer should it break.

The presence or lack of presence of the handles may yield a clue. Some families have taken the handles off to prevent young children from opening windows, especially on upper floors. But it would be of interest to find the windows open and the handles missing.

An Engineer's Observation

The structure was large, even for the neighborhood. It was a two-story L-shaped structure, with a two-car garage with rooms over the garage on the shorter wing. The front of the house was classic in design with evenly spaced casement windows on both the first and second floors. However, the back of the house was almost all sliding glass doors looking out over a short backyard and into a scenic bay with houses lined up on the other side of the bay.

The first 9-1-1 call came in for smoke in the neighborhood and a single engine was dispatched. Before the engine arrived, 9-1-1 calls started coming in for a structure on fire with fire coming from the roof. As the dispatcher reported this information, the first engine arrived and reported fire through the roof. A supply line was laid and an attack line was stretched through the front door. Because the door was locked, forced entry was required, which took time for the three-man crew on the engine to accomplish.

The driver/engineer on the first fire truck had everything lined up and took a moment to look around. It was then that he noticed that fire was coming from the room on each end of the L-shaped structure, but there did not seem to be any fire in the middle of the structure. The flames that appeared to be coming through the roof were really flames coming from the two rooms on the second floor level. The roof was still intact. This structure was more than 6,000 square feet, so there was a lot of room between the two ends of the building.

Something just did not look right. Then, it hit him. He was looking through the windows directly through the house at the lights from the homes on the other side of the small bay. He then took further stock of the situation and saw that the casement windows were all opened, on both the first and second floors. This was unusual because it was December and the temperature was just above freezing. Even more important was that he could see into the windows on the second floor in the center of the house and there was no smoke. He pointed this out to another firefighter who arrived with the second alarm. After getting attack lines in place, the firefighter took a quick check and found that all the casement windows on the first floor had no handles and could not be closed.

Before additional attack lines could be laid, the entire second floor became involved in fire and it was decided to do an exterior attack. No one was able to confirm whether handles were in place on the second-floor windows, and there was too much fire damage to verify one way or another.

Needless to say, the observations of the engineer on the first engine proved invaluable to the investigator. All evidence pointed toward the homeowners being involved in setting the fire. Although an arson charge could not be established, the insurance company denied the insurance claim based on insurance fraud.

arrow patterns
as described in the NFPA 921, the patterns on wooden structural members that show a direction or path of fire travel

WOOD FRAMING

Once the wall coverings have burned away, they expose to the fire the structure's wood framing. The wood framing may provide a pattern that can help to determine the direction of fire travel or the fire origin. NFPA 921 uses the term **arrow patterns** to describe the patterns left on wood structural elements. When examined, arrow patterns point pack toward the area of origin. Wall studs farther from the area of origin have damage at a higher level than do those studs that are closer to the area of origin. As the pattern gets lower and lower, it simulates an arrow pointing back toward the area of origin, as shown in **Figure 11-11.** The studs themselves may show more rounded edges on the side facing the area of origin.

■ **Note**
As the pattern gets lower and lower, it simulates an arrow pointing back toward the area of origin.

As the wood studs are subjected to flames and heat, they start to char. There is no exact science that can definitively state that the fire burned for a certain length of time with a certain amount of char. But comparison of the depth of the char on wood framing can help to give a direction of fire travel. The depth of the char can be compared from stud to stud and from wall to wall. The investigator is cautioned that fuel load may cause heavy charring in an area that is not the area of origin. Also, ventilation can create a situation where there is more char

Figure 11-11 *Arrow patterns pointing back toward the area of origin in the corner. As shown, there may be more than one arrow pattern in any given area.*

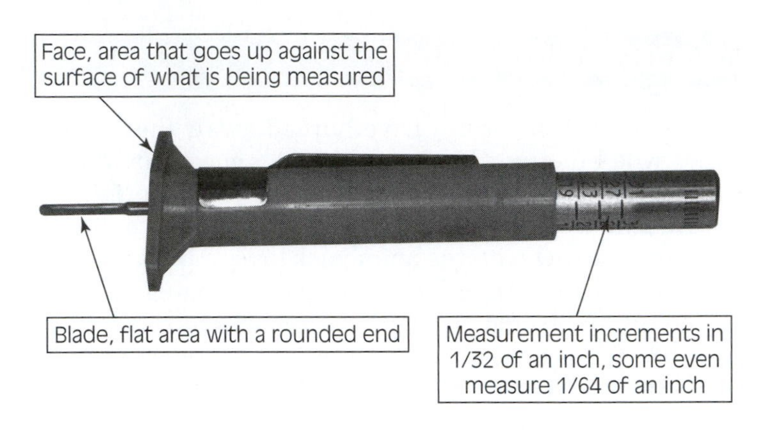

Figure 11-12 *A depth gauge showing the nomenclature. This device is more commonly used in the automotive tire industry to measure tread depth.*

because the fire burned more intensely near an opening such as a failed window or door.

Note
The depth of the char can be measured with a simple, inexpensive device that you can find at most automotive hardware stores.

The depth of the char can be measured with a simple, inexpensive device that you can find at most automotive hardware stores. A tire depth gauge in increments of 64ths or 32nds of an inch is an ideal tool with its measured increments and blunt blade on the end. (See **Figure 11-12.**) One person should take all measurements because the blade must be pushed into the char with the same pressure each time a measurement is taken. If the studs still maintain their original shape and size, then the tool can measure with the face against the char on the stud. The measurement can be read on the upper side of the device.

However, some studs may have experienced some partial consumption as a result of the fire, and you must compensate for that missing material. You can cut an unburned section of a stud, and then place it flush (even with) and parallel to the stud being measured. This will then show where there is missing char. Take the measurement from the edge of the char to the edge of the sample stud; this measurement is the missing char. Then, push the depth gauge into the char as before, and take the reading. Add the two measurements together and this will be the total char. Using this same method, go around the rest of the area. Measurements can then be charted and compared to find the heaviest damage. (See **Figure 11-13.**)

Note
Taking into account excessive fuel load and ventilation, the variations of the depth of char should show a direction of fire travel. This may point back to the area of origin.

By comparing the measurements around the room, taking into account excessive fuel load and ventilation, the variations of the depth of char should show a direction of fire travel. This may point back to the area of origin.

FLOORS

Patterns on the floor can be of great value depending on the circumstances. Before you use these patterns, you must consider certain conditions. If the room went to flashover, then any previous pattern may be destroyed as the result of the massive heat when combustibles in the room reached their ignition temperature.

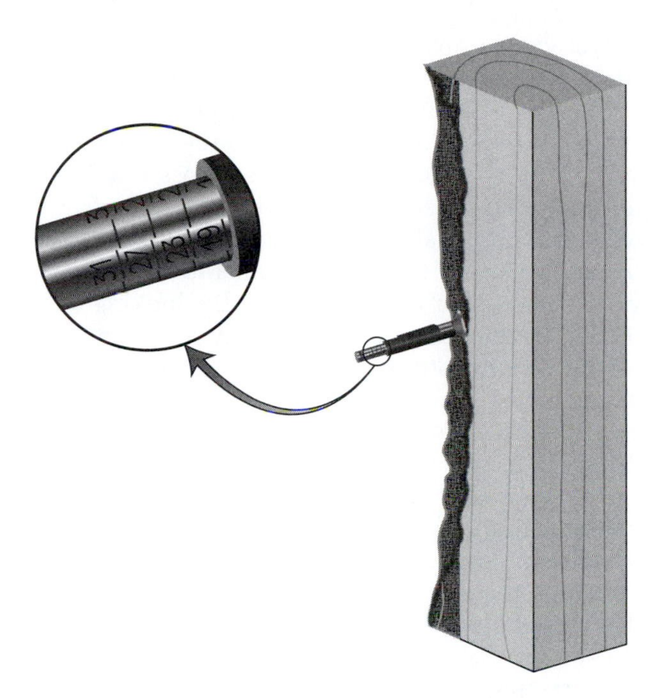

Figure 11-13 *Diagram showing the use of a depth char gauge. To use this gauge, hold the face of the gauge against the char, push plunger in with an even pressure, and allow blade to sink into soft char. Take the measurement. This is the depth of the char. Then, take an unburned 2 × 4 stud and line it up where the front of the burned stud would have been. Take a measurement to identify how much of the stud burned away and is missing. Add the two together and that will be the total impact on the stud. Do the same in other areas of the room where there is char for comparison.*

■ Note
Patterns left after flashover can be misinterpreted for something that they are not. Previous studies have shown that postflashover floor patterns can resemble ignitable liquid burn patterns.

Patterns left after flashover can be misinterpreted for something that they are not. Previous studies have shown that postflashover floor patterns can resemble ignitable liquid burn patterns.

Other cautions include situations involving worn wood floors in very old country stores. The common areas in the aisles and at the register can wear down over the years. This may not be noticeable on a normal day. But these same wear patterns have exposed wood fibers and created a rough surface that can char differently from areas that have not worn away. In some instances, the charring of these wear patterns may look similar to a pour pattern of an ignitable liquid.

Floor patterns are great indicators in many circumstances. In small simple fires starting near or at floor level, the pattern may clearly indicate the area of origin because it is the lowest burn. There may be circumstances, however, in which there is more than one low burn point. This can be the result of multiple points of origin, or there is one area of origin and other low burns from burning materials falling down, creating subsequent low burn points. By taking into account fuel load and other evidentiary clues, the investigator should be able to discern which low burn was the area of origin.

However, not all fires are small and simple. The vast majority leave the floor area covered in debris. When this happens, the first action that must be taken is to ensure there is no more damage to the structure. Barricades should be established to stop anyone from walking across the floors in and around the area of origin unless it is absolutely necessary. If the area of origin has not been identified, then all potential areas of origin must be protected.

■ Note
After the scene is documented, the investigator can then start removing debris in layers, like an archaeology dig, looking for evidence.

After the scene is documented, the investigator can then start removing debris in layers, like an archaeology dig, looking for evidence in each of the layers. Once the debris has been carefully removed, the investigator can document and examine any patterns on the floor. One mistake an investigator can make is not to remove all the debris in and around the area of origin. Leaving any area unexamined will most likely lead to faulty conclusions.

Holes in the floor are always a safety concern. As debris drops onto the floor, burning embers may continue to burn even as additional debris continues to fall, covering the burning embers and protecting them. As a result of the continued combustion process of the coals, there is potential that the coals may burn all the way through the floor over time.

Holes can also burn down through the floor as the result of flame impingement from ventilation or from the underside of contents burning such as a couch or bed, creating radiation or flame impingement on the floor. Although this is not a rare event, it is one that would need to be explained to ensure an accurate determination of the fire and its progress is made.

Holes can be created by heat and flame impingement on the underside of the floor as well. This may give an indication that the fire started on the level below the floor where the hole is located. This would be a natural progression of the fire and would be expected if the fuel load below the floor can warrant enough energy to create the damage.

■ Note
The direction of fire travel can usually be determined by looking at the burn pattern around the hole itself.

The direction of fire travel can usually be determined by looking at the burn pattern around the hole itself. Provided the hole was protected and was not damaged by suppression operations, there will be an indication that there was more fire on one side or the other. When the damage around the hole is more on top than on bottom, there will be a slope on the top side, like the beginning of a crater (see **Figure 11-14**). The edges will look just like someone used a scoop to scoop out the upper layers of the floor with the same slope on the edges. When the fire burns from below, it has the reverse effect: there will be more damage on the under side. The slope of the sides of the hole will be toward the center of the top layer, with the underside surface being wider than the top side, just like someone used a scoop from below.

Different types of floor finishes can create some unique situations. The following subsections discuss some examples of different floor finishes and what you might see in your investigation.

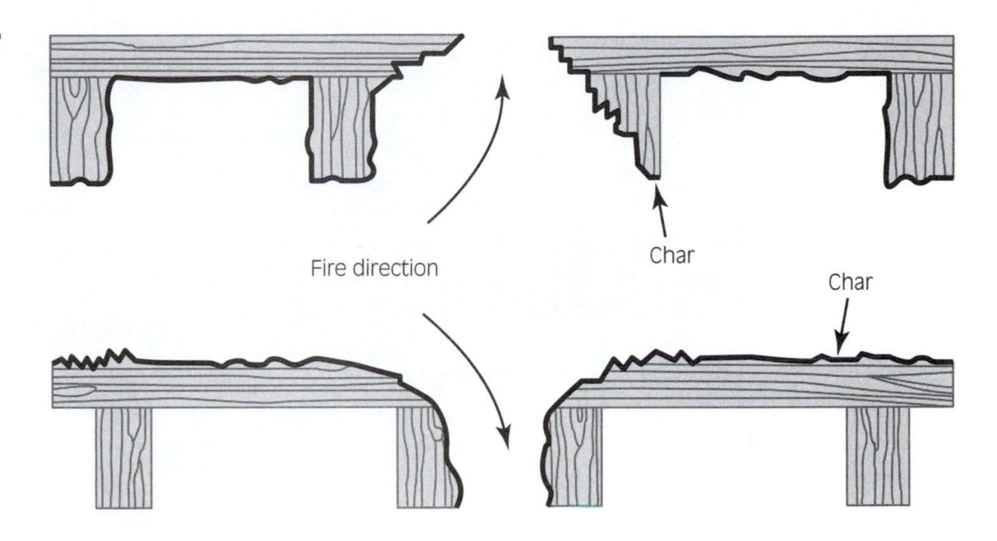

Figure 11-14 *Drawings showing the fire damage from the fire breaching the floor from above and from below.* (Reprinted with permission from NFPA 921-2008: Guide for Fire and Explosion Investigations, Copyright ©2008, National Fire Protection Association, Quincy, MA. This reprinted material is not the complete and official position of the NFPA on the referenced subject, which is represented only by the standard in its entirety.)

Carpets

Even though a carpet may have a fire-resistive rating, it will eventually burn. More important, if an ignitable liquid is present, the carpet will more than likely hold the liquid by absorbing it into the carpet backing or the foam padding. Then, the carpet fibers may act as a wick, aiding in the burning of the liquid fuel. A common indicator of the presence of an ignitable liquid is the sharp line of demarcation present on the carpet between the burnt and unburned areas, as shown in **Figure 11-15.** These liquids may have been poured on the floor intentionally to create a trailer allowing the fire to travel from one point in the structure to another.

Note
Ignitable liquids are not the only thing that can cause a sharp line of demarcation.

However, ignitable liquids are not the only thing that can cause a sharp line of demarcation. Ordinary combustibles can cause a line of demarcation on carpet and other floor finishings. The line of demarcation usually is a concentrated area and not so much a pour pattern leading out of a room or area. If there is any question, the investigator should take samples and submit them to a laboratory for analysis. A positive result for ignitable liquids still requires the investigator to determine whether this was caused by an accident or was an intentionally set fire. The negative result for ignitable liquids from the laboratory may only mean that there was not enough ignitable liquid in the sample or there was no ignitable liquid in the first place.

Wood Flooring

As with all floor surfaces, charring or burning indicates a low burn, which can be associated with the area of origin or drop-down. The duration of the burn and the condition of the flooring and its finish can make a difference as to the type of

Figure 11-15 *Burn pattern on a carpet resulting from the presence of a poured ignitable liquid.*

Rushing Out of the House

The engine company arrived at the residence of a reported fire to find no smoke but found a burn pattern, quite narrow, running the length of the carpet that led from the center of the house to the back door. The doors were open and it was winter. Checking with a neighbor, they were told that another neighbor took the homeowner to the hospital for burns.

The investigator was dispatched automatically and arrived just minutes after the engine company lieutenant had talked to the neighbor. Looking at the pattern, they found it went from the entrance of the kitchen to the back door, on the carpet only. In the backyard, they found at least a clue to the mystery: a large 3-inch-deep cast-iron frying pan sitting in the yard with a burnt potholder. They also found one burnt slipper.

The investigator interviewed the homeowner at the hospital. She had gone to the kitchen that morning and was going to do some deep-fat frying with the iron skillet. She got distracted by a phone call until she heard the smoke detector. Upon hanging up, she found that the grease in the pan, about 2 inches deep, had ignited. She grabbed the pan, and holding it behind her with just one hand (it was a large pan), she ran out of the house. When she got outside the fire in the pan was out. But the carpet in the hall was burning where she left a trail of grease from the kitchen to the back door.

(Continued)

(Continued)

She at least did one right thing by going to a neighbor and reporting the fire. After calling the fire department, the neighbor took her to the hospital for burns to her left foot and right hand. When she went to the neighbor's house, the grease in the carpet was still burning. Fortunately, by the time the engine company arrived the fire had burned out.

Although there was a sharp line of demarcation and an ignitable liquid, the fire was accidental in nature.

pattern that may be seen. A small amount of an ignitable liquid with a relatively short burn time on a sealed hardwood floor may show whisping and sooting. However, a similar floor without the tight finish, where there are actual minute cracks between the boards, will allow the liquid to seep into these cracks and then burn as the vapors are released. This, in time, could create rounded charred edges on the individual boards.

Concrete

Concrete flooring is more commonly found in basements and garages. However, in warmer climates where frost is not a problem, it is not unusual to have homes built on concrete pads. Concrete that has any foot traffic usually is sealed. You might find this in areas such as basements and garages. This sealant prevents chemicals and other liquids from seeping into the concrete. Even if there is a sealant, it can eventually be worn away by high traffic.

The sealant does not play a role in the investigation other than assuring the investigator that any sample that may be taken for laboratory examination has not been contaminated. Under most circumstances, you want to take a comparison sample of the concrete from an area that you know was not contaminated.

■ Note
Spalling of the concrete floor is not an indication of anything other than the fact that there was moisture in the floor and that moisture was heated to 212°F.

Spalling of the concrete floor is not an indication of anything other than the fact that there was moisture in the floor and that moisture was heated to 212°F. (See **Figure 11-16.**) You might question why the concrete reached that temperature in one place and not another. This can result from the burning of the underside of a vehicle such as in a garage. The burning of the sides or underside of furniture can raise the temperature at floor level as can the presence of ignitable solids and liquids on the concrete floor itself. Unburned items on top of the concrete protect the floor, but if they ignite, they may be the source of heat that causes the spalling to occur.

If an ignitable liquid was poured on the floor as a trailer, the pattern may be evident. (See **Figure 11-17.**) In the process of removing the debris to examine the floor, you may have to wash the floor. You might be concerned that if there is any residue of an ignitable liquid on the floor, it will be washed away as well. As a precaution, while removing debris and after the debris is removed, use a

Figure 11-16 *Typical spalling on concrete as the result of moisture in the concrete expanding as it turns to steam, forcing the surface of the concrete to separate from the rest of the floor.*

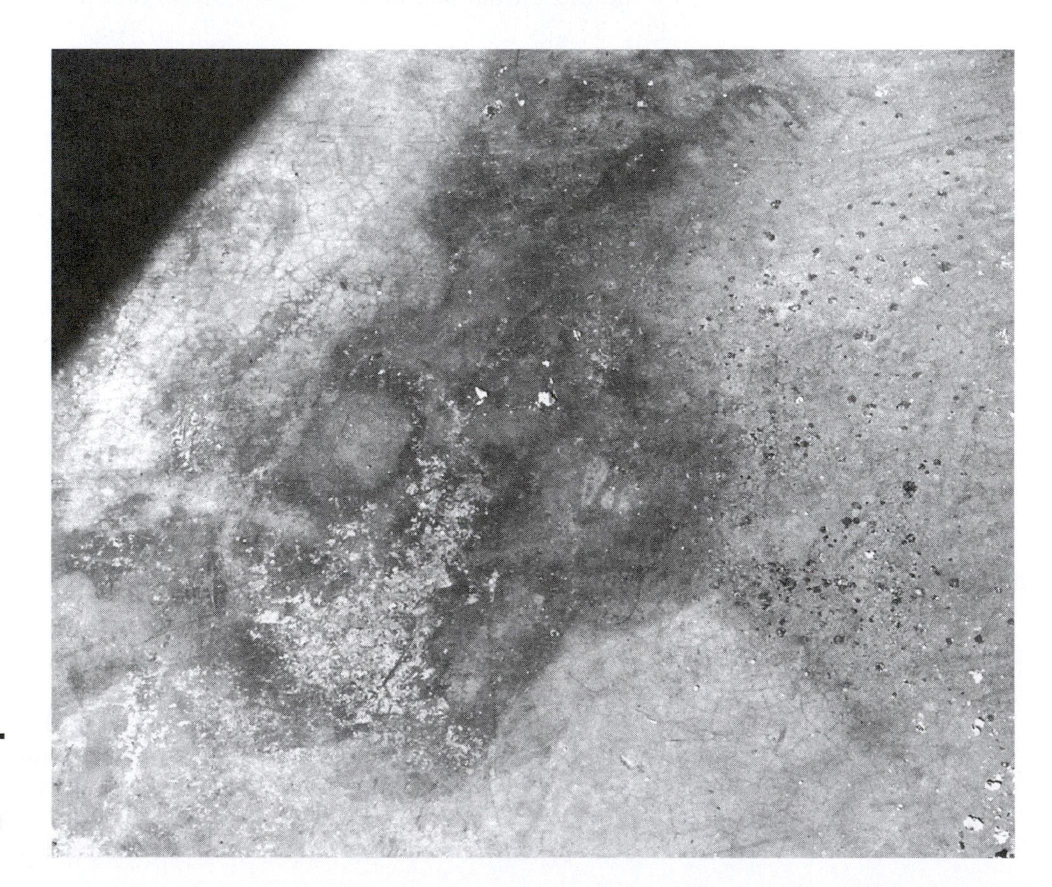

Figure 11-17 *Whisping pattern on the floor from an ignitable liquid.*

Figure 11-18 *Pouring clean water onto shallow, murky standing water on concrete can sometimes allow the investigator to see the pattern on the concrete surface.*

hydrocarbon detector to see whether there is any measurable sample in the area. An accelerant dog is ideal because it has the ability to detect residue in smaller amounts than any electronic detector. Using these detectors can ensure that you get the best sample possible to send to the laboratory for analysis of the presence of a liquid accelerant.

Sometimes, in such cases as in basements, it may not be an easy task to remove the murky standing water. Depending on the depth of the standing water, clean water may be poured on the area, as shown in **Figure 11-18,** to temporarily force away the murky water long enough for a quick snap of a photograph. This allows the investigator to see any patterns that may be on the concrete surface.

PROTECTED AREAS

As the fire burns and patterns from fire damage are left on the floor, there are other things to look for such as the lack of fire damage. As the floor is cleaned off in the examination of the debris, you will notice areas that have been protected from the fire and that are relatively or completely undamaged. **Figure 11-19** shows that books strewn on the floor protected that part of the floor from fire damage. More important, these protected areas will show the proper placement of furnishings and may help to tell the story of the fire. In the case shown in Figure 11-19, you might wonder why several books were on the living room floor. In this particular situation, a high school student was doing her homework on the floor when the fire occurred, and she left her books where they were as she escaped from the structure.

■ Note
Protected areas will show the proper placement of furnishings.

Anything sitting on the floor will most likely leave an indication of its placement as the fire damages the floor surface. During firefighting and the overhaul

Figure 11-19
Protected areas where schools books had been left on the floor.

phase of the event, some of the furnishings may be moved or even removed from the structure. The protected areas can help the investigator place the contents back where they were at the time of the fire. This will provide additional information and additional patterns to be observed.

Protected areas occur not just on the floor. As the heat moves through a structure, furnishings can block and sometimes limit the amount of heat absorbed in various areas. The back of a solid chair near a wall can somewhat protect the wall behind the chair, creating a shadow. This imprint on the wall, should it survive the fire, shows the placement of the chair. Even photos hanging on walls can create a protected area on the wall. When the hanging picture falls from heat damage, the shadow it created on the wall may survive. The shadow can be the result of lesser heat damage or limiting the collection of soot in that specific area.

PATTERNS ON FURNISHINGS

With the floor cleaned off, you can see the protected areas. You can place the furnishings back where they belong, or if they are already in place, you can verify their location thanks to these indicators. With the surviving contents of the structure in place, the overall pattern from the ceiling, walls, floors, and contents can be compared. Sometimes the patterns on the furnishings can provide a much better picture of the direction of fire travel. If nothing else, they may confirm or refute the patterns documented on ceilings, walls, and floors.

It may be obvious that the most damaged corners or sides of the furnishings will be on the side from where the fire came. If the right side of a chair is heavily

Figure 11-20 *Light damage on a kitchen chair showing the direction of fire travel is from the right to the left.*

damaged but the left side of the chair has only moderate damage, then the fire came from the right and traveled to the left. The furniture may also be in the area of origin and can even be within the V pattern.

Still at times it may not be so obvious. There may be minor differences in the pattern such as on a wood chair. The chair may look blackened by the fire but is still intact. By looking at the squared edges, you may find the edges on the right are slightly rounded but on the left side of the same piece of wood they are still sharp and pristine, as shown in **Figure 11-20.** In such cases, then, you have a pattern that shows the fire moving from right to left. (See also **Figure 11-21.**)

The melting of various objects also gives an indication of heat in that particular area. Aluminum objects can melt at temperatures of 1,100°F to 1,400°F. Some other pot metals can melt at even lower temperatures. If the drywall has failed, the wiring in the walls can even give an indication of temperature. Where both aluminum and copper wires travel in walls, you may find the aluminum wire melted but the copper is not melted. This indicates that the fire was at least 1,100°F and below 1,981°F.

annealing
the collapse of coil springs, loss of temper, as the result of heat

Steel springs today are most commonly found in bed mattresses, although not in all mattresses. In homes with older furnishings, steel coil springs might be in large stuffed chairs as well as in mattresses. These steel springs have been tempered to give added strength, allowing them to spring back to their original shape. When heated excessively, coil springs lose their temper and collapse on themselves, flattening out. This process is called **annealing.**

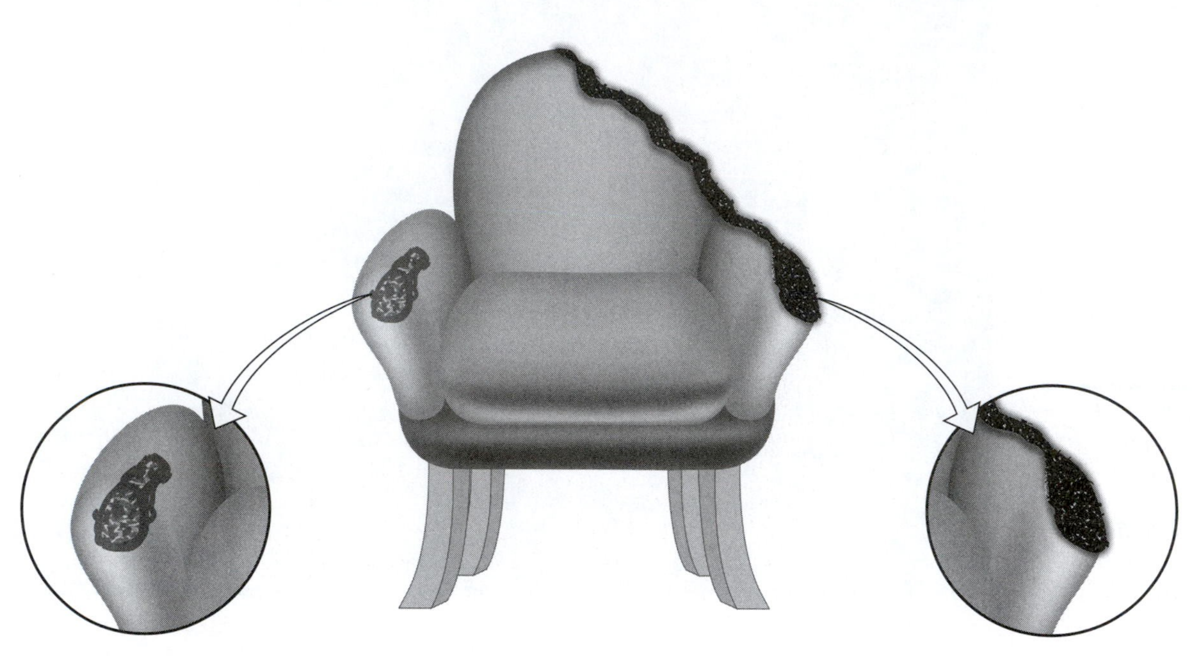

Figure 11-21 *Diagram showing the rounded edges of the arms of a stuffed chair along with damage to the back of the chair indicating direction of fire travel.*

Note
No one thing can conclusively solve the riddle of how the fire started.

Steel springs also flatten out from constant use, such as the sagging in the center that occurs in an old heavily used mattress or couch. During a fire situation, the springs will flatten out more prominently if there is weight on the springs, such as a basket of clothes on the bed or even a body.

Occasionally, annealing can give an indication of direction of fire travel. If only one portion of the bedsprings are collapsed—annealed—then there is a probability that it may be an indication of the direction of fire travel. For example, if the foot of the bed on the east side has collapsed, this may be an indication that the fire came from the east and moved toward the west. This is a result of the heat having a longer impact on the east corner of the bed than on the rest of the bed. Of course, as the heat builds up in the room, the entire mattress may collapse as a result of annealing. When this happens, it essentially eliminates your ability to determine fire direction from this one piece of evidence.

The loss of the mattress coverings can help as well. By the time the entire mattress anneals, the coverings most likely have been consumed. With just one corner of a mattress being consumed, it is a pattern on its own and may indicate the direction of fire travel. Annealing by itself should not be a defining indicator so much as it is one more fact to add to the culmination of evidence that leads to the final analysis.

SUMMARY

Finding and recognizing burn patterns aid the individual doing the cause and origin investigation. These tasks can best be accomplished by working from the area with the least amount of burn patterns to the area that has the most burn patterns. The char on the walls and ceilings will point back to where there is more char, eventually leading the investigator to where the fire started. This chapter covers a lot of material on burn patterns at the floor level, but you must remember that the fire can just as easily start on or in walls, on countertops, and in ceiling areas.

Along the path to the area of origin, the investigator may encounter heavy char and conflicting patterns. This area may be where the fire burned more intensely as a result of a heavy fuel load or a fuel that had a higher heat release rate than in the area of origin. Ventilation created by the fire, such as the breaking of a window, may have allowed the fire to burn more intensely in that area, which creates a deeper char at the ventilation source than in the area of origin. This ventilation or the ventilation created by suppression sources may provide air currents that could even change the direction of the fire travel. Working past these other patterns, recognizing them as secondary issues, the investigator should then be able to continue working back toward the area of origin.

In the event the fire went to flashover, the investigator is cautioned about burn patterns, especially those on the floor. In these situations, it has been observed that floor patterns are distinctly similar to ignitable liquid patterns when in fact there were no ignitable liquids present at all.

Other indicators that help the investigator to work back toward the area of origin are the patterns left on the furnishings. Protected areas on the floor can help ensure that the furnishings are properly placed where they were at the time of the fire. The char on furnishings, such as a chair, may show more damage on one side, which will be the side from which the fire traveled. The melting of certain metals such as aluminum will show the temperature that specific area reached during the fire. Comparison of this with other areas may help lead back to the area of origin.

Patterns that look like they indicate an intentionally set fire must be thoroughly examined to ensure that this is the case. Yes, there may have been an ignitable liquid, but that fact alone does not absolutely mean the fire was intentionally set. There may be a sharp line of demarcation, but some ordinary combustibles can create similar patterns.

This reinforces a most important tenet of fire investigation: no one thing can conclusively solve the riddle of how the fire started. The culmination of all the facts gathered as the result of a systematic analysis of the entire scene eventually leads to the successful final determination, hypothesis, of the fire origin and cause.

KEY TERMS

Annealing The collapse of coil springs, loss of temper, as the result of heat.

Arrow patterns As described in the NFPA 921, the patterns on wooden structural members that show a direction or path of fire travel.

Calcination The process of driving off the moisture in the gypsum, discoloring and softening it in the process. The process includes the driving off of moisture that is chemically bonded within the gypsum.

Demarcation A distinct line between a fire-damaged area and nondamaged areas; not a gradual charring but an abrupt end.

Drywall Also called sheetrock; a wallboard of gypsum with coating of paper on each side.

Oxidize The effect that heat has on a metal surface, in which it consumes any covering, resulting in the rusting of the metal surface.

Spalling When the surface of concrete pops off as a result of water in the concrete reaching boiling temperature and turning to steam. When this happens, the steam expands 1,700 times in volume, creating the energy to pop off the concrete from the surface.

REVIEW QUESTIONS

1. Define *burn pattern* as it relates to the investigation of fire origin and cause.
2. Describe two types of events that can affect the fire pattern and possibly give false leads as to the direction of fire travel.
3. What is a cone pattern?
4. Describe drywall and calcination.
5. What is the natural flow of fire and associated hot air and gases?
6. Define *spalling.*
7. Define *annealing.*
8. What is an arrow pattern?
9. What type of evidence might be found in relation to glass both inside and outside a fire-damaged structure?
10. What simple tool can be used to measure the depth of char?

DISCUSSION QUESTIONS

1. What new household products are on the market today that could affect the growth and development of a fire, possibly skewing burn patterns?
2. Patterns indicate that there may have been an ignitable liquid involved. Lab results will not be available for 2 weeks. Should interviews with occupants take place before or after the laboratory results come back?

ACTIVITIES

1. Search on the Internet to find articles on the crazing of glass. Document what temperatures are necessary for the crazing to occur when cold water hits the glass. At what temperature will glass melt, and what impact will heat have on the breaking of glass?
2. What scientific articles are available on the Internet that document the process of the spalling of concrete?

EXAMINING THE SCENE AND FINDING THE ORIGIN

Learning Objectives

Upon completion of this chapter, you should be able to:

- Apply what has been learned in previous chapters to see how it all comes together to complete a fire scene examination to search for the fire origin.
- Describe what to look for on the exterior of the structure that would indicate the area of origin of a fire.
- Describe what to look for when examining the interior of a structure that will lead you to find the area of origin of a fire.
- Describe the steps taken to do a fire scene reconstruction and the benefits obtained from such an effort.
- Describe events or conditions that may create greater char, skewing the path of fire travel.

CASE STUDY

No two fire scenes are ever alike, but there is one consistency: For the most part, the investigator will run into conflicting burn patterns that lead one direction and then another, only to wind up with a scene that must be thoroughly analyzed and presenting a challenge to the investigator. When the call came in for a fire in a resort-type home in an exclusive neighborhood, there was little thought that this fire scene would be different.

On the scene, there were evident burn patterns on the outside of the structure, coming from the side door and heading up and out. A clear V pattern pointed back down toward the door. Regardless of this clear arrow pointing back to the area of origin, it was necessary to do a clockwise search of the property, heading from the farthest point not impacted by the fire, circling around the structure, and working from least damage to most damage. In the process, plenty of photographs were taken of both damaged areas and potential heat sources that were not involved in the fire for the cause phase of the investigation. No evidence was discovered other than the exterior burn seen on arrival, so it was time to look inside the structure.

Many high-end log cabins will sheetrock the interior or put up standard interior wall partitions. This was not the case at all. Every wall was a true log cabin wall. Every surface both inside and out was the same—no deviation on finish or shaping, and each log was slightly rounded as one expects to see in a log cabin.

The interior furnishings were sparse and were rustic as would be expected in a cabin in the woods: heavy wood chairs with light padding, very few wall hangings, and plastics were at a minimum, essentially limited to the TV, stereo, a few kitchen appliances, and a trash container.

Working from the least damaged area to the most damaged area was hard to do, not because of debris or problems, but because on the left was the kitchen and at floor level in the kitchen was a pile of ash and debris. Moving directly up from this point was a V pattern. This pattern continued up the wall to the door, which had a glass window. The glass was laying on the floor and the pattern continued up the wall to the ceiling and outside up the wall in a continuation of the burn pattern.

Regardless of this obvious area of origin, we examined the entire structure and found there was only sooting in the remainder of the house. The area of origin was official then: floor level in the kitchen on the east wall. For this chapter, this is where this story should stop. But for the curious—and you should be if you want to be an investigator—sifting through the debris we found the base of the plastic trash can in the center of the debris pile. In the structure are full ashtrays, cigarettes, and matches in most rooms. Interviews with the occupants divulged that when they left the house they emptied the ashtray they had been using into the trashcan. That was about 90 minutes before smoke was first reported in the area.

■ **Note**
Although we frequently use the terms *origin* and *cause* together, you cannot find the cause of the fire until you have located the area of origin.

INTRODUCTION

This is where you take all the knowledge you have obtained from previous chapters and apply it to the physical examination of the scene in search of the area of origin. Although we frequently use the terms *origin* and *cause* together, you cannot find the cause of the fire until you have located the area of origin. To find the area of origin, investigators must work their way through the fire scene interpreting the burn patterns. By doing this properly, these patterns will lead back to the area where the fire originated.

It should again be mentioned that neither one text nor one class could ever make someone a fire investigator. To attain the level of competence necessary to make accurate determinations of the area of origin, you must have a strong knowledge of building construction. You must also have experience with fire in knowing the natural progression of a fire and experience in knowing how that process can be altered by fuel load, heat release rate of materials, ventilation, and suppression activities.

■ **Note**
National standards clearly indicate that the fire origin and cause determination is the responsibility of the fire department.

THE ASSIGNMENT

officer in charge (OIC) the fire officer who has ultimate charge and control of the overall fire or emergency scene; usually but not always the senior officer on the scene

As discussed in Chapter 3, "Role of First Responders," there must be a comprehensive policy and procedure on how fires are to be investigated by the locality. National standards clearly indicate that the fire origin and cause determination is the responsibility of the fire department. The National Fire Incident Reporting System (NFIRS) has an entire section on the origin and cause of the fire to be filled out by the fire **officer in charge (OIC)** or the designated person doing the report. Although not all fire departments use NFIRS, it is hoped that someday every fire department in the United States will participate in this process. A common phrase is that we fight fire with facts; the facts come from the NFIRS data. The more participants, the better the data. The better the data, the more we can accomplish.

■ **Note**
A common phrase is that we fight fire with facts; the facts come from the NFIRS data. The more participants, the better the data. The better the data, the more we can accomplish.

Even if the local department does not use the NFIRS, it should create some type of report. When a citizen loses a home or even a vehicle as a result of fire, the insurance company usually requires a fire report from the locality or the responding fire department. Without this report, there could be a delay in handling the claim and in turn a delay in the suffering family getting their lives back together.

To ensure consistency and accuracy, the department must decide what role it will take in the determination of the fire origin and cause. Under most circumstances, for the average single-engine fire, the fire OIC will identify the area of origin and take the report. Even for larger events where an assigned investigator may be automatically dispatched, the OIC of the first engine arriving may do the preliminary report and leave the details on the area of origin to be updated in the NFIRS by the assigned investigator.

■ Note
The term *fire investigator* refers to the person who examines the scene to determine the origin and the cause of the fire. This may be a firefighter, fire officer, or the assigned fire investigator such as a fire marshal.

Remember, the term *fire investigator* refers to the person who examines the scene to determine the origin and the cause of the fire. This may be a firefighter, fire officer, or the assigned fire investigator such as a fire marshal. The assigned investigator usually does the more detailed investigation. An investigator may be assigned to an incident as the result of an automatic dispatch because of the size of the incident or because the engine company officer needs assistance finding the area or origin or cause.

When the assigned investigator arrives on the scene, regardless of rank or responsibility, the OIC of the fire is still the leading official until the fire is placed under control. The investigator must report to the OIC to advise of his or her arrival and planned activities. This is all explained in the National Fire Protection Association standards for proper incident command. If the investigator enters the structure before the fire is placed under control, it must be with the OIC's knowledge and consent. (See **Figure 12-1.**) This is predominantly a safety issue that must come before all else.

SYSTEMATIC: A PROCESS

■ Note
By using one methodical approach on most, if not all, fire scenes, the investigator can prevent overlooking key indicators and evidence.

■ Note
Above all else, the investigator must be a seeker of truth.

Whether it is the fire officer, looking at the scene after it is marked under control or the assigned investigator, the approach must be systematic and consistent. It must follow a logical path that is conducive to identifying the fire area of origin. By using one methodical approach on most, if not all, fire scenes, the investigator can prevent overlooking key indicators and evidence. Above all else, the investigator must be a seeker of truth. The best tool to accomplish this is the use of scientific methodology to come up with the proper hypothesis as to the area of origin.

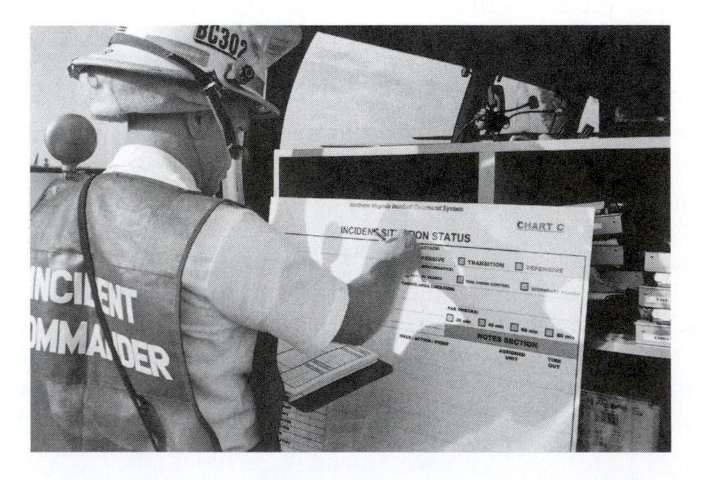

Figure 12-1 *The OIC with an accountability board, tracking the location of all personnel working at the emergency scene.*

A common approach is working from the least damaged area to the most damaged area. By doing so, the investigator should, under most circumstances, be working from the extension of the fire to the area where the fire originated. Smoke patterns and burn patterns can be the path to the area of origin and are indicators that assist the investigator.

■ Note
Each fire is unique; in other words, no two fires will ever be the same.

Each fire is unique; in other words, no two fires will ever be the same. This is because of outside influences such as fuel load, ventilation from suppression activities, or from the fire burning through to the exterior of the structure. Regardless, there should be consistency in the burn patterns that will allow the investigator to follow the path back to the area or origin. But before you get there, consider what must be done first.

SAFETY ISSUES

As the fire is being knocked down, even before the last ember is extinguished, the investigator should be thinking about the first and primary concern: safety. Everything from the condition of the structural elements to the presence of unsafe electrical lines or even exposure to chemicals must be considered. After the fire OIC leaves the scene, it is the fire investigator's responsibility to ensure that the scene is safe not only for those who may be investigating but for the curious public as well.

Structural Stability

Structural stability is the first concern. Knowledge of proper building construction is essential. If there are areas that appear unstable, then actions must be taken to render those areas safe before entry. If there is any question as to the safety of the structure, the local building official can be contacted to ask for assistance.

Other resources can possibly assist as well. Larger departments may have *heavy technical teams* who have the necessary tools and supplies to shore up walls and overheads. If the insurance company representative has arrived, he or she too has an interest in looking at the area of origin and possibly even doing an inventory. The insurance representative may be willing to provide resources in the form of contractors and supplies. Most insurance companies recognize that leaving an unstable structure is a liability and will want to assist. A key concern is to make sure the investigator stays in control of the scene and guards it against contamination from equipment and supplies.

■ Note
Sometimes damage may be so severe that the only recourse is to demolish those structural elements that pose a hazard.

Sometimes damage may be so severe that the only recourse is to demolish those structural elements that pose a hazard. If this is the case, a crane or other heavy equipment can be called in to lift damaged items away from the structure. However, efforts must be taken to make sure that the heavy equipment does not contaminate the scene should it be necessary to take debris samples for ignitable liquids.

Utility Hazards

The first utility that is of most concern is the presence of electricity. The department or the electrical utility company may have already disconnected the meter. But that is no assurance that all electrical sources have been eliminated. Some investigators have found extension cords running to neighbors' homes for one reason or another, legally or illegally. With today's technology, battery systems and solar units can still provide quite a jolt if encountered without protection.

■ **Note**
Some investigators have found extension cords running to neighbors' homes.

■ **Note**
Every investigator should know how to use an electrical multitester properly.

Every investigator should know how to use an electrical multitester properly. They are quite inexpensive and easy to use. These devices can determine whether there is energy on any given electrical circuit; if there is not, then the wires should be safe to examine. Testers generally have two probes, one going to the wire or post and the other to a ground. They then indicate the amount of voltage present. There are some devices on the market labeled as noncontact voltage detectors, as shown in **Figure 12-2,** that activate when the plastic tip is placed near an energized line. Of course, the investigator should always test all equipment before relying on the readings.

On occasions, there are overhead power lines that could create a hazard, even if not damaged. The electricity stopped at the meter base may still be supplied by the power company through an overhead line from the power pole to the structure. Every effort should be made to check it to ensure that it is safe and that there will not be a need to get anywhere near this line. If there may be a potential

Figure 12-2
A noncontact detector that will activate in the presence of an electrical current.

of having to get near this line or if the line suffered any heat damage, the power company should be contacted to have the line deenergized.

The investigator should look to see whether the structure had gas lines. Merely checking records or asking neighbors if public gas lines are available is not enough. Looking in the back or side yard is also insufficient because some liquefied petroleum gas tanks are buried; if a tank is buried, there is a control head cover that could be visible if not covered in rubble. A close examination of the scene should help to ensure that the hazard is nonexistent or, if gas lines are found, that the supply is shut off at the source. The atmosphere should be checked with a multigas detector for combustible gases, oxygen, and carbon monoxide.

■ Note
The atmosphere should be checked with a multigas detector for combustible gases, oxygen, and carbon monoxide.

Even water lines can be a hazard. If the structure is a multistory building, leaking water lines in the upper areas could add additional weight to lower floors, weakening the structure. Water continually leaking into a basement can create problems by hiding other hazards under the water or potentially drowning someone if that person was to fall into the water-filled basement. Depending on the weather, water can create additional safety problems with icing, adding weight to the structure and creating slipping hazards. If the structure was supplied with well water, then cutting the electricity should solve the problem, unless there is a secondary source of electricity for the well pump. If the water comes from a public water source, that can be cut off at the street by the fire department or local water authority.

■ Note
Water can create additional safety problems with icing, adding weight to the structure and creating slipping hazards.

If water is collecting on the floor, steps should be taken to allow it to drain off the floor and away from the structure. Water that collects in the basement may need to be pumped out should the basement need to be examined by the investigator. Some fire departments have portable pumps that can assist in this process.

SECURE THE SCENE

During the suppression operations, efforts should have been made to cordon off the area for the safety of the general public. When the assigned investigator arrives on the scene, one of the first duties is to check the fire line and expand it if necessary to cover the entire fire scene. Depending on the situation, the investigator may want to find personnel from the fire department or police department to man the lines to keep everyone out of the fire area. If the fire officer is to be the investigator, then as soon as the fire is under control the fire officer should check the fire line or assign this duty to one of the firefighters.

■ Note
Depending on the laws of the locality, even the building owners or tenants should not be allowed to enter the scene unescorted until at least the preliminary scene examination is complete.

Security must be maintained throughout the investigation. Arrangements must be made for someone to monitor the exterior while the investigator is inside the structure. Depending on the laws of the locality, even the building owners or tenants should not be allowed to enter the scene unescorted until at least the preliminary scene examination is complete. Many times, an engine company may stay on the scene to render assistance and provide scene security.

EXAMINING THE SCENE DURING SUPPRESSION OPERATIONS

For smaller events where the fire officer takes the role of investigator, not much will be done upon arrival on the scene. Depending on the situation, the fire officer, having set plans in motion such as giving the crew instructions, may then have a chance to look around and consider the area of origin. If this is an active burning fire, at this time investigation is not a primary function. The most important task at hand is safety first, not just of the crew working at the scene but of the officer. Wandering off alone is not a wise course of action. The fire scene examination as a whole can take place as soon as the fire is knocked down.

■ Note With the number of camera phones in use in the United States, someone may have taken a photo of the fire in the early stages.

■ Note The investigator should take as many photographs of the overall scene as possible.

imminent danger to life and health (IDLH) an atmosphere in which anyone entering endangers themselves unless proper protection is taken, which could include full turnout gear with self-contained breathing apparatus

accountability tag an identification tag with the holder's name and sometimes unit number or assigned company; see also, end of chapter definition

Under most circumstances the engine company officer will not have the equipment or the opportunity to take photos of the scene until the exigencies of the situation are over. Even then, most engine companies do not have sophisticated camera equipment. However, today's technology does offer the ability to get some photographs. It has been common practice by some departments to issue disposable cameras for each engine. Once the engineer sets up the apparatus, gets hose lines hooked up and a water supply established, there may be time to snap a few photographs.

The camera can use a standard 35-mm film that requires processing, or the camera can be digital where the photos can be downloaded onto a computer. Most cameras are also available in waterproof containers to protect them in harsh climates. Of interest is that with the availability of the camera phone, many photos are taken by the engineer on personal camera phones. The same can be said of the witnesses around the scene. With the number of camera phones in use in the United States, someone may have taken a photo of the fire in the early stages.

For larger events when an assigned investigator is dispatched, upon arrival the investigator will report to the OIC. After checking the security of the scene, the investigator can start making observations about the fire and suppression activities. Where is the fire burning now? Where are there signs that the fire was knocked down? Tracing the fire hose may be a way of seeing where the attack on the fire began, which can come in handy when interviewing the firefighters about the scene later in the investigation. During this observation phase, the investigator should take as many photographs of the overall scene as possible. Photographs should include the structure, surrounding area, and the crowd.

The investigator may find it beneficial to go into the structure even before the fire has been completely extinguished. This should only be done if the investigator is completely qualified to work in an **imminent danger to life and health (IDLH)** atmosphere with full turnout gear and self-contained breathing apparatus (SCBA). Once all equipment is donned and a second check is made that all is ready, the investigator should then take his or her **accountability tag** and report to the OIC to ascertain where he or she can or cannot work at that time. If there

■ **Note**
If personnel are available, it is always best to work in pairs.

are areas that can be examined with relative safety, then the OIC will place the accountability tag and assign the investigator to that sector of the structure. If personnel are available, it is always best to work in pairs, but this is not always possible or practical at this stage of the emergency event. One important rule that applies during the entire time while on the scene is that the investigator must know at least two paths of escape at any given time. Everyone on the scene must be aware of their surroundings so that in case one avenue of escape is blocked, there is always the second path to get out safely.

It must be understood that if the OIC feels that it is not safe for the investigator to enter, it is prudent for the investigator to comply. In many jurisdictions, the OIC and the investigator in charge have equal and concurrent jurisdiction at any emergency scene. Regardless, safety must come first, and until the fire is under control, the OIC should be the lead in all safety rulings.

If there is sufficient personnel and with the approval of the OIC, the investigator and a partner may be able to go anywhere in the structure, provided they do not hamper suppression operations. With this ability, the investigator may see firsthand the fire's progress and its extinguishment. Observations may be limited in areas because of smoke and steam, but as the atmosphere clears, various patterns may be visible. This may enable the investigator to identify sections or areas of the structure to protect and where overhaul should be held off until the area can be properly documented.

As with any accountability system, should the investigator change from one sector of the fire scene to another, the OIC should be notified, usually in person, so that all personnel on the scene can be accounted for at any given time.

EXTERIOR EXAMINATION

Once the fire is under control the overall scene can be examined in earnest and in its entirety. If it is still in the dark of the night, the investigator may choose to wait until sunrise to begin the investigation. This may be for the sake of safety or to ensure that no clues or burn patterns are overlooked. Regardless, the Supreme Court has defined (*Michigan v. Tyler*) that under most circumstances this may be acceptable. If there is a delay in the scene examination, the scene must be protected to ensure there is no alteration or damage to potential evidence.

■ **Note**
If there is a delay in the scene examination, the scene must be protected to ensure there is no alteration or damage to potential evidence.

The patterns on the outside of the structure can tell a story in themselves. Documenting the patterns is essential. Photographs can show burn patterns that indicate wind direction or more damage in one location than another. Later in the investigation, it will be necessary to explain why there was more burning in any one opening than in another. Even if you cannot explain this now, you may need to explain it later in court.

Sometimes there may be conflicting patterns that will also need explaining. In **Figure 12-3,** the patterns from the door show the wind going from left to right, but

Figure 12-3 *Conflicting burn patterns on the exterior of the structure showing wind direction different for each opening.*

the pattern in the adjacent window shows a pattern indicating a wind direction from right to left. In this case, statements from suppression forces advised that when they showed up on the scene, fire was coming from the door and the smoke was going away from the street into the backyard, to the right in the photo. As they knocked down the fire in that room, the fire continued burning in the adjacent room and the wind changed direction, sending the smoke toward the street and engulfing the engine and the engineer. Thus, the pattern in the door was first, and when the fire finally breached the window, the wind had changed direction.

■ Note **It is a great tool to be able to see the overall scene from above.**

One beneficial tool to a fire scene examination is the truck company apparatus. Whether it is a ladder truck or an aerial scope, it is a great tool to be able to see the overall scene from above. Photographs from this area will be of value as well. So, catch the truck company before they leave the scene or convince the suppression OIC of your need for a truck company to assist.

As you work toward the structure, examine the debris to see whether there is any evidence of an explosion. Examine any natural ventilation areas or areas where there was forced ventilation by suppression forces, and again, document and take photographs as the investigation progresses.

BUILDING SYSTEMS

The topic of building systems could fill volumes of books. As an investigator, you may not need to know everything, but you will need to know your resources, where you can obtain help to assist with the various building systems questions as they arise. However, every investigator must know the basics on systems such as building construction, utilities, and fire safety and suppression systems built into the building.

Compartmentation is a component of the building system. The construction of the walls and their ability to resist fire travel, concealed spaces where fire can travel, and the control of openings into these spaces are components that can affect fire growth or can be key elements in resisting fire growth.

Walls should have a fire rating in the form of sheetrock. The room should be completely enclosed, with no holes leading to concealed spaces to prevent extension of fire. The doors to the room should have automatically closing doors or fire doors that will seal the room when activated by a fire. The fire investigator must have a good working knowledge of building construction and safety systems such as fire doors. However, this is in a perfect world. The investigator will also need to know about the balloon construction building with wood paneling, open dumbwaiter, and decorative features that are far from fire resistive.

Fire investigators must be familiar with fire suppression systems or have known resources to assist with the examination of such systems. These can include all forms of sprinkler systems, fire-extinguishing foam or dry chemical extinguishing agent delivery systems, or total flooding systems such as carbon dioxide systems. These types of systems must be examined in both damaged and undamaged areas to ascertain their condition as well as determine if they reacted properly and performed as they were designed. Detection systems must also be examined to ascertain if they properly activated and performed as required for that type of occupancy.

Of course, any examination of building systems will include the examination of utilities such as the heating, ventilation, and air conditioning (HVAC) system as well as the type of finish on the walls.

The key issue is the impact building systems had on the fire growth and the extension of the fire to other areas of the building. A detailed exam and documentation of the structure are instrumental to the future outcome of any investigation. This is an area where computer fire modeling has been instrumental by giving visual demonstration of different impacts that building design can have on fire growth and travel.

INTERIOR EXAMINATION

Note
Before entering the structure, the investigator must assess the safety of the structure.

Before entering the structure, the investigator must assess the safety of the structure. This assessment must continue the entire time that occupants are within the structure. If at any time there is any indication that the structure is not safe, all occupants should exit until the structure can be returned to a safe condition.

As you complete the exterior exam there is a lot to remember, so good note taking is important. When you enter the structure, go to the least damaged area first and work your way to the least damaged area again.

■ Note Each pattern must be taken in its entirety with the surrounding area.

■ Note The lowest burn is of importance because fires naturally burn upward and outward.

■ Note The investigator must never forget that the building construction can have a dramatic impact on how the fire grows.

■ Note The investigator should use tools such as the depth of char gauge on wood char and check calcination to help to determine the relative burn times.

Examine burn patterns you encounter within the structure to interpret the relative length of time the fire burned in each area. Each pattern must be taken in its entirety with the surrounding area. The volume of the fuel load or the heat release rate of the available fuel in each area must be examined to determine whether that may have caused a heavy char rather than the length of time the material burned, which leads back to the area of origin. The potential for ventilation generated charring must be examined in areas where either the fire self-ventilated or ventilation was created by suppression forces.

The lowest burn is of importance because fires naturally burn upward and outward. However, some low burns may be caused by dropping curtains that ignited at the top and fell down when they burned away from the curtain rod. Some fires can burn downward depending on how the fuel is positioned, and one should never forget that fires can start in a multitude of areas, so the investigator must keep an open mind, applying logic to all patterns observed.

The investigator must never forget that the building construction can have a dramatic impact on how the fire grows. In a balloon construction home, the fire could start in the basement only to extend through the interior of the walls all the way to the roof. The entire upper level of the structure could burn away, leaving a floor relatively undamaged between the basement and attic.

A fire in a structure with concrete floors and walls with a metal roof could start in an area that contains relatively little fuel load. After smoldering and burning for hours, it may reach an area with massive amounts of fuel with high heat release rates. Then, the fire grows in intensity, breaching the structure and alerting those passing by of the fire. The intensity of the fire where the fuel was located may throw off the investigator at first. However, examining the burn patterns should lead the investigator to question why the fire damage occurred in the remote location and whether the fire had burned longer in that area.

In following the burn patterns back to the area of origin, the investigator should use tools such as the depth of char gauge on wood char and check calcination to help to determine the relative burn times. Along the way, a 60-watt light bulb may have been distorted, pointing back to where the heat originated. Melting can assist as well. Look for similar items that may or may not have melted along the fire's path to give an indication of where the hottest temperature occurred.

Debris Removal

Fire suppression personnel must be taught to not hamper the accurate determination of the fire origin and cause, the debris must not be disturbed any more than absolutely necessary. An example of why debris removal could be a problem to the investigator is shown in **Figure 12-4,** where debris has been thrown out the window, collecting on the porch roof and on the ground with the debris from the first floor. There is a fine balance in that the fire suppression personnel

Figure 12-4 *Debris thrown out of the second-floor window, hampering a proper investigation of the fire scene.*

■ Note
Fire suppression personnel must be taught that to not hamper the accurate determination of the fire origin and cause, the debris must not be disturbed any more than absolutely necessary.

must conduct overhaul operations to put out each and every last ember and at the same time disturb the evidence as little as possible. Failure to suppress the fire entirely could result in a rekindle, which may end up destroying any evidence remaining from the first fire.

To truly make the final determination of the area of origin, the debris must be carefully removed from in and around all suspected areas. Any area where debris is not examined may lead to incorrect conclusions. Just as bad, if the case ends up in court, the jury may doubt your conclusions because you failed to look at all the evidence. Either way it may end up being a miscarriage of justice for all concerned.

The debris needs to be removed in layers and carefully examined for any potential evidence. After the debris is completely removed, all patterns can be more easily seen and evaluated. The debris itself may give indication as to what may have been in the area. A pile of newspapers will burn but usually only on the surface and with relatively low heat release rate. However, a bean-bag chair filled with polystyrene beads will have a tremendous heat release rate in comparison and in turn may create more char in the area. Residue will usually be left from each, giving the investigator clues as to what happened.

■ Note
Spoliation is always a concern even with the unnecessary removal of debris.

Spoliation is always a concern when there is unnecessary removal of debris. If debris removal is not necessary to prevent rekindle or lessen the weight on a weakened structure, it may be spoliation if it indeed hampers the scene examination and eventual determination of the fire origin or cause.

EXAMINING CONTENTS

■ Note
With the careful removal of the debris in the area, the investigator will be able to see any protected areas on the floor that will match up with the furniture.

The patterns on contents can be vital to the discovery of the area of origin. They can support patterns on walls, ceilings, and floors and can help resolve any confusing patterns in the structure. The key issue is to make sure that any furnishings that were removed are placed back in the correct position. With the careful removal of the debris in the area, the investigator will be able to see any protected areas on the floor that will match up with the furniture. Shadows on walls can indicate the placement of the furniture as well.

■ Note
The burn patterns on furnishings can show the direction of fire travel.

The burn patterns on furnishings can show the direction of fire travel. More damage on one side than on the other may indicate direction of fire travel. However, it could also mean that there was a sufficient fuel load on that side of the chair to cause damage and have nothing to do with the travel. Searching through the debris will identify any potential of a secondary fuel affecting the patterns in that area of the fire scene.

THE AREA OF ORIGIN

■ Note
It is only after a systematic search of the entire fire scene, using a scientific methodology that a successful hypothesis can be tested to identify the area where the fire started.

With the thoughtful and careful examination of all burn patterns throughout the entire structure and with the examination of the debris and patterns on all furnishings, the area of origin can be identified. It is only after a systematic search of the entire fire scene, using a scientific methodology, that a successful hypothesis can be tested to identify the area where the fire started. Any conflicting patterns as the result of ventilation, fuel load, heat release rate, or fire suppression activities must be identified and taken into account before making the final hypothesis.

With the area of origin identified only part of the job has been completed. The next step is to identify the first material ignited, the heat source, and the act that brought them all together, which is the cause of the fire.

SUMMARY

As with everything dealing with a fire investigation, safety has to be the first concern of everyone on the scene. An accurate determination of the area of origin of the fire is the first step in the final determination of the fire cause and identification of how to prevent future fires. With the application of a systematic search along with scientific methodology, the investigator can search the property from the exterior, working toward the structure. Once inside the structure, with all safety concerns handled, the investigation can continue from the least damaged area to the most damaged area, which under most circumstances leads the investigator directly to the area of origin.

KEY TERMS

Accountability tag An identification tag with the holder's name and sometimes unit number or assigned company. More sophisticated tags also include the individual's medical history, including blood type and medical allergies. Used in the accountability system to track the location of all personnel at an emergency scene.

Imminent danger to life and health (IDLH) An atmosphere in which anyone entering endangers themselves unless proper protection is taken, which could include full turnout gear with self-contained breathing apparatus.

Officer in charge (OIC) The fire officer who has ultimate charge and control of the overall fire or emergency scene. Usually but not always the senior officer on the scene.

REVIEW QUESTIONS

1. Why does the investigator need to know two routes out of a structure at all times?
2. What is NFIRS and how can it help to prevent fires?
3. Why should investigators have to report to the OIC and let him or her know where they will be working?
4. Give three outside influences that can affect or change the fire burn pattern.
5. How could a building official be of any assistance at a fire scene investigation?
6. Why would the investigator photograph a gas line or electrical panel that was not involved in the fire and not even damaged?
7. How could water be a hazard to the fire investigator?
8. What tool could be used to determine how long the wood had burned in one area of a room in contrast to another area of the room?
9. Why would it be essential to not have suppression clean out a room, shoveling out the debris so that the investigator can look at the floor?
10. How can burn patterns on furniture help point back to the area of origin?

DISCUSSION QUESTIONS

1. The National Fire Incident Reporting System collects fire investigative information from fire reports filled out by the fire companies. What type of statistical information on fires do you feel would be beneficial to help the community or the fire department?
2. Once the lead investigator arrives on the scene, she reports to the OIC. They are both of the same rank within their department and have been in the department for the same length of time. Who do you feel should be in charge from that point on? Why? What if circumstances were that the fire was an obvious high-impact arson? Would that change who you think should be in charge?

ACTIVITIES

1. Contact the local fire department and ask if they have an accountability system in place. If yes, ask if you can get a copy of it. Does it clearly address the fire investigator when the investigator arrives on the scene?
2. On the Internet, search for melting temperatures of some common household materials. Make a list for yourself. The list should include but not be limited to copper, aluminum, lead, steel (as used in I-beams and as used in coil springs), glass, and various plastics.

FIRE CAUSE

Learning Objectives

Upon completion of this chapter, you should be able to:

- Describe what needs to be done in conjunction with the search for the area of origin that will help lead to the fire cause.
- Describe evidence found on the exterior of a structure that indicates whether an explosion occurred before or after the ignition of the fire.
- Describe the importance of identifying the first material ignited and the act that brought the source of ignition together with the first material ignited.
- Describe the four different classifications of fire causes.
- Describe the process of identifying the product or person responsible for the ignition sequence.

CASE STUDY

The oil furnace received its semiannual servicing by a local oil and heating contractor. This process had taken place in the spring and fall each year for the preceding 10 years. The next day was the first cold day in the fall; the family was in the basement in a small den that was located adjacent to the furnace room. The thermostat was turned up, starting the furnace for the first time after the summer months.

It was not long before the family discovered smoke coming from the basement area. Fortunately, they thought fast and cut off the furnace, but that did not stop the black smoke from permeating the structure, coating expensive original paintings and antique furniture with black, oily soot. The insurance company was faced with damages in excess of $150,000 from sooting alone.

The investigation started like any other systematic search of the property exterior, going inside and working from the least damaged area to the most damaged area. The smoke patterns clearly led to the basement area. In the basement, the patterns led back to the furnace area. The wall adjacent to the furnace was burned clean all the way to the floor, as shown in **Figure 13-1.**

Figure 13-1 *The furnace in the area of origin. The wall adjacent to the furnace is burned clean.*

(Continued)

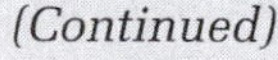
(Continued)

Figure 13-2 *This is the burner assembly removed from the furnace by the investigator. In the center of the photo is a small 1-inch-square hole at the base of the furnace. This is the air intake, but by all accounts, it appears that this is where the flames escaped the furnace.*

Burn patterns indicated that a flame came out of the furnace and impacted the wall, spreading heat and fire upward and outward. The only opening where the flame could have escaped from the furnace was the small 1-inch-square hole, which is designed for air intake (see **Figure 13-2**). Under normal circumstances, this opening has a cover with one screw. The cover can be adjusted to allow just enough air into the burn chamber for a proper efficient flame. For flames to come out of this small hole, flames in the burn chamber would have had to go up and over the burn chamber and then down to the hole to escape; this was not usually likely.

burner assembly
the assembly houses the electrodes in a furnace to ignite the oil that is pumped out of the jet nozzle; this unit also contains the fan, which blows air into the chamber, allowing for a bright and efficient flame

Pulling out the **burner assembly** proved the oil **jet nozzle** had not been replaced as should have been done the previous day during the service call. The nozzle showed considerable age and wear. In the removal of the assembly, an excessive amount of soot and **scale** came out with the unit. If the furnace had been properly serviced the previous day, this area would have been vacuumed clean as would the area where the heat is exchanged just above the burn box area.

Pulling out the burn chamber provided the largest shock of all. The firebox is a cylindrical steel chamber with the top open allowing the heat to escape upward. As you can see in **Figure 13-3,** the chamber had melted away on both sides of the box. This allowed flames to escape the chamber and come out the air intake. In the chamber is a bed of soot and scale, as shown in **Figure 13-4;** it defies the imagination as to how long it would take to get that much damage and debris.

(Continued)

(Continued)

Figure 13-3 *Steel burn chamber showing where fire burned through the steel as a result of debris clogging the oil jet and skewing the flame to the side. Over time, more debris in the jet skewed the flame to the other side so that holes were burned on both sides of the chamber.*

Figure 13-4 *The area where the burn chamber sits, the designed vacant space is filled with soot and scaling from the destruction of the burn chamber itself.*

(Continued)

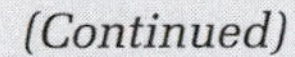

(Continued)

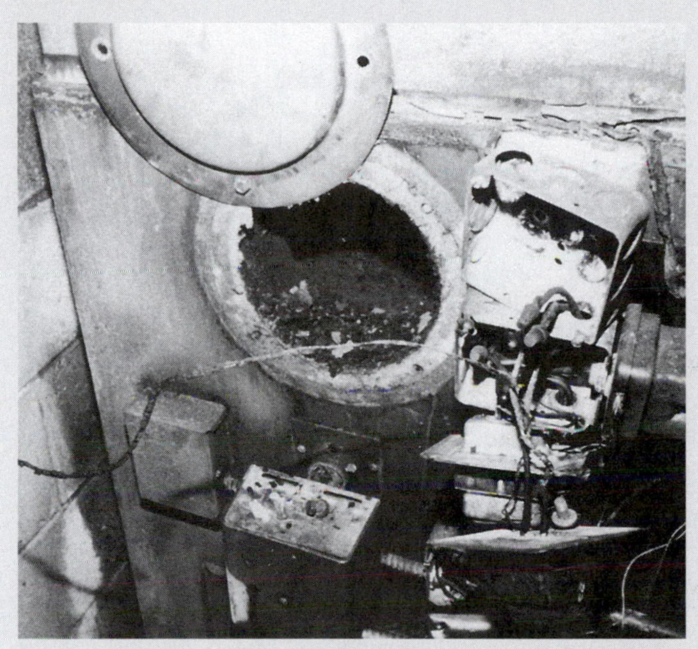

Figure 13-5 *The access area to the heat exchange with the plate turned up out of the way. This area had allegedly been cleaned the day before, but soot and scale are evident, piled up at the face of the opening.*

Above the burn chamber is a round opening to access the heat exchange area. Remember, this too should have been vacuumed out. To get into this area, the cover plate had to be removed; however, one screw broke off and a second would not come out, so it was used to pivot the plate. The plate itself had to be pried away from the face of the furnace. It had not been opened in some time. The first look showed scale and soot piled up at the plate, as shown in **Figure 13-5.** By using an external, remote flash, the investigator could see the soot inside the exchange area, as shown in **Figure 13-6,** which obviously had not been cleaned in several heating seasons.

The area of origin of the fire was the area at floor level, just outside of the furnace, in the area of the air intake. The first material ignited was a cardboard box adjacent to the furnace, in the area of origin, against the cinderblock wall.

The homeowner was able to show 10 years' worth of service calls, twice a year, with a cost of just under $200 for each call. Each document had marks on the form showing the jet was replaced, the electrodes were aligned, the burn chamber was inspected and vacuumed, as was the heat exchange area. The furnace was then adjusted to give an efficient flame; this would have been the movement of the plate over that 1-inch-square hole. One more interesting note was that the small metal plate was never found in the debris.

This was a situation in which the furnace was at fault for failing. But, because the service company accepted money for a job they never did, they created the

(Continued)

jet nozzle
an oil furnace has a nozzle with an extremely small opening; the oil is pumped up to this nozzle, where it squirts out and is ignited by the electrodes, providing a proper flame for the burn chamber; nozzles come in varying sizes

scale
a flaking of metal as the result of oxidation

(Continued)

Figure 13-6 *With a remote flash you can see into the heat exchange area, which shows heavy soot buildup.*

situation. Even more interesting was when the civil suit (subrogation) was initiated, the insurance company obtained a subpoena for the service records from the oil company. The average service call takes between 1 and 2 hours, depending on the experience of the technician. The average technician made anywhere from 12 to 20 service calls in an 8-hour day. The case was settled and a part of the settlement was that the oil company shut down their service department and only deliver oil. The second part of the settlement was that the oil company would pay to have each home that was serviced that fall revisited by another company to properly service those furnaces to ensure that a similar incident would not occur.

INTRODUCTION

Note
Safety is highest among all the issues to be considered.

As in Chapter 12, you can use the knowledge gained from all of the previous chapters to seek out the facts that will lead to a successful conclusion as to how the fire started. Safety is highest among all the issues to be considered. Then, the legal issues of properly being on the scene along with potential spoliation are of concern. Knowledge of chemistry and science principles is an invaluable tool when you examine all the burn patterns at the scene of the fire. Even with an absolutely accurate determination of the cause of the fire, if the evidence is not collected and preserved properly, it is all for nothing. Not only must you, as the investigator, know how the fire started, but you must also prove such to your peers, and there is the potential that you must prove such in a court of law.

The first vital necessity in the search for the cause of the fire is locating the fire origin, as discussed in Chapter 12. During the search for the origin, you might find many pieces of evidence that will assist you in determining the cause of the fire. Burn and smoke patterns can give some indication of the fire cause, especially if there was an unusual amount of charring that would not be expected with the normal type and amount of contents in the structure. Documenting is critical along the path to the fire origin. Once at the area of origin, a detailed examination must be undertaken to discover the material first ignited along with the heat source that brought that material to its ignition temperature.

■ Note The search for the cause must be done systematically using scientific methodology.

As with the search for the origin, the search for the cause must be done systematically using scientific methodology to come up with the hypothesis that explains your opinion as to how the fire started. This process may result in a hypothesis that does not stand the test, requiring you to go back and reexamine your notes or the scene itself. The fire may require a more in-depth investigation to come up with an accurate determination of the cause.

■ Note If at any point, the investigator uncovers evidence that the fire may not be accidental in nature and may in fact be an intentionally set fire, the investigation changes focus. If the investigator is a fire officer, this is the time to call for an assigned fire investigator.

LEGAL ISSUES

The reason the fire service is allowed to stay on the fire scene is to handle the emergency and, for the good of the public, to identify the area of origin and the cause of the fire. All portions of the structure should be examined, even those areas without fire damage. However, you are only looking for the extension of the fire or the elimination of potential heat or fuel sources. The investigator has no right to look in furniture such as drawers. If there is no connection to looking for extension of the fire, then it would constitute an illegal search. In addition to being ethically wrong, if any evidence is found in this process, it may not be admissible in a court of law.

If at any point, the investigator uncovers evidence that the fire may not be accidental in nature and may in fact be an intentionally set fire, the investigation changes focus. If the investigator is a fire officer, this is the time to call for an assigned fire investigator. The legality issues may be different from one jurisdiction to another, but it would be prudent for the investigator to place the scene under guard, keeping all out of the scene, and approach the proper authority, such as a magistrate, for a search warrant to continue the examination. Based on the evidence collected, the investigator would take an oath, stating the facts in the case that confirm that probable cause exists that a crime has been committed. The investigator identifies the remaining areas to be searched and what is being sought in the process. Only then can a valid search warrant be issued. This process protects any further evidence collected from being thrown out of court because of the possibility that it was illegally obtained.

■ Note If evidence is out in the open, if there is a chance that it may disappear or be damaged, then it is acceptable to document, collect, and preserve that evidence.

One major caution: If evidence is out in the open, if there is a chance that it may disappear or be damaged, then it is acceptable to document, collect, and preserve that evidence. Most important is that the investigator must consult legal counsel in advance to ensure that investigative procedures are up-to-date and appropriate to the laws of the jurisdiction.

Protecting the Scene

Investigators must protect the scene using barrier tape and guards to ensure that all the evidence is protected. You are just as obligated to protect evidence from an accidental fire as you do a criminal fire case. The evidence for an accidental fire is just as crucial in a civil case. Proving a product failure may well prevent the failure of similar devices or products in the future. This in turn can save lives and property.

EXTERIOR EXAMINATION

■ Note
Although the cause of the fire can only be determined after the area of origin has been located, plenty of supporting evidence of the cause may be uncovered along the path to the fire origin.

When you start the actual examination for the fire origin, you are also searching for evidence of the fire cause. Although the cause of the fire can only be determined after the area of origin has been located, plenty of supporting evidence of the cause may be uncovered along the path to the fire origin.

Start your examination from the farthest point away from the structure where there is no damage at all. Working in a systematic way, examine the surrounding area for any indications of debris or evidence that may indicate a cause of the fire. Although many items identified outside as evidence may be evidence of the fire being incendiary in nature, the cause of the fire is still undetermined during this phase of the investigation.

In this process, it does not matter whether the investigator goes clockwise, counterclockwise, or sets up a grid system. The important fact is that the examination is conducted in a consistent manner from one fire to the next, as much as practical.

All evidence should be photographed in place as it is discovered and before it is moved. It should be documented, and if it is to be collected, it requires further detailed documentation, such as cataloging and taking the various steps necessary to ensure the evidence is preserved for further examination by a laboratory. Items found in the outside area could be glass from the windows. Broken glass that is clean with no soot found some distance from the structure may indicate an explosion before the fire. Sooted glass may indicate the fire occurred first, and then the explosion was secondary or was caused by the fire.

Other items may be indicators of a potential incendiary nature of the incident. For example, smoking materials discarded in an unusual area, especially if they appear unweathered and no one in the family smokes, can be suspicious. The presence of refreshment bottles or cans, both soda and alcoholic, can be suspicious, especially if they are not weathered and there is no history of anyone drinking anything in that area or discarding such containers in that area. A container that may have contained an ignitable liquid could be in the area and of interest if it does not belong to the occupants of the structure. A caution: Do not place your nose over the container to detect its contents; this is an unnecessary potential exposure to hydrocarbons. The use of a simple hydrocarbon detector is the best test to see whether there is residue of an ignitable liquid.

■ Note
The use of a simple hydrocarbon detector is the best test to see whether there is residue of an ignitable liquid.

Other incendiary indicators may be ignitable liquids used as a trailer leading from the structure to a hiding place outside the structure. This type of trailer is not always successful. The success of this type of trailer depends on the type of fuel, the amount of fuel, the type of soil (some soils may absorb the liquid quickly), and the temperature (which may dictate the amount of vapors given off by the fuel). Regardless, the trailer leaves a telltale residue that will make it visible and obvious, as shown in **Figure 13-7.** This specific trailer of diesel fuel

Figure 13-7 *A trailer of poured diesel fuel that leads from the kitchen, out the door, down the steps, across the backyard, and behind an above-ground pool.*

Figure 13-8 *An evidence tag sitting on top of a bucket that protects a footprint in the soft earth.*

was a failure; however, the perpetrator reentered the structure and used gasoline inside to get ignition.

Fresh footprints in and around the area may be of interest, as would tire tracks not expected to be at the scene. Even though they may not be of any known value, it is best to protect these items until the entire scene can be examined. This can be accomplished by placing items such as boxes or buckets over the print or track and then marking it with a tag as evidence. For example, if you find a footprint in moist soil, place a bucket over the print to protect it, and then place an evidence marker on top of the bucket, as shown in **Figure 13-8.** If the examination of the scene indicates an incendiary nature and there is a possibility that the footprint may be of value, then a plaster cast of the print may be made prior to leaving the scene.

■ Note
It is also important to photograph all fuel and heat sources entering a structure.

It is essential to photograph all sides of a structure regardless of whether there are patterns. It is also important to photograph all fuel and heat sources entering a structure. The electrical meter base, the overhead wiring, or the conduit from underground lines should be photographed and the condition documented. Any gas lines entering the structure should likewise be examined, documented, and photographed. In addition, any chimneys or flues coming through the roof should be photographed, at least from the ground.

INTERIOR EXAMINATION

The path to the area of origin inside the structure will also render potential evidence. The investigator must keep an open mind and yet analyze what is found along the way. It should be mentioned that more often than not evidence items located outside the area of origin are potential incendiary indicators. This does not imply that evidence of arson is all that the investigator is looking for or all that will be found. However, evidence of a fire accidental in nature is more often found at or near the area of origin.

Clues that may assist in the investigation can be found in many areas. For example, an abundance of spliced extension cords is not evidence by itself, but an indication that the investigator should take a close look at electrical problems near or at the area of origin.

■ **Note**
An abundance of spliced extension cords is not evidence by itself, but an indication that the investigator should take a close look at electrical problems near or at the area of origin.

What to Search for and Document

The investigator must search all rooms and areas of the structure. This is not a detailed search of any furnishings but a search of the structure for extension of the fire and to eliminate any potential fuel or ignition sources, as well as unsuccessful fire starts. As heat or fuel sources are located, they must be photographed and documented to either show potential involvement or no involvement in the fire. All too often, an investigator goes to court on a product liability or incendiary fire and the opposing counsel brings up the problem with a furnace or faulty electric heater in the bedroom and how that had to be the cause of the fire. With the proper documentation, the investigator can easily refute with a photograph of the heater or furnace in question, clearly showing that it did not start the fire and in fact suffered no fire damage at all. This could certainly help the jury or judge render an opinion on the case.

Accidental Indicators

What comes to mind most when thinking about indicators of an accidental fire are clues that may indicate what to search for around the area of origin. For example, when ashtrays around the house are all full of cigarette butts or when burn marks from smoking materials are found in areas not damaged by the fire, this indicates that there was carelessness with smoking materials. This is not evidence but clues on what needs either to be confirmed or eliminated. It must also be stressed that the process of elimination is just as important as confirming a potential correlation between the clues and actual proof of the fire cause.

Electrical wires and connections in the undamaged areas of the structure require the investigator to, again, confirm or eliminate structural electricity as being involved in the incident. When the investigator finds open containers of paint thinner or other chemicals in areas of the structure unaffected by fire, the investigator must confirm or eliminate their involvement in the ignition sequence in the area of origin. As with all these indicators, the investigator must keep an open mind and not become blind to other sources of ignition. It is essential that these potential hazards be eliminated in the overall report should the cause of the fire be something different. The scientific methodology does require that the investigator include all hypotheses that were eliminated in the process of making the final determination of the cause of the fire.

Note The investigator must keep an open mind and not become blind to other sources of ignition.

Incendiary Indicators

Should the fire have any indication of incendiary nature, the investigator may consider having the local police department on the scene to help with security. Depending on the local policy, the police may already be on the scene if it is their responsibility to examine any fire scene where a potentially criminal act took place.

Note Should the fire have any indication of incendiary nature, the investigator may consider having the local police department on the scene to help with security.

Burn patterns, such as pooling patterns and sharp line of demarcations on the carpet, may be indicative that ignitable liquids were used to get the fire to start burning. These liquids can also be used as trailers to help the fire travel from one area to another. Many materials can be used as trailers. Sometimes it is not so much their composition as their form. Stacking magazines on the floor from one room to the next may not result in travel of the fire. Yet, by taking those same magazines, tearing out the pages, and crumpling them up can result in a faster moving fire that will travel wherever the trailer leads.

Common household items such as fabric softener sheets or wax paper streaming from one piece of furniture to another will act as trailers. These items will almost disappear, except in areas where they were protected from burning, such as under cushions. However, the crumpled paper from magazines will leave behind carbon residue that, if not destroyed by suppression forces, will still appear as a trailer; some pages may even hold their crumpled shape after burning.

Note The mere presence of an ignitable liquid pattern or even a positive sample is not an indication by itself of an incendiary nature of the fire.

The mere presence of an ignitable liquid pattern or even a positive sample is not an indication by itself of an incendiary nature of the fire. Unusual as it may seem, there may be a legitimate reason for that flammable liquid to be in the structure at that location. Another caution is flashover conditions where post-flashover burn patterns can be very similar to ignitable liquid pour patterns.

When taking samples, it is best to use either K-9 detection or an electronic hydrocarbon detector to identify the best location to obtain a sample. If possible, find a similar product such as the wood flooring or carpeting that has not been contaminated to use as a comparison sample.

SOURCE AND FORM OF HEAT OF IGNITION

In Chapter 9, we discussed the many sources of heat that would be substantial enough to bring the first material ignited to its ignition temperature. Everything from smoking materials to faulty electrical wiring and escaping sparks from a fireplace or chimney can provide sufficient energy to start a fire. The role of the investigator is to identify the specific source of heat that came in contact with the first material ignited within the area of origin.

■ Note
Many items once burned are hard to identify, and the investigator needs to know what to look for in the debris.

Recognizing residue is a major skill of the investigator. Many items once burned are hard to identify, and the investigator needs to know what to look for in the debris. Some of this experience comes from being a firefighter and looking at scenes after a fire. By picking up debris during overhaul and examining it, firefighters can discover what the object was, for example, the base of a coffee maker. Other investigators get experience from their training program, which may require them to investigate a room after a fire. They can examine the debris and then watch a video or view photos of what was in the room before the fire to gain "before and after" perspectives on burnt items. Even better training is when the team that put the room and content training fire together sees firsthand what a hairdryer or curling iron looks like after being subjected to 1,000-degree heat.

Many times, just examining an item in its burnt and misshapen form can reveal its original form and use. The item in **Figure 13-9** was found adjacent to a furnace. In fact, 10 of these were found surrounding the furnace. By looking at the bottom, it became evident that the object was a 1-gallon milk jug. An additional clue was that there were small pockets of liquid preserved as the sides melted down on themselves. Laboratory confirmed the liquid was gasoline. The homeowner attested to having a serious problem with the furnace. However, the

Figure 13-9 *Careful examination of the scene revealed melted plastic that was the residue of a 1-gallon milk jug that contained gasoline at the time of the fire.*

oil tank valve was in the off position and weeds had grown into the valve handle, making it impossible to open or close without killing the weeds. When confronted, the homeowner admitted that he set the fire using the milk jugs, hoping to make it look like a malfunctioning furnace.

■ Note **Sometimes the evidence of the source of ignition may not be in the area of origin.**

Sometimes the evidence of the source of ignition may not be in the area of origin. It could have been a lighter carried away by the arsonist. A fire alongside a road in the leaves could be from a spark from the muffler or the heat from a catalytic converter that set the leaves on fire as the vehicle drove away. A brush fire may lead back to a burn barrel, but the burn patterns may begin about 15 feet from the barrel. In this case, perhaps a spark from the barrel started the fire, but the likelihood of finding that burning ember may be slim. The point is that in some circumstances, the heat source may not be present at the site, but through eyewitness accounts, the investigator may be able to label the heat source with a reasonable amount of certainty even when it is not in the area of origin.

FIRST MATERIAL IGNITED

The first material ignited is just what it implies. There are a few ground rules to make sure that there is consistency. During a training session for filling out the forms for the National Fire Incident Reporting System (NFIRS), a student was given the scenario of a child playing with matches while in a closet. In the process, the flame came in contact with the plastic film covering clothing that had just returned from the cleaners. The student entered the first material ignited as wood. Well, he was correct to a certain extent—the match head chemicals were attached to a wood stick, the matchstick. In his mind, the match chemicals ignited the wood match and everything else was secondary to that. At the same time, another student argued that the gas in the butane lighter was the first material ignited in this scenario.

Although technically correct, neither explanation meets the intent of the information fire investigators want to gather for a report or an investigation. The first material ignited tells the story of what happened. The heat source could be a gas burner on the stove, and the first material ignited was the dish towel left near the burner. This is the beginning of the explanation of how the fire starts. This tells us that more training on keeping combustibles away from gas range burners is a preventative solution. The fact that the pilot ignited the gas serves no value other than to prove the appliance operated as it was designed.

The first material ignited has a form or shape, and this too is important. To say that the first material ignited was wood has limited value. You must be more specific. If it was a wood log, 12 inches in diameter, and the heat source was a pilot light, the wood would take some time to ignite and that fact is of importance. However, if the first material to ignite were wood shavings, they would ignite rapidly in contrast to the log.

Some items need a little more clarification. Plastics, for instance, take many shapes and sizes and have varying characteristics. Plastic can be the soft sponge foam in a couch cushion or a hard plastic toy. The plastic in toys can be hard and brittle or it can be soft and bendable. Each has its own burning characteristics that need to be examined. Of particular concern are the soft foam plastics that can turn to liquid upon exposure to heat, and then ignite and flow. Sometimes the telltale signs of flowing plastics are the same as those of flammable liquid pours, but not always. If plastic is a potential first material ignited, then it would be prudent to obtain similar material so that it can be examined and tested to see whether it just chars or is capable of spreading the fire.

■ Note
Sometimes the telltale signs of flowing plastics are the same as those of flammable liquid pours.

IGNITION SEQUENCE

If you are ever in a position to be a public safety investigator, you may realize one day that it is amazing that we don't have more fires than we do. There are many times when an appliance or device malfunctions, but circumstances were just not right for the proper material to be present for ignition. Often, this is by design when manufacturers recognize a potential heat source and insulate or isolate it from ignitable items. At other times, manufacturers ignore this fact and as a result there have been devastating losses, recalls, and civil suits.

But when a heat source and a fuel do come together, there can be ignition. It is the role of the investigator to place these two items at the proper location and time to explain in the hypothesis what happened that resulted in a fire. When considering the heat source and the fuel coming together, there is no implied movement at that time. The two could have been together over a long period of time before flaming combustion took place. The two could have been in contact with no action until an external force interceded and caused ignition, such as a strong wind current fanning a pilot flame.

■ Note
You must apply science that will prove that the available heat source was sufficient to ignite that specific first material ignited.

■ Note
The final hypothesis should identify and explain the heat source in detail. It should also identify the first material to be ignited and explain why the two were in close proximity.

To satisfy this part of the investigation, you must apply science that will prove that the available heat source was sufficient to ignite that specific first material ignited. Sometimes the ignition sequence is so obvious that verification need not be documented—such as a wooden kitchen match dropped onto crumpled writing paper. But it is not always this simple; when there is any doubt that the heat source was capable of causing ignition, there are NFPA texts and documents that can verify the ignition temperature of various materials. The final hypothesis should identify and explain the heat source in detail. It should also identify the first material to be ignited and explain why the two were in close proximity. You might need to explain why that material was in that area. In some instances, such as a gas leak, it may be necessary to explain the location of the leak that lead to the event. It is a matter of common sense when writing the report.

ELIMINATION OF ALL OTHER CAUSES

Before solidifying the final hypothesis, the investigator should be prepared to answer questions prior to them being asked. In particular, the investigator should check each and every heat source and potential fuel to ensure that these items could not have been involved. You might have already done this in the scene examination process, but you should verify and clearly document in the notes, including the reason why that heat source or fuel could not have been involved in the fire. The photos taken during the scene examination of the various heat sources can support your elimination determinations. In some situations, even a fuel source has been removed, such as shown in **Figure 13-10** where the liquid propane tanks had been removed from the property prior to the fire.

Sometimes it may be difficult to eliminate some heat sources such as smoking materials. If no one in the house smoked and there were no guests in the preceding 24 hours, it may be logical to eliminate smoking materials as being involved. To be sure, your interviews should include questions about whether someone used to smoke and is in the process of quitting. If there are teens in the house, you want to believe they were not smoking, but this is a call you must make based on your opinion of the answers given in the interview process. More important is the fact that you have a good logical heat source at the

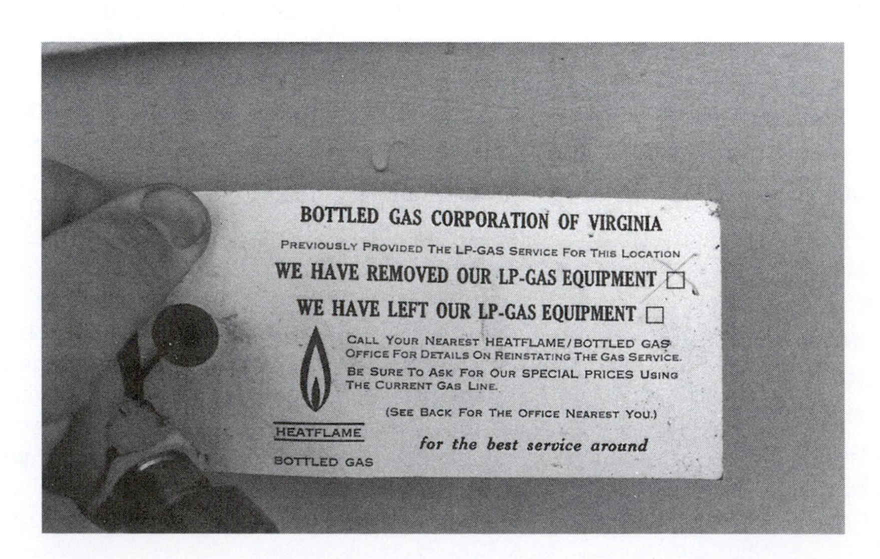

Figure 13-10 *Tag left on the gas supply line outside the structure showing that the liquid propane gas bottles had been removed prior to the fire. If there is no date on the tag, the gas provider can give exact dates of removal and reasons why.*

area of origin that can stand the challenge test from peers and opposing counsel in a trial.

CLASSIFICATION OF FIRE CAUSE

There are variations on the listings of fire cause, but they are essentially all the same. The first tenet of fire cause is that all fires are considered accidental in nature until proven otherwise. Some states mandate this through case law. However, it is good practice to adopt this attitude if you are to be a true seeker of truth. Going into a fire scene with a clear mind and no preconceived notions of the cause keeps your mind open to new ideas, thoughts, and concepts as you uncover piece after piece of evidence about the cause of the fire.

■ **Note**
Going into a fire scene with a clear mind and no preconceived notions of the cause keeps your mind open to new ideas, thoughts, and concepts.

Natural

Some texts or articles may say the fire was an "act of God." Depending on one's beliefs, this could be accurate. To clarify, the term *natural* implies that the fire was not started by the actions of a human. This category includes fires caused by lightning, lava flows, or flooding that creates electrical shorts. As long as there was no act by a person that created the cause of the fire, the fire should have been labeled as natural.

Accidental

Accidental is just as it implies: The fire was started because of an accidental act. Yes, someone made a mistake and some action of a human resulted in the heat source coming in contact with the first material ignited. This includes product liability cases where the fire was the result of a design error, a manufacturing error, or an error in how the device was used.

For example, a steam iron is a good tool, but left face down on a fabric while the person doing the ironing runs to answer a phone can result in a fire—an accidental fire. Electrical wiring installed using the wrong components that cause an electrical short and then a fire is considered an accident. It could be argued that the electrician consciously used the wrong connector, knowing that it could start a fire. But unless there is proof that the electrician intended for the fire to start, the fire is accidental in nature. However, labeling the fire as accidental does not remove blame for the electrician's actions. Civil suits are available for those seeking recourse in a loss resulting from an accidental fire.

■ **Note**
When a child plays with matches and starts a fire, the fire is considered in most locations as being accidental in nature.

When a child plays with matches and starts a fire, the fire is considered in most locations as being accidental in nature. Because the child is too young to know the consequences of his or her actions means that the fire was accidental. In many jurisdictions, the courts have ruled that children under the age

Pint-Size Problem

The engine company pulled up on the scene of a small brush fire. As they got off the engine, the officer spotted a small child running into the house next door to the open lot with the fire. The door slammed loudly. As they pulled off the booster line to extinguish the small fire, three other young boys pointed to the house next door and said Jimmy started the fire. Jimmy, of course, was the boy the officer saw slamming the door.

An investigator was called because it was a potential juvenile case. The investigator knocked on the house's door and little Jimmy opened the door a crack. The investigator identified him and asked Jimmy if his parents were home. The young boy looked up at the investigator and shouted that he was 5 years old and that there was nothing the investigator could do about it, and then Jimmy slammed the door shut. As tragic as this may seem, this young man is destined for more trouble in the future. But the engine company personnel got quite a chuckle when the 6-foot-plus investigator had a door slammed in his face by a 5-year-old.

of 7 years could not have understood their actions. However, there are 5-year-olds who clearly demonstrate that they know what they are doing is wrong. But the courts are unlikely to prosecute unless it is a compelling case to get assistance for the child.

Incendiary

Many fires are intentionally set, but that does not make them incendiary. Setting fire to the logs in a fireplace or igniting the trash in a legal burn barrel are both intentionally set fires. The key difference between these fires and incendiary fires is the intent. It is as simple as determining the frame of mind of the individual who starts the fire. The key component of this frame of mind is *malice*.[1] If someone sets anything on fire with malice, then it is against the common good of the public. It doesn't need to be hate; *malice* also includes being reckless toward the law and the rights of others. Malice can also imply an evil intent, such as committing insurance fraud by burning a car for the insurance money.

■ Note

Case law in many jurisdictions has dictated that it is up to the courts—not the investigator—to decide whether a fire is arson.

The term *arson* means essentially the same thing as *incendiary*. But case law in many jurisdictions has dictated that it is up to the courts—not the investigator—to decide whether a fire is arson. The definition of arson and incendiary are similar in *Black's Law Dictionary*, but how those words are used in your own court system makes the difference as to which term you should use in your report.

Undetermined

No matter how good an investigator's skills, there are fires that will go undetermined as to their origin and cause. The investigator may find the origin, but the

If You Are Going to Do It, Do It Right!

The fire almost totally consumed the two-story multiroom house. The entire structure was reduced to a 3-foot-high pile of debris. The local investigator labeled the fire as electrical in nature because the homeowner mentioned he had been having problems with the light. One week later, a fire investigator for the insurance company looked at the scene. Only one small corner had been disturbed by the local investigator. To conduct a thorough examination, the insurance investigator brought in labor in the form of off-duty firefighters and they dug the fire scene. When they got to the bathroom, which was 3 feet lower than the rest of the house, a 55-gallon drum was located on its side with holes in areas that were against the floor. Further investigation gave clear indication of an incendiary fire. Although additional evidence of an incendiary fire was found and documented, the homeowner maintained his innocence. His only defense during depositions was the report created by the local investigator that said that the fire was electrical in nature. Charges were never brought, but the insurance company denied the insurance claim.

cause may elude him or her because of conflicting patterns or because the evidence sustained too much damage. This can be the case when the fire burns too long or the fuel load was such that it caused extensive damage.

■ **Note**
There should be no shame in having an undetermined fire.

There should be no shame in having an undetermined fire. However, there will be senior officers in municipal departments who do not understand investigations and who will insist on having a cause for each and every fire. This is dangerous because it can cause false data to be created and create problems later if the true cause of the fire is discovered. Worse yet, to make a guess as to a cause can result in the imprisonment of an innocent party.

DETERMINING RESPONSIBILITY

Once the origin has been located, you have a place to look for the cause. With the cause identified, there may be sufficient evidence, after interviews, to identify *who* or *what* caused the fire. This determination must meet the requirements of a systematic investigation using scientific methodology. The *what* can be a product failure, which may allow the victim or the victim's insurance company to recover their funds through subrogation. The *who* may be an individual that either intentionally set the fire with or without malice. If it was with malice, there may be pending criminal charges. A no-malice situation might be something like a child playing with matches or the homeowner accidentally using gasoline in a kerosene heater.

Temporary Repairs

The responsibility could be a third party's. A homeowner found flames shooting out of her gas water heater and had the presence of mind to immediately shut off the fuel supply. She called the gas company, which sent out a service repair person to look at the water heater. He advised the homeowner that the valve assembly leading to her pilot light was faulty. He advised her that he would put a temporary fix on the valve and get her an estimate on the repair before ordering the part.

In less than an hour after the repair person left, she smelled something burning and found that the flame from the pilot was as much as 3 feet tall and had ignited some nearby combustibles. She called the fire department, and the engine company officer who did the preliminary, first responder investigation labeled the cause as faulty equipment. The insurance company sent an investigator. In the presence of the gas company representative, the homeowner, and the repairman who worked on the heater, the investigator disassembled the valve for the pilot light. He found a wad of aluminum foil, what appeared to be a gum wrapper, as shown in **Figure 13-11.**

The homeowner said she had noticed the repairman take out a stick of gum and wad up the wrapper when he was doing her repair. The repairman confessed it was a common practice for him to push foil in the valve, limiting the gas flow to save his customers' money.

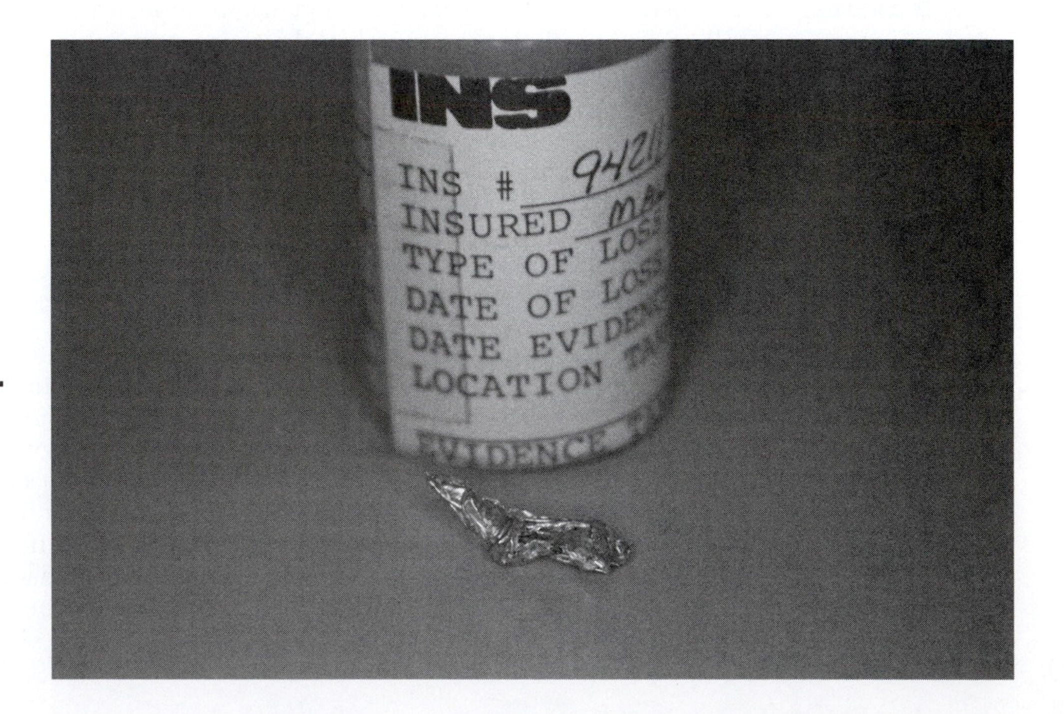

Figure 13-11 *Small foil from a stick gum wrapper improperly used to limit the flow of gas in a water heater pilot light. Notice the use of a film canister as an evidence container.*

Product

■ Note
Product failure is not unusual in this age of technology.

Product failure is not unusual in this age of technology. A number of kitchen appliances have been recalled because of their propensity to overheat, creating a fire hazard. The investigation into the failure of an appliance or any other device is to eventually identify whether the failure was a result of design or the fact that it was improperly manufactured.

A third product issue is when the product is proper, but the instructions were lacking or the individual did not follow the instructions. Many instructions may appear ridiculously obvious. Such as "coffee is hot" or "do not use clothes iron on clothing being worn." The reason these instructions are in place is because someone did hurt themselves by doing something foolish, yet a jury awarded the individual a settlement.

When fire reports are filled out properly, they may eventually indicate a noticeable number of fires with one specific product. That could result in further research and in turn could result in a recall of those products. The end result is the reduction or elimination of fires caused by those specific products, which definitely saves future damages and possibly lives.

Person or Persons

■ Note
The typical arsonist is not usually a dirt bag, low-life scum or deviant bum; most arsonists are people not much different from anyone you may see on any street.

Identifying the person who maliciously set a fire, making the arrest, prosecuting, and having them incarcerated prevents that person from setting another fire, at least while he or she is behind bars. The typical arsonist is not usually a dirt bag, low-life scum or deviant bum; most arsonists are people not much different from anyone you may see on any street. They are businesspeople, homeowners, and people with families; they may even live next door to you. The arrest made in an arson will not only put away the arsonist, but it will act as a deterrent for those who may be thinking of turning that insurance policy into a fiscal profit. It just may make that homeowner with the cracked walls caused by water in the basement, which is not covered by insurance, to think hard before burning the house down to collect on the fire insurance policy.

Deterrence is a tool in the fire prevention toolbox. People will think twice before taking an illegal action of setting a malicious fire if they risk going to jail. Deterrence is not just the fact that the fire department made the arrest so much as it is the fact that the department is looking at fires and making determinations as to the cause of fires. This may provide the deterrent necessary to keep someone from committing a crime.

SUMMARY

It is essential to determine the area of origin. Without the origin, there is no possibility to determine the exact cause of the fire. In the process of identifying the cause of the fire, all other potential heat sources must be eliminated. All this can be accomplished by conducting a systematic search using scientific methodology.

The cause of the fire can fall in the category of natural, accidental, incendiary, or undetermined. It is hard to leave any fire scene with an unknown cause of the fire. However, as a result of destruction of evidence and other indicators, this may be the only solution.

Once the area of origin and cause have been confirmed, then it may be necessary to identify the persons or things responsible for the fire. There is a multitude of heat-producing appliances in any home. The improper placement of these items, misuse, and poor design or manufacture could be at fault. Depending on the outcome of the investigation, the investigator could be testifying in civil court on his or her findings.

Should the investigation reveal that a person was responsible for the fire, the investigator must determine whether there was malice. If someone made a mistake and stacked the wood in the fireplace so that it rolled out onto the floor and ignited the rug, it is human error but not an incendiary act because the fire was an accident and there was no malice. The same is true for someone placing gasoline in the kitchen sink to clean auto parts. This is not what an average person would do, but it is done from time to time. The vapors could reach an ignition source and ignite. There was no malice, but there was stupidity. Arrests are usually not made for stupidity.

Then, there is the person who sets the fire intentionally either with spite, revenge, or for the insurance money. In such cases, the fire is considered to be set with malice and it is a crime. Solving such crimes and prosecuting the people involved are a deterrent for others considering similar acts.

Everyone has a job to do on the fire scene. The overall focus is to save lives and property by conducting searches to find those trapped and by extinguishing the fire as soon as possible to limit the fire damage. But the responsibility of the fire service does not stop when packing up the hose to leave the scene. To prevent future events, we need the tools to prevent fires. To do this we need complete and accurate fire incident reports. The fire officer or an assigned investigator must make a determined effort to find the area of origin and the cause of the fire. Transferring this information to the NFIRS ensures that others can benefit from this data, eventually providing guidance for future fire prevention efforts.

KEY TERMS

Burner assembly The assembly houses the electrodes in a furnace to ignite the oil that is pumped out of the jet nozzle. This unit also contains the fan, which blows air into the chamber, allowing for a bright and efficient flame.

Jet nozzle An oil furnace has a nozzle with an extremely small opening. The oil is pumped up to this nozzle, where it squirts out and is ignited by the electrodes, providing a proper flame for the burn chamber. Nozzles come in varying sizes.

Scale A flaking of metal as the result of oxidation.

REVIEW QUESTIONS

1. What is the primary reason why fire departments are allowed to stay on a fire scene and search for the origin and cause of the fire they just extinguished?
2. If the investigator discovers evidence that the fire is incendiary in nature, what is being risked by not obtaining a search warrant at that point?

3. What is the best device to use to obtain a good sample of a potential accelerant to send to the laboratory?
4. All fires are considered_____until proven otherwise.
5. Why take a photograph of an electric hot water heater if it had no involvement in the fire?
6. Give an example of a fire that could be classified as a natural cause.
7. A fire resulting from a 3-year-old child playing with a lighter should be classified as what type of fire?
8. What key component must be present to accuse a person of a criminal intent in setting a fire?
9. Why must you have an area of origin before you can find the cause of the fire?
10. Under what circumstance, mentioned in this chapter, would it be necessary to obtain a search warrant in the process of the investigation?

CLASS ACTIVITY

1. In the case study at the beginning of this chapter, the area of origin was labeled as just outside the furnace. If the furnace failed on the inside, should that be the area of origin instead? Why?
2. In the case study, the cause of the fire was the malfunction of the furnace. The person responsible was the technician. Was he the only problem?
3. In the case study, the oil company settled the case by agreeing to shut down its service section. It also had to pay a third party, another service company, to visit all of the clients that had a service call that fall to ensure that the furnaces were properly serviced. Was this fair or should the case be continued in the courts in hopes of a stronger punishment?

ACTIVITIES

1. In a dictionary, look up the definitions of the terms *arson* and *incendiary.* Discuss the similarities and differences in definitions.
2. On the Internet, enter the term *fire investigator* and explore the sites listed including InterFire Online, National Fire Protection Association, International Association of Arson Investigators, and National Association of Fire Investigators.

NOTES

1. National Fire Protection Association, *NFPA 921, Guide for Fire and Explosion Investigations* (Quincy, MA: NFPA, 2008), Section 11.5.6.

VEHICLE FIRES

Learning Objectives

Upon completion of this chapter, you should be able to:

- Describe the safety precautions that must be taken before starting the investigation.
- Describe the fuel and electrical systems of most common vehicles.
- Describe the potential heat sources and fuels available in most vehicles today.
- Describe the potential for arson and which indicators may point to an incendiary nature of the fire.

pro bono
doing the work free of charge; at no cost

CASE STUDY

School was quite a commute, more than 50 miles one way. But because of a physical disability, staying in a dorm was out of the question. The student had suffered a major injury, losing all sense of feeling below the arms. The family was able to find a used car that had hand controls that enabled him to at least be mobile.

On this particular day, he got in his vehicle after class and even before getting on the interstate he could smell something burning. It was an unusually cold day, so he had to keep the windows closed. The smell did seem to dissipate, so he kept on his travels home. When he arrived home, he got in his wheelchair and went into the house. Almost immediately, his mother asked him what was burning. The smell had not dissipated so much as his body had continually ignored it and compensated and adjusted, essentially not sending the message to his brain that something was burning.

He leaned forward and his mother noticed a blackened area on his shirt. He moved to his bed and laid on his stomach. Pulling the shirt up, she found blisters and a very reddened area. He had experienced first- and second-degree burns. She called 9-1-1.

He was hospitalized to treat the burns and as a precaution to prevent infection. His mother called local authorities only to find out that there was no official fire investigator in the town where they lived. Going through the yellow pages, she finally found a fire investigator in the private sector who would at least come and look at the vehicle. After hearing the story and visiting the scene, the investigator offered to work the fire **pro bono,** free of charge.

It was obvious at first glance that the seats were heated. There was some amazement in looking at the upright portion of the seat, as shown in **Figure 14-1;** the fabric was not even burned. However, there was a small char in the area where the back of the seat met the bottom of the seat; it was slightly smaller than the size of a dime.

The design of the seat included a zipper on the back, which allowed the cover to be pulled off. Inside was the heating element. Other than having to pull the fabric off, the melted plastic element came out freely. As shown in **Figure 14-2,** the element shorted out and continued to burn until the supply wire melted off, which was where the small hole burned through the fabric.

Not all stories have a happy conclusion. Because the vehicle was purchased second hand, the manufacturer would not even consider doing any repairs or even discuss the issue. The family could not even find an attorney to take on the case—at least with their available, limited funds. Because the fire department did not respond, there was not even a fire report to document the potential danger. Copies and photos were sent to the manufacturer and to Consumer Product Safety, National Highway Traffic Safety Administration, and the National Transportation Safety Board.

(Continued)

(Continued)

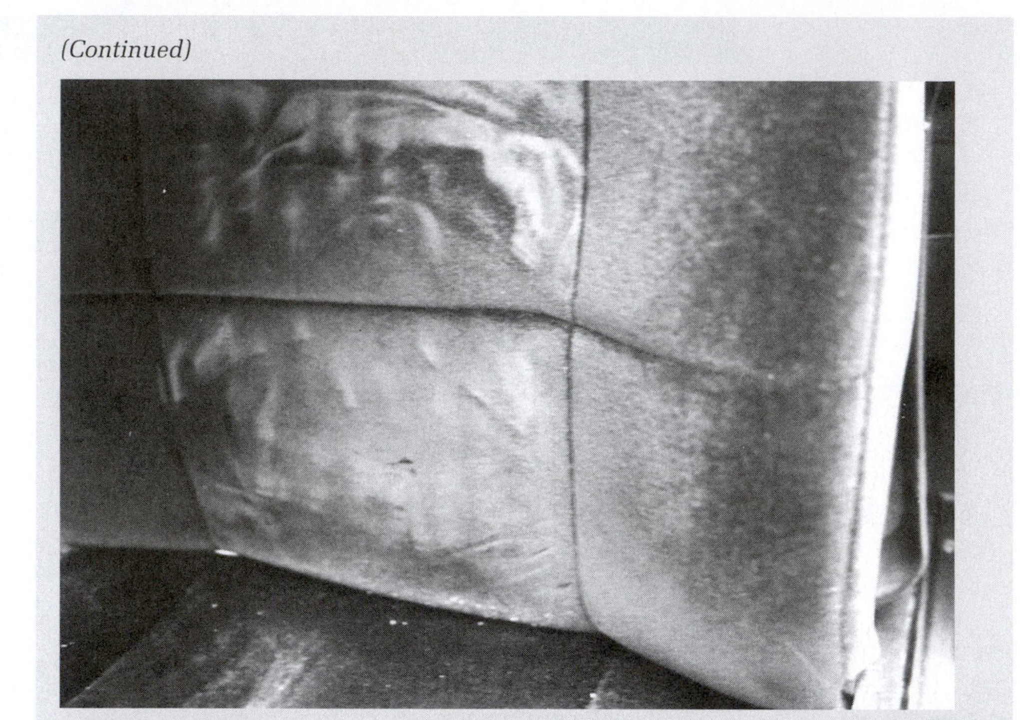

Figure 14-1 *Front seat driver's side showing heat damage of seat fabric.*

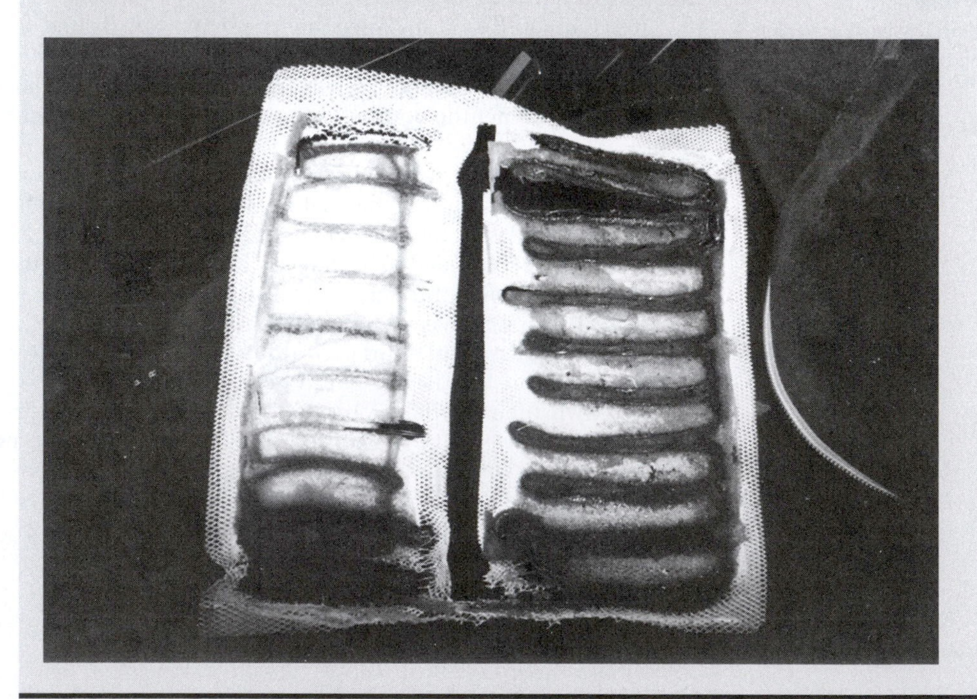

Figure 14-2 *The heating element that was located directly behind fabric on the seat's upright.*

INTRODUCTION

■ Note Vehicle fires are undoubtedly some of the most confusing and difficult fires to work.

Vehicle fires are undoubtedly some of the most confusing and difficult fires to work. Some believe they are the easiest, probably because they are not exactly sure what they are looking at. For all, the vehicle is a mode of transportation; for some, a means of stature in the neighborhood. The key to investigating a vehicle fire is not just determining the origin and cause, but discovering how the vehicle looked before the fire.

We teach investigators throughout their career that in investigations they need to do fire scene reconstruction—to remove all the debris and place the contents back where they were prior to the fire. This can help them determine a more accurate origin and cause. This rule applies to both structure fires as well as vehicle fires. As with structure fires, the investigator needs to work the scene from the least damage to the most damage.

■ Note As with structure fires, the investigator needs to work the scene from the least damage to the most damage.

Investigators must not become complacent in investigating a vehicle fire. They must take the next step to make sure they have covered all the basics, and do not treat these fires as minor incidents. Vehicle fires are just as important as any structure fire. The investigator needs to ask for assistance in the areas he or she is not familiar or comfortable with. The Internet is a great resource; you can find dealerships and manufacturer sites for specific vehicles to obtain information on components or even the particulars of the vehicle's construction features.

The passenger vehicle today is an extremely complex device. Yet it is very simple to use. It is full of hundreds of feet of wires, tubing, and combustible material, and for the most part uses ignitable liquids for combustion. These characteristics add to the fuel load of most vehicles, enhancing the burning process. It is the intent of this chapter to give the first responder investigator the tools to investigate the vehicle fire initially and to recognize when an assigned investigator needs to respond.

National Fire Protection Association (NFPA) statistics from 2006 show that there was a decrease in vehicle fires during that year. The report shows that 278,000 vehicle fires occurred in 2006. This number is down 4 percent from the rate in 2005; however, because of the rising costs of vehicles, the dollar loss was up. The report shows that $1.5 billion in property damage occurred related to vehicle fires, and that a vehicle burned once every 113 seconds. The interesting statistic is that more than 20,000 vehicle fires were intentionally set. Even though this number decreased from 2005, it shows that these types of fires still occur.[1]

SAFETY AND INITIAL EXAMINATION

Of the utmost importance in the investigation is safety. All precautions should be taken to protect yourself and others from injury. Fire investigators tend to get lazy when it comes to vehicle fires, probably because they are usually outside of a structure in the open air. The mind-set is that open air equals clean air. This is far from the case.

The materials in modern vehicles contain numerous toxic fumes that are released from burning components. The foams and plastics in vehicles give off numerous toxins when subjected to heat. These toxic chemicals range from hydrogen cyanide to carbon monoxide. **Off-gassing** is a term used in the fire service that means that materials, fabrics in particular, absorb the by-products of combustion during the fire and release the gases after the fire. Carbon monoxide is a colorless, odorless gas, so you cannot see it. But absorbed in quantity in the lungs, it can be detrimental. Off-gassing occurs as well in your fire gear from the smoke and particulate matter produced by fires. It is for this reason that self-contained breathing apparatus (SCBA) or respirators must be used by all who investigate fires, even when dealing with cold scenes. Debris that may be cold on top may be holding heat within or underneath.

■ Note
Off-gassing is a term used in the fire service that means materials. Fabrics in particular, absorb the by-products of combustion during the fire and release the gases after the fire.

off-gassing
the release of gas or vapors in the process of aging or decomposing, also refers to vapors or gases that were absorbed by fabric, carpet, and so forth during a fire and then released after the fire was extinguished

The first responder investigator should not take for granted that the initial fire crews have secured the vehicle. Is the vehicle in a position that it may roll or move during the investigation? One interesting note about the electric hybrid vehicles is that the vehicle may engage and start up if the accelerator is pressed, even though the motor is not running. Chocking the wheels and making sure that the vehicle's battery has been disabled by fire crews are musts. If the vehicle's electrical system has been damaged, shorting may occur, causing the vehicle to move or even worse to burn again.

■ Note
One interesting note about the electric hybrid vehicles is that the vehicle may engage and start up if the accelerator is pressed, even though the motor is not running.

Airbag deployment is a huge area of concern in almost all vehicles today. There are side bags, front passenger, driver's side, and a multitude of configurations to protect the passengers from impact. These bags contain a chemical propellant. One component most commonly used is sodium azide. A rapid chemical reaction takes place within the inflator that creates mostly nitrogen gas, which inflates the airbag. Bags can inflate in 1/20th of a second.[2] This is not enough time for the investigator to move out of the way. Typically, the airbag will deploy during the fire if the canister containing the pressurized gas reaches temperatures over 300°F.[3]

It is not recommended that you crawl under the vehicle during your investigation. Not only are you possibly contaminating your protective clothing by lying in the unidentified liquids that have leaked from the vehicle onto the ground surface, but you are endangering yourself from the vehicle. Shifting weight within the vehicle could cause it to move or collapse, causing crushing injuries, or jagged steel items cut by fire crews or ripped and torn by the vehicle accident impact may protrude from the underside of the vehicle, causing harm. Instead, one idea is to have the towing service raise the vehicle with all safety precautions in place. This will give you a brief chance to look at the underside of the vehicle. At no time should you crawl under the vehicle while it is lifted. All observations can be made from a safe area off to the side. By having the vehicle raised, you can observe smoke and soot stains or flame impingement from a safe

■ Note
It is not recommended that you crawl under the vehicle during your investigation.

distance. Any items of evidence that may be recovered will remain after the vehicle is removed.

■ Note **Always be aware of sharp edges and broken glass, and wear protective equipment at all times. Latex gloves do not afford the investigator protection from sharp edges and glass.**

Always be aware of sharp edges and broken glass, and wear protective equipment at all times. Latex gloves do not afford the investigator protection from sharp edges and glass.

TYPE OF VEHICLE AND IGNITABLE FUELS

The manufacturing divisions of the automotive industry are constantly changing. Popular demand, safety innovations, and new technology keep the vehicles we use daily in an ever-changing mode of appearance and functionality. There are diesel, biodiesel, gasoline, electric hybrids, ethanol, methanol, compressed natural gas (CNG), and liquid propane–driven vehicles, both passenger and commercial. All have their own distinctive characteristics but contain very similar components. Basic makeup of various vehicles is the same, framework, wheels, interior, and so forth.

The engine area is where there are major differences, with the fuel storage the second area of concern. One item that stays the same on most vehicles, with the exception of a fully electric vehicle, is the cylinder in the engine, the combustion chamber, where fuel and air come together, to form an ignitable mixture, which is ignited by a spark. This mixture is explosive and drives the pistons that turn the crankshaft. Vehicles have a transmission that attaches to the engine that when engaged transfers the turning of the crankshaft to the gears and makes the vehicle move.

Diesel Engines

Diesel engines operate on a compressed combustion chamber (cylinder). Fuel is sprayed into the cylinder prior to the pressure stroke. The fuel and air mixture are compressed in the cylinder and an explosion occurs, thus pushing the piston back down to the exhaust stroke. This action makes the revolution of the crankshaft, driving the piston down and back up the cylinder to the intake stroke of the motor. Air is drawn into the cylinder and the process starts over.

The diesel engine needs diesel fuel to run. Diesel fuel is a combustible liquid with a flash point above 100° F. Biodiesel is an addition to diesel that has a flash point above 300°F, although it will combust in the engine cylinder.[4] The fuel is stored in a sealed tank and is pumped from a fuel pump through lines and filters, which run along the frame rails of the body of the vehicle. The fuel is then pumped under pressure—upward of 78 pounds per square inch (psi)—into the engine's cylinder to begin the combustion process.[5] Failure of the system at any given point may be a possible point of fuel release in some vehicle fires.

Maintenance issues come into play here with the upkeep of the vehicle. Were the fuel lines tightened correctly? Are there leaks that supplied the fuel to an ignition source? There may even be an issue that the wrong fuel was used in the vehicle. It is not uncommon for someone to make the mistake of pumping gasoline into a diesel fuel tank. The nozzle of most gas pump dispensers will fit into the neck of the diesel tank. However, the nozzle on most diesel dispensers will not fit into the neck of a gasoline-fed vehicle.

Gasoline Engines

Gasoline, being more volatile and having a lower ignition temperature, causes concern about it being the first fuel ignited. With the fuel tank usually in the rear of the vehicle, the fuel lines run up the frame of the vehicle to the engine. The force to get the fuel from the tank to the engine is the fuel pump. The majority of fuel pumps in today's vehicles are located within the fuel tank. However, older vehicles and vehicle kits may use a fuel pump that can be located anywhere between the tank and the engine.

Gasoline engines operate on flammable liquids that have a flash point of less than 100°F. These vehicles can operate on a multitude of additives with the gasoline, including ethanol and methanol.

Ethanol, also known as ethyl alcohol, drinking alcohol, or grain alcohol, is a flammable, colorless, slightly toxic chemical compound with a distinctive perfume-like odor and is the alcohol in alcoholic beverages. In common usage, it is often referred to simply as alcohol. Ninety-five percent of ethanol is produced from corn; 11 percent of the U.S. corn crop went into ethanol production in 2004. In that year, the U.S. consumed approximately 140 billion gallons of gasoline and produced a record 3.4 billion gallons of ethanol. In August 2005, the National Energy Bill became law, mandating production of 7.5 billion gallons of ethanol per year by 2012.[6]

Ethanol fuel is a biofuel alternative to gasoline. It can be combined with gasoline in any concentration up to pure ethanol E-100. As fuel, ethanol is primarily used in two forms. E-10 is a blend of 10 percent ethanol with 90 percent unleaded gasoline. In this form, it can be used in any vehicle. E-85 is 85 percent ethanol blended with 15 percent unleaded gasoline. In this form, it can be used only in especially built vehicles. This means that ethanol can be found in quantities of 10 to 85 percent in gas pumps and 95 percent pure with 5 percent gasoline added in rail cars, tank trucks, and barges.

■ Note

Ethanol can be found in quantities of 10 to 85 percent in gas pumps and 95 percent pure with 5 percent gasoline added in rail cars, tank trucks, and barges.

Physical Data on Ethanol[7]

- Boiling point: 173°F (75°C)
- Melting point: −179°F (−117°C)
- Specific gravity: 0.7893
- Volatility: 100 percent
- Vapor pressure: 40 mmHg at 13°C
- Evaporation rate: (carbon tetrachloride = 1) 1.4

- Solubility in water: Complete
- Appearance: Clear, liquid
- Odor: Pleasant odor
- Flash point: 55°F

Both methanol and ethanol are water soluble. Both have characteristics that enhance the burning process, creating less exhaust emission. This is the trade-off for a cleaner environment: a vehicle that will burn hotter and faster, with less emission gases.

Gasoline is what the consumer has been most accustomed to. It is the most common fuel used today, with the addition of alternate fuels as supplements. Gasoline has a flash point of –43°F to –45°F. It is not soluble in water, and its ignition temperature is 536°F.[8] It is stored in the same manor, in a sealed tank, attached to the vehicle frame. Fuel is pumped from the tank through a fuel filter and to the injectors. This process is very similar to that of the diesel engine. The difference is that the fuel is sprayed into the cylinder and followed by a spark to ignite the mixed air and fuel, thus pushing the piston downward on the power stroke. This process is called electronic fuel injection (EFI). The EFI system is under great pressure, some more than 100 psi.[9] Just as with diesel engines, lack of maintenance or worn and loose parts can cause the fuel to be sprayed, thus atomizing it. This could be a possible cause for the spread of the fire in the event it came into contact with an ignition source.

Other Liquids

Other liquids in a vehicle include transmission fluid, windshield washer solvents, ethylene glycol, motor oil, brake fluids, and other hydraulic fluids. All of these liquids have different flash points but may become ignitable under extreme conditions, either contributing to or causing a fire.

Commercial vehicles in particular carry large amounts of these other liquids, as well as fuel. Some carry as much hydraulic fluid to run different equipment attachments as they do fuel. For example, trucks with snow plows, tailgate lifts, and farm equipment for implements carry large amounts of hydraulic fluid. Care should be given to conducting your investigation around these items because they may still have pressure built up in the hydraulic lines for the attachments. If you attempt to push or close a valve or knob, without knowledge of the vehicle, the implement could be lowered or dropped, causing injury.

■ Note
Windshield washer solvent, if sprayed onto a hot surface, may ignite.

HOT SURFACES

The manifold, catalytic converter, and exhaust pipe operate at sufficient temperatures to ignite a wide range of fuels. For example, windshield washer solvent, if sprayed onto a hot surface, may ignite. This could occur if the rubber line between the windshield reservoir and the wiper becomes detached. The liquid may then spray onto the hot engine manifold instead of the windshield, possibly igniting.

Leaking brake fluid is most often the first fuel ignited when it leaks onto a hot surface. Transmission fluid, especially if heated from the vehicle being overloaded, leaking onto the catalytic converter of associated exhaust pipe is also capable of ignition.

However, gasoline will not usually ignite when sprayed onto a hot surface. There are many variables such as ventilation or humidity that can influence whether the gasoline will ignite or not. The investigator should not dismiss this as not probable, but should continue the investigation, seeking any possible heat source and fuel that may have been involved in the ignition sequence.

ELECTRICAL SYSTEMS

■ Note Some commercial vehicles such as tractor trailers, farm equipment, and boats operate on a 24-volt system.

■ Note The voltage in hybrid vehicles can reach more than 300 volts of alternating current (AC).

Most vehicles operate on a 12-volt direct current (DC) system. But this is not always the case. Some commercial vehicles such as tractor trailers, farm equipment, and boats operate on a 24-volt system. The newest vehicles are hybrids, vehicles with an electric motor to supplement the gas power engine. Hybrids bring a whole new realm to the investigation of vehicle fires. The voltage in hybrid vehicles can reach more than 300 volts of alternating current (AC).

DC 12-volt vehicles operate with a lead acid battery that feeds circuits and fuses in the vehicle. The voltage is supplied to the circuits through wires that run from the battery to the fuses and switches. The energy is then distributed throughout the vehicle to various applications. The wires coming from the battery to the main fuses and switches are typically 2 to 6 AWG in size, often called the primary wiring. The wires from the fuses to the distribution areas can run in size from 8 to 18 AWG, and these wires are often referred to as secondary wiring. All of these wires are insulated with a coating to insulate and to protect them.

The secondary wiring, the wires that run from the fuse panel to other switches, outlets, and lights, carry the load of the vehicle's applications that the driver of the vehicle uses every day. The wires operate the lights, radio, windshield wipers, heated seats, and anything the vehicle uses, besides ignition. The secondary wiring can come into play in fires from wear and tear and misuse. Overloading these wires can occur with the addition of cell phones, computers, DC/AC converters, and so forth. Wiring to these accessories are typically 16–18 AWG wires, designed to carry no more than 100 to 200 watts. Operators of the vehicle sometimes disregard the maximum capacity of these outlets and strain the wire. This strain, which is called resistance

Figure 14-3 *Cracking of wires from age and continuous heating and cooling of the insulation. (Photo courtesy of Bobby Bailey.)*

heating, can cause the insulation around the wire to break down or melt. This cracking or melting of the insulation may expose the wire to a short or arc, which is an illuminous discharge across a gap that can cause a fire. In **Figure 14-3,** the wires show signs of cracking as a result of age and heating and cooling of insulation.

HYBRID VEHICLES

Hybrid vehicles combine a smaller internal combustion gasoline engine with an electric, battery-powered motor. There are basically two types, the series and the parallel. The series hybrid system switches between gasoline and the electric power source. The parallel system supplements the engine.

These vehicles contain a battery pack within the interior of the car. The batteries are high-voltage nickel-metal hydride (Ni-MH). The packs are made up of individual battery cells that contain potassium hydroxide (KOH). KOH has a pH of 13.5 and is highly alkaline.[10] KOH liquid is absorbed in each cell by a thin membrane paper; the result is a gel, or a dry cell battery. The battery packs may range from 100 volts to upward of 300 volts and are considered high voltage. The Honda Insight has approximately 54 C size cell batteries

Figure 14-4 *Battery pack and components of the Honda, located in the rear passenger area under the seat. (Photo courtesy of Bobby Bailey.)*

■ **Note**

In a series hybrid, the gasoline engine will automatically stop and start as the vehicle is running. The gasoline engine will stop when the vehicle comes to a stop, such as at a traffic light. Pressing the accelerator pedal restarts the gasoline engine.

in each battery pack, with 4 to 6 packs in each vehicle battery system. (See **Figure 14-4.**) Care should be taken when working in or around these vehicles after a fire. The battery system is located in the rear of the Honda under the passenger seat. The battery packs are tied together in series, in which the main feed wires run to the DC/AC converter located in the engine compartment area of the vehicle.

In a series hybrid, the gasoline engine will automatically stop and start as the vehicle is running. The gasoline engine will stop when the vehicle comes to a stop, such as at a traffic light. Pressing the accelerator pedal restarts the gasoline engine. The gasoline engine may run to recharge the battery pack. This may cause an issue for the investigator when investigating the scene. If the vehicle is not shut off and the gas pedal is accidently touched, it may engage the engine. In a parallel system, the electric motor runs in conjunction with the gasoline engine. The vehicle does not operate on the electric motor alone. The gasoline engine only shuts down when the vehicle is at a complete stop.

Figure 14-5 *Shut-off switch under hood of Honda Insight. (Photo courtesy of Bobby Bailey.)*

Note
If the vehicle is not shut off and the gas pedal is accidently touched, it may engage the engine.

Both Honda and Toyota battery packs are well insulated and protected from impact damage during a collision because of their location above the rear axle. Potassium hydroxide (KOH) reacts violently when it comes in contact with metals such as aluminum, zinc, and tin. The battery packs themselves have several shut-off switches located on the vehicle. Just because the investigator shuts off one of the switches does not mean that the system is secure. In the photo in **Figure 14-5,** the switch located on this vehicle is under the hood in the engine compartment. It shuts the power off to the electronic box to the engine but does not shut the power off from the batteries. There is a second switch that needs to be turned off and secured.

Remember that if the vehicle is partially or more submerged, do not touch any high-voltage components because of the electrocution hazard. Keep in mind that components of the high-voltage system have or may have in excess of 300 volts. Most manufacturers use a bright orange cable to identify the system. The wires are anywhere from 0 to 4 AWG in size. The wires carry high voltage from the battery pack to the DC converter of the vehicle. These cables can run inside (**Figure 14-6**) or under the vehicle (**Figure 14-7**) depending on the manufacturer. In addition, there will be a 12-volt DC battery in the vehicle that operates the accessories.

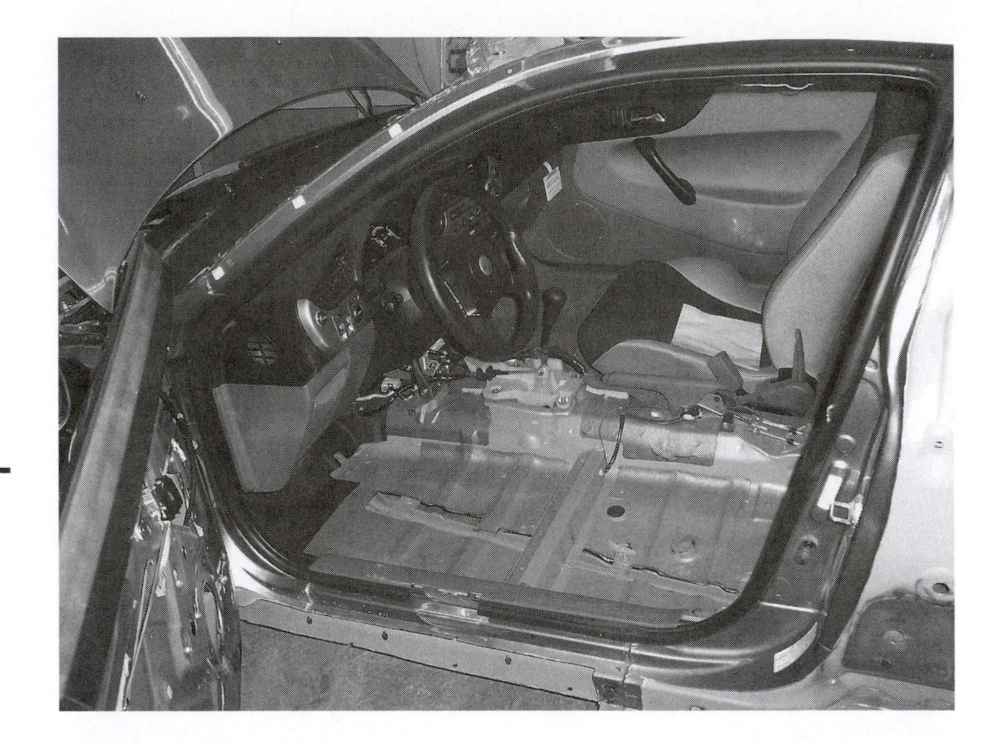

Figure 14-6 *Cable (orange) running inside Honda Insight; cable runs under carpeting under driver's seat. (Photo courtesy of Bobby Bailey.)*

Figure 14-7 *Underside of Toyota, showing cable (orange) from battery pack. (Photo courtesy of Bobby Bailey.)*

OTHER FORMS OF HEAT

■ Note
Just because you may find a bead on some wires, that is not a definite that the vehicle had a short there to cause the fire; instead the bead could have been created as the result of the fire.

The vehicle's heater, fans, and wires attached need to be looked at by the assigned investigator. These items need to be looked at by someone who is familiar with the vehicle's components—looking at all these wires can be a specialty in itself. Just because you may find a bead on some wires, that is not a definite that the vehicle had a short there to cause the fire; instead the bead could have been created as the result of the fire. Basically, the only thing that it shows initially is that there was current on the wire at the time of the fire. Heated seats, power mirrors and seats, and other electrical applications built into the vehicle need to be eliminated by the investigator as sources of heat of ignition.

A properly operating catalytic converter, located on the underside of the vehicle, can reach temperatures of more than 700°F externally.[11] The converter is located in close proximity to the floorboard. The interior covering on the floorboard includes an insulation barrier, and then the carpeting is placed over that. Heat can build up on the underside of the vehicle from a clogged system that has not been maintained, possibly causing heat transfer through the metal floor to the combustibles in the interior. Obtaining owner or driver feedback on the maintenance and how the vehicle was running prior to the fire is important to verify the damage assessment that is noted at this point.

One incident that occurred with a catalytic converter was when the driver, weary from travel, parked the passenger car next to the curb in the rear of a shopping mall. The vehicle was backed into a parking space where leaves and debris had accumulated. The vehicle was running for approximately 6 hours to keep the occupant warm on the cold night. The vehicle's catalytic converter and the exhaust system ignited the dry leaves under the car and ignited the vehicle. Leaking fluids from lack of maintenance fed the flames as the fire consumed the vehicle—unfortunately with the passenger inside.

Turbo chargers are located within the engine compartment and add as much as 40 percent horsepower to the vehicle's engine. The turbo charger collects air from the exhaust and turns a turbine at over 100,000 rpm.[12] The turbine then forces collected clean air into the engine in addition to the fuel to increase combustion and power. These appliances create tremendous amounts of heat and may be a source of ignition. The air is pulled in from the exhaust and pushes the turbine that is on a shaft. The shaft turns an opposite turbine that forces air into the combustion chamber.

Smoking-related incidents can be difficult to find. However, looking at the ashtray of the vehicle and talking with the occupants of the vehicle may lead to discussions of how the fire began. It will very possibly be a slow smoldering fire that will begin low inside the vehicle or in the seats. The smoking-related items will have to have combustibles to ignite or the fire will most likely burn out. Information from the occupants about smelling smoke for a while may indicate that a lit cigarette or match was dropped in the floor or seat area.

HOW DID THE VEHICLE BURN AND WHY?

Most vehicles are extremely safe and reliable. But just as with any mechanical device, they must be maintained and kept up. If the vehicle was kept in clean and running condition and there is a fire, it may have been arson. If the first responder investigator looks at the vehicle and cannot come to a conclusion by using deductive reasoning and a systematic search of the fire, an assigned investigator needs to be called.

■ Note **The owner may have burned the vehicle for the purpose of insurance fraud or to get out from under a lease agreement.**

■ Note **The vehicle may simply have been burned to escape a high auto loan payment or even high fuel prices.**

The owner may have burned the vehicle for the purpose of insurance fraud or to get out from under a lease agreement. It may be that the vehicle was burned by another person as a vendetta against the owner. The vehicle may simply have been burned to escape a high auto loan payment or even high fuel prices. With the run of big SUVs built over the last 10 to 15 years, this may become more of a reality over the next couple of years, if gas prices rise substantially. Either way, a more experienced investigator needs to be called because this determination is well above the first responder's role in the investigation. First responders need to recognize that this is the time to ask for assistance.

One thing to remember at all times is that the investigator has to work the scene from least damage to most damage. Investigators miss so much when they jump into the scene before the outside perimeter is checked. There is nothing worse than having potential evidence stepped on and crushed into the ground never to be recovered because the investigator went directly into the interior of the vehicle.

The scene first needs to be photographed from all sides. Photos of the area from the vehicle looking outward need to be taken as well. This is particularly important when a stolen vehicle is found, showing egress path and direction to the scene. Are there tire tracks into the scene? On your exterior walk-around, did you see evidence located nearby that may be associated with the fire? All of these items need to be photographed and documented before conducting the examination of the vehicle. All evidence needs to be documented and photographed before moving and collecting.

WHAT DOES THE EXTERIOR OF THE VEHICLE TELL YOU?

The extent of damage may show direction of fire travel. Burn marks or demarcation lines may show path of travel. The patterns left behind will show the direction from where the fire began. Fire burns typically up and out. A fire that begins in a vehicle will show lines on the metal left on the vehicle and upward. As it burns upward, the fire will spread from side to side. These patterns will show what appears to be a V pattern, the same as seen in a structure fire.

This V pattern will show the possible direction of fire. Looking at the vehicle's damage will indicate whether the fire originated inside or outside the vehicle.

If there is no damage from flame impingement and/or smoke on the underside of the vehicle, it is very possible that the fire originated inside. Door jambs need to be checked as well to see if any damage shows, possibly indicating that the door was left open or someone tried to remove the vehicle identification number (VIN) plate.

The VIN plate is usually located in the driver's side door jamb or on the lower corner of the inside windshield on the driver's side. The VIN plate is usually made of aluminum alloy and may melt at very low temperatures. The VIN contains information about the vehicle, such as year, make, model, engine, and paint color. The VIN may burn partially or completely away on the dash; however, law enforcement officers are able to retrieve this information from several other secret VIN plates on the vehicle, and they can give the investigator this valuable information. **Figure 14-8** shows a VIN plate on a vehicle located on the driver's side of the dashboard. The VIN plate has been cleaned using a small piece of steel wool to remove debris so that the number can be read.

Burn patterns may not be as obvious on total burn vehicles. These cases will be harder to solve because of the extent of damage. The metal remaining on these will be distorted and buckled. The distinction of demarcation lines may not be

Figure 14-8 *Vehicle identification number (VIN) plate after it has been cleaned. (Photo courtesy of Bobby Bailey.)*

visible to the first responder. However, there may be clues. Remaining paint left in areas and plastics that may not have melted may give clues to direction. If the inside and the outside of the vehicle is totally burned, but the bottom shows no damage, it will be probable that the fire originated inside.

One interesting facet to look at is the vehicle's glass. The windows on most vehicles are tempered glass, meaning that they will shatter into small pieces upon impact or fire. This includes the side and rear glass. The front windshield is laminated safety glass. The tempered glass will fall into the vehicle or just outside in the event of a fire. This gives the investigator a guide to go by to see if the glass was intact with the window in the up (closed) position prior to the fire. If there is a large percentage of glass inside the door, the window may have been down at the time of the fire. If still intact, the window mechanism will indicate the position of the window at the time of the fire. However, most newer vehicles use plastic parts for this assembly whereas older vehicles may have metal construction. The glass will show soot or smoke stains that may have been left prior to breaking from the fire.

Fire will cause long cracks on the front windshield where the laminate separates on the side of the area of higher heat. These long cracks can appear to point to the area where the laminate in the safety glass has separated as a result of the heat and flame. The glass, if found intact still in the vehicle, may give a clue as to the side and location of the fire. For example, a fire that began in the driver's-side floorboard will break out the lower driver's-side area of the windshield, and the fingers or cracks will run from this area all the way across to the passenger side. This shows that the fire was low in the vehicle. Flame impingement was through the dash and just over the dash.

On the other hand, a fire that occurred in the rear of the vehicle results in the cracks on the windshield running down toward the hood of the vehicle as well as side to side. In the following photos, note that the glass has cracked and where flame is coming through.

Figure 14-9 shows the fire coming through the windshield high near the roof line. The fire originated in the rear seat of the vehicle. Note that the laminate separates half way down the windshield, with flame coming through on the top. This idea that the vehicle's windshield will provide direction and travel is only a tool that works if used early enough after fire suppression activities. It is a potential tool that when added to the systematic approach of the investigation may add to the deductive reasoning and a hypothesis.

Look at the tires remaining on the vehicle. Do they show consistent damage? Are the front tires destroyed and the rear intact, or vice versa? Some newer vehicles have a plastic fuel tank that may rupture and spill fuel contents below the vehicle, which enhances the burning process on the underside and the potential ignition of the tires. This can give the investigator a false reading on where the fire occurred because of additional V patterns on the vehicle.

In **Figure 14-10,** the vehicle was set on fire in the seats. Notice the burn marks left on the doors. They are up several inches from the bottom of the door.

Figure 14-9 *Photo of laminated windshield with flame from top down, showing fire originated above the floor. (Photo courtesy of Bobby Bailey.)*

Figure 14-10 *Vehicle that has burned with indicators showing fire originated inside passenger area and not in engine compartment. Note the remaining materials on the front bumper and the paint on the bottom of the driver's-side door exterior. (Photo courtesy of Bobby Bailey.)*

The front bumper is still intact as are the plastic headlight covers. The roof area is buckled over the center of the vehicle. The area over the driver's-side door shows flame impingement because rust is already developing. This is not on the other doors. Your investigation should also show whether the window was up or down prior to the fire. Note that the tires are flat on the driver's side from heat. However, they did not help fuel the fire, as exterior burn patterns show. The tires may also show direction of flame, in particular on the inside side wall of the tire. The rubber in the tire will begin to melt in the area of flame impingement from the outside in.

INSIDE THE VEHICLE

Look at the vehicles' contents. Glove box information may or may not be present. Did the owner leave the keys in the vehicle? Can the keys be located? The metal components of the steering column will sometimes fall, including the key assembly, onto the floorboard on the driver's side. The keys, unless initially visible, may not be seen without the use of an X-ray machine. The carpeting and materials need to be collected to see if the keys are in fact there.

Most of the interior of the vehicle is combustible, composed of either plastics or fabrics. These items melt or burn very fast, leaving melted globules or the metal frame to look at. However, they may tell direction as well. What does the hood and trunk area look like? If the burn patterns are exterior and not on the inside, there is probably nothing in the engine compartment that could have been the cause of the fire. The trunk usually will not show any impingement inside unless the fire traveled through the rear passenger seat area into the interior of the trunk. But there are potential heat sources inside the trunk. The vehicle owner may install items in trunks such as large audio equipment. There are also vehicles with factory-installed equipment located in the trunk such as retractable antennas, and some even have the batteries located in the trunk.

■ Note

There are potential heat sources inside the trunk. The vehicle owner may install items in trunks such as large audio equipment. There are also vehicles with factory-installed equipment located in the trunk such as retractable antennas, and some even have the batteries located in the trunk.

The hood will show burn patterns from the heat and flame impingement if the fire was inside the engine compartment. The paint will begin to melt and burn away, spreading outward across the hood. Also the hood's metal will start to bubble or distort, and shortly thereafter, the hood will begin to rust very quickly. This is caused by oxidation and introduction of moisture from the hose streams during fire suppression, along with rapid cooling of the metal. Hood damage will possibly show the side of the engine area involved.

MOTOR HOMES

Motorized homes on wheels, a typical term, bring a structure to life on the road. The motor home contains components of a vehicle combined with the combustibles of a structure. Furniture, plastics, electronic equipment, to name

Figure 14-11 *Motor home fire, which occurred in the kitchen area. (Photo courtesy of Bobby Bailey.)*

a few, are what make up the motor home. Add a fuel tank and an engine, and if not properly maintained, the motor home could burn with massive damage. In **Figure 14-11,** in this motor home that burned, the fire began on the inside from food on the stove. The close proximity of cabinets and combustibles does not allow for a lot of room for error with a heat source nearby.

■ Note
Motor homes are unusual and must be investigated with care. They may contain liquid propane gas for heat and fuel for the cooking equipment.

Motor homes are unusual and must be investigated with care. They may contain liquid propane gas for heat and fuel for the cooking equipment. The motor home may have a built-in generator that may use a flammable or combustible liquid. Sometimes the tanks that contain the fuel are shared with the engine. Thus, the tanks are very large. Electrical panel boxes located on the unit tie into AC current with an electrical cord attached to a power source.

For motor home fires, everything involved in a vehicle fire applies, and these additional items of interest need to be addressed as well. In motor home fires, working from the least damaged to most damaged areas and the exterior to the interior is of the utmost importance. You must find all sources of ignition and look at all the burn patterns.

BOATS

Depending on the boat, they are not much different from a vehicle or a motor home. However, there are inherent differences. When the boat uses gasoline, these fumes are heavier than air and can collect in the lowest part of the hull. These boats

Figure 14-12 *This boat burned, sank, and was raised. With caution, most of the debris remained within the hull of the boat.*

usually have built-in ventilation to ensure that the hull is free of ignitable vapors, but this system can fail and there could be a potential ignition source nearby.

■ Note
Depending on the size of the boat, it can be very similar to a motor home. All the comforts of home along with all the associated fuel load and ignition sources.

Depending on the size of the boat, it can be very similar to a motor home. All the comforts of home along with all the associated fuel load and ignition sources. A full kitchen is not uncommon and even elaborate entertainment systems are available.

There are other unique situations in boat fires, including the possibility that the boat sank, such as the boat shown in **Figure 14-12,** was raised, and now much of the contents and some of the evidence may be at the bottom of that body of water. When larger boats are involved in fire and have either been sunk by the fire department to put out the fire or sank as a result of the fire, the insurance company may even hire a specialized investigator that has scuba diving skills, enabling the first phase of the investigation to take place before the boat is raised.

With all fires, look from the least damaged area to the most. Check the overall condition of the boat as you would a vehicle. Where you check the tires on a vehicle, look at the hull of the boat for damage or lack of upkeep. Is the propeller (sometimes referred to as a wheel) pitted and worn, bent, or out of round? This could be an expensive repair that may have been on the horizon.

With inboard motors you should look into the engine compartment as you would a vehicle fire. These areas are quite similar to an automobile. Fuel tanks and associated fuel lines will be present. **Figure 14-13** shows the motor mounts and the area surrounding where the two engines were located (motors removed for examination). This area clearly shows less fire damage than the living quarters do.

Working from the least damaged area to the worst damaged area, the investigator moves into the living compartments. This area of the boat would be examined the same way as you would examine a home. This section was covered in debris and required searching and removal of debris, same as a house, as shown in **Figure 14-14.**

If anything is out of the ordinary or if the area or origin or cause is not immediately apparent, then the assigned investigator should be notified. If

Figure 14-13 *The engine compartment has little fire damage and much less than the crew quarters.*

Figure 14-14 *Debris removed from galley area, revealing char patterns.*

the assigned investigator arrives and is not comfortable with boat fires, he or she may want to ascertain whether the insurance company is going to send a specialist to investigate. If so, it may be in everyone's best interest to protect the vessel while awaiting the arrival of that investigator, who has specialized knowledge in marine fires. However, the investigator should at the very least take a complete series of photographs to document the entire scene, including a diagram and any notes. The area where the boat was moored should be closely examined, documenting all damage. **Figure 14-15** shows the burn pattern on the dock, which will match the patterns on the boat. Also of interest on the dock is the electrical shore connection. This needs to be examined to see if it was energized and the condition of the connection both before and after the fire.

■ Note
Working with an experienced marine fire investigator is the first step in adding to the sum of the assigned investigator's experience so that he too will be able to conduct full marine fire investigations in the future.

Although there are benefits of conducting interviews after the examination of the fire scene, in this scenario it may be best to at least get preliminary information from any witnesses and suppression forces.

Working with an experienced marine fire investigator is the first step in adding to the sum of the assigned investigator's experience so that he too will be able to conduct full marine fire investigations in the future.

Figure 14-15 *Examining the area adjacent to the location where the boat was moored may provide additional information and evidence and is no different from doing an exterior exam of the property surrounding a house fire.*

SUMMARY

First responder investigators need to remember that all cases might not be solved. Investigators don't have to be auto mechanics but must understand vehicles. The information from a local mechanic or automobile dealership may provide valuable insight into the investigation of a vehicle fire. All possibilities in an investigation must be exhausted before concluding that the fire is undetermined. For every action there is a reaction. Something, either accidental or malicious, caused the vehicle to burn. An individual's incompetence or lack of knowledge is not a crime, but the intent to destroy personal property or the property of another is.

The investigator must look at all facets of the case. Work the fires from the exterior to the interior and from the least damaged to most damaged areas. Take into account how the vehicle arrived at the location where it burned. The scene needs to be documented and photographed on all sides. This should include the surrounding area as well. The investigator needs to sift debris to find evidence that will aid in using a scientific methodology, which will lead to an accurate and credible hypothesis. Lighting is a key issue when investigating these scenes. Darkness hampers the location of evidence needed for making an accurate analysis, not to mention it is a safety issue.

Has the vehicle been secured from movement or engaging itself? The battery should be disconnected and all shut-off switches locked out. Make sure that you are wearing protective clothing because the sharp edges of metal and glass, along with the contaminated liquids that remain, can cause injury to you and bystanders.

If I could leave you with one thing, it would be this: If you don't know exactly how the fire occurred or the fire brings more questions than answers, get a second opinion, or bring in another investigator to validate and give an honest second look at exactly what happened. Make sure this investigator is not unduly influenced by the initial opinion if it does not make any sense. Deductive reasoning needs to be used. Basically, if it does not make any sense, it may not have happened.

Both motor homes and boats have the same characteristics as other motorized vehicles. Likewise, they will also have similar characteristics of a structure and should be investigated no differently from any other structure fire. The similarities will aid the investigator in the search for the area of origin and fire cause.

KEY TERMS

Off-gassing The release of gas or vapors in the process of aging or decomposing, also refers to vapors or gases that were absorbed by fabric, carpet, and so forth during a fire and then released after the fire is extinguished.

Pro bono Doing the work free of charge; at no cost.

REVIEW QUESTIONS

1. Two different types of window glass are discussed in this chapter. What are they and how can they assist the investigator in determining the direction of fire travel and cause and origin?

2. Discuss the differences between parallel and series electric hybrid vehicles.
3. What temperature will the exterior surface of a catalytic converter reach when it is operating properly?
4. What is the flash point of ethanol? How does it affect the investigation when used in conjunction with gasoline in a vehicle fire?
5. Discuss the use of self-contained breathing apparatus (SCBA) and why it is important to use them in vehicle fires.
6. When should a first responder investigator call for the assigned investigator and why?
7. Describe how to properly secure a vehicle (electric, diesel, or gasoline) prior to investigating the incident for safety.
8. List a flammable or combustible fluid used in the motor vehicle and how it may be involved with the cause of a fire.
9. Discuss how an engine with no spark can operate on a combustible liquid.
10. List five potential and credible heat sources and how their failures may lead to an accidental fire.

DISCUSSION QUESTIONS

1. One larger fire department decided they needed to cut back on services because of budget cuts. As such, they decided not to investigate vehicle fires. What problem might this create in that jurisdiction? What problem might it create in neighboring jurisdictions?
2. Hybrids are creating some unique problems for the firefighter and fire investigator. What actions can be taken by manufacturers to alleviate some safety problems?

ACTIVITIES

1. Using the Internet, identify all of the vehicles with high-voltage electricity.
2. Using the Internet, identify all vehicles that have plastic or fiberglass bodies in contrast to metal.

NOTES

1. Michael J. Karter, Jr., *Fire Loss in the United States During 2006* (Quincy, MA: National Fire Protection Association, Fire Analysis and Research Division, September 2007).
2. How Stuff Works, "How Airbags Work: Air Bag Inflation," http://auto.howstuffworks.com/airbag1.htm.
3. Dr. Rich Hollins, "Airbag Inflator Explosion," *Fire Engineering* 149, no. 12 (December 1996). http://www.fireengineering.com/issue/toc.html?publicationId=25&issueNumber=12&volumeNumber=149.

4. Economy Watch, "How Biodiesel Works," http://www.economywatch.com/renewable-energy/biodiesel-functions.html.

5. Lee S. Cole, *Investigation of Motor Vehicle Fires*, 4th ed. (San Anselmo, CA: Lee Books, 2001), 53.

6. Energy Justice Network, "Fact Sheet: Ethanol Biorefineries," 2007, http://www.energyjustice.net/ethanol/factsheet.html.

7. National Fire Protection Association, *Fire Protection Guide to Hazardous Materials*, 11th ed. (Quincy, MA: National Fire Protection Association, 2001).

8. National Fire Protection Association, *Fire Protection Handbook*, 20th ed. (Quincy, MA: National Fire Protection Association, 2008), Table 18.13.18, page 8–165.

9. Anthony Schwaller, *Total Automotive Technology*, 4th ed. (Clifton Park, NY: Thomson Delmar Learning, 2005), Table 14-4, page 188.

10. Environmental Health and Safety, "Material Safety Data Sheet, Potassium Hydroxide," MSDS Number P5884, effective February 1, 2007, http://www.jtbaker.com/msds/englishhtml/P5884.htm.

11. National Fire Protection Association, *Guide for Fire and Explosion Investigations* (Quincy, MA: National Fire Protection Association, 2008), 25.4.3.1, page 921–212.

12. Schwaller, *Total Automotive Technology*, 458.

FATAL FIRES AND FIRE INJURIES

Learning Objectives

Upon completion of this chapter, you should be able to:

- Describe the documentation of victims (live persons) and potential evidence.
- Describe the documentation of the fatality and the surrounding area.
- Describe the process of identifying the body.
- Define who establishes the cause and mechanism of death.

CASE STUDY

It was about four in the morning when Engine One was toned out for a vehicle fire. Upon arrival, they found that a four-door sedan had been completely consumed but was still smoldering. The fire had been discovered by a paper delivery person who was driving through the area on his route. The interior was almost completely consumed, but on the front seat framework was a lump that could have been a body. Rather than investigate further, the fire was extinguished with the minimum of water and a fire marshal was summoned.

As the debris was carefully removed, it became evident that it was a body. The victim had been lying across the front seat with his head against the driver's door, torso across the seat, and his feet on the floor on the passenger's side. The legs were almost totally consumed, only bones remained, but one shoe, still containing a foot was only slightly burned. However, it only provided limited information at the time.

There was no head; it had disintegrated, and bones were lying on the floor of the car under the seat. After documenting, measuring, and photographing, a sample was taken to test for accelerants, and then the remains of the body were lifted out and placed in a body bag. Any item that could possibly be related to the body was carefully collected and placed in a smaller bag and also placed in the body bag. In this process, on the floor, in the area where the head would have been, there were several charred items that resembled candy corn. These items were collected and retained by the investigator.

The medical examiner could do little to identify the body, and DNA analysis would not be done until there was a sample for comparison. There had been no missing persons filed in any of the surrounding jurisdictions. The registration on the vehicle came back to a local address. In checking, they found the mother of the vehicle's owner. She said that her son was out of town; he was an independent truck driver and would not be back for several days. He had called her that morning from another state. She also said it was not unusual for him to loan out the vehicle to his old army buddies when they came to town.

In the meantime, the investigator took the teeth to a nearby major hospital that had several teaching curriculums, one of which was medical forensics and the other was the school of dentistry. Great fortune was with the investigator because the first person to greet him was a professor of dentistry who also taught identification techniques for the forensic school. An X-ray confirmed they were teeth, and the professor advised that if dental records could be located, he could attempt an identification of the victim.

Two days later, a mother called the police to report that her adult son was missing. The date of disappearance matched the date of the fire. The investigator was dispatched to the woman's home for an interview. She confirmed that her son was an army veteran. She also confirmed that he had borrowed a car from an old army buddy on the night that he disappeared. She was not sure where he went that night but had heard that a bunch would get together in an old abandoned house to drink and play cards.

(Continued)

(Continued)

When she described the location of the house, it matched the location where the car was found. Now the investigator had a body and a potential identification. The mother said that her son never would go to the dentist and didn't really need to, but when he went in the army she knew he had lots of work done on his teeth.

She also described his friends; all were interviewed and confirmed that they were playing cards at the old house and left sometime after midnight. They had all been drinking; the drunkest was the victim. One of his friends had taken the keys of the car to keep the victim from driving in his condition. The explanations were logical and matched evidence collected to date. In the meantime, the medical examiner called and said the cause of death was smoke inhalation and the victim was definitely under the influence of alcohol.

The Department of the Army was contacted and within 2 weeks they provided a set of dental records. These were taken back to the teaching hospital along with the teeth. By the end of the day, it was confirmed that the teeth recovered matched the army records; this was the woman's son that she reported missing.

The fire investigator found no indications of an incendiary fire and neither was there any evidence that the fire was caused by any malfunction. Friends and family reported that the victim was a heavy smoker, three or more packs a day, and his mother reported that her son did smoke in bed almost every night. The fire remained undetermined because the data gathered gave a hint to the situation but did not provide sufficient evidence to label the fire's exact area of origin within the vehicle and there was not any definitive evidence of the cause of the fire.

It would have been so easy to just presume identification of the victim because there was only one person missing. But it was critical to take the extra steps and follow the scientific methodology to ensure that the identification for the family was accurate and to ensure that no other crime had been committed. The fire service owes it to the victim, the victim's family, and to society to confirm the identity of every victim. We need to close the door to seek the truth.

INTRODUCTION

It is impossible to cover all aspects of conducting a death investigation in one lone chapter. Instead, this chapter provides an overview of the different functions necessary for investigating a fire involving injuries or death. The first responder investigator and the assigned investigator may find themselves in a different world when such a tragedy is associated with their investigation. There will be addition pressures from senior leadership, the press, and even from elected officials. These people want answers and may be impatient about the time it can take to work the scene and develop a proper hypothesis on the cause and origin of the fire as well as the cause of death. Pay heed to these pressures but never let them compromise the integrity of the investigation.

A fire with injuries or a fatality requires more documentation than others. And all of this is demanded at a time when you must put the needs of the injured victim and the concerns of the family as a priority. Just as important as finding the area of origin and cause of the fire is the need to identify how the victims were injured or how and why they died.

■ Note
The purpose of identifying the cause of the fire is to prevent future fires.

The purpose of identifying the cause of the fire is to prevent future fires. The reason to identify the cause of the injury or death is to take steps necessary to prevent future injuries and fatalities. This process is done the same as with the fire investigation. It needs to be a systematic process involving scientific methodology.

The path leading to the cause of the fire may lead to a new avenue as the result of the injuries or deaths. The fire may now be a homicide scene, and even if the fire was accidental, there may now be a higher probability of future litigation that may require the involvement of the fire investigator.

The essential focus needs to be on safety as always, and a more determined effort must be made with evidence identification and collection. The investigator will now be exposed to more challenges: interviewing victims and family members of the deceased. It will be more difficult and require more time and effort to achieve the same objective.

PREPARATION FOR THE INVESTIGATOR

Before dealing with the emotions of the survivors, relatives, loved ones, or witnesses, the investigator must be prepared for his or her own emotions. Those who have been in public safety for any period of time have faced death, not just from fires but vehicle accidents and illnesses when doing emergency medical response. Most are stoic at the time, but others may use humor to cover their unease with the situation. Even back at the fire station the attitude continues. Some put their emotions under control and others truly are not bothered; they are hardened to the facts of the job. But no matter how hardened or how much emotions are kept in control, most are bothered by the loss of a child. Some can witness fatal fires for years with no problem, but then a case comes up that touches them in a way so that their own emotions come flowing out.

■ Note
No matter how hardened or how much emotions are kept in control, most are bothered by the loss of a child.

The public safety field, fire, police, and rescue, have recognized the need for peers to get these emotions in control in a positive way. As such, many public safety organizations have created programs to handle this type of stress: critical incident stress management (CISM). Most often you hear of these organizations when they are conducting Critical Incident Stress Debriefing (CISD) for larger events when there are multiple fatalities. However, they do have recommendations that can help the individual on how to address these emotional stresses for both the psychological and physical health of emergency personnel.

A part of the fire investigator's job requires in-depth involvement with the deceased. Knowing this upfront, it is always a good idea for you to contact the CISM team if for no other reason than to refresh your memory on stress-relieving exercises that can help you in your job.

Most important, you want to evaluate your emotions and how you will interact in a positive way with the emotions of others associated with the case being worked.

THE SCENE

■ Note The fire scene with an injury or fatality is no different from any other fire scene when it comes to the process and procedures to follow in the investigation.

■ Note Even though all fires are considered accidental until proven otherwise, the fatal fire scene may need to be treated as a homicide scene in regard to protection and evidence collection.

The fire scene with an injury or fatality is no different from any other fire scene when it comes to the process and procedures to follow in the investigation. However, it provides a different intensity for those working the scene. There may be a more concentrated search of the scene in dealing with a fatality, and there are different types of evidence, which in turn may require different methods of evidence collection and preservation.

Depending on the injury, and in all fatal cases, the investigator should notify other law enforcement authorities. The presence of these other agencies will most likely bring additional resources and expertise to the scene, enhancing the capabilities of the investigative team as a whole.

Even though all fires are considered accidental until proven otherwise, the fatal fire scene may need to be treated as a homicide scene in regard to protection and evidence collection. The cause of death will not be confirmed until after the autopsy—all the more reason for the scene to be protected and searched for any and all evidence at the time of the initial investigation.

DEALING WITH INJURY VICTIMS

The first responder investigator may be the first person to have any contact with an injured victim. However, depending on the severity of the injury, the victim may already be on the way to the hospital. This occurs before there is time enough to set up suppression activities, let alone start an investigation. When there are injuries or fatalities, the first responder investigator should call for an assigned investigator.

Once it is known that there is a victim associated with the fire, every step should be made to secure the scene. Every piece of potential evidence must be preserved. The scene should be treated as if a crime scene, setting up security and keeping overhaul to a minimum. In addition, until the assigned investigator arrives, the first responder can identify available witnesses of the fire or the injury, witnesses that heard comments made by the victim, emergency medical services (EMS) personnel who transported the victim, and those first responders who may have arrived prior to the engine arriving on scene. This information will aid the assigned investigator by providing a clear list of who can be interviewed and where to locate them.

The assigned investigator should interview the victim as soon as possible. If the victim is still at the scene, the first concern is to make sure the victim does not need further medical attention. If the victim is already at the hospital, it is prudent to get to the hospital just in case injuries were serious enough to lead to a coma or, regretfully, death. If the victim is hospitalized, the fire scene should be treated as a fatal fire scene in case the victim succumbs to his or her injuries.

■ Note **Anything the victim was wearing is evidence. If anything, jewelry or clothing, is removed by EMS or hospital personnel, it needs to be preserved for the investigator.**

Anything the victim was wearing is evidence. If anything, jewelry or clothing, is removed by EMS or hospital personnel, it needs to be preserved for the investigator. The investigator, first responder or assigned, should ask everyone who came in contact with the victim if they have anything that may have come from the victim. In the case of an ambulance, it is always a good idea to look in the waste can because if the victim was seriously injured, the emergency medical caregiver may not have even realized he or she removed something during treatment.

■ Note **Even gauze or pads used in the treatment of the victim are evidence.**

Even gauze or pads used in the treatment of the victim are evidence. They may have absorbed residues other than body fluids that may be evidence. If the victim was the arsonist, there may be trace residue of the accelerant on the victim. If there was an explosive device, trace residue may have been on the skin and then picked up by the gauze. All materials that come in contact with the patient should be collected, properly preserved for future examination, and documented for submittal to the laboratory for analysis. The exact type of test to be run on the items depends on what is discovered at the scene or from interviews.

At some point during the interviews, steps must be taken to document any injuries. The entire burn or injured area must be photographed, within reason. All clothing should be photographed, both damaged and undamaged. It may also be beneficial to have a signed medical information release form. If this is not a criminal case, there will be no precedent for this information. If information is received, remember that there are regulations and statutes that address the release of an individual's medical information.

FATALITIES

It is not unusual to have suppression personnel to do their job by rushing in, doing a search, finding a body, and pulling the person out to safety only to find that the person is deceased. If there is any hope at all, life-saving procedures may be started and the patient transported to a medical facility and only then pronounced dead at the emergency room. When this happens, the investigator needs to contact the hospital to ensure that nothing is cleaned up until the body and belongings are examined for evidence. Protocol in some jurisdictions dictates that EMS providers transporting a victim are to not clean out their unit until it is examined by the investigator.

In fire situations, some bodies when found leave no doubt that the person is dead. When this happens, the scene must be secured immediately. Nothing should be moved unless absolutely necessary, such as extinguishing any hot spots in the area. The first responder investigator's biggest challenge will be to keep out other emergency responders. After securing the scene, it should be photographed thoroughly. The first responder investigator should take steps to photograph the body and surrounding area, especially if there is any possibility that the scene may be disturbed because of an unstable nature of the building.

■ Note

The first responder investigator's biggest challenge will be to keep out other emergency responders.

Even though the scene is photographed by the first responder, the assigned investigator must take a complete series of photographs, too. This includes any photos that relate to the fire cause as well as a complete series of photos of the body and surrounding area. All aspects of the room in which the body is found must be documented to the minutest detail.

Measurements of the room need to be as accurate as possible. Additional measurements of the body in relation to furnishings should be documented to allow proper placement of the data on a diagram for future purposes.

Once the scene has been properly documented, a body bag must be brought in and the body placed in the bag. Any parts of the body must be identified and put in the bag as well. A badly burnt body may have fingers and toes burned off. In these cases, it is important to uncover the debris carefully as in an archaeological dig, removing layer by layer until all body parts have been accounted for and they are all in the body bag.

Then, the scene must be thoroughly examined; look at all debris to identify the room contents and anything in or around the body. Remember that the debris under the body may be permeated with fluids from the body. Depending on how long it has been after the fire, there may be bacteria that are of concern. As with your entire investigation, always wear work gloves with latex gloves underneath. One side note: It is essential that any cloth or leather gloves be thoroughly cleaned and sanitized after the scene examination. In some instances, it may be just as easy and prudent to discard the work gloves instead of reusing them.

■ Note

It is essential that any cloth or leather gloves be thoroughly cleaned and sanitized after the scene examination. In some instances, it may be just as easy and prudent to discard the work gloves instead of reusing them.

If the examination of the structure reveals pet bowls or belongings, it is recommended that a search be made for that pet. It may be found in the same room as the victim. It is cruel to leave pet remains to be found by the victim's loved ones as they search the scene later for personal belongings or valuables. If found, the remains can be wrapped in an opaque plastic bag, and a responsible party can be notified of the location for future disposal.

Impact with the Ground

In a multiple fatality, knowing the placement of the bodies can be critical at a later date. The case in point was a tragic small twin-engine plane crash. The night was unusually foggy and when the plane was coming in for a landing at a small municipal airport, it was descending and losing altitude as well as speed. Too late, the pilot realized that he had made a mistake as to the location of the runway and he pulled the plane up to avoid collision with structures, but the plane's speed was insufficient. The plane rolled and immediately pointed nose down toward earth, crashing straight down into the ground at approximately 200 mph.

The location was a small patch of woods between the airport and a main highway. A small fire started after impact in one of the engine nacelles. This proved a blessing because it allowed a local state trooper to see the crash site from the road. Otherwise, in the fog it would have taken quite a search. The trooper took his small extinguisher from his vehicle and was able to extinguish the fire.

First responders found the plane nose down and tail straight up in the air. The front of the plane was buried and crushed into the ground. There was some obvious buckling, but the plane was still intact. After some effort the engine crew was able to get the side door open. Looking inside they found all the passengers piled on top of each other at the front of the plane. No pulse could be found on any of the victims. As they were removed, care was taken to document each body as it was pulled from the plane.

A few months later the senior officer on the scene, who happened to be the local fire investigator, was contacted by two different law firms. A meeting was set up at the local fire station. One of the attorneys led off by explaining the situation. The pilot was a wealthy, successful businessman, but he had no will. His first marriage had failed, but he had two children from that marriage who were now young adults. The businessman had remarried and his second wife was in the plane. They did not have children, but she had two grown children from her previous marriage.

The attorneys simply wanted to know who died first in the plane. The investigator made it very clear that he would have no way of knowing that information and would not even speculate. What happened next was an interesting exchange. One attorney claimed that the pilot was the first to die because he was in the pilot seat and was the first to impact the ground. At that point, his entire estate went to his current wife. As such, she died second because she was in the back of the plane. As a result, she was the holder of the estate when her husband died, and when she died the estate went to her children, not the children of the businessman.

How this was finally settled is unknown to anyone in the fire service. The argument presented caused a lot of shaking of heads around the fire station. The point is that documentation of the scene did show that the businessman was in the pilot seat. The wife was in a seat directly behind her husband. Not only was this documented in the after-action report, it was thoroughly photographed.

MEDICAL EXAMINER

With the body at the morgue, it will be scheduled for an autopsy. This is necessary not only to help with the positive identification of the body but also to determine the cause of death. The role of the medical examiner (ME) in this process is already established and followed. As for policy, some states and localities even have the medical examiner show up on the scene to examine the body and to supervise the collection of the body.

■ Note Some states and localities even have the medical examiner show up on the scene to examine the body and to supervise the collection of the body.

The ME will run a series of tests, from a detailed physical examination of the body to a series of X-rays of the full body. Samples will be taken for analysis to determine the carbon monoxide levels, alcohol content, drugs, and other toxins. A sample will be kept for future tests if necessary.

The throat and mouth will be examined to see if sooted. There may even be heavier soot just below the nose, where the victim breathed in the sooted air and exhaled, concentrating it on the skin below the nose. The investigator on the scene would have looked in the mouth to look for soot. No soot indicates that the person died prior to the fire. Soot in the mouth indicates that the person was breathing during the fire and was alive when the fire started. The ME will take that one step further by examining the throat as well. Smoke and soot can permeate an open mouth giving false indications, but will not do so in the throat or lungs.

The ME will document the extent of the burns for the report. There will also be a search for any other damages or injuries. The body will be documented for dehydration of the muscle tissue as a result of exposure to heat, which causes the flexing of the muscle and puts the arms in what is referred to as *pugilistic attitude,* which makes the victim look like he or she was fighting or fending off blows. This stance can sometimes be misinterpreted. What can also be misinterpreted at the scene is the presence of blood coming from the ears, nose, or mouth. This may be the result of the burn damage to the body and not necessarily trauma damage before the fire.[1]

■ Note In addition to the DNA that can be collected, fingerprints may still identify the victim. Lightly burned hands may still have ridges that can be identified.

In addition to the DNA that can be collected, fingerprints may still identify the victim. Lightly burned hands may still have ridges that can be identified. Today with electronic fingerprint readers there is an even greater probability of getting a readable fingerprint. For identification purposes, in addition to dental records, previous X-rays can be of assistance, especially if there were previous bone injuries.

CAUSE, MANNER, AND MECHANISM OF DEATH

Cause of death is the event that leads to the loss of life. The victim may have died of asphyxiation as the result of inhaling carbon monoxide, but the *cause of death* was the fire. The asphyxiation was the *mechanism* of death, that medical event that led to loss of life.

The *manner of death* relates to how the entire event was created. It could be a homicide, where the fire was intentionally set to harm the victim; a suicide, where the victim set the fire to end his or her own life; or accidental, where a device failed to work properly, producing sufficient heat that unintentionally started the fire and lead to the victim's death. A fourth category is natural, as in a fire started by lightning.

For a multitude of reasons, from time to time the medical examiner may be faced with the same dilemma the fire investigator is dealing with. There may not be sufficient evidence to support any accurate determination as to the manner in which the victim died. In these cases, the manner may just be undetermined.

There may be times when the victim lives but later dies. The *cause* and *mechanism* are separate but still related. A patient who survives the fire but is severely burned may live days or a few weeks, only to die as a result of complications related to his burn injuries. When the patient dies in the hospital 2 weeks after the fire, the mechanism may be pulmonary edema or the failure of a vital organ. However, the cause of the death will relate back to the fire such as homicide from the setting of the fire with the intent to kill the victim. The cause, manner, and mechanism of death are the purview of the medical examiner. However, as a result of these findings, the fire investigator may have more work to do to help solve any potential crime.

■ Note
The cause, manner, and mechanism of death are the purview of the medical examiner.

DELIVERING THE NEWS: DEATH NOTIFICATION

Most frequently, the police can assist in delivering the news of the loss of a loved one, even from a fire. But in some departments, when it is a fire fatality, the fire department is assigned this task. The reason may be as simple as being able to answer the questions of those who have just lost a loved one. Before carrying out such duties, you should be trained because doing this task wrong may cause even more harm in such a dreadful situation. The first responder investigator will most likely not be assigned this duty. However, the first responder investigator may be the only investigator and the lead officer on the scene when the next of kin arrives. Understanding the following information may help the first responder cope with the task of notification.

■ Note
The task of death notification should never be taken on alone.

The task of death notification should never be taken on alone. There should always be a pair: the investigator or fire officer who has knowledge of the incident, and the other can be the chaplain from the fire or police department, who has experience in delivering such messages. If the religious leader of the victim's faith is known, it would also be beneficial to have that person accompany the team. You may notice that this concept implies a visit; such messages should be delivered in person, never over the phone.[2]

There are a few rules to follow for the carrying out of such duties. First is to be honest in the delivery, without false platitudes. Carole Moore in her article

■ Note
Words have to be chosen carefully because they will be remembered for a lifetime by those receiving the message of the loss of a loved one.

■ Note
The family needs to know the truth to a certain extent—not the gory details, but the facts as known about the incident. All too often the imagination can be so much worse than the truth.

"Breaking Bad News" gives good advice about being truthful. Words have to be chosen carefully because they will be remembered for a lifetime by those receiving the message of the loss of a loved one. She states that words such as *your husband is dead* may seem harsh, but there is no way to sugar coat such a message. Trying to soften the blow by using words such as *passed on* or *expired* do nothing but cloud the issue and make it more difficult for the family to understand; don't leave them guessing.[3]

Karen Dewey-Kollen, in the *Law and Order* magazine, concurs that choice of words is important. Saying that you are sorry for their loss is sincere and to the point. But saying that you know how they feel is not exactly true for most individuals; besides, even if true, it would not be believed at such a terrible time in their lives. She also recommends that the family needs to know the truth to a certain extent—not the gory details, but the facts as known about the incident. All too often the imagination can be so much worse than the truth.[4]

Sometimes it is best to state the obvious. When making the notification, verify that you have the right family and the right person. You don't have to ask for identification, but a simple question such as, "Are you the wife of ..." can be sufficient. Always ask to enter the residence or someplace not so public. If at all possible, arrive before the news hits the media. Show your credentials even if in uniform, and then be straightforward. Your presence has most likely put them on notice that this is not a good visit.[5]

If the loss of life was not from an accidental fire or if there is reason to believe that the cause of death was not accidental, this may not be the time to ask detailed questions related to your investigation. Watch the body language and responses of the family members for future note. Don't judge or make rash assumptions at that time. People exhibit a wide range of emotions for numerous reasons that may not be associated to your case.

SUMMARY

The evidence to help determine the cause of the death may well be found on or around the body. For this reason, there needs to be a thorough examination of the body, the area surrounding the body, and the room in which the body is found. This area needs to be documented by measurements and inventory as well as a complete set of photographs showing the entire scene. The investigator must have a keen eye and the persistence to go over each and every piece of debris to determine whether it is or is not evidence. Experience is a valuable tool in identifying what to look for at the scene.

The medical examiner should be able to provide the identification of the victim as well as the cause, manner, and mechanism of death. Based on these findings, the fire investigator may need to conduct a more in-depth investigation.

REVIEW QUESTIONS

1. Why would the fire investigator be assigned to be part of the team to deliver the death notification of a fire victim?
2. What is a CISM?
3. If the burn victim survives, what needs to be collected and documented for your investigation?
4. What tools or resources are available to make an accurate body identification?
5. What is the difference between the cause of death and the mechanism of death?
6. Why should the fire scene be treated as a fatal fire if the burn victim is still alive but at the hospital?
7. Why would the fire investigator want to look in the ambulance that just transported a fire victim?
8. Why should the gauze used by EMS to treat an injured victim be recovered by the fire investigator?
9. What would be apparent in the nose and mouth that would indicate that the victim was alive at the time of the fire?
10. What is the pugilistic attitude of the body of a fire victim?

DISCUSSION QUESTIONS

1. The facts surrounding the opening case study prevented the investigator from labeling the exact origin and cause. Should the fire have been labeled accidental and the cause listed as smoking materials? Explain your reasoning.
2. Why would the investigator want to interview the victim as soon as possible? If the department's policy is to always examine the fire scene before doing interviews, should the investigator keep to this policy and interview the burn victim after completing the scene examination? Why?

ACTIVITIES

1. Research the policies in your locality as to the role of the medical examiner. Does the ME respond to the scene or only work in the laboratory?
2. If there is a critical incident stress management (CISM) program in your locality or a neighboring locality, approach them to get a copy of their procedures and see if there is an opportunity to learn about their activities.

NOTES

1. National Fire Protection Association, *NFPA 921: Guide for Fire and Explosion Investigations* (Quincy, MA: National Fire Protection Association, 2008), 921–192.
2. Wayne Bennett and Karen Hess, *Criminal Investigation*, 8th ed. (Belmont, CA: Thomson Wadsworth Publishing, 2007), 261–262.
3. Carole Moore, "Breaking Bad News," *Law and Order,* May 2003, 106–109.
4. Karen Dewey-Kollen, "Death Notification Training," *Law and Order,* May 2005, 12–14.
5. Douglas Page, "Death Notification: Breaking the Bad News," *Law Enforcement Technology,* March 2008, http://www.officer.com/print/Law-Enforcement-Technology/Death-notification--Breaking-the-bad-news/1$40885.

ARSON

Learning Objectives

Upon completion of this chapter, you should be able to:

- Describe the three elements of arson.
- Describe the necessity of proving motive.
- Describe and give examples of the different types of motives.
- Describe various techniques that could be arson indicators.
- Describe characteristics that may indicate that the fire was intentionally set.

CASE STUDY

A local homeowner left for vacation on Sunday night. He and his family had a 12-hour drive to get to the cabin they had rented on a lake in another state. He arranged for a neighbor to check his home but gave specific instructions to go over only on Tuesday evening. A few hours after the homeowner left for vacation, a strong thunderstorm came through the area and knocked down several trees. It took almost 20 hours for the power to come back on. When the neighbor came home from work on Monday and found that the power had come back on, he felt he should not wait until Tuesday to check on the home next door. When he opened the back door, he could smell something burning. When checking the house, he finally made it down to the basement and found a 4-foot by 8-foot model railroad set on fire. It took three buckets of water, but he finally put out the fire.

Once the fire was out, he realized that an electric soldering iron had its trigger switch taped into the on position. The soldering iron was hanging from a hook in the ceiling by its power cord and was plugged into a 24-hour electric timer, which was attached to an extension cord plugged into an outlet. On the table were remnants of

Figure 16-1 *The electrical timer allows for a time delay of just under 24 hours.*

(Continued)

(Continued)

what was burning, and he realized it was crumpled up toilet paper that was stacked up about 2 feet high, with the soldering iron stuffed in the middle of the paper. He immediately called the sheriff's office, and then called the homeowner at his vacation home.

When the homeowner arrived back at his house the following day, he was met by the sheriff's office personnel. The homeowner's prints were taken as a process of elimination of prints found in the home. The homeowner's prints matched the prints found on the sticky side of the electrical tape that was used to set up the time delay device. The homeowner set up the family trip to take suspicion off himself and had arranged for his neighbor to check the home, but only after he knew the home would have burned down.

The timing device, as shown in **Figure 16-1,** could be set for just under 24 hours, giving the homeowner plenty of time to be far from the house in hopes of escaping suspicion. The power failure certainly messed up his plans, as did the good neighbor.

INTRODUCTION

Under most circumstances, the fire officer is not the one to complete the investigation on an incendiary or intentionally set fire. An assigned investigator who is either part time or full time is assigned to handle such cases. Depending on the jurisdiction, the assigned investigator could be someone who works for the fire department such as the fire marshal or an investigator for the local police department. Depending on the state, the fire investigator may be assigned to the state fire marshal office, the state police, or the state insurance bureau. However, the initial investigation may be started by the fire officer and it is beneficial for this investigator to know what happens in the next stage of an investigation of an intentionally set fire.

In previous chapters, we have dropped little hints and tidbits about arson or incendiary fires. As a general rule, we use the term *incendiary* in our reports because in many jurisdictions it is up to the courts to decide whether a fire is considered arson. The report must document the facts, such as that Mr. Jones set the fire using a match to ignite balled-up paper in the northeast corner of the garage.

This chapter discusses legal issues and the elements of the crime of arson. It discusses the motives for arson and gives special consideration to juveniles involved in such incidents. Various indicators are discussed including where the arsonist may set the fire and the various tools of the trade they may use in their desire to destroy.

IMPACT ON THE COMMUNITY

■ **Note**
The criminal act of arson has an impact on each and every person in the community.

criminal act
as stated in *Black's Law Dictionary*, an external manifestation of one's will which is prerequisite to criminal responsibility; there can be no crime without some act, affirmative or negative

The **criminal act** of arson has an impact on each and every person in the community. It can even have a far-reaching impact at the national level when funding is taken from other worthy projects to support the investigation and suppression of arson. What you pay for vehicle insurance or for your home or renter's insurance policy is predicated on the anticipated losses, including those losses that result from arson or insurance fraud associated with a fire loss.

A more direct impact in your community is the potential for injury or loss of life as the result of an arson fire. This includes not only direct losses from the fire but when emergency units are tied up with an arson fire and are not available for other calls for assistance such as for vehicle accidents, engine response on accidents and illnesses, or even from other nonarson fires.

■ **Note**
The firefighters are at an unnecessary risk from the arson fire.

The firefighters are at an unnecessary risk from the arson fire. They are not only endangered by the flames, but sometimes they are at risk from the different tactics arsonists use to keep firefighters from extinguishing the fire, such as cutting holes in floors at the door entrance or arranging for structure failure just as the firefighters are making their attack.

■ **Note**
Often, arson is directed at older structures that may seem of less value to some but that may be irreplaceable pieces of a community's history.

Often, arson is directed at older structures that may seem of less value to some but that may be irreplaceable pieces of a community's history. Even if fire does not destroy everything, there is additional damage resulting from the suppression activities, such as water damage. Inside some of these structures are irreplaceable artifacts that may seem like junk, but they too could be pieces of the past that someone had cherished.

perpetrator
the one who commits the criminal act

One of Our Own

One of the facts we have to face is that sometimes one of our own may be the **perpetrator** of the crime we so detest in the fire service. It had been quite a week with no fire runs at all. The older firefighters tend to enjoy these types of duty tours, but the young ones are anxious to hone their skills and use their newly learned talents of fire suppression techniques.

One of the younger firefighters came running into the station shouting that there was a fire next door in a shed. The alarm was called in as the engine pulled out and set up in front of the small structure. The fire was knocked down quickly, but a road flare was quite visible in the window where the structure was set on fire.

Investigators spent the rest of the night interviewing each firefighter on duty. The last to be interviewed broke down and told the truth: that he set the fire because he was bored. The tragedy is that the structure in question was a historic building, the last standing one-room schoolhouse in that jurisdiction. Inside were artifacts from the lessons taught in the previous century. What was a shack to one person was a treasure to the community.

■ **Note**
To have the crime of arson there must be a corpus delicti, in other words, the body of the crime.

corpus delicti
the body of the crime; the physical proof that the crime did occur

ARSON: THE CRIME

To have the crime of arson there must be a **corpus delicti,** in other words, the body of the crime. If nothing was damaged, there was no crime of arson. For example, an inmate got himself locked up in solitary confinement. While there, he tore open the mattress, pulled out the stuffing, and stacked it in a corner. Federal law requires the outside of an institution mattress to be fire resistant; however, it does not regulate the interior of the mattress. Once the stuffing was stacked in the corner, the inmate pulled out a match and lit the stuffing. It burned poorly but put off a massive amount of smoke. (Yes, the inmate was searched before being sent to solitary, but let's not discuss where he hid the match.)

His motive was to escape when the jailers came to rescue him. He knew there were only four jailers on duty, yet there were more than 50 inmates that particular night. His plan did not work, but the inmates were dumped into the fenced-in compound, not out the front, which is where he thought he would be taken because the compound was at the other end of the building. Both the inmate and the guard doing the rescue took in quite a lot of smoke getting to the other end of the structure, but they were given a clean bill of health by the attending physician at the scene.

■ **Note**
There are three elements to the crime of arson:
1. There must be clear evidence that something burned.
2. The burning had to have been an intentional act by the perpetrator.
3. There must have been malice in the setting of the fire.

The fire did not damage the structure. There was one blister of paint on the wall, but the only other damage was to the mattress and massive smoke damage throughout the structure. It was finally decided to charge the inmate with destruction of government property and reckless endangerment of the staff and inmates in the jail. It carried more punishment then arson of a mattress. It added 2 years to his time to serve. However, his fellow inmates had to clean the jail following the incident. Needless to say, he had a somewhat uncomfortable stay for the rest of his sentence—there was no outright violence thankfully, but he lived to regret his actions. It was surprising how the inmates made sure he did not set any further fires.

For most jurisdictions, there are three elements to the crime of arson:

1. There must be clear evidence that something burned.
2. The burning had to have been an intentional act by the perpetrator.
3. There must have been **malice** in the setting of the fire.

It must be stressed that this text cannot answer the legal questions about your jurisdiction. It is essential that your department research the legal issues and establish a policy and procedures manual to address legal issues.

malice
intentionally, willfully, and knowingly doing a wrongful act with the intent to inflict harm on another; recognized by law as an evil intent

Arson versus Incendiary

Although covered earlier in Chapter 13, it bears mentioning again here in this chapter: *Black's Law Dictionary* includes an extensive definition for arson that states that arson is the malicious burning of the house of another.[1] This is the earliest form of the charge of arson. Over the years, the law has changed, expanding what is covered under arson. States with model laws place different penalties for the different types of structures or items that are set on fire. For years, arson has been a crime against property, and now it is recognized that it is also a crime against people, which allows for greater penalties.

Black's Law Dictionary includes a lesser definition of *incendiary,* which is when one sets a house or building on fire.[2] *Webster's Dictionary* goes into a little more detail, saying that *incendiary* means the willful destruction of property, and the definition mentions incendiary devices.[3] As a general term, an incendiary fire is one that is intentionally and willfully set.

The bottom line is to set a department policy on the proper terminology that will be approved by your jurisdiction. The policy can be reviewed by legal counsel who may be part of the county staff, but the final approval should be given by your local prosecutor because this person will be the one to argue your department's cases in a court of law, after an arrest has been made.

MOTIVES

In many states, it is not necessary to have a motive for the crime to get a conviction. However, with today's jury, the motive can be a vital part of the prosecution's case in court. Today, more than any time in the past, the general public is well trained in legal issues. There has been a progressive number of TV shows dedicated to crime and solving crimes. This trend started many years ago, in particular with *Dragnet* and *Adam 12,* and progressed to *Hill Street Blues* and the various Murder Mystery shows. Today, with the many variations of the Law and Order and the CSI series, the public is better educated than ever before. This education includes knowing why the perpetrator committed the crime.

■ Note
When the jury deliberates, they may have a hard time coming up with a verdict without knowing why the individual committed the crime.

These educated public will serve on juries. When an arson case comes forward, a conviction is legally possible without ever knowing the motive or reason the individual committed the crime. However, when the jury deliberates, they may have a hard time coming up with a verdict without knowing why the individual committed the crime. Regardless of the legal minimums, always strive to answer all potential questions that would come up with any case. In this instance, always strive to identify the motive of the individual who may have committed the crime for which he or she has been charged.

■ Note
Research conducted by the National Center for the Analysis of Violent Crime (NCAVC) has identified six motive categories: vandalism, excitement, revenge, crime concealment, profit, and extremist.

A study by the Department of Justice (DOJ), the Federal Bureau of Investigation (FBI), and the Federal Emergency Management Agency (FEMA) and research conducted by the National Center for the Analysis of Violent Crime (NCAVC) has identified six motive categories: vandalism, excitement, revenge, crime concealment, profit, and extremist.[4]

The NCAVC, along with the American Psychological Association, has concluded that the use of **pyromania** as a motive is an oversimplification of a more complex issue that may be indicative of other disorders or problems. As such, the

pyromania
the tendency or impulse to set fires

■ **Note**
Pyromania is an impulse that can be ignored if the individual so desires.

■ **Note**
Vandalism is usually done by those individuals who have not mentally matured.

■ **Note**
Vandalism fires can be associated with petty theft and sometimes with breaking and entering by the perpetrators. Vandalism fires are also associated with gangs, who may even deface property with paint.

NCAVC has taken it off the list of motives for fire-setting.[5] Studies have indicated that pyromania is an impulse that can be ignored if the individual so desires. The investigator must have supporting documentation such as fiscal discovery, interview comments, or data collected from the fire to present in court when giving testimony as to the motive of the individual. If not supported by tangible evidence, the motive should not be introduced.

Within each category of motives is a wide range of reasons for why the person set the fire. Consider the elderly widow who is destitute because her husband died and left her no insurance money except Social Security. She has debts that will soon see her out of her home. She does not drive but has three vehicles. She tries to sell them to no avail and doom is closing in around her. In a moment of weakness, she sets one car on fire one night. Her attempts were extremely poor; the vehicle burns, but the evidence is overwhelming that she committed the crime. Clearly, the motive was profit. But the investigator, now knowing the facts, must decide whether he will arrest a grandmother just before the holiday season.

Vandalism

This type of crime is usually associated with juveniles and adolescents. It is hard to place an age on this type of senseless crime. Instead, vandalism is usually done by those individuals who have not mentally matured.

Vandalism fires can be associated with petty theft and sometimes with breaking and entering by the perpetrators. Vandalism fires are also associated with gangs, who may even deface property with paint. These types of crimes are usually, but not always, done in groups. The targets may be limited to those that the vandal is associated with, such as a school. What may start out as a petty crime can change quickly.

Teen Prank

Three young teens were not happy with their shop teacher for the bad grades they got on a recent project. With nothing to do on a Sunday afternoon, they decided to go to their junior high school, which was just around the corner. With no plan of exactly what to do, they did find a door that could be opened. They went to the shop and found three 1-gallon gasoline cans, full. The students picked up the gasoline cans and emptied them, and then they grabbed one road flair from a tool box and went out the rear door. The boys managed to strike the flare and threw it in the room, slamming the door behind them as they ran. Why their clothing did not ignite, we will never know. The room did not ignite because the atmosphere was too rich—too much ignitable vapor and too little oxygen.

(Continued)

(Continued)

The shop teacher arrived Monday morning and found gasoline still puddled on the floor and the residue from a burnt road flare by the back door. At the back door there was a smooth concrete pad on the outside, and on that pad investigators found the perfect outline of three sets of sneakers, you could even read the sneaker manufacturer on two pairs of imprints. The students had walked through the gasoline, which had broken down the soles of their shoes, and while they stood still, it left the evidence behind.

When asked, the shop teacher said that only three students had failed their project this quarter. However, he was quick to add that they were good boys and would never have done anything like this. Each of the boys' parents was visited and when asked they all allowed the investigators to look in their son's rooms. In each instance, investigators found the sneakers that matched the imprints at the school. Faced with the evidence in hand, three confessions followed. What started out as a small rebellious vandalism event could have burned down the school and, worse yet, could have taken the boys' lives.

In the past, we used to say that vandals did not use devices. However, in this age of computers young people have learned how to make bottle bombs and often place them in mailboxes for the best effect. These devices don't result in fires, but they contain the energy to cause serious bodily harm.

One more event worth mentioning was a grandfather calling the fire marshal to report his grandson's acts of vandalism. Seems the boy and three of his friends stole a case of the grandfather's antique Coca-Cola bottles, the thick green glass returnable bottles from the 1950s. The kids poured in gasoline and stuffed a rag in the top, making **Molotov cocktails.** They lit the rag and threw them at mailboxes as they drove around their subdivision. Fortunately, the bottles did not break. However, the burning rag did create small grass fires that, because of the moisture on the ground, went out on their own. Grandfather got all the bottles back, still intact, but a little sooted. The kids got counseling and community service, part of which was to replace the damaged mailboxes. Although the bottles did not break, the impact dented up about a dozen mailboxes.

Molotov cocktail
a bottle filled with an ignitable liquid that is ignited using a wick in the bottle, reactive chemical coating, or wrapping in a fuel-soaked rag

Excitement

The firefighter who set the one-room schoolhouse on fire mentioned earlier in this chapter is an example of someone with an excitement motive. He fit the description by being bored and wanted the thrill of putting out the fire and the attention he would get in the process. It is an embarrassment to all of us when one of our own commits arson. It goes against the grain of our senses and gives us a feeling of betrayal when this happens. This is usually an act committed by one person, unless alcohol is involved, and then there may be more involved.

■ **Note**
Individuals with this motive of excitement usually start with small fires and work up to dumpsters, vacant buildings, and possibly occupied structures after time.

Other individuals with this motive of excitement usually start with small fires and work up to dumpsters, vacant buildings, and possibly occupied structures after time. They usually stay in areas they are familiar with, such as around their home or work. They have been known to stay around and watch the firefighters, even offering to help, to live vicariously through the firefighters' actions. They like the thrill of the lights and sirens and they like the fact that they know how the fire started and no one else knows that yet. They may even seek out recognition and attention by helping to pull hose lines or even make suggestions of what they saw before the fire department arrived.

Most of the time, firefighters are too busy to recognize familiar faces in the crowd. But the engineer, operating the apparatus, may have the opportunity from time to time to check the crowd to see if there is a familiar face. Some departments issue disposable cameras for each engine so that the crew can take pictures of the fire and the crowd.

Revenge

■ **Note**
A fire set with the motive of revenge can be set to harm an individual, a group, or an institution. Sometimes the revenge is directed at none of these but at society itself.

A fire set with the motive of revenge can be set to harm an individual, a group, or an institution. Sometimes the revenge is directed at none of these but at society itself. For some, things have gone so wrong in their lives that they just strike out at anything that represents society as a whole. They just want to get even for what life has dealt them and the life they lead. There usually is no pattern; the strikes will be random.

Person When the revenge is aimed at a person, it is more obvious. In addition to the fire set, there may be vandalism of personal property. When pictures are broken, clothing is torn up, or holes are cut in pants or dresses, these actions indicate a broken relationship. Anything that may have been a favorite of the victim may be the target of the perpetrator's outrage. These types of fires usually have been planned, but time delay devices are not normally used. The fire setter may lay in wait for the right opportunity to strike at the individual, and destruction and fire are primary tools.

In these particular types of fires, interviews are key to discovering the motive. During the discussion with all the parties involved, the investigator may get an indication of what is happening. Checking into relationships between couples, both heterosexual and homosexual, may be revealing. A betrayal could have caused someone to lash out, and it may not be their partner they attack. Instead, the victim may be a third party who is presumed to be breaking up the relationship. Thus, the third party could be the one who has their possessions destroyed and burned.

Looking at office relationships may reveal that there was an actual or perceived hardship suffered by the perpetrator. That individual may have been laid

Obvious Evidence

A 9-1-1 call resulted in an investigator being sent instead of an engine company. The English teacher at the local high school went out to get into her Volkswagen Rabbit in the morning and found the vehicle's fuel door open and a rag stuffed into the fill tube. The rag was burned as was the plastic trim that came in contact with the burned rag.

After photographing the rag, the investigator pulled it out of the fuel tube. The part stuffed into the fuel fill had been tightly wound up and had no damage. The rag turned out to be a towel, and embroidered onto the towel was the name of the local restaurant that just happened to be next door to the high school. With no other incidents occurring overnight, the investigator asked the teacher if anything happened at school the previous day. She said that she had to fail one of her students, and as a result, that student will not be able to graduate as planned.

The investigator checked with this student, and she had an alibi that night. When asked if she had a boyfriend, she said yes but that he graduated last year. Sure enough, he worked as a dishwasher at the restaurant next door to the high school, the same restaurant identified on the towel. When interviewed, the boyfriend eventually confessed after failing a lie detector test. He said he was upset because he wanted to get married, but his girlfriend's parents would not consent until she graduated from high school. One point of humor about the case was that it was below freezing weather and it was a diesel fuel vehicle. There was no way this attempt would burn any vehicle.

off or did not get a promotion he or she felt was deserved. Thus, the perpetrator may strike out against the person who got the job or the supervisor who failed to promote them or laid them off. Even if perpetrators brought their situation upon themselves, they still want to strike out. The target could be the personal items in the office, the individual's vehicle or even their home.

Note Churches and synagogues are common targets for revenge types of crimes.

Group A perceived injustice against a group can manifest to the point that the perpetrator lashes out at something that represents that group. Churches and synagogues are common targets for revenge types of crimes. The Bureau of Alcohol, Tobacco, Firearms, and Explosives heads up task forces to examine these types of crimes against places of worship.

The target for this type of revenge is not aimed at an individual so much as an organization. It doesn't even have to be an organized group. However, this is where we see the type of fires set by gangs when they get territorial or lash out at anything that can threaten their life or existence. It is also a crime that can be committed by an unorganized crowd where they, as a whole, at that time, strike out against what they perceive is a common cause.

Although the crowd may not advocate committing a revenge crime, there are those in the crowd who may decide to take things into their own hands. Commonly, those seeking revenge lash out at symbols of what they perceive as wrong—a cross, the Star of David, or corporate logos. These are usually well-planned events and may use a more sophisticated means of setting the fire.

Concealment of a Crime

■ Note
Whether the crime was planned or was a crime of passion, the criminal will seek to cover up the crime just committed.

Whether the crime was planned or was a crime of passion, the criminal will seek to cover up the crime just committed. This is especially true when perpetrators believe that they have left evidence behind that may indicate their guilt. For this reason, some criminals turn to arson in the hopes of destroying the crime scene and allowing them to go undetected. Because of this, the assigned fire investigator must be properly trained on the investigation of other crimes, if for no other reason than to be able to recognize evidence and process it appropriately.

Usually when fire investigators find evidence that suggests that other crimes may have been committed at the scene, they call their counterpart in the local police or sheriff's office so that they can work as a team. Should they make their case and find the culprit, there will be two experts testifying in court on the facts surrounding the incident. This teamwork atmosphere must be worked out in advance so that there is a smooth flow of information sharing and cooperation from the very beginning. If for some reason there are going to be territorial issues or command questions, the fire scene is not the place to work this out; it must be done in advance.

Almost any serious crime can result in the perpetrator setting a fire for concealment. In the past, before DNA analysis, there were cases in which the criminal actually did get away with murder, but not so much today. New forensic technology has brought us a long way and there is new science just around the corner to help even more.

burglary
the entry into a structure or part of a structure with the intent to commit a crime

robbery
feloniously taking any item of value from another directly or in their immediate presence and against their will while using force or fear

Other crimes that may lead to setting a fire for cover-up are **burglary, robbery** that leads to murder when the suspect realizes the victim recognizes him, fraud, and embezzlement. For example, the owner of a bar kicked out a patron because he was behind in paying his bar tab. That night, the bar experienced a break-in and rather damaging fire. In the center of the area of origin was the ledger, opened to the page on which was listed the name of the patron who was thrown out earlier in the evening. The book was stored by the owner before leaving that night, but it was found on a tray, doused in gasoline and ignited. The gasoline may have been the first material ignited, but the bar tab was the motive for setting the fire. Eight cases of beer with delivery labels for the bar and coin boxes with serial numbers from the bar video games were found in the perpetrator's trunk when his vehicle was searched.

Those committing embezzlement or fraud can sometimes use arson to cover their crimes. Selling parts out of the company's inventory and pocketing the money only to find out that an annual company inventory is going to be taken may be a reason to set fire to the stock. These types of crimes are only limited by the imagination.

Check Cashing

One particular company paid their staff every other Friday. The payroll clerk signed the checks using her employer's signature stamp. She would then take the checks to the staff in the workshop. Many dealt in cash, so she got them to endorse their checks, took the checks to the bank, and returned with the correct cash for each employee. What may seem like a benevolent act was a perfect part of the rouse. Although the next Friday was not payday, the clerk created payroll checks for the cash-based employees, stamped the checks with the employer's signature, endorsed the checks, went to the bank, and kept that cash for herself.

One Friday, an auditor showed up and looked at the books for accounts receivable. He told the payroll clerk that he would look at the books for accounts payable on Monday morning. Saturday night the business suffered a major fire. The fire started in the accounting office. All the desks were in proper order but the desk drawers of the payroll clerk were left wide open. The books, which are usually left locked in the desk drawer, were in the center drawer and gasoline was poured over the desk, books, and surrounding area. Interesting enough, the books survived; they were charred, but legible.

No trial took place; a warrant was issued for the clerk's arrest on charges of embezzlement. Once she was arrested, she could be interviewed and hopefully charged with arson as well. But she was nowhere to be found. Ironically, about 2 months later a similar business about 5 miles from the first business went to the police with an amazing story that they thought the payroll clerk was creating unauthorized payroll checks and cashing them. By the time investigators arrived at her office, she was gone. By the time they got to her home, they had just missed her again. Last known location was somewhere in the Caribbean. So, even if you find the area of origin and the cause, even if you identify who is responsible, you may not always get that person to court.

Profit

Sometimes the profit to be gained is obvious and sometimes not so obvious. The obvious are such situations as when businesses under hard times with outdated stock are set on fire for the insurance money or the person with a gas-guzzling vehicle responds to a sharp rise in gas prices by burning the vehicle to get the insurance proceeds to buy a more economical vehicle. Or the vehicle owner just wants to get out from under a high vehicle payment.

This motive can sometimes cause one business planning the demise of a competitor by burning it out of business. Or the crime of extortion is committed where a business is threatened with a fire if it does not pay a certain amount to the crime boss or local gang each week.

Some types of fraud can be obscure at first. One property owner intentionally burned his neighbor's barn because it blocked the view of the water from his house. It wasn't that he wanted to see the water; he just wanted to be able to seek a higher value for his house because it would have a scenic view. In the city, land is at a premium. It is not unusual for empty lots to be more valuable than the lots with existing structures because demolition costs are high. A developer must figure in demolition costs for structures sitting on the land. But if the structure burns, the property owner realizes lower demolition costs and a more valuable lot.

■ Note
For every type of insurance policy, there is a way to commit fraud.

Insurance Fraud For every type of insurance policy, there is a way to commit fraud. In the aspect of a fire loss, the fraud can be the setting of the fire with the intent to defraud the insurance company. In a second scenario, the fire really was accidental, yet the property owner decides to take advantage of the situation and put in a claim for losses that did not occur, such as by overstating the value of items or putting content items on the list that were not in the fire.

■ Note
Insurance fraud has been so costly for the insurance industry that the major companies have special investigative units to look into some fire losses.

Insurance fraud has been so costly for the insurance industry that the major companies have special investigative units to look into some fire losses. These units either have their own investigative staff or they hire fire investigators in the private sector to look at fire scenes. The intent is to see whether there are any reasons to deny the claim based on the policyholder being involved in setting of the fire or defrauding the insurance company after the fire.

■ Note
There is a thin line between cooperation and collusion.

An insurance investigator, either from the insurance company or a private investigator on contract, has no legal right to enter the scene before the government investigator is done with his or her examination. However, depending on the situation and the individuals involved, there is nothing prohibiting the investigators from working together. There is a thin line between cooperation and collusion. Cooperation is essential to the point that many states have arson immunity laws that allow the insurance company to share information without fear of civil or criminal prosecution. Caution must be exercised to ensure that there is no collusion, that the government investigator and the insurance investigator do nothing to abridge the rights of the insured in any way. The two should not set plans in motion to prove the person's guilt. The investigator is a seeker of truth and should be just as interested in evidence of innocence as well as evidence of guilt.

■ Note
Most policies do not mention arson. Instead, they adopt a line such as "Concealment of Fraud."

Insurance Policy States have an insurance bureau or agency that regulates the insurance industry. They or the legislative body of the state may adopt a standard policy language that is to be used for the sale of all policies within the state. Interestingly, most policies do not mention arson. Instead, they adopt a line such as "Concealment of Fraud." Essentially, the policy is null and void if the insured should misrepresent any fact or circumstance or misrepresent themselves by committing fraud or falsely swearing of any material fact.

After a fire, the insurance company requires a written proof of loss. This is essentially a signed statement that the fire occurred, along with a list of all the contents lost in the fire and the value of each item.

deposition
obtaining testimony of witnesses that is recorded, authenticated, and reduced to writing and that can be used in future court testimony; used as a discovery device by either side in a civil or criminal trial

Under the requirements of most insurance policies, the insured must, if requested, present themselves for an interview with the insurance representative for an examination under oath (EUO). An EUO is essentially a **deposition,** where the insurance company, with its legal counsel, is able to ask a series of questions about the incident. A court reporter will be present and the insured may bring their own counsel if they so desire. Keeping this in perspective, it is an opportunity for the policyholder to explain any unclear circumstances surrounding the event. Should the policyholder refuse to participate, and after reasonable attempts are made to obtain cooperation, the insurance company can deny the claim and not pay the policyholder for the claim.

■ Note **Should the policyholder refuse to participate, and after reasonable attempts are made to obtain cooperation, the insurance company can deny the claim and not pay the policyholder for the claim.**

To make sure that the insurance companies do not abuse this ability, the courts have imposed penalties in the form of bad faith, where penalties can be imposed if the insurance company does not act in good faith toward their policyholders.

It is important for the assigned investigator to have knowledge about the insurance industry because the investigator may be subject to a deposition should testimony be required in a civil case. The deposition is an opportunity for all parties to hear what will be in their testimony if they go to court. Sometimes the evidence presented in deposition is so compelling that the case is settled out of court. If the case does go to court, the deposition and the investigator's answers can be used to verify what is said while the investigator is on the stand.

All witnesses are subject to a deposition, including the municipal investigator who may be compelled by the courts to testify. If subjected to a deposition, the deponent, the investigator in this instance, should never waive the right to review. You should always get a copy of the deposition and read it carefully. If there was a mistake in the recording, insist on having it verified and corrected if necessary. If you misspoke and the statement was not correct, then make correction in writing and submit to all sides in the case. This might be quite embarrassing but is necessary because investigators are seekers of the truth, so should also always speak the truth.

■ Note **Because investigators are seekers of the truth, so should also always speak the truth.**

Extremism

■ Note **Just as the name implies, any person or group of people who possess an extremely different outlook on life in comparison to the world around them or those who take extreme measures to make a point (their point) or to get their way are considered to be extremists.**

Just as the name implies, any person or group of people who possess an extremely different outlook on life in comparison to the world around them or those who take extreme measures to make a point (their point) or to get their way are considered to be extremists. Terrorism is the best example of this type of fire setter. Today more than ever before, you can see examples of extremism on the nightly news. This type of attempt at social change has been around since recorded history in one form or another. The Provisional Irish Republican Army (IRA) took it to new heights when they formed in 1969 to oppose British rule.[6] Today there are radical groups around the world that are against one form of organized society or another. These extremists are willing to use whatever means necessary to force their ideals upon others. Sometimes it is not the ideals so much as the destruction and terror they can create. Today, extremists from the Taliban and al Qaida have taken terror to a new level, and regrettably, the world has not seen the limits of the horrors that can be unleashed.

Extremist groups don't have to come from other lands or from other cultures. The United States has its own share of extremism in various organizations. They speak for the rights of animals, the rights of the wilderness, and the right not to be taxed or to do as they please in or around their compounds. Some extremist ideals are based on religion and others are based on a philosophical theory.

Some upset about the damage to the environment have lashed out and set SUVs on fire in one neighborhood. In another state, a set of high-dollar homes under construction in a cul-de-sac represented humanity's growth intruding on the wilderness, and those homes were set on fire. The list goes on and on.

OPPORTUNITY

A time line should be established for the fire event. It can only be as precise as the data available. Reliable times that can be documented are the first call to the emergency communications center and the time that the first engine arrived. The next time to be documented are any eyewitness accounts, which can vary and may not be 100 percent reliable but at least it will provide something to work with when placed on the time line. The fire scene can provide some time indicators, but reliability will depend on the circumstances. For example, electrical analog clocks that stopped either as a result of fire damage or of the circuit being interrupted by overcurrent protection devices can provide a time, as shown in **Figure 16-2** and **Figure 16-3**. A word of caution: Ascertain from the homeowners whether the clocks were actually working and properly set prior to the fire. In this day and age, analog clocks are becoming scarce with the advent of newer digital liquid crystal displays, which regretfully will not provide such evidence.

■ Note
The fire scene can provide some time indicators.

Although not used much except as travel alarm clocks, mechanical clocks can also provide a time of failure as well. Some homes may even have antique or reproduction mechanical mantel or grandfather clocks that can also be of assistance. When subjected to excessive heat, the mechanics of these clocks will also fail, providing a time when that occurred. Because these clocks need winding, it will be necessary to ascertain from the owner if the clock was working and if it had recently been rewound.

NFPA 921, 2008 edition, cautions about using the depth of char for establishing a time line.[7] Laboratory test results with pine did indicate a burn ratio of 1 inch of char in 45 minutes. But there are so many variables and unknowns that to use this in an actual burn scene would be unreliable. There may be times where the investigator can make a generalization based on experience, especially if there is a gross disparity in actual times and alleged times, such as when the homeowner says he left only 5 minutes ago and exposed 2 × 4s at the area of origin are almost consumed.

When the time line is complete, the fire has been determined to be incendiary in nature, and a suspect has been identified, the investigator must then

Figure 16-2 *A clock from the bedroom nightstand. Always check to see if it was plugged in, and ask the homeowners if it was in working order.*

Figure 16-3 *An analog clock located on the kitchen oven. A check with the occupants is necessary to ascertain whether it was accurate or set for the correct time.*

verify whether the suspect had an opportunity to set the fire. In addition to documenting the scene as to when the fire occurred, you need to collect evidence on any person of interest in the event to show whether they had the opportunity to commit the crime.

INCENDIARY: ARSON INDICATORS

■ Note
The world of fire investigation is not black and white.

The world of fire investigation is not black and white. There are no indicators that emphatically prove that something is this or that. It takes a culmination of all the facts to eventually have an accurate hypothesis of the fire origin and cause. The following indicators may be signs of an incendiary fire, and through a thorough investigation of each situation they may or may not be part of the facts that support your final determination of the cause of the fire and persons responsible.

Location and Timing of the Fire

The arsonist wants the fire to burn as long as possible before being discovered. Thus, the time to set the fire would be when there are fewer people around to discover the fire. After midnight and in the early morning hours most people are in bed and not out and about. So, this would be the arsonist's preferred time. The location and time have a correlation as well. If the fire location is an industrial complex, like an industrial park that is quiet with few people around on Saturdays or Sundays, then this too would be an opportunity to set the fire to delay its discovery.

The location on the property is also a key giveaway. It has to be a location that will allow the fire to spread. Fire burns upward and outward, so setting the fire at the lowest point in the building provides a better opportunity to cause more damage. This does not eliminate the roof of the building as a location. Attic spaces usually have no windows, which will delay discovery. This type of location is also hard for the firefighters to get to, so this delay will also add to the desired destruction.

Another decision on the location of the fire is available fuel. To make the fire look accidental the fire may be set in the area with the most volatile fuel. The fuel load and heat release rate of the fuel may make it the best bet to get the fire going and cause as much destruction regardless of who sees it and how fast the fire department gets to the scene.

■ Note
When considering the location, also consider the ability of the fire to spread.

The most obvious of location issue is multiple set fires. If there appears to be two or more area of origins that cannot be explained, the investigator must consider that the fire was set in more than one location.

When considering the location, also consider the ability of the fire to spread. Check to see whether any of the suppression systems such as sprinkler systems were shut off, fire doors were blocked, or detectors were disconnected. It would be typical for an arsonist to disarm anything that could slow the fire or give away early detection of the fire.

Fuel, Trailers, and Ignition Source

■ Note **The intent of most arsonists is to make the fire look accidental while doing the most damage.**

The intent of most arsonists is to make the fire look accidental while doing the most damage. The way to make this happen is to use the fuels that are normally found in the structure. The debris left behind is then what someone would expect to find in an accidental fire. Using some of these fuels to move the fire from one part of the structure to another takes creativity and may go undetected depending on the amount of destruction and the amount of damage done by the use of suppression hose streams.

An arsonist who is in a hurry and needs rapid damage and does not care that it will look incendiary uses an ignitable liquid such as gasoline. Of course, there is that element in society that will use the gasoline anyway, not even thinking about being caught or being identified. Other individuals will use unique methods that involve mixing chemicals to allow for a delayed ignition. Some of these chemical mixtures have an extremely high heat release rate that can ignite any common combustible.

■ Note **The investigator needs to learn to recognize the residue of various items that could be possible heat sources.**

This means that the investigator needs to learn to recognize the residue of various items that could be possible heat sources. This includes ordinary items typically found in any residential or industrial occupancy. It is also beneficial to recognize where these items were found. Residue of a road flare would not be unusual in a garage toolbox or the back of a vehicle but is unusual if found on the kitchen floor.

This is another situation when a strong background of working fires comes in handy. The average firefighter has learned to recognize different items in their burned state. To confirm or refresh your knowledge it is always good to be involved in training sessions where items are set on fire in various scenarios, and then you can do a scene examination. For firefighters, following such an exam, they get to look at a video and still shots of what the room looked like before the fire. Another tool is to simply subject certain items to fire and examine the residue.

Cigarettes around the fire scene can be found by the filter residue, but the filter itself at the area of origin with any disturbance of the debris may disappear altogether. The match, paper or wood, may be difficult to find, but it has not been unusual to find residue of either. Again, a lot depends on the amount of debris at that exact site and how much it has been disturbed by fire streams or salvage.

The evidence of an electrical arc from a copper conductor could survive a typical residential fire, but aluminum will be more apt to melt. The condition of the wire may indicate an electrical short, but what must be determined is if the short caused the fire or the fire caused the short.

Several common materials when mixed together can cause sufficient heat to act as a heat source. Depending on the mixture, they can act as a delay device as well. Each leaves various residues. These chemical mixtures should not be listed in texts such as this book but can be made available to investigators by the state or local training authority or from specialized training events put on by federal agencies such as the ATF or the FBI.

Some explosive items, such as black powder, merely burn when not confined and the fire spreads rapidly making it good as a trailer. They make a burn pattern across the surface on which they are poured. Fuses for model rockets will burn (like the fuse on a firecracker) and leave a scorch mark. In contrast, safety fuses used in modern blasting leave a waxy residue where they are stretched across a surface.

■ Note
What you find at the scene will dictate the additional field investigation required.

What you find at the scene will dictate the additional field investigation required. If there are traces of what is believed to be an ignitable liquid such as gasoline, as shown in the pattern in **Figure 16-4** and **Figure 16-5,** the investigator may want to visit local gas stations to see if anyone recently purchased fuel in any containers. It would be even more beneficial to the investigation to find the persons who purchased the gas using credit cards. Videotapes from the station may be of assistance. But because these videotapes are reused on a regular basis, it is beneficial to visit local stations as soon as practical.

If chemicals are used, then a visit to other local stores is recommended depending on the chemical. If there is a purple residue, possibly potassium permanganate may have been used. This can be bought in large containers from a

Figure 16-4 *An ignitable liquid pour down a set of steps. The separate char on the left side of the lowest step is from a cardboard box that was left on the steps.*

Figure 16-5 *Gasoline splashed onto walls from 1-gallon milk jugs. Notice the rundown patterns in the center, and note the whisping of smoke from the burning fuel on the pattern on the left of the photo.*

local pharmacy. If there is evidence that a model rocket motor was used, then a visit to the local hobby shop may prove interesting as well.

Black powder can be purchased at the local gun shop or sportsman store. Pyrotechnic cord and safety fuses are available from various suppliers, but it may be just as useful to check to see whether there have been any thefts or vandalism at local facilities such as explosive contractors or rock quarries. Although there are no guarantees on any of these ventures, not to visit would be an opportunity lost for gathering potential clues.

Sometimes looking for the unusual is important. For example, while sifting through the debris in a trashcan located at the area of origin, the investigator noticed that a burnt soda can was heavier than normal. Peeling back the metal revealed residue of solid resin, as shown in **Figure 16-6,** typical of what is used in fiberglass work. The business that experienced the fire had nothing whatsoever to do with fiberglass. Of interest was that the owner of the facility that burned was the last one in the room of origin. Even more interesting was that when the owner showed up, he was in his truck for his second business, which advertised fiberglass repairs on the side of his truck. Resin when mixed with a catalyst produces an exothermic reaction that is capable of self-ignition as well as igniting nearby combustibles such as a trashcan of mostly paper products.

Figure 16-6 *A soft drink can that contained the residue of activated fiberglass resin. The use of the resin and the catalyst, methyl ethyl ketone, created an exothermic reaction that was the heat source for the fire.*

CHALLENGE THE UNUSUAL

■ Note
Keep an objective and open mind on what to look for in any fire situation.

Investigators should make an extra effort to see what is there and not what they expect to see. They must be able to judge the situation and not compare it to the norms of their life. Both fire and police personnel should have no problem with this factor. In their work, they have seen how many people live in many different ways. Bottom line, keep an objective and open mind on what to look for in any fire situation.

■ Note
One thing to look for is what is *not* there.

One thing to look for is what is *not* there. Sometimes it can be obvious, such as shown in **Figure 16-7.** As you can see, the smoke produced a great V pattern and the flame impact on the wall, which burned clean, is an inverted V. This is typical in the beginning stages of a fire. This was exactly what the engine company found upon arrival. It was alleged that the young man at this house had a metal cooking pot full of gasoline and he was ladling gasoline and pouring it into wine and beer bottles. Neighbors were watching, but no one bothered to call this strange behavior in to the police. According to a neighbor, the young man stopped his work to sit back and light up a cigarette. The neighbor thought for sure there would be a big explosion, but nothing happened at first. Then, the young man, with the cigarette between his fingers, picked up the ladle with the same hand and dipped it into the metal pot. There was a flash and the pan was on fire. The neighbor ran in her house and called 9-1-1.

Figure 16-7 *Burn pattern on the wall where a pot of gasoline ignited. Notice the V pattern from the smoke and the inverted V from the flames that burned the wall clean.*

When she came back outside and looked across the street, the man was gone, the pot was gone, and so were the bottles. The fire department arrived and checked around but found no hazards. However, there was a strange sound as they walked across the porch. Most responders on the scene were ready to clear up, but the mystery was too much to not look further. Someone finally looked under the porch and there was the young man, hiding in a space he could just barely fit in. When someone stepped on the porch, the boards depressed enough so that it was pushing the air out of his lungs and he was making a small gasping noise. Surprisingly enough, he only singed the hairs off his hand and arm. He was cuffed and hauled away for making explosive devices.

Figure 16-8 *Sooted back door with dead bolt. This door requires a key for entry from both sides; it was locked at the time of the fire.*

It is also good to listen intently to the stories given by victims. In one case, the homeowner went to great lengths to describe how he just barely got out alive by running out the back door. But, as shown in **Figure 16-8,** the door has a key deadbolt and it is still secure as is the lock on the door knob. There is one front and one back door on the structure; no one went out the back door during this fire. Such suspicious stories should lead to another line of questioning by the investigator.

One structure in a rural area burned to the ground and was not investigated until a claim was filed with the insurance company. Seems the house was purchased for $30,000, but the claim produced by the policy holder asked for $100,000 in damages because the homeowner had done so many improvements. Listed was the fact that all iron pipe was removed and replaced with copper, and the plaster walls were removed and all new sheetrock installed and painted. The appliances were all replaced and new floors installed.

Figure 16-9 *The building owner tried to defraud the insurance company by claiming the structure more than doubled in value as a result of new pipes, walls, floors, and so forth. But evidence on the scene proved otherwise.*

When the investigator arrived, the two-story structure was on the ground. The tallest portions of the structure were the stove and refrigerator, which were clearly made in the 1950s. The debris consisted of plaster from plaster and lath walls; not a single piece of sheetrock residue was found. Just as important, as shown in **Figure 16-9,** all the pipe was iron. A detailed investigation revealed that the fire was accidental: When the plumber's assistant started to redo the plumbing, he accidentally set the interior walls on fire with his torch. He could not put it out, and by the time he drove to a phone, and the delay allowed the house to burn to the ground. The owners wanted to cash in on their policy, but their claim was denied by the insurance company and they were out the value of the structure. They were fortunate that the local authorities did not want to charge them with insurance fraud.

■ Note
Make sure you are legally on the scene and have a valid reason for looking throughout the structure; just because you have a badge is not enough.

CONTENTS

The contents of a burned structure hold a wealth of knowledge. The first priority is to make sure you are legally on the scene and have a valid reason for looking throughout the structure; just because you have a badge is not enough. In searching the structure, you may find some interesting facts that are not supported by the occupants of the home.

Figure 16-10 *Empty cabinets in an occupied home is suspicious. The fact that all the doors were left open prior to the fire may suggest that the cabinets were emptied prior to the fire.*

For example, in one case the family said they had just finished breakfast when the smoke detector went off, and they just barely got out of the house with their lives. During the interview, they described that they had breakfast in the kitchen, yet there was no fire damage in the kitchen and not one ounce of food could be found in the structure. As shown in **Figure 16-10,** the cabinets are empty and the doors were even left open before the fire started. In contrast, at another fire scene, the closet shown in **Figure 16-11** was full at the time of the fire, but the contents were removed prior to the investigator taking his photos. As a seeker of truth you must look for all evidence to prove guilt and innocence.

Contents in closets that experienced fire damage may not still be hanging but may have dropped to the floor. Always check the floor debris to determine the closet's contents. In closets with just smoke damage, it is a little easier to tell whether there were hanging contents at the time of the fire. The sooted bar will have protected areas from the hangers. In **Figure 16-12,** the three hangers have been moved aside to show the protected area. This by itself has no value as to the nature of the fire, but it is one piece of data that will be used in the final determination of the fire cause and origin.

■ Note

The contents of the structure may indicate other activities that may be unrelated but that may show that an occupant has a propensity to commit other crimes.

The contents of the structure may indicate other activities that may be unrelated but that may show that an occupant has a propensity to commit other crimes. A small fire occurred in a back room that by all accounts was accidental in nature. However, the contents of this residential back room included more than 50 traffic and property signs that were stolen. The theft of the street signs,

Figure 16-11 *Closet with clean areas showing that it had contents at the time of the fire but that they were removed after the fire.*

Figure 16-12 *Closet rod showing that only three hangers were in closet at the time of the fire. The hangers have been moved aside to show the protected area.*

Figure 16-13 *Some of the stolen street signs found in the home of a fire victim.*

some of which are shown in **Figure 16-13,** constituted a felony for this 20-year-old tenant.

Some of the contents can tell a possible story in relation to the fire. A restaurant fire was clearly incendiary in nature. There were no food stocks for the following day and even staff had questioned why no food deliveries were made that would be necessary to open the next morning. At the scene, the office was clearly meant to burn because gasoline had been poured over the open drawers in the desk and file cabinet, but for some reason the office door was closed. There was plenty of soot from the fire in the kitchen, and it was obvious suppression forces had been in the office because things were moved around a little from when they went in to open a window for ventilation.

On top of the desk the material had been moved around a little because the sooting was uneven. In plain view was a book titled *How to Beat the Bill Collector,* as shown in **Figure 16-14.** This book by itself did not mean much, but several other papers were overdue notices. The investigator on this particular fire was a private fire investigator contracted by the insurance company. As such, he was able to take photographs of all the contents of the structure. In addition to the rights of the insurance company, the owner had given permission to do the investigation.

To give some closure to this particular case, there was a lot of evidence that the owner set his own place on fire. But it was just not enough to deny the claim or for local authorities to make an arrest. However, the key witness was the young

Figure 16-14 *Material on the desk covered with gasoline included a book titled* How to Beat the Bill Collector. *Having this book is not evidence by itself but does indicate further investigation may be necessary.*

wife of the owner. She eventually divorced him and contacted the county about 5 years after the fire, stating that she was willing to produce additional evidence and testify against her ex-husband. It was a very short deliberation by the jury, who found him guilty as charged.

Sifting a Scene

■ Note
Sifting a fire scene is a daunting task but can be handled well enough with time and a good hardworking staff.

Sifting a fire scene is a daunting task but can be handled well enough with time and a good hardworking staff. The first decision to make is the size of the openings of the screens. It will completely depend on what you expect to find. Screens can be commercially purchased or they can be made on the site with basic, simple supplies. Intricate models can be made that will swing back and forth, but the basic, simple plan calls for screens that are 4 foot square set on two sawhorses. The screens can be picked up and sifted by a team of two, as shown in **Figure 16-15,** or the sawhorses can be balanced on just two legs and rocked back and forth so that they take the weight of the sifting screen. A third person can shovel the debris onto the screen. The fourth person in charge documents and tags everything that is found in the process.

The reason for sifting is to look for evidence, such as a weapon or a device. The process is no different than what was discussed in Chapter 4 where the insurance investigator will be searching for contents for a possible fraud case. In either case, the entire scene must be sifted to clearly indicate that the job was done properly. The first concern has to be for safety. Although this may be a cold scene, dust will rise as the debris is disturbed. Thus, it is critical that everyone on the scene is properly protected by respirators that have been properly fit

Figure 16-15 *Sifting a scene: two investigators handling the screen, while a third shovels in the debris. The fourth person documents, photographs, and packages any evidence found. Although this is not a task undertaken by the first responder, first responders are often drafted for this duty because of their basic knowledge and to provide training opportunities as well.*

tested as per OSHA regulations. Crews must be dressed for the environment. The presence of hazardous materials will dictate whether specialized clothing is needed. For most residential structures, any clothing that covers most of the body is acceptable, especially a good pair of gloves.

Special precautions must be undertaken. Any time food is to be ingested, the workers must be in a clean environment, and outer clothing should be taken off. Hands must be thoroughly cleaned. In addition, the clothing should not be washed in an ordinary washing machine, which will become contaminated from the dust from the fire scene. Scrubbing with a water hose and brush and washing clothing in a commercial turnout gear washer are safe ways to handle the clothing.

The final result is a scene that has clear signs of the sifting process with various piles of fine sifted material. In adjacent piles is material that was examined and set aside. As an area is cleared off, the floor is examined, and then the screens are moved forward and the next area is sifted in a grid fashion. The process can make it possible to identify any items that survived the fire, such as firearms as shown in **Figure 16-16** or tools and other metal furnishings as shown in **Figure 16-17.** Sometimes jewelry may be found in the debris, as shown in **Figure 16-18.** However, not all jewelry will survive a fire. It depends on the temperature to which the items are exposed, whether they were protected and insulated, and the melting temperature of the materials from which they are made.

Figure 16-16 *Items such as firearms are still identifiable in the fire debris.*

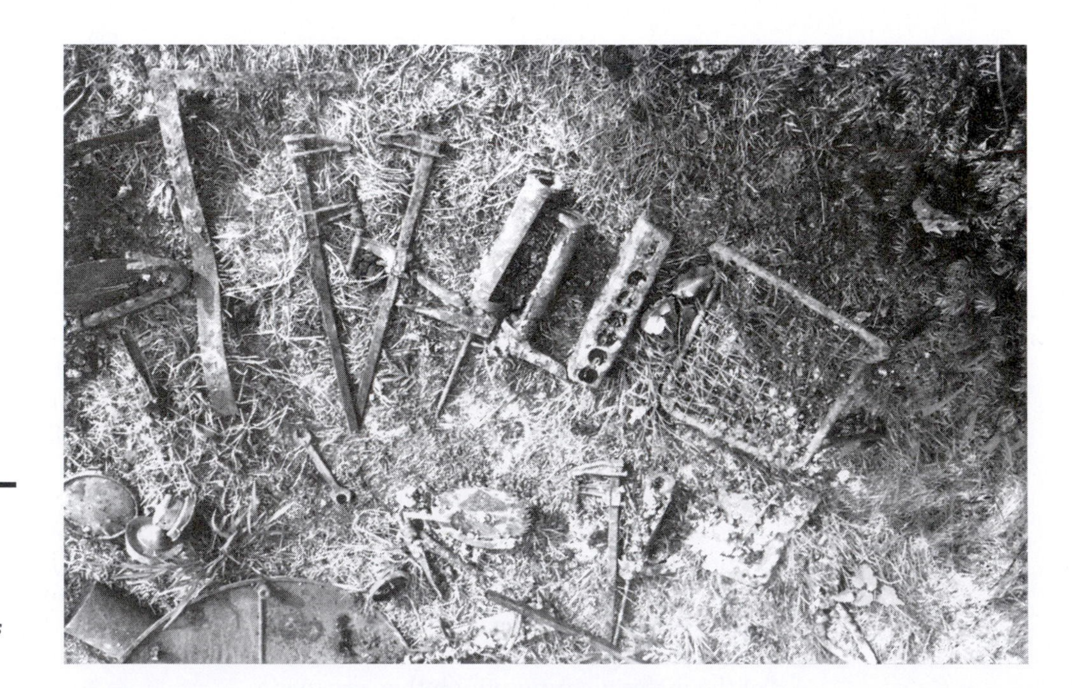

Figure 16-17 *Metal items recovered from the sifting such as tools and kitchen appliances and utensils.*

Figure 16-18 *Ring found in the debris as a result of the sifting process.*

SUMMARY

Arson is a horrendous crime that affects the entire community and taxes the emergency services. There are three elements to the crime of arson. First, something must have burned; second, the burning must have been intentional; and third, the element of malice must be present. Malice is the key indicator to determine whether the fire was an illegal act. As for who sets the fires, it can be anyone you meet on the street, sit next to in the café, or pray beside in church. For whatever reason, people have turned to using fire as a means to an end even though they would never want to hurt anyone.

Motives for setting the fire are far ranging but fall within the six categories: vandalism, excitement, revenge, crime concealment, profit, and extremist. Even though the law may not require the motive to get a conviction, it may well be required to get the jury to understand why an individual would do such a horrible thing.

Many indicators give various clues as to how the fire was started or why the fire occurred. These indicators aid the investigator in finding the evidence not only to determine that the fire was incendiary but also to identify who may have committed the crime.

KEY TERMS

Burglary The entry into a structure or part of a structure with the intent to commit a crime.

Corpus delicti The body of the crime; the physical proof that the crime did occur.

Criminal act As stated in *Black's Law Dictionary,* an external manifestation of one's will which is prerequisite to criminal responsibility. There can be no crime without some act, affirmative or negative.

Deposition Obtaining testimony of witnesses that is recorded, authenticated, and reduced to writing and that can be used in future court testimony. Used as a discovery device by either side in a civil or criminal trial.

Malice Intentionally, willfully, and knowingly doing a wrongful act with the intent to inflict harm on another; recognized by law as an evil intent.

Molotov cocktail A bottle filled with an ignitable liquid that is ignited using a wick in the bottle, reactive chemical coating, or wrapping in a fuel-soaked rag.

Perpetrator The one who commits the criminal act.

Pyromania The tendency or impulse to set fires.

Robbery Feloniously taking any item of value from another directly or in their immediate presence and against their will while using force or fear.

REVIEW QUESTIONS

1. What are the three elements of arson?
2. Define *malice.*
3. What are the six motives for arson as identified by the National Center for the Analysis of Violent Crime?
4. Give an example of all six motives.
5. Describe in detail five indicators that the fire may be incendiary in nature.
6. What are the differences between the terms *arson* and *incendiary?*
7. What is the difference between cooperation and collusion of the government investigator and the insurance investigator?
8. When the contents are missing in a structure fire, does that mean it is an incendiary fire? Explain in detail.
9. What safety precautions should you take if you are to take part in sifting the debris of a residential fire?
10. Why is pyromania not a motive now?

DISCUSSION QUESTIONS

1. Discuss the impact on the fire service when one of its own commits an act of arson. If convicted and after completing any punishment, should that person be allowed to come back as a firefighter?
2. Earlier in this chapter is a story about a grandmother who set a vehicle on fire. Should she be arrested? What good would it serve?

ACTIVITIES

1. The Bureau of Alcohol, Tobacco, Firearms, and Explosives (ATF) has a National Response Team for arson. Go to the ATF website and research the team's duties, abilities, and responses.
2. Research local, regional, or state associations that offer training on fire investigations where you live.

NOTES

1. *Blacks Law Dictionary*, 5th ed. (St. Paul, MN: West Publishing Company, 1979), 102.
2. *Blacks Law Dictionary*, 5th ed. (Eagan, MN: West Publishing Company, 1979), 685.
3. *Webster's New World College Dictionary*, 2nd ed. (New York: Simon & Shuster, 1984), 78, 709.
4. Gordon P. Gary, Timothy G. Huff, David J. Icove, and Allen D. Sapp, *A Motive-Based Offender Analysis of Serial Arsonists.* (Washington, DC: Department of Justice, Federal Bureau of Investigation, and Federal Emergency Management Agency, n.d.), http://www.interfire.org/res_file/fuab_mb.asp (accessed August 27, 2008).
5. Gordon P. Gary, Timothy G. Huff, David J. Icove, and Allen D. Sapp, *A Motive-Based Offender Analysis of Serial Arsonists.*
6. BBC News, "Provisional IRA: War, Ceasefire, Endgame?" http://news.bbc.co.uk/hi/english/static/in_depth/northern_ireland/2001/provisional_ira/1969.stm (accessed August 27, 2008).
7. National Fire Protection Association, *NFPA 921, Guide for Fire and Explosion Investigations* (Quincy, MA: NFPA, 2008), Section 6.2.4.4, pages 921–41.

INVESTIGATIVE RESOURCES

Learning Objectives

Upon completion of this chapter, you should be able to:

- Describe the various resources to keep the investigator safe.
- Describe various investigative tools that are necessary to work a fire scene.
- Describe two types of cameras that can be used on the fire scene.
- Describe the benefits of creating a local fire investigative association.
- Describe additional resources available to the investigator.

CASE STUDY

An entire city block gone. One building, built at the turn of the century, had burned to the ground. It was open on the inside, there were shafts from the basement to the top floor, and it was made of heavy timber construction with most timbers exposed. In the debris, there were indications that some of the timbers were as much as 16 inches wide.

This fire was a great example of the use of outside resources. The investigation division of that locality consisted of two full-time investigators. That area had one of the older fire investigator associations in the state, and they had created a mutual aid pact between each of the surrounding jurisdictions. But even with the mutual aid it was a daunting task. One member of the mutual aid was a Bureau of Alcohol, Tobacco, Firearms, and Explosives (ATF) special agent who was able to offer his agency's assistance.

Within hours, specialized units and personnel showed up from that state and from surrounding states. They were organized into an interview team to blanket the area to search for witnesses with information, a scene team to assist and relieve a tired team of local investigators who had been digging the scene by hand, a document team to photograph and diagram the scene, and a team to concentrate on the business with a specialist in accounting.

A forensics specialist was on scene, and Mattie the K-9 was on a plane on the way from Connecticut. A neighboring police department put their helicopter team on call for the ATF, who used it to document the scene and surrounding area by air. The insurance company sent a contract fire investigator with authority to provide resources to the government if it was necessary and relevant to the discovery of the fire origin and cause.

During morning team meetings, reports were given on the previous days' findings and new assignments were given out. A request was made of the insurance company for heavy equipment to move some of the larger debris from the area, making the scene more secure and giving access to areas that had not been examined.

A crane and additional trucks were on the scene by midmorning. Under the direction of the ATF scene workers, the debris was removed and the scene investigation continued. Mobile command vehicles from a neighboring jurisdiction provided both a command position and a rehab location for those working the scene. The insurance company investigator and an ATF agent were assigned to photograph the scene and the process.

That afternoon, the ATF K-9, Mattie, and her handler showed up on the scene. The dog was put through her off-scene testing phase in which she had to identify which metal can contained the miniscule sample of a hydrocarbon—less than a drop, more of a mere touch of the liquid inside the can, not even enough to hardly be seen. With the command, a wave of the hand, Mattie went into work mode, sniffing and examining each area indicated by the handler. When she came to the can with the hydrocarbons, she immediately sat down, looked at the can and then up at her handler. The handler asked her to show him the sample, and Mattie again put her nose on the can and looked back up at the handler while still in the sitting position. She got

(Continued)

(Continued)

her reward of a few kibbles; Mattie was a food reward dog. She got to eat if she found the sample.

For most of the local investigators that was the first dog they had seen work a fire scene. To make sure Mattie was not taking cues from the investigators on the scene, everyone was pulled back to the outside edge of the building. What the investigators had been working for days now turned into a less-than-1-hour search when Mattie was allowed to start in one corner and given her command. As she approached the area that had already been designated as the potential area of origin, the team maintained their silence, but you could see everyone being a bit more attentive. Mattie stopped, went back about 2 feet, sniffed that area again, and then sat down. Asked by the handler to show her again, she sat down again and pointed her nose to the floor. Samples were taken by the forensic specialist and rushed to the laboratory.

Next morning's meeting revealed that the sample was positive for gasoline. The area where it was found was the office, and per the owners gasoline had never been in that area. The investigation continued, but the perpetrator was never identified. There were a few good leads, but none led to an arrest. Regardless, it showed that ATF was exemplary to have on anyone's scene, and the team building resulting from the local mutual aid program provided invaluable lessons that yielded benefits for years to come.

INTRODUCTION

■ Note **There may be times when the first responder investigator and the assigned investigator may need to work together.**

■ Note **Teamwork is more efficient and productive if the first responder understands all aspects of the full investigation.**

Even though the fire officer may not be able to follow through with the full investigation, it is important for this individual to know about the tools and resources available to the assigned investigator. There are certainly benefits of working with the assigned investigator from time to time, if for no other reason than for the first responder to hone investigative skills. Moreover, there may be times when the first responder investigator and the assigned investigator may need to work together throughout the entire investigation process. This teamwork is more efficient and productive if the first responder understands all aspects of the full investigation.

Resources include safety supplies, tools, and outside agencies available to assist with various aspects of the case. Outside contractors and specialists may also be able to provide resources and expertise otherwise unavailable.

PROTECTING THE INVESTIGATOR

The first resource to be considered is those items used to protect the investigator in carrying out investigative duties. For the fire suppression officer, who is the first investigator on the scene, the available safety resources are limited to full turnout gear used in suppression activities. If a filter mask is necessary and none are available, investigators should always move up to the next level of protection and use the self-contained breathing apparatus (SCBA).

The assigned investigator has several levels of protection available, from coveralls to fully encapsulated suits. The investigator may already have the specialized knowledge necessary to know when to upgrade the level of protection. If not, the department or locality may have hazardous and environmental specialists who can identify the safety requirements of various scenarios and can assist in creating a safety policy that dictates when all investigative staff are to don the various suits or pieces of equipment.

To ensure that all guidelines and regulations are met, it is always best to approach those who enforce the Occupational Safety and Health Administration (OSHA) regulations. Some states have taken on that role and their staff may be found in the state Department of Labor. The location of these divisions can be varied. For example, in California it is located in the Department of Industrial Relations, and in New Mexico in Environmental Department, but the majority use the title Department of Labor. A little less than half of the states and territories have enforcement at the state level. The other local fire investigators may be contacting the regional OSHA offices under the U.S. Department of Labor.

Those who enforce safety regulations include specialists who can assist in the design of a safety plan as well as policies and procedures on fit testing of any breathing respirators or apparatus. They can also provide assistance on using various coveralls and other types of safety clothing. One note on the use of fire helmets during investigations that will be 8 hours or longer: If the risk warrants the use of fire helmets, then by all means use them. But over a long period of time, they add to the fatigue of the wearer. If the risk warrants a lighter type of head gear, such as a hard hat, it may be worth the extra purchase of such equipment.

■ Note
Because the investigator may be digging in debris of unknown contents, the first layer of protection is latex gloves.

■ Note
On the outside of the latex gloves the investigator should wear a good pair of leather work gloves.

Gloves

Gloves must first protect the investigator and the evidence. Because the investigator may be digging in debris of unknown contents, the first layer of protection is latex gloves, which today are available in most fire departments. Spares should be on each engine to be used on emergency medical services (EMS) calls. (See **Figure 17-1.**)

The investigator should wear one layer and, depending on the situation, may want to consider two layers. The reason for wearing two layers of latex gloves is first to protect the wearer; should one layer fail, there will still be protection. The second reason is that if the outer layer becomes contaminated by liquid soaking through the leather work gloves, it can be removed to enable a protected but clean hand to operate the camera. Then, the hand can go directly back into the leather glove. On the outside of the latex gloves the investigator should wear a good pair of leather work gloves that fit tight enough for dexterity but that are not too thin as to fail to protect from broken glass and such.

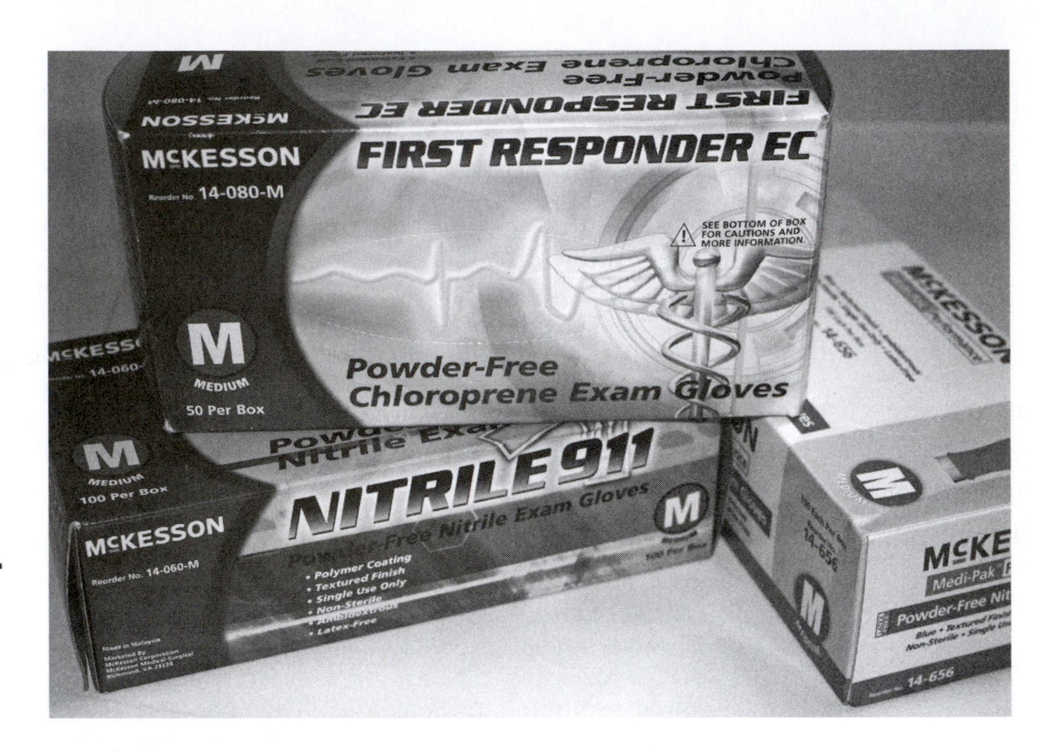

Figure 17-1 *Every investigator should have an ample supply of gloves at each investigation.*

Testing the Environment

The suppression forces should have been testing the environment before leaving the scene. If the investigator does not have the equipment, the investigator could either borrow it from the engine or ask the engine to stay on site, monitoring the environment until the investigator is finished. Depending on staffing, this may not be practical or possible.

Something that should be done, for a multitude of reasons, is to develop a relationship with other area investigators. Many areas have gone so far as to form local investigative or prevention associations to share ideas and resources. If this is the case, then ask for assistance from a neighboring locality.

As soon as the budget allows, it is best for the investigator to obtain a three- or four-gas monitor/detector. At the very least, the detector needs to measure available oxygen and the carbon monoxide (CO) levels. Preferably, it is best to have an atmosphere of 21 percent oxygen, and the CO level threshold should be as dictated by the local labor laws.

TOOLS

Some of the tools discussed in this section have been previously mentioned but are placed here to put into perspective the overall tool needs of the investigator. A key issue is to make sure that the items in the investigator's toolbox are what will be needed at the scene and are in a condition that will not hamper the case but enhance the investigator's capabilities. Much of what is available to the first responder investigator from the engine is inappropriate for use in an investigation with one exception: The engine company officer may want to keep a multitool so that it can be used at an investigation site should it be necessary. This does not preclude it being used elsewhere as long as it is properly cleaned and uncontaminated.

■ Note
The engine company officer may want to keep a multitool so that it can be used at an investigation.

Safety Tools

On occasions, it may be necessary to enter a hazardous environment involving potential ignitable vapors. A word of caution about the tools being used. Even though the tools purchased say they are *nonsparking,* this is not always 100 percent true. Any metal tool is capable of creating a spark,[1] especially if it comes in contact with oxygen-bearing materials, such as rust. It is an extremely risky venture to enter such an atmosphere just to collect evidence. There must be an overwhelming reason to put anyone at risk in such a venture. Instead, maximum effort should be made to make the atmosphere safe.

■ Note
Any metal tool is capable of creating a spark, especially if it comes in contact with oxygen-bearing materials, such as rust.

Nonsparking tools are usually made of brass, bronze, or other metal compounds. They are soft metals, and that means that they can bend or wear excessively. Some safety operations recommend using stainless steel tools, which are less apt to create a spark but maintain their structural integrity.

All tools need to be thoroughly cleaned and dried. Do not use any oil-based lubricants because of their potential of contaminating a fire scene. Even if you do not actually contaminate the scene, using such lubricants can raise doubts as to that potential, negating any value of a positive sample from the scene. As with all tools, it is a good practice to test each tool with a hydrocarbon detector just before using it to prove that the tool will not contaminate a scene or sample being taken.

■ Note
Test each tool with a hydrocarbon detector just before using it to prove that the tool will not contaminate a scene or sample being taken.

The Investigator's Toolbox

Investigators must keep their own separate toolbox for fire scenes. The tools may not look in good condition—there will be rust and corrosion—but these tools will not contaminate the scene. The toolbox from the engine traditionally will have well-oiled tools, which is exactly what they are supposed to be when properly cared for and maintained. The first responder investigator must use extreme caution and not use any tool that may contaminate what may be future evidence.

■ Note
The first responder investigator must use extreme caution and not use any tool that may contaminate what may be future evidence.

Standard tools are needed. An assortment of screwdrivers of various sizes is handy along with pliers and locking wrenches. A socket set in both metric and American standard sizes along with extenders and handles are also useful. The toolbox should also contain a set of Allen wrenches and possibly some open-ended wrenches. Various saws and cutting tools will be handy as well. A good hammer, hatchet, and a small maul will certainly come in handy at several fire scenes. Tape measures and a folding ruler can come in handy as well. Of course, most firefighters like to have a good multitool available as well. Included in the multitool is a knife, which also is a valuable tool; a separate pocketknife can be of assistance as well.

■ Note
When the tools are purchased, they should be washed in a good detergent, and then completely dried.

■ Note
Don't worry about a rusted saw blade—it will still cut and won't contaminate your scene or the sample you are taking.

■ Note
The best tools of all: mason's trowels, one large and one small.

Sometimes it takes a gentle touch. A small whisk boom and even a dustpan can be good tools to have from time to time. Smaller bristle paintbrushes and even a half-inch artist's round paintbrush can come in handy. Be prepared for the brushes to be disposable. If they cannot be adequately cleaned, they must be replaced from one fire scene to the next.

When the tools are purchased, they should be washed in a good detergent, and then completely dried. Same with the toolbox, even though it may be plastic. Because these tools will not be lubricated, they will require more attention than usual; constant care can make them last a long time.

Battery-operated tools are also a great assistance, from power screwdrivers to small skill saws. They can help collect evidence and open electrical panels with ease. These items must also be cleaned and not lubricated. They should have sealed bearings that do not require oiling. The blades on the saws must be scrubbed clean and dried. Don't worry about a rusted saw blade—it will still cut and won't contaminate your scene or the sample you are taking.

The best tools of all: mason's trowels, one large and one small. These tools can act as a miniature shovel to clear away debris, layer by layer. They have pointed ends that you can use for exploring and picking out small items. Trowels can be used in a way that is similar to an archaeological dig, removing each layer to see what fell last, until you get to the bottom, which will identify what fell first, at that location.

Larger Tools

hux bar
one brand name of a firefighting multitool that comes in different sizes but that averages 36 inches in length; different tool heads on each end can be used for prying, poking, striking, and opening hydrants or closing gas valves

This list will be obvious. Shovels: Flat blade, pointed blade, and a large scoop shovel are all convenient tools to dig a fire scene. If working the area of origin, do not use the shovel from the engine. Not only could it carry contaminates from a previous fire, but it may have been stored near the gasoline-powered tools located in the storage compartments. This could contaminate the scene, rendering any evidence inadmissible in a court of law.

A fire axe is necessary for larger tasks, and the pick end is good for making drain holes to run off water from a floor. A **hux bar,** or similar tool, can be handy for prying and forcing and has been known to be used as a prop from time to time. Rakes and hoes may sound like farm tools, but they can come in handy for searching larger areas and will cut down on the fatigue of the investigative team if it is a larger scene.

■ Note
A good, strong, dependable flashlight is absolutely essential.

LIGHTING

A good, strong, dependable flashlight is absolutely essential. But one flashlight will not provide all the lighting necessary for a thorough investigation. A generator with floodlights will certainly come in handy to light up the entire scene.

■ Note
If a gasoline generator is to be used, it must be kept well away from the potential area of origin.

This will also be necessary during the day unless the scene is under open skies. If a gasoline generator is to be used, it must be kept well away from the potential area of origin. All cords and light fixtures should be examined and tested with a hydrocarbon detector to ensure that they will not contaminate the scene.

LOCATION

Technology can help identify the exact location of each piece of evidence. If this is an important case where the location of multiple pieces of evidence needs to be documented, there are tools that may be available from the local law enforcement agency to do this.

Today there are specialized investigators who carry specialized tools to give an exact latitude and longitude of each point at a fire scene. Vehicle accident investigators have had these tools available for marking items for future testimony in court. Depending on the need for this same type of testimony, they may make themselves available to come on your scene, take each piece of evidence you tagged, and create a computerized report showing the location of each and in some cases a diagram with each item posted. To maximize their time, prior to their arrival make sure you have identified everything you want plotted.

■ Note
A compass is essential to make sure the investigative diagrams are oriented correctly to magnetic north.

At the very least, a compass is essential to make sure the investigative diagrams are oriented correctly to magnetic north. Simple GPS equipment can help as well to give an approximate position. GPS equipment can be found at many retail stores with handheld units sold at most sport, hunting, and outdoor enthusiast retailers.

CAMERA

The use of photographic equipment is discussed in the next chapter. The type of equipment can only be described in general terms and the intent of this section is not to make someone an expert on camera use. Many forensic schools have as much as 30 to 40 hours of classes dedicated to photography alone, so really a text unto itself is needed to cover all the topics.

■ Note
The first responder can make use of a good quality point-and-shoot camera with a minimum of 8-megapixel photos and at least 3× optics.

The age of print film is mostly behind us now with the advent of new technology of digital photographs. This is primarily a result of the boom in technology in this area. By the time we list the latest and greatest, an even better model with more features will most likely be on the market.

The first responder can make use of a good quality point-and-shoot camera with a minimum of 8-megapixel photos and at least 3× optics. Let's break this down. The photos from most cameras at 8 megapixels, set on the best quality, will produce photographs that can easily depict discernable burn patterns and take close-up shots that will be crisp and clean. They can also be blown up large enough to display in court with no discernable degradation. The 3× is a standard zoom that will magnify the shot sufficiently to help get photos from across a room that may not be safe to enter. The memory disc should be at least 1.0 gigabyte to allow for sufficient photos at most fire scenes. Photos can then be downloaded to a computer and saved.

■ **Note**
The most limiting factor for any point-and-shoot camera is the flash.

The most limiting factor for any point-and-shoot camera is the flash. Under normal circumstances, the flash will hit most objects and reflect back to the camera to give a quality reproduction. However, most fire scenes are soot black. Not just that, sooted surfaces have a tendency to absorb light rather than to reflect it, yielding darks shots in many cases.

■ **Note**
The assigned investigator should have a digital single-lens reflex (SLR) camera.

In contrast, the assigned investigator should have a digital single-lens reflex (SLR) camera with a larger flash attachment that accepts various lenses and has manual settings that allow the photographer to change the depth of field and lengthen the amount of time the shutter is open, allowing in more light. This camera is essentially the same as a sophisticated film camera with digital components added.

■ **Note**
Fire investigation photos should have as much of the scene in focus as possible.

The depth of field is important. Fire investigation photos should have as much of the scene in focus as possible. The depth of field is how much the camera will have in focus for any given shot. If taking portraits of an individual, it is best to have a narrow depth of field where the tip of the nose to the earlobe is in focus and the background is slightly blurred (out of focus) so that the viewer can concentrate on the person in the picture, not the background.

The simple rule is the smaller the opening of the lens, the more the depth of field, but the less light that will get in. Thus, the attached flash will most likely flood the area with light to adjust for the smaller opening in the lens, giving a well-lit photo with everything in focus.

One of the best parts of using good quality digital cameras today is that the investment in quality equipment will save money over the use of film type cameras. The number of photos to be taken is limitless. Whereas before the number of photos was dictated by the budget for film canisters for the camera and the only way to see the results was to pay to have the film processed.

A major metropolitan department took their entire budget for film and developing for one year and used it to replace all film type cameras with midpriced digital cameras with extra memory cards, a dedicated computer for viewing photos and choosing which to print, a high-end professional printer, and enough photo paper for 3 years. The next year they were able to replace all the midlevel cameras with top-quality cameras, and they still saved money. Today their fiscal need for photography is a fraction of what it used to be for film and processing in the past.

Bottom line, digital cameras provide better quality of photographs, unlimited quantity of photos, and storage of photos with no degradation so that they can be used for years to come. Printing of photos is only necessary for court or press events. Moreover, money might be left over to spend on safety equipment.

■ **Note**
The National Incident Management System is a template for all aspects of emergency services, including the investigative scene.

■ **Note**
Individuals should not enter an unsafe area, but if they must, then it should never be when alone on the scene.

INCIDENT MANAGEMENT

The National Incident Management System is a template for all aspects of emergency services, including the investigative scene. Accountability is essential to safety because it allows for the tracking of all personnel at the scene. Even when the emergency units clear and leave the scene, the lead assigned investigator should track the location of all personnel who stay to assist. Individuals should not enter an unsafe area, but if they must, then it should never be when alone on the scene. There must be a communications system on the scene to call for help should that be necessary.

It is a fact that the investigator may end up on a fire scene alone from time to time. Actually, because of most localities' limited workforce, this happens more often than not in most jurisdictions. When this happens, it is essential that the investigator let someone know the location of the scene and his or her anticipated time of return.

ADDITIONAL EXPERTISE

Because prevention offices are historically understaffed, it is more important than ever that an association be created with neighboring jurisdictions. Some of these associations are limited to strictly those with police powers. This allows for the sharing of information that is allowable from one law enforcement agency to another.

Another type of association opens their membership to both assigned investigators and those in public safety agencies interested in investigations. This brings in the engine company personnel who help to make the first determination of the fire origin and cause. Some associations also include personnel from special investigation units of the insurance industry.

■ **Note**
The purpose of most fire investigation associations is the sharing of information.

The purpose of most fire investigation associations is the sharing of information. It can be surprising when two jurisdictions find that they have similar incidents and the same names start popping up on suspect lists. Arsonists do not abide by jurisdictional lines and jurisdictions working together can tear down those lines and create an atmosphere that is more apt to catch these individuals and put them away.

■ **Note**
A secondary benefit of having a fire investigation association is the creation of mutual aid between the jurisdictions.

A secondary benefit of having a fire investigation association is the creation of mutual aid between the jurisdictions to be there to help each other when needed. There has to be a balance so that one jurisdiction is not tagged to do more than their share of the work. Sharing can be a team showing up to work a large scene or just one investigator being called to assist the first responder investigator in a neighboring jurisdiction.

The third benefit of a local fire investigation association is training opportunities. With a team, formal training events can be established. Depending on the size

■ **Note**
The third benefit of a local fire investigation association is training opportunities.

of the association, these training events can even take place monthly. Instructors can be from each of the localities or can be other investigators from around the state, from state investigative agencies, or from federal agencies such as ATF. In many instances, depending on whether there is a local office, the ATF agents assigned to fire investigations may even join the local association to be part of the team.

Forensics

The assigned investigator will be adequately trained in the use of fingerprint tools for the discovery and collection of latent prints. This will be in the form of using powder for dusting the print, fuming to develop the print, as well as chemicals to spray on prints located on paper.

There is also new technology being studied. One such process being researched by the Virginia Fire Marshal Academy is the use of liquid latex in soot removal to uncover potential prints. With an oily soot covering walls and furnishings, it is impossible to use any existing method to identify or lift the prints. The process is to spray liquid latex on the walls or furnishings, let it dry, and then lift the latex, taking the soot along with the latex. Then, conventional powders and chemicals can be used to locate and lift the prints. To date the tests have been successful.

■ **Note**
The Federal Bureau of Investigation has a great laboratory and personnel who can assist with a major investigation especially if it involves other crimes.

There are many other forensic tools and processes at the investigator's disposal: collection of residual blood for serology, tool mark impression examination, paint chip examination, and of course the debris examination for trace residue of a petroleum distillate. The local, regional, or state lab can provide many services. It is incumbent upon the assigned investigator to become familiar with the appropriate lab, its resources, abilities, and procedures.

STATE AND FEDERAL RESOURCES

■ **Note**
The Bureau of Alcohol, Tobacco, Firearms, and Explosives is an outstanding resource. They have a National Response Team that comes with specialized equipment and plenty of resources for any major fire event.

The state may have resources that can be of benefit. Investigators from the state police may be able to assist on the scene or with follow-up interviews, background checks, or polygraph testing. Specialized forensic technicians can assist with the evidence as well.

The state forensic laboratory can provide a wealth of knowledge and assistance. Most important, they can give advice and recommendations for taking better samples in a manner that will aid them in their analysis. They are also a training resource for the local association. However, they do not often have the resources to send personnel to a scene to lend assistance.

The Federal Bureau of Investigation has a great laboratory and personnel who can assist with a major investigation especially if it involves other crimes. From time to time, they also offer specialized training at their facilities.

The Bureau of Alcohol, Tobacco, Firearms, and Explosives is an outstanding resource. They have a National Response Team that comes with specialized equipment and plenty of resources for any major fire event. ATF agents assigned to work fires and bombings have their own training academy that follows the national standards. Agents completing their training are referred to as CFIs, Certified Fire Investigators. Not only are they trained to look at a scene by themselves but as a team, each knowing the others' duties and each having the ability to trade off responsibilities as necessary.

Other Resources

■ Note
Polygraph examiners can be a great resource to assist with the primary suspect.

In addition to specialized forensic and fire investigative resources, there is a wealth of resources that can assist with building a case when working an arson scene. Both ATF and FBI have specialist, such as certified public accountants (CPAs) who can look at the fiscal records of a business to see if there was a financial motive for the fire. Briefly mentioned earlier, polygraph examiners can be a great resource to assist with the primary suspect. Another way to use polygraphs is to do an elimination process if there are multiple suspects, but this should be a last resort because it takes considerable time and effort.

■ Note
If hazardous materials were involved in the fire, many states have environmental agencies with enforcement divisions that can assist.

If hazardous materials were involved in the fire, many states have environmental agencies with enforcement divisions that can assist with identification of unknown chemicals or substances. They can also identify whether that particular business was targeted by state or federal agents for environmental violations that could result in substantial fines, penalties, or criminal charges. Such potential actions could be a motive for the fire, to cover up or eliminate evidence. Engineers can assist by giving information on structural integrity of the remaining fire scene for safety issues. They can also evaluate the structure's condition prior to the fire. Engineers are part of some building official offices, state university departments, or may be hired by the insurance company and agree to be of assistance.

The investigator may develop other resources for future reference. If there is a college in the locality, the chemistry teaching staff can be a wealth of knowledge. If there is an arts program at the college, it may provide training or be a resource to help with the use of photographic equipment. Local sprinkler contractors can provide information on both standard and specialized sprinkler systems.

■ Note
A local building contractor can provide training opportunities for the investigator and first responder investigator to see structures under construction to understand what is normal and what may be unusual.

A local building contractor can provide training opportunities for the investigator and first responder investigator to see structures under construction to understand what is normal and what may be unusual at a future fire scene. The same can be said of a local electrician, who can help train investigators on basic electricity as well as show investigators electrical components as they are installed in both residential and commercial properties.

DETECTING ACCELERANTS

The term *accelerants* needs to be better defined. An accelerant is, well, something that accelerates the fire, speeds up the burning process. That is a simple definition, but the point is that the presence of an accelerant is not proof of an intentionally set fire. A playing child knocks an oil lamp off a shelf, where it breaks and spills the oil, an accelerant, on the floor. If the lamp had been lit, it may ignite the oil on the floor. The fire would be an accelerant spread fire but not necessarily arson. On the other hand, an arsonist, seeing an oil lamp, picks it up and intentionally throws it on the floor, where it breaks, spreading the oil on the floor. Now the accelerant is part of an intentionally set fire and part of the evidence that will be used to hopefully get a conviction. If other indicators, clues, evidence point to an intentionally set fire, you need to collect a sample of the debris to prove the accelerant was present. To obtain the best sample to send to the laboratory for analysis, you can use an accelerant detector.

■ Note The presence of an accelerant is not proof of an intentionally set fire.

■ Note To obtain the best sample to send to the laboratory for analysis, you can use an accelerant detector.

Even though the investigator gets a positive indication from a hydrocarbon detector or an accelerant dog at the scene, the laboratory results may come back as negative for the presence of an accelerant. It does not mean that an accelerant was not used, but that no residue could be detected in the sample provided. The K-9 and many field hydrocarbon detectors are more sensitive to the presence to an accelerant than the laboratory equipment. However, the detectors cannot identify what it is detecting. The patterns themselves can be introduced as part of the investigator's process in coming up with the final hypothesis.

Hydrocarbon Detector

■ Note The hydrocarbon detector is a device that can discern the presence of a hydrocarbon fuel at a minute level.

The hydrocarbon detector is a device that can discern the presence of a hydrocarbon fuel at a minute level. (See **Figure 17-2.**) By running this device over an area, you can pick up on the presence of hydrocarbon vapors. The detector gives an alert or provides a metered display to enable you to find the strongest sample. The unit needs calibration on a regular basis but is a readily available device to ensure that tools are not contaminated before using at the scene and to ensure that evidence cans have not been contaminated before using.

There are several on the market today. As mentioned before, it is advantageous to use a multi device that can scan for the presence of hydrocarbons as well as read the atmosphere for the lower explosive limit (LEL) should a gas mixture be in the air. This type of unit can then be placed on the engine for the use of the engine company officer, who may also be the first investigator on the scene.

More sophisticated units are available but somewhat out of reach of the municipal investigator because of their high price. The price of these units can range in the tens of thousands to more than a hundred thousand dollars, whereas the ones commonly used by localities range from one thousand to four thousand dollars.

Whether it is the normal model used by local investigators or the high-end expensive models, all hydrocarbon detectors can and will hit on false positives. Only a laboratory can make a definitive determination of the presence of accelerants from the debris sample.

Figure 17-2 *A typical hydrocarbon detector in use by the local fire investigator. This particular unit is mounted on a hand light and is designed to use one or two sensors, which can be plugged into the device as needed.*

■ Note
A properly trained dog for locating accelerants can be a great resource for obtaining the best debris sample possible.

Accelerant Dogs

A properly trained dog for locating accelerants can be a great resource for obtaining the best debris sample possible. (See **Figure 17-3.**) These dogs are presently being provided to localities by one insurance company at no cost or from training institutions for a relatively low cost. There will be additional expenses for cages, modifications to the investigator's vehicle, veterinarian fees, as well as food and other supplies. But, the real cost is in time. This is not a pet but a working dog—*work* being the operative word because the dog will need to be worked

Figure 17-3 *The accelerant dog is a great resource for locating the best sample to send to the laboratory.*

constantly, most times, daily. Even though they are not pets, most handlers take their dogs home and they do become part of the family.

Dogs can be a bonus on the scene. It is interesting to see a dog work a crowd only to have her sit down and point with her nose at someone's shoes. This is a normal occurrence, and the handler understands that hydrocarbons can come from other places other than the scene. It is up to the assigned investigator to do the work necessary to determine whether the individual has anything to do with the investigative scene.

Dogs can make an impressive show by being able to search a room and find that one minute sample placed by the handler before the dog enters the room. This is a test the handler uses to make sure the dog is working and that she is accurate. That is all it is. Some may want to put the dog on the stand to testify that she hit on a hydrocarbon residue in a corner in the room and that that is where the fire started. The problem today is that we do not truly know what the dog identified until a sample is taken to the laboratory and the sample is tested.

■ Note

There is no doubt that dogs are a great asset, one of the best. But this cannot take the place of a thorough, systematic fire scene investigation.

There is no doubt that dogs are a great asset, one of the best. But this cannot take the place of a thorough, systematic fire scene investigation that considers all the facts at the scene, including the sample identified by the dog or an electronic detector.

INSURANCE COMPANIES

Throughout the text, the role of the insurance company has been discussed in one manner or another. An insurance company may send one of its own investigators or contract a private fire investigator to look at the scene to determine the cause of the fire. The reason for making such an investment is to determine whether the fire was accidental, and if so, who or what was the responsible person or thing. The second part of that assignment is to see whether there is a possibility of recovering the funds they will pay to the insured through subrogation against the responsible party.

Of course, if the insured was the one to accidentally start the fire, then there is no such recovery action. People have insurance sometimes to protect them from their own mistakes. The same goes for a child who accidentally or out of curiosity sets a fire; the insurance company in a standard policy is usually there to cover the loss should that be the case.

At the same time, the insurance fire investigator may determine that the fire was incendiary in nature, and of course the insurance company would want to know if its insured was responsible for this act; if so, the insurance company can then deny the insurance claim based on the terms of the policy.

■ **Note**
As a resource, the insurance company can be of value because it has funding not normally available to the locality.

As a resource, the insurance company can be of value because it has funding not normally available to the locality. Should the structure be on the ground with the scene completely covered by the metal roof, the insurance company investigator has to get it uncovered to work the scene, so the insurance company may pay for the heavy equipment to uncover the scene. Depending on the situation and the local rules, the insurance investigator and the assigned investigator may then work together to identify the area of origin and the cause. But each individual must make his or her own determination based on the facts discovered in the investigative process.

■ **Note**
Under arson immunity laws, the insurance companies usually must turn over any information uncovered if a fire is determined to be incendiary in nature.

Under arson immunity laws, the insurance companies usually must turn over any information uncovered if a fire is determined to be incendiary in nature. They furthermore will be able to turn over additional information from their own investigation, such as finding the contents of the home not in the debris but in the local storage unit, where it was placed prior to the fire.

SPECIALIZED INVESTIGATIONS

Situations may arise that will require specialists to assist with the investigations. Bomb devices or postblast scenes may require the assistance of the bomb unit from local or state police agencies. Federal agencies such as ATF also have bomb specialists who can assist along with their laboratories that can examine debris residue from the blast.

Wildland fires create unique scenes. If the assigned investigator is not experienced with locating the area of origin or examining the wildland fire scene, there are resources available to assist. Most state forestry agencies have specialists in wildland fire investigations. The National Wildlife Coordination Group (NWCG) has created training programs and documents for those wishing to be certified as a wildland fire investigator.

Hazardous materials also provide unique situations that require specialized attention. Most investigators will be familiar with the chemicals and hazardous materials found in residential occupancies and small businesses. But they may not have the expertise to handle situations where large drums of an unknown chemical are found at the area of origin. The local or regional Hazardous Materials Response Team could be of great value in taking a sample and processing it for laboratory analysis to identify the product. They can also provide other services, from decontamination of the investigators entering the scene to acting as safety officers at the event.

SPECIALIZED INFORMATION

Using local resources and personnel with expertise is another tool for the investigator. However, even specialists may need to look up information that is not readily available in their resource manuals. That is where the Internet is the next valuable resource investigators have at their disposal. Every investigator should develop a list of Web sites that can provide specialized information, and investigators should be familiar with search engines and various parameters for making an efficient search on the Internet.

SUMMARY

The first resource to consider are those that help protect the fire investigator on the scene. This starts before the fire occurs. Policies and procedures must be created to ensure that proper equipment and proper fitting of respiratory equipment are available. Several resources are available to make sure this is done properly. The same goes to make sure there are proper tools including electronic devices to monitor the atmosphere.

Setting up a local fire investigation association allows neighboring jurisdictions to band together and pool their resources for larger fire events. The association can include assigned fire investigators, first responders, local police, and investigators from the insurance industry.

Federal resources (such as ATF support) can be assets in the form of providing trained certified fire investigators or helping with training events for local investigators. A major response from a full investigative team such as the ATF National Response Team can be invaluable. Not only can they help with the scene, but they can offer other expertise in the form of fiscal experts and engineers.

KEY TERMS

Hux bar One brand name of a firefighting multi-tool that comes in different sizes but that averages 36 inches in length. Different tool heads on each end can be used for prying, poking, striking, and opening hydrants or closing gas valves.

REVIEW QUESTIONS

1. What federal regulations address the safety of the firefighter in the work place?
2. Are brass tools 100 percent effective in never creating a spark?
3. Describe the protection that should be donned to protect the fire investigator's hands while conducting a fire scene investigation?
4. What type of cameras are best for the assigned investigator to use at the fire scene and what features make certain cameras better adapted for taking fire scene photographs?
5. What are the three benefits of creating a local fire investigation association?
6. What federal agency has a National Response Team for fire investigations?
7. What resource could a local college or university provide the fire investigator?
8. What resource could a local contractor provide the fire investigator?
9. Why would the fire investigator use the hydrocarbon detector at the fire scene?
10. What assistance can an insurance fire investigator provide the local municipal fire investigator?

DISCUSSION QUESTIONS

1. A K-9 accelerant dog hits on a spot in a structure that has already been labeled an incendiary fire. A sample was taken, but the laboratory said it was negative. Should the dog and its handler be allowed to testify in court to show that an accelerant was at the scene?
2. K-9 accelerant dogs have a reward system that is their incentive to find evidence of an accelerant at the fire scene. One system gives them kibbles; the only way they feed is to detect an accelerant. The second reward system is a play reward system in which, if they find an accelerant, they are given a toy, such as a rolled-up towel or a ball, and the handler plays with the dog briefly. Which system do you believe is the better system to use in training dogs? Why?

ACTIVITIES

1. Research the arson immunity laws in your area to see if they exist, and if so, what they actually allow.
2. Hydrocarbon detectors use electronic sensors to detect the hydrocarbon molecule. Research on the Internet to see what new features are already being developed and marketed in the area of the detection of hydrocarbons.

NOTES

1. Canada Centre for Occupational Health and Safety, "Non-Sparking Tools," http://www.ccohs.ca/oshanswers/safety_haz/hand_tools/nonsparking.html (accessed August 28, 2008).

DOCUMENTING THE SCENE

Learning Objectives

Upon completion of this chapter, you should be able to:

- Describe when and what to photograph to document the fire scene adequately.
- Describe the different types of diagrams and how they can assist the fire investigator.
- Describe symbols used to represent common items used in most diagrams.
- Describe the importance of notes, proper note taking, and final disposition of notes.

CASE STUDY

It was a simple fire, no injuries and the damage could be easily repaired, which was accomplished within a month after the fire. The report was short and simple. The fire was caused by a malfunction of the dryer. Burn patterns were clear and moved upward and outward away from the dryer. No other heat source was in the area and there were no other low burns.

Notes from an interview with the homeowner said that she had washed some sheets and placed them in the dryer. The sheets just came off the bed, so there were no chemicals on them and only laundry detergent was used in the washing machine, again no unusual chemicals. Approximately 30 minutes later, the smoke detector in the hall activated. In searching for the source of the smoke, she found the laundry room on fire. She closed the door, called 9-1-1, and left the house. The notes even quoted the investigator telling the homeowner to go to the neighbor's next time to call 9-1-1 rather than risk her life by staying in a house that was burning.

The typed report said the investigator searched the dryer and found only debris that was consistent with cotton. The laundry room contained one wicker basket that was only partially heat damaged, and burn patterns on the washer clearly came from the dryer. The vent was metal and did not show any burning below the floor where the vent ran under the house to the outside.

Final determination was that the fire started in the laundry room, approximately 18 inches off the floor, and that the cause of the fire was a malfunction in the dryer. The only other heat sources in the room were the washer and electrical service to the washer, dryer, and two outlets. All electrical was examined and found not to be involved in the ignition sequence. Simple, short, and concise.

Five years later, the investigator received a summons to appear for a deposition in the civil suit of an insurance company against a major appliance manufacturer. The investigator was asked to supply a copy of all material on the case before the deposition. So far, this is normal.

At the deposition, the investigator brought the file and was questioned by two attorneys. An engineer from the appliance manufacturer fed questions to the attorneys during the 2-hour deposition. Every word and every note was covered multiple times. At the end of the deposition, the investigator correctly did not waive signing of the transcript but requested a copy for review before signature.

After 6 months, the investigator received a letter of appreciation for the time spent in the deposition and, more important, for supplying the notes and report on the fire scene. Even though the investigator did not find the exact cause of what failed in the dryer, the manufacturer decided to settle. According to the insurance company's attorney, the fact that the investigator could not work off memory was to be expected after all that time. The fact that all answers during the deposition came from the notes and report were sufficient for a settlement.

INTRODUCTION

■ Note
Every fire scene must be adequately documented to allow recall for the investigator.

■ Note
To test your hypothesis you must document all the facts of the case.

Every fire scene must be adequately documented to allow recall for the investigator and, just as important, to allow others to share in the information gathered at the time of the fire. This is critical because it may be several months and sometimes a year or more before a case goes to court.

To test your hypothesis you must document all the facts of the case. To stand the tests of others, you must be able to present all the facts of the case to them for them to make a fair assessment. All the facts of the case include all notes gathered, reports written, photographs taken, and diagrams created. They also include all dates and material that was proven not to be part of the case but that was used in a previous hypothesis that was disproved and rejected. Only with all this information can someone assess all aspects of a case.

The first responder investigator most likely will not make diagrams or keep detailed notes other than what is needed for the fire incident report. If the fire ends up being turned over to an assigned investigator, the first responder's notes may also be turned over to that investigator for the event file. However, first responder investigators should know what is needed for a full investigation so that they can be of assistance to the assigned investigator.

PHOTOGRAPHS

In the past, arguments have been made against digital photography in fire investigation for logical reasons. But times have changed and the basis of the arguments have disappeared. This and the advances in technology have proved that digital photographs are equal and sometimes better in detail as their film counterparts.

Those older arguments were correct in their time. Digital photographs in the early days did not have the clarity or the details that would allow any duplication into larger prints. In addition, there was always the apprehension that photographs could be fixed (doctored electronically) to depict something other than fact. Today the fire investigator as an expert witness can testify that photographs used in the investigation, digital or film, depict exactly what the investigator saw during the scene examination and that they are true and accurate.

■ Note
A major benefit of going digital is the fact that the photo can be reviewed immediately on the camera to make sure it is exactly what is needed.

A major benefit of going digital is the fact that the photo can be reviewed immediately on the camera to make sure it is exactly what is needed. If there is insufficient lighting, the camera can be placed on manual and the aperture opened wider, allowing in more light and getting a clearer shot. When you return to the office, you can refer to photos immediately to write the report rather than waiting for film processing. Of course, this is only possible if your courts allow digital photography. To be sure, contact the prosecuting attorney for your jurisdiction to make sure that digital photographs are accepted in your particular courtroom.

In the past, a photographic expert could magnify a photograph to tell whether it had been altered. This was possible based on the chemical process that created photos, which were essentially series of dots of varying shades and colors. When the photo was scanned electronically, the dots were converted into pixels, which are tiny squares of color. When the photo was enlarged enough, these squares or dots could be identified. Today this is a moot point. It is not so much that the courts have accepted the use of digital photographs as much as it is that the courts have not banned them.

Photography for the First Responder Investigator

While pulling up on a fire scene, taking photographs should be the last thing on the mind of the fire officer. First and foremost, the safety of the crew and the attack must be the priorities. The engineer will be setting up lines and working on water supply. Once water supply is established and all lines are at proper flow and pressure, there may be an opportunity for the engineer to stay at the engine and take a couple of overall photos of the scene, surrounding area, and the crowd while at the same time making sure all things are proper at the pump panel.

■ Note There may be an opportunity for the engineer to stay at the engine and take a couple of overall photos.

Once the fire is under control and the crew is taking care of overhaul, rehab has been established, and all victims have been cared for, including comforting if necessary, the officer can then use the camera to take a few photographs that may come in handy later if the fire scene warrants further investigation.

Overall shots of the property should again be taken because the first ones may have been obscured by smoke and lighting may have been skewed by flames. A point-and-shoot camera can only get so much with the small built-in flash. But even if a photo is dark, keep it. Next, a photo framing each side of the structure is appropriate along with any debris found outside the structure. Moving inside, the first responder investigator can take photographs of each room, even if there is limited damage. Because the lens will not allow a full shot, the first responder can take two photos, each from two opposite corners. Anything that may shed light on the cause of the fire should be photographed.

■ Note A series of photos showing the path of travel should be taken from the furthest point leading up to the area of origin.

If the direction of the fire is obvious to the first responder investigator, then a series of photos showing the path of travel should be taken from the furthest point leading up to the area of origin. If the path of fire travel is not obvious, then the first responder should call for an assigned investigator. While waiting for the investigator, the first responder can take additional photographs of all recognizable heat sources such as the furnace, electrical entrance, range, and portable heaters. Also beneficial are photographs of recognizable fuel sources such as oil tanks, gas cut-off valves, and propane tanks.

■ Note The assigned investigator should start from the beginning and take all photographs needed for the case.

Photography for the Assigned Investigator

Even though the first responder investigator took a series of photographs, the assigned investigator should start from the beginning and take all photographs needed for the case. The photographs from the first responder will be reviewed and added to the report, especially if they show the scene early on and they have some value. All photographs not used still stay with the file.

When using film cameras, the investigator used photo log sheets to document each shot taken; however, log sheets may not be necessary with digital photography. The investigator's policy and procedure book should state whether log sheets are required by the locality.

■ Note
The investigator's policy and procedure book should state whether log sheets are required by the locality.

The first photo taken should depict the area where the fire occurred. It should be taken from farther back, showing the type of neighborhood. A second shot should be nearer to the scene, showing the street signs of the closest intersection, if street signs exist.

On the scene, the next shot should show the structure and the surrounding property such as the limits of the yard if in a subdivision. There should also be a photo showing the distance of the structure from property lines, and front, back, and side yards. There should also be photographs of each side of the structure. If there are exterior patterns from the fire, it may be beneficial to take corner shots showing the burn patterns on two exterior walls.

All evidence found on the exterior should be photographed, if necessary from various angles. A photo should show the location of the evidence in relation to the building or another landmark. If the evidence is sitting in grass and hard to see, an evidence marker can be placed next to the evidence, and then a photograph taken in relation to the building and/or other landmark.

Inside the structure, the photos should be taken just as the investigator should work the scene: from the least damaged area to the most damaged area. Each indication of the area of origin or the cause should be documented with clear, in-focus photographs.

Each piece of crucial evidence should be photographed in close-up detail. In addition, there should be a photograph showing its placement within the structure. Any photographs of the area of origin should include one shot from a distance, a shot about half as close, and then a series of close-up shots from at least two sides, and if possible four sides.

Any item identified as the first fuel ignited should be photographed. The heat source should also be photographed. These photographs should show as much detail as possible. Any device involved should be documented with photos, including the product manufacturer's name, the model name, and if possible a model and serial number.

■ Note
All potential heat sources should be examined and photographed to show that they were not involved in the ignition sequence.

All potential heat sources should be examined and photographed to show if they were or were not involved in the ignition sequence. This includes the electrical entrance, electrical panel, furnaces, range, dryer, hot water heater, and any other major appliance.

As evidence is collected, it should be photographed, and then photographed again once it is in the evidence container. Each scene requires some digging through the debris. Photograph before doing any digging and after the digging before leaving the scene.

SPECIALIZED PHOTOGRAPHY

A benefit of using a single-lens reflex (SLR) type camera is the ability to change lenses and use specialized flash units. The one lens length that best depicts what is actually seen by the eye is the 50 mm lens. Investigators also need a wide-angle lens to capture as much of one room as possible. A wide-angle lens should be at least 28 mm or lower. A zoom lens is beneficial to take close-ups of those items that are too far away to get a clear shot. The minimum length should be 200 mm, and if funds are available, 300 or 400 mm will aid in surveillance cases as well.

Instead of purchasing three lenses, the investigator can purchase two lenses and have the best of all worlds. First, purchase a lens that is 28–135 mm or something close to those numbers. This includes the 50 mm that can give the best true picture, and it provides the wide angle without switching lenses, and for most close-up shots the 135 mm will be an added bonus. The second lens to purchase is a zoom lens. These units also come in combination lengths such as 70–300 mm. A shot could be taken of a close-up detail and the lens can pan back to give a wider shot of that detail in perspective to other things in the room, without changing lenses. (See **Figure 18-1.**) A note of caution to anyone familiar with 35-mm cameras: a third-party (off-brand) 2× converter will not always work on

Figure 18-1 *Digital cameras can range from a simple pocket camera to SLR units with specialized lenses and flash units. The flash unit on the right is a macro flash that allows close-up photography lighting that will softly light up the area and not wash it out with a strong flash of light, great for fingerprints, blood splatter, and tool mark impressions.*

digital cameras. Before purchasing such a device ensure that it will work with your equipment.

The next part of the camera that is crucial to good photography at a fire scene is a high-quality flash unit. If possible, a second flash unit with a cord will give the flexibility of using both flashes, which can help reduce shadows. Another flash unit that can come in handy is a ring-flash. This flash unit fits over the lens and plugs into the camera. It will help with those macro shots taking close-up photos with a soft flash of fingerprints, arc patterns, and small hard-to-read serial numbers on small appliances. The ring flash is also handy in taking photos of electrical circuit boards because the soft flash will not wash out the etched circuitry and allows a good shot of any failure on the board itself.

If shots are to be taken at night, a tripod may be necessary. *Photo painting* is a process of putting the camera on "blank," which locks open the shutter, and using a flash to manually light up a building, wall, vehicle, interior, or any other item. This is necessary only if there is any possibility that the scene will change before sunup, such as when the wall with burn patterns threatens to fall. Situations such as a wildfire in the area that endangers all structures or when the investigator cannot make it back to the structure at sunup are reasons to do a photo-paint.

With a digital camera, it is a lot easier to ensure a proper photo-paint shot is taken because each shot can be reviewed after it is taken. To ensure a good shot, all emergency lights must be turned off, personnel at the scene must be aware of what is being done, and they must stand clear. It may be difficult to get all emergency lights off because some may be needed for safety, but limit as many as possible from reflecting on the area being photo painted.

The person holding the flash should hold it away from the body and slightly forward. If a flat wall is being photographed, one shot about every 10 feet should be taken. Look at the results, and then do more flashes or less accordingly. Depending on the equipment, distance from the wall to camera, and the type of flash, there is no one formula to let you know how many flashes to take for each shot.

Photo painting can also light up the interior of a garage. Use the flash to paint the front of the garage. Then, enter the garage and flash each interior wall. Flash the sides of any vehicle, and then, standing in the back of the garage to one corner, flash the top of the trunk and the roof of the vehicle. Always let the flash hit the inside of the vehicle as well. Check the camera then to see if you need to do more flashes or less. If the structure had a great room, a basement, or if it is a commercial property, this process helps you get a good image. Yes, there will occasionally be some shadows, but they should not interfere in showing the patterns. Occasionally, you will get ghosting, which is a slight image of the person holding the flash, but practice will help get rid of this in the future. Examples of photographing a scene are shown in **Figures 18-2 through 18-6**; all shots were taken using a Canon 30D (SLR) digital camera with a Canon Speedlite 580EX flash.

Figure 18-2 *Using the standard flash attached to the camera, this photo was taken on an automatic setting.*

Figure 18-3 *Photo painting. Keeping the shutter open and making multiple flashes while walking around the vehicle lights up the vehicle and shows details of vehicle, ground, and surrounding area. Also notice by keeping the shutter open the sky seen between the trees lightens up as well.*

Figure 18-4 *Photo painting. This photo was taken while walking across the front of the structure and firing the flash 15 times, which gives moderate results on this 72-foot house. The occasional shadow in the image is caused by the photographer holding the flash too close to the body and blocking some of the light reflected from the face of the structure to the camera.*

Figure 18-5 *Standard single-flash photo of a carport with two vehicles.*

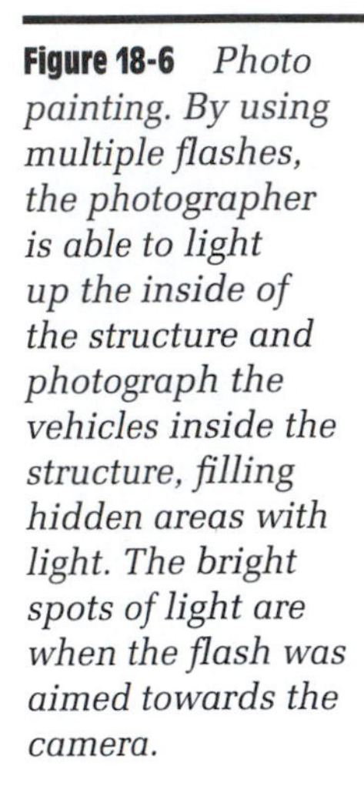

Figure 18-6 *Photo painting. By using multiple flashes, the photographer is able to light up the inside of the structure and photograph the vehicles inside the structure, filling hidden areas with light. The bright spots of light are when the flash was aimed towards the camera.*

DOCUMENTING WITH DIAGRAMS

Diagrams are drawings of the property that depict the property boundaries, the position of the structures on the property, and that use an arrow to depict the direction north to orient the viewer as to the relative position of structures with neighboring properties. The investigator places pertinent information specific to the case on a diagram(s) to help tell the story of what happened on the property during the event.

■ Note
The investigator places pertinent information specific to the case on a diagram(s) to help tell the story of what happened on the property.

The first responder will most likely draw only a sketch to assist in the fire report. If an assigned investigator has been requested, the first responder investigator may start taking measurements, which will certainly be appreciated.

Sketching and Measuring

To document a fire scene properly, a diagram must be created. A sketch is a simple rough drawing, which is the first step in creating a detailed diagram. Sketching starts with drawing the area, and then a sketch of the structure. Both can be roughly drawn, but because all notes, including rough sketches, should be included in the investigative file be sure to include only material for the diagram on these papers. Simple graph paper can help in making as accurate a depiction as possible in the field.

The final diagrams need to be as accurate as possible. To help with this, all walls, roads, landmarks, key points, evidence, or other data placed on the diagram should be measured. The actual measurement can be included in the diagram. Distances can be measured with a walking tape, which is a wheel on a stick that will properly measure from one point to another. This is especially important if any evidence was discovered outside the structure. Inside the structure, a tape measure will help to get measurements under 20 feet. Longer distances can be taken as well, but anyone who has used a tape beyond that distance knows that from time to time it takes a couple of attempts.

A product that has been on the market for some time is an ultrasonic measuring device that can give measurements from wall to wall or from any given point to a wall. This device bounces a signal off a solid wall or other object and back to the device to give a measurement. If walls are gone, this limits the use of this type of device. If you plan on using electronic measurement, the device should be tested using a series of tests to ensure its accuracy. Always test the device on a regular basis to make sure it still works properly.

Before leaving the scene, the investigator should verify all measurements and that all evidence locations have been captured on the rough sketch.

The Diagrams

The sketch needs to be rendered into a diagram. Some investigators prefer to create their diagrams on drawing paper, and this is certainly acceptable. Others—the rest of us with limited talent for such ventures—purchase software that enables you to create two-dimensional floor plans. These floor plans will be necessary to add to the report. In addition, many of these same software programs, for a relatively small cost, can also create three-dimensional diagrams that allow for the accurate depiction of the burn patterns.

The basic completed diagram needs to be attached to the final report. It needs to be extremely accurate, but must state "Not to Scale." If this comment is not added and a mistake is found on the diagram, it can give the opposing counsel an open door to challenge what else may not be accurate, such as the determination of the fire origin or cause.

There are several pieces of information that should be considered for placement on a diagram. This data can be placed on one diagram, or special diagrams can be created to show the data separately or in groups. One way to show multiple sets of data on one form is to create an overlay. Clear acetate sheets can be purchased at most art and graphic supply stores in 8½ × 11 sheets, and usually by the ream. These sheets can be laid over the initial diagram, attached on one side by tape so that they can hinge back onto the sheet or off to show the original floor plan. Data can be placed on the clear sheets that when placed over the diagram show how those data relate to the structure.

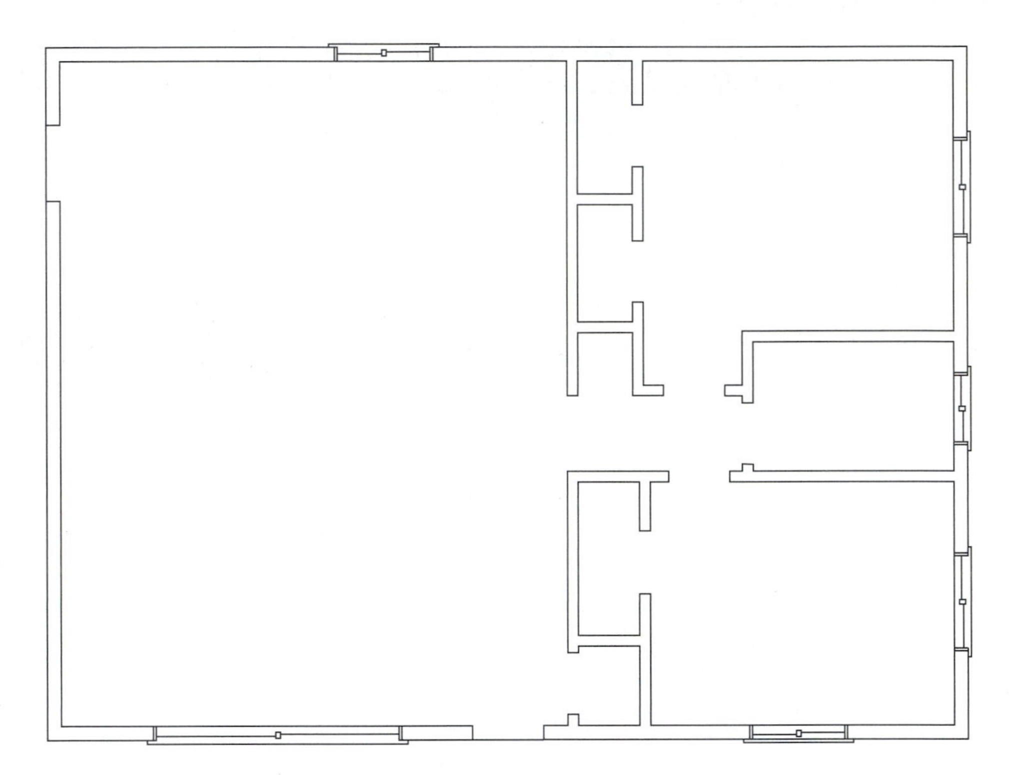

Figure 18-7 *A simple diagram showing all spaces within a structure as well as all openings such as windows and doors.*

Pre-Fire Floor Plan The first diagram should show the layout of the rooms and spaces as they would have appeared prior to the fire. Closets, hallways, and any hidden spaces should be included. (See **Figure 18-7.**) From this template, you can create the other diagrams. If created electronically, it can be an easy task to make additional diagrams based on this initial one. If created by hand, the basic floor plan can be duplicated and used for the other necessary diagrams.

Note
Diagram details must include all windows, doors, and any other openings that may have an impact on the fire's spread.

Diagram details must include all windows, doors, and any other openings that may have an impact on the fire's spread. It may be beneficial to show the electrical and gas meters as well.

Pre-Fire Floor Plan with Known Furnishings Furnishings whose locations prior to the fire can be verified should be placed in their proper position on a diagram. Include only furnishings that have enough mass to contribute to the fire or that can provide indications such as the fire's path.

Burn and Smoke Patterns In addition to showing which areas had char, the diagram could also reflect the depth of the char. It should show the direction of fire travel from the area of origin. Additional diagrams can show smoke damage as well as the other extreme, areas of total consumption. Diagrams need to support the report; as the report is written it will identify which diagrams are needed to help the reader understand the fire growth as well as the final hypothesis identifying the fire origin and cause.

Evidence Diagram Depending on the amount of evidence, it may be beneficial to mark it on a diagram. A graphic representation showing the location of each piece of evidence identified and collected is a good tool for others reading the report and for presentation in court, if this is ever necessary.

Each piece of evidence that was identified should be depicted on the diagram. Each location should show two measurements from structural elements that will enable a third party to understand the location in relation to the rest of the building.

Photo Diagram The final report should include a collection of photographs that will complement the report and allow the reader to follow the investigation as it progressed. Photographs can show the various patterns and indicators that the investigator used in the creation of the hypothesis. They can show where and how evidence was collected, the area of origin, and if possible, physical evidence that indicates the cause of the fire.

Each photograph should be numbered and mounted or printed on paper, either one or two per page. Each page should have a header, page number, and a clear description under each photograph. Each photograph should be referenced in the report.

Photo information should also be placed on a diagram sheet. The photograph number should be inserted in the diagram in the location where the photographer was standing when the shot had been taken. For identification's sake, it is beneficial if the number is placed in a small circle with an arrow pointing in the direction that the camera was pointing when the photograph was taken, as shown in **Figure 18-8,** which is one way to document a photo on a diagram.

Suppression Activities If for any reason the actions taken, or not taken, by suppression forces will be in question, then it is important to capture all possible information before leaving the scene. Placement of apparatus, water supply lines, location of hydrants used or not used, ladder truck placement, and attack lines should be placed on a sketch and transferred to a diagram for the incident file. One diagram needs to encompass the scene from the longest water supply line to the scene itself. A series of diagrams or overlays can show the arrival of apparatus and where they set up to operate. Fire attack lines can show which door they entered, and master streams can be documented as well. (See **Figure 18-9.**)

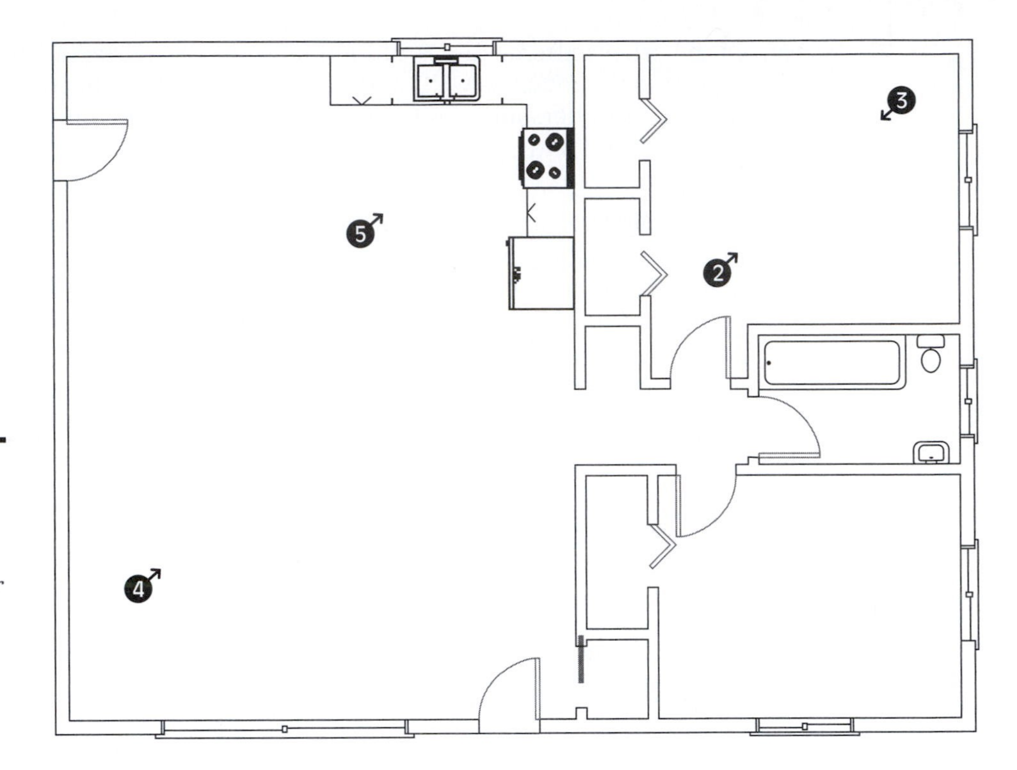

Figure 18-8 *A number in a circle with an arrow shows the photograph number and the direction the photographer was pointing the camera when the photo was taken.*

■ Note
A full set of plans may be available from the building official, the contractor, or architect.

Commercial Building Plans Many commercial structures have plans that were used to build the structure. Depending on the building's age, a full set of plans may be available from the building official, the contractor, or architect. Many years ago, it was a common practice for a set of plans to be created on the building as it was completed. That does not happen now; in fact, the plans now may not even reflect how the project was completed. During construction or after occupation, a structure can change. Thus, those plans at the building official's office may not reflect the structure floor plan as it was on the day of the fire. But close examination can enable the investigator to discover any differences and compensate accordingly.

NOTES

■ Note
"The dullest pencil is better than the sharpest mind"

It may be an old cliché, but the saying that "the dullest pencil is better than the sharpest mind" is just as true today, in our age of technology, as it was when it was written. Notes are an essential part of any investigation. In fact, they are critical to the success of the investigation. A good investigator should be able to identify who to interview, what to ask in that interview, and listen to the

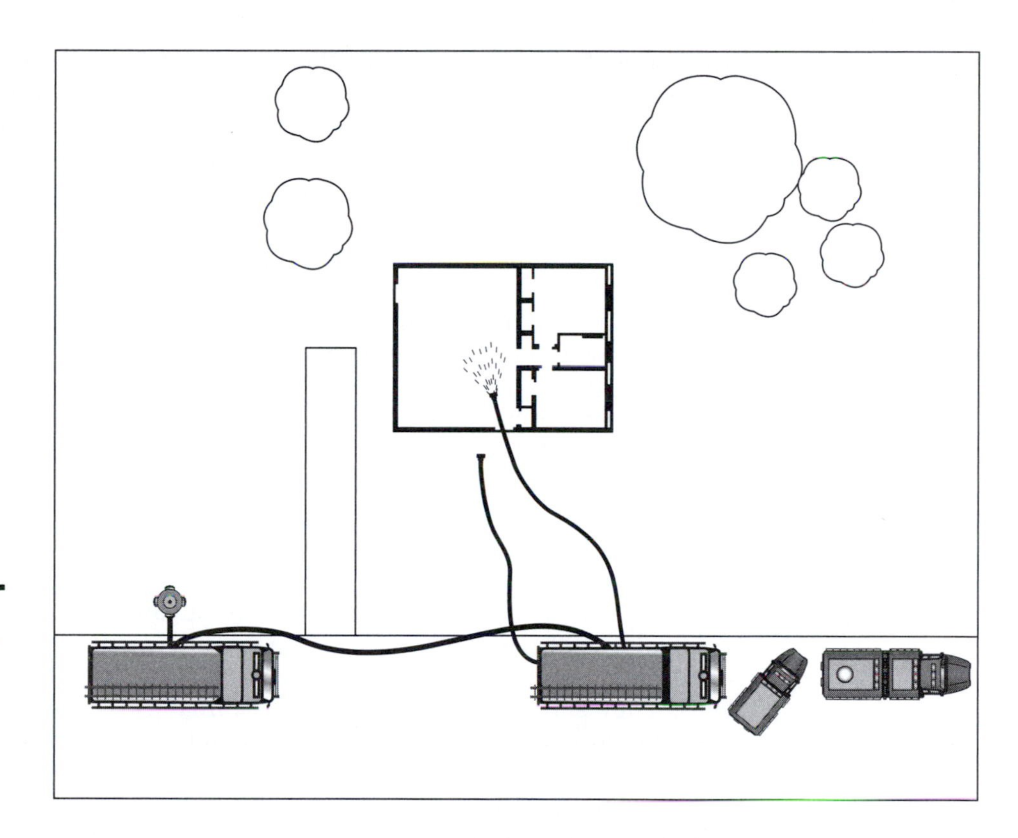

Figure 18-9 *A typical suppression diagram showing apparatus placement, hose lines, and water source.*

answers. If you do not write down what was said in the interview, then the interview will be of limited use.

This argument could be countered by saying that the interview can be tape-recorded. This is true; the conversations should be recorded for future use. But, during the interview, an issue may arise that needs more clarification, and investigators cannot excuse themselves to back up the recording and review it to ask additional questions. Investigators need to learn how to listen and what to write down. They need to identify key points that can be used in the scientific process to come up with answers to questions that must be asked to narrow down all hypotheses.

■ Note
Writing the report is much easier and more efficient if written notes are used to get the information compiled and in the correct order.

Writing the report is much easier and more efficient if written notes are used to get the information compiled and in the correct order. Written notes are also a good tool to use in answering questions during depositions and at trial. Again, they are there to refresh your memory and to aid in providing the best answers possible about anything involving the case.

Notes need to be legible; there is nothing wrong with using abbreviations or shorthand as long as you know what they mean now and in the future. Some investigators have gone so far as to memorize the shorthand used by stenographers for common words or sayings. The notes need to be kept in a logical order, and if possible, use a method that will keep them all together. Some departments use preprinted forms that can be filled in as the investigation proceeds. These are handy to ensure that everything is covered. Some preprinted forms can be 40 or 50 pages in length. If a page does not apply to that specific case, it is marked accordingly. Other investigators use spiral-bound note pads with case numbers marked on the cardboard cover. There is no right or wrong way to handle notes as long as all data are properly captured and documented.

There are two theories on keeping and destroying notes. Some departments destroy their written notes once they have typed their report and supervisors have approved it. This is done only for the sake of file size because over time written notes can be quite extensive and most municipal departments are limited in office and storage space. Other departments keep their notes in the file so that all information is available for review by staff, prosecuting attorneys, and opposing counsel. Some feel this sends a message that there is nothing to hide. As for which way is best, that depends on the recommendations of the prosecutor's office and the policy of the department.

Notes are the key to correlating the investigator's thoughts, ideas, and eventual final hypothesis. The notes are the tools that will be used to create the final report. The investigator's notes reflect directly on the investigator taking the notes. They should never contain discriminatory or inflammatory comments. They should not be speculative in nature but should contain facts. If the notes are framed in a clear, concise, logical, and professional manner, the same will be thought of the investigator.

A warning: Because notes are used for cases and may be reviewed by others, it is critical to avoid putting anything in notes that does not involve the case at hand. Be cautious about editorializing, and write only those things that would not result in shame or embarrassment if printed in the local paper.

SUMMARY

Diagrams and notes are key tools to creating the final report. They are the tools that will be referenced to come up with the hypothesis as to the fire area of origin and cause. From these diagrams and notes will come a final report that conveys to the reader all the actions of the investigator that lead to the final determination of the fire origin and cause. The first responder investigator most likely will not create diagrams or take notes in volume enough that they will be kept after creating the fire incident report. But assigned investigators use their notes throughout the entire investigation.

Diagrams are valuable tools and a vital part of a good, concise final report. A variety of item can be captured with a diagram by using overlays or multiple diagrams. No one way is better than another as long as they provide accurate information that supports the final hypothesis. To make sure the diagrams are not misunderstood, each and every diagram should be labeled as "Not to Scale."

The notes reflect on the investigator and should be legible, logical, and concise. They should be written in such a way as to reflect positively on the investigator. Always take notes as if they were to show up in the local paper the next day and so that they include nothing that would be embarrassing or inappropriate. Professionalism is a total package. The investigator is a seeker of truth and this should always be reflected in all work products.

REVIEW QUESTIONS

1. Give three major benefits of using digital 35-mm SLR cameras for fire investigations.
2. Why should a first responder have a digital camera?
3. What are the three lens sizes that are beneficial for the fire investigator to have?
4. What is photo painting and how is it accomplished? Why would the investigator want to use the photo painting process?
5. Make a list of what can be included in the final diagram for your report.
6. Make a list of what should be photographed at a fire scene.
7. Why would the investigator create a diagram depicting placement of suppression equipment?
8. What should be put on all diagrams to be sure the diagram is not misunderstood?
9. Under what condition should the investigator be able to use shorthand in field notes?
10. Can an investigator use his or her notes when testifying in a deposition?

DISCUSSION QUESTIONS

1. Should investigators destroy their notes after writing the report or keep them in the investigative file?
2. Which note-taking practice is better for the fire investigator: use of preprinted blank forms bound together or use of a narrow spiral-bound note pad?

ACTIVITIES

1. Contact court services or look on the Internet to obtain symbols that can be used by the fire investigator in note taking.
2. On the Internet, search for software programs that are capable of creating diagrams for the fire investigator. In particular, are there any that can provide an exploded view?

INTERVIEWS

Learning Objectives

Upon completion of this chapter, you should be able to:

- Describe and understand the process of obtaining sufficient information from the scene before conducting in-depth interviews.
- Describe the creation of an interview plan in which the questions and their answers will further the cause of the investigation.
- Describe the protection of privileged information and who is entitled to such protection.
- Describe the Miranda ruling and the Constitutional rights associated with the Supreme Court decision, including when it is necessary to advise a suspect of his or her rights.
- Describe the various types and methods to document an interview and the benefits of each.

CASE STUDY

Nothing brings out the natural instinct of parents to protect their children like when a person of authority wants to interview a child about a fire scene. Every investigator has stories about dealing with parents and children. There was one father who insisted his child would never set a fire, accidental or otherwise. He made the perfect figure of a dad, protecting his young, holding the boy close to his side. Once dad was convinced by the insurance investigator that his insurance company would pay for the damages regardless of whether the fire was set by the son, that dad reared back his hand and smacked the boy on the back of the head. At the same time asking the boy what he was thinking when he was playing with the propane torch, a fact the investigator didn't even know about.

But the most shocking event took place on a quiet summer afternoon. The fire marshal got a call from a father asking if a fire engine could come to his house and talk to three families who had young boys who had been playing with fire. A call to the closest engine company was made and both the fire marshal and the engine arrived at the house at the same time. While walking up to the front door, they observed three obvious burnt circles in a well-manicured front yard. Each circle was about a foot in diameter. The father answered the door and nodded to the burnt areas and said there were six more in the backyard.

Seems dad had been out of town, and when he came home he found the burnt areas in his lawn and his wife had no idea what had happened. It didn't take long in talking with the neighbors for them to find out that the four boys, sons of the three parents, had seen old reruns of *Wagon Train* on the television and they decided to make their own campfires in the front and back yards. One boy was 9, two were 7, and the youngest was 6 years old.

Every attempt was made to keep this a friendly atmosphere, and it worked. Everyone crowded into the screened-in front porch, three sets of parents sitting in chairs or on the arms of chairs, the kids sitting on the floor, and the firefighters leaning up against a post or wall here and there. The fire marshal sat on the floor with the kids. Taking a tip from his chief, who had been his mentor, the fire marshal asked the kids if they ever wanted to be firefighters. Of course, there was a resounding yes. The fire marshal then told the boys that if they ever wanted to be firefighters, they could never lie to the chief or the fire marshal. Then, he asked them if they knew that. Again, a resounding yes, yes, we know that. The fire marshal then asked the boys to explain what happened to the yard. They explained about the campfires and it was obvious that they were remorseful.

As in all interviews with kids—another tip he learned from his chief—the fire marshal asked if the boys had set any other fires. Four boys anxiously told of setting some fields on fire because they saw a show that said it would get rid of snakes. They burned an old shed down, again a tip from another movie, and the list went on. All the smiles were gone from the parents' faces, and shocked looks were on the firefighters' faces; after all they remember most of those fires. The fire marshal felt that at any minute he would have to protect the kids from their parents. Actually, shock set in all around except for four small boys who were absolutely oblivious to the danger at hand.

(Continued)

(Continued)

Then, the atmosphere changed, and each child was interviewed separately with his respective parents and data were collected. In all, there were almost 70 fire events over a 2-year span, 32 of which had fire department responses. Reports for those incidents were changed because most had an unknown cause and others had notes that they thought it was kids playing. The fire marshal's office had looked at only six of the events, and those files were now listed as solved. Almost all of the fires were quite small. Almost half of the fires were extinguished by the boys themselves. Thankfully, there appeared to be no malice in any of the events. Counseling sessions were set up as an educational issue and the parents fully agreed that they held some responsibility to more closely watch their own children.

Of course, the chief wanted to know why the engine company didn't pick up on that many calls. With a station that had 70 volunteers at the time and ran almost 2,000 fires a year, it just did not get noticed. One minor note: One of those boys did end up being a career firefighter with an outstanding record. Most important, a valuable lesson was learned in interviewing techniques—don't ever hesitate to ask the big question because you might be surprised by what you may find out.

INTRODUCTION

There are various types of interviews to be conducted at any scene. The first responder investigator does informal interviews to collect data from firefighters as well as to obtain any information on what emergency medical services (EMS) or local law enforcement might have seen when they arrived or as they were on the scene. More formal interviews of victims still at the scene or occupants are also obtained if they are available.

The assigned investigator most likely handles the in-depth interviews or follow-up interviews. These are planned events, especially if they involve a criminal case. There are laws that affect what can or cannot be done and even a few legal rulings in favor of the investigator that you may find surprising.

In this chapter are references to reading the Miranda Warning to individuals. The courts have ruled that this is a function of law enforcement personnel. An assigned investigator such as a fire marshal is a law enforcement officer. Even if the locality has not administered police powers to the individual but that individual does in-depth interviews, the fire marshal, as a representative of the government, is still responsible for administering—reading—the Miranda Warning to the person being interviewed.

The key point is not if the investigator really has police powers so much as how the person being interviewed views the situation. If the fire marshal's actions make the interviewee feel that he or she is not free to go, then it is important that the individual be advised of their rights. After all, the fire investigator is acting as an agent of the government and any information obtained is intended to be used in a court of law should it be discovered that the person being interviewed was a party to a crime.

The converse to this is the fact that the fire officer on the scene, who is the first responder investigator, should not be put in a situation where he is seeking anything more than facts surrounding the fire. The first responder investigator should not be in a questioning mode that would remotely put him in a position as a government agent in a criminal matter. Thus, under these circumstances, first responder investigators should not read the Miranda Warning to individuals. The minute that there is any doubt that the interview is going beyond seeking cause and origin information, the interview should be brought to a close and the case should be turned over to an assigned investigator.

INTERVIEW BEFORE DIGGING OR INTERVIEW AFTER DIGGING

There are two separate schools of thought about when it is appropriate to interview the victim/property owner. Whether to interview before or after examination of the fire scene has long been a subject of debate. The former allows the investigator to get as much information as possible about the scene prior to physical examination; however, this may taint the way the investigator looks at the scene. Having listened to victims and witnesses, the investigator may consciously or unconsciously start looking for clues to support or refute what was heard.

Doing the scene examination first allows the investigator to enter with a clear mind, no preconceived notions, as to how the fire started. This provides two benefits. The scene will be examined based on the patterns and the physical evidence. The second is a very defendable stance in court should opposing counsel infer that you looked for arson because the neighbor said the property owners were arsonists.

Each process, digging before and digging after the interviews, has merit. However, it is largely the preference of the investigator as to what course of action to pursue. Investigators who choose to interview the owner or witness prior to examination will almost always wish to reinterview them after the examination. Likewise, facts brought forward from the interviews may lead to a reexamination of the scene to confirm or dispel the information brought out in the interview.

■ Note
Even the most seasoned investigator requires and wants the information that is told by another person about the fire.

■ Note
An investigation is not complete until all witnesses and any persons of interest are interviewed.

WITNESSES

Even the most seasoned investigator requires and wants the information that is told by another person about the fire. Although a good investigator will be able to determine much information on the scene, an investigation is not complete until all witnesses and any persons of interest are interviewed. This gathering of information is as much a part of the investigation as the examination of the fire scene.

Of vital importance to the investigator is the location of witnesses who may have seen the fire in progress. This information can be invaluable in supporting the hypothesis that the investigator develops as the investigation progresses. The victim of the fire may also be a witness and perhaps the most informed witness if that person discovered the fire. Likewise, in the event of arson where the victim is also the perpetrator, the interviewee may use the interview as an attempt to persuade the investigator that the fire was an accident.

■ Note The victim of the fire may also be a witness and perhaps the most informed witness if that person discovered the fire.

Witnesses may make themselves known to the investigator or suppression personnel early in the event, or they may have to be located. The most logical place to look for witnesses is among the people who live in the neighborhood. Also of consideration are persons who are in the neighborhood such as postal carriers, service workers, or delivery personnel and newspaper carriers.

■ Note Victim and witness interviews occur on the fire scene; steps should be taken to allow for privacy, comfort, and confidentiality.

The place setting of the interview can be an important part of the success of the interview. Although most victim and witness interviews occur on the fire scene, steps should be taken to allow for privacy, comfort, and confidentiality of the interview. It is difficult to interview a victim over the idling engine of a fire truck; therefore, it may be advisable to select a spot away from the engine, where the witness is warm, dry, and comfortable. This may be at your vehicle or in the home of neighbor or even standing outside away from the crowd if the weather is conducive. (See **Figure 19-1.**)

Figure 19-1 *Interviews sometimes may have to be done on the scene, away from the noise and bustle.*

RECORDING INTERVIEWS OF SUSPECTS

Arson is a crime of stealth, and often the findings and the hypothesis of the investigator can be supplemented and verified by the statements made by witnesses and the victim. There may also be a time when the investigator finds that the fire has been intentionally set and that there is a person responsible for the setting of the fire. In such cases, the investigator must confront the suspect and attempt to solicit as much information as possible.

Interviews of suspects or persons of interest are likely to be much more structured. If the person is to be interviewed at the station or office, it is advisable to conduct the interview in a private place with little or no distractions. Drab, barren, windowless interview rooms are not as encouraging to suspects as a business-like environment is, if such environment is not cluttered and presents a pleasant atmosphere. If this interview area is equipped with recording apparatus, it should be tested prior to the interview to make sure that all parts are in working order.

If the investigator desires to record the interview, he or she should have this prepared in advance. The use of recording devices was sanctioned in *United States v. McKeever* and *United States v. White*[1] where the United States Supreme Court affirmed that recording of a conversation is legal as long as one party of the conversation is aware that the conversation is recorded. Several states, however, have requirements that persons being recorded be notified or they prohibit recording without a court order. It is advisable to consult the legal counsel for your department prior to any recording to ensure the admissibility of such evidence or any special requirements which must be met.

■ Note **The United States Supreme Court affirmed that recording of a conversation is legal as long as one party of the conversation is aware that the conversation is recorded.**

Regardless of whether the interview will be recorded, the investigator must be prepared to document the interview. Documentation of the interview is not only necessary to prepare a report of the interview but also important to guide the interview. Also, if the interview is recorded, there is always the possibility that the recording could fail or parts or all of it may be lost in transcription, and then the only documentation of the interview will be the investigator's notes. If the suspect or witness makes a statement that needs clarification, the investigator will need to have notes available to confirm what was said the first time. This is important to track inconsistencies in the statements and to clarify points of interest. To accomplish this, notes need to be taken during the entire interview. It is important that note taking does not distract from the interview process. The investigator should note facts in bullet statements and not attempt to write down every word that is spoken, except for such statements that would require a word-for-word transcription to the report.

■ Note **It is important that note taking does not distract from the interview process.**

THE INTERVIEW

It should be remembered that victims may present many and varied reactions to the situation. In residential fires, the sum of interviewees' lives may have just gone up in smoke, and in the most extreme situations a loved one may be dead as a result. In between, there may be feelings of desperation caused by financial loss, fears that insurance may not be adequate (or in some cases insurance may not have been purchased), loss of prized or sentimental possessions as well as feelings of hopelessness that they cannot recover from such a catastrophe. They may also be concerned over a killed or missing pet, which in many cases appears much the same as the loss of a loved one. Finally, a person who started the fire intentionally or negligently may be nervous and will attempt to conceal that fact in a deceitful interview.

■ Note The investigator should work to develop rapport with the victim prior to any questioning.

■ Note It helps to remove your hat or helmet and be sure you are on the same plane as the victim.

■ Note Interview witnesses separately to avoid getting a composite view of what happened and to ensure that each witness is allowed to express his or her own individual recollection.

■ Note Let the witness terminate the interview and allow him or her to talk as long as the person has meaningful information to give.

Any fire can traumatize the victim and the investigator should be prepared for any number of reactions when interviewees are approached for questioning. They may be frightened or angry, withdrawn or confused, but in any case they should be treated with understanding. These feelings of anger and fright can often be misinterpreted by the interviewer as deception; therefore, it is prudent to conduct several interviews with the victim of the fire to establish norms and patterns that may be useful in the determination of truthful behavior.

Interviews conducted after the scene has been examined should be done without revealing to the interviewee the findings of the scene. Rather, allow the interviewee to tell the story without interruption, and then ask questions to clarify the facts of the investigation.

Planning

Planning the interview should allow for a settling in period at the beginning, when questions center around matters to confirm the layout of the place, the arrangement of furniture and fixtures, and the use of various areas. This will allow the investigator to develop in his or her mind the prefire conditions that were present to account for fuel load and placement of combustible material, and to provide supporting data to confirm or dispel the findings of the scene examination. At this point in the interview, the questions should not pose a threat to the individual being interviewed.

The investigator should work to develop rapport with the victim prior to any questioning. First consideration should be the well-being of victims. Relate your concern and sorrow for their loss. Reassure them they are safe. It helps to remove your hat or helmet and be sure you are on the same plane as the victim. If they are seated, kneel down so that you are not talking down to them. (See **Figure 19-2.**) Carefully discuss the matter at hand and be prepared for a variety of reactions. Victims may be very stoic or break down with emotion at the loss they have experienced. Ensure there is support lined up for them and offer to assist them in getting that support as needed.

Interview witnesses separately to avoid getting a composite view of what happened and to ensure that each witness is allowed to express his or her own individual recollection. Allow the witness to relate observations in his or her own words and avoid leading the witness with suggestive questions. Ask open-ended questions, and then guide the interview with questions for clarification. Let the witness terminate the interview and allow him or her to talk as long as the

Figure 19-2 *The investigator may not always have a seat. In these instances, always be sure to be on the same level as the victim, kneeling, with hat off. (Photo courtesy of Karl Mercer.)*

person has meaningful information to give. Be sure not to overload interviewees with questions and give them ample time to respond. Conduct your interviews in private with minimal distractions.

THE COMMUNICATIONS PROCESS

■ Note
The investigator needs to be a good listener and focus on what is being said and how it is being said.

■ Note
It behooves the investigator to maintain an open mind and evaluate each answer on its own merit.

The key to a successful interview is to understand the communication process. Communication is a two-way process. The investigator needs to be a good listener and focus on what is being said and how it is being said. Focus on the present question and do not think ahead to future questions. Avoid anticipating what the interviewee will say or analyzing the answer before the person is finished speaking. The investigator needs to cover the basic elements of any story, which include the basic principles of who, what, when, where, and how the event occurred. It behooves the investigator to maintain an open mind and evaluate each answer on its own merit.

Witnesses recall the event as they saw it. Their perceptions may be different from other witnesses, and they will recognize things that have the most significance to them individually. For example, if a person is small in stature, then that person may report others as being bigger than they actually are, or vice versa. If someone has a particular interest in automobiles, they may be able to identify makes and models, which other people would otherwise overlook. This may be true of any life experience. It is advisable to learn as much about the witness as can be done in the short time of the interview.

In preparing for the interview an understanding of the communication process is important. The actual spoken word accounts for less than 10 percent of the entire communication process, while vocal nuance and nonverbal body movements comprise the majority of the process. The spoken word is the most controllable aspect of communication, whereas the nonverbal is much more difficult to mask. For this reason, pay full attention to the interpretation of the communication that is exchanged.

Communication starts with the source (sender), who encodes the message and selects the words to send as the verbal message. Nonverbal communication is more difficult to encode and often betrays the true meaning of the spoken word when one is being deceitful. The receiver then decodes the message and attempts to understand it. The receiver provides feedback in the form of facial expressions, verbal response, and body language. The process then reverses as the sender reads the feedback and sends more information. When words and the nonverbal communication do not match, it hinders the communication process. This may be a result of deception or that the words being sent are not correct.

NONVERBAL COMMUNICATIONS

Watch for facial expressions because they are the hardest for a person to control. In most situations, the expression on a person's face will reveal true feeling for an instant before the person can then force a smile or other expression to mask the true emotions. Normal facial expressions are the most universal of all nonverbal communications. Expressions of fear, surprise, disgust, happiness, and guilt/nervousness tend to transcend most ethnic and cultural barriers.

■ Note
Body language or nonverbal communication can be significantly influenced by a person's culture, ethnic background, and physiologic makeup.

Nonverbal communications are often referred to as *body language*. The study of such is a key to the veracity or motives of the speaker. Normal reaction to fear may be to close down and withdraw, which manifests as arms folded and a desire to create distance between the fearful person and the person who is causing the fear, such as the interviewer. Conversely, it is held that people who are truthful will be open and maintain contact with the person they are talking to.

However, it must also be noted that body language or nonverbal communication can be significantly influenced by a person's culture, ethnic background, and physiologic makeup. The most acceptable way to look at nonverbal communications is to study the normal posture and position of the person while he or she answers nonsensitive questions where there is no need for deception. The change to a more defensive body posture in many cases is an indicator of deception. The interviewer may want to bear in mind that in a fire-related interview signs of deception by the victim may be a result of traumatic experience such as the escape from the fire or the loss of their property, pet, or perhaps even a loved one. It may also figure into the picture if the fire was a result of some negligence that the owner or victim is uncomfortable in discussing.

LISTEN CAREFULLY

The words spoken can provide a link to the veracity of the statement the person is making. The investigator should listen closely to what is being said as well as the choice of words because most people are uncomfortable lying and they will substitute words or use of words that make the telling of a lie easier. This may provide insight into the incident and the person being interviewed. Statement analysis follows a two-step process. First, investigators determine what is typical of a truthful statement, referred to as the *norm.* They then look for any deviation from this norm. Truthful statements differ from fabricated ones in both content and quality.[2]

Usually, this is a two-step process. Listen closely for the normal use of words and key in on changes. Be prepared because this may occur in the same statement. For most people, it is difficult to lie overtly, and the true feeling of a person can show though in his or her choice of words. People being interviewed may not realize that they are betraying their true feelings. In addition to coping with the fact that they are being interviewed, they are also dealing with their guilt for something they may have done.

Singular pronouns such as *I, me, you, he,* and *she* show responsibility, whereas *they, it,* and *we* transfer or share responsibility. The use of *I* creates a personal responsibility whereas using *we* shares that responsibility. The use of *we* denotes a partnership whereas *he* or *she* denotes another party entirely. Failure to use *we* in a statement involving multiple people may reveal a desire to separate the party from the group or other person.

In a similar manner, the use of possessive pronouns may also give an indication of the attachment or desire to separate from the subject at hand. Pronouns can indicate either a separation or an attachment to the person or thing. Pronouns can be used to avoid saying a person's name. The use of *my* or *our* denotes a personal ownership or relationship, whereas the use of *the* creates a barrier and depersonalizes the attachment. Note the statement if changes occur in the same general area; for example, "I bought *my* house when my wife and I were first married, and tonight *the* house burned down." Although the use of "the house" in the latter part of the sentence may indicate a separation from the house, it could also be a defense mechanism to help the healing process begin, to help the person come to terms with the fact that "my house," as the subject remembers it, no longer exists.

Use of past and present tenses can be an indication of true feelings. For example, describing a person in the past tense shows a complete break for that person in some cases, or it may show a significant change in the way that person is viewed at this time. Something else that may concern the investigator is when an interviewee describes a past event in the present tense, which may indicate the interviewee is altering the event as he or she speaks.

Note
Sometimes too much information can be an indication that something is amiss.

Sometimes too much information can be an indication that something is amiss. The overassertion of information that is not relevant may tend to show a justification of action or an attempt to reinforce a position where the subject erred or was imprudent. It can also be a stalling tactic or whitewash to put the event in a better light.

Lack of conviction or words that do not match the seriousness of the situation can cause the investigator to take notice. If the subject has been through a substantial trauma as a result of the event, and then passes it off as if nothing happened, this may cause the investigator to doubt the veracity of the person's statements. However, it may also be a coping mechanism to deal with the trauma.

PROTECTED COMMUNICATIONS

The investigator will want to interview any and all persons who may have a connection in any way with the fire. This may include neighbors, relatives, medical personnel if the victim was injured, and any other persons who have knowledge to share on the circumstances of the fire.

Note
There are times when communications are protected under the rules of evidence as privileged communications, which limit the extent of questioning and the use of the results of such questioning in legal proceedings.

There are times when communications are protected under the rules of evidence as privileged communications, which limit the extent of questioning and the use of the results of such questioning in legal proceedings. Such a privileged communication normally extends to statements and information exchanged between clergy and parishioner, doctor and patient, attorney and client, and husband and wife.

To examine privileged communications there are certain common denominators to be considered. Such communications apply only to incriminating statements or admission made to the party that may be claimed as confidential. The measure is to establish whether there was an expectation of privacy in the communication and whether that privacy was dependent upon a precedent of law. Such communications must apply to crimes past and not be a part of any conspiracy between the communicating parties. For example, if a husband were to tell his wife in the confines of the marriage that he has committed a crime in the past, the admission of such information may be excluded based on the privilege that the marriage bears a bond of confidentiality. This is a premise of common law that is often recognized in modern courts. However, if the husband includes his wife in the planning of a future crime, then the wife becomes a co-conspirator and the bond of confidentiality is no longer related to the sanctity of the marriage.

Communications between clergy and parishioners may be claimed to be privileged on the basis that the desire to make a settlement with the spiritual well-being cannot be introduced into the secular world. In some religious organizations, there is a covenant of the religion to confess the sins to the clergy to seek absolution. It may be argued that the desire to cleanse the soul and ask for forgiveness must take precedence over the court of law.

Attorney and client relationships are normally absolute as a rule of law so that the attorney can provide for the defense of a client under the Fifth and Sixth Amendments unobstructed from discovery. Most courts recognize this sanction of law.

Relationships between physician and patient are centered around the desire to allow a patient to disclose to the doctor all factors that might influence his or her recovery, either physically or mentally, with the interest of the patient in mind.

■ Note
In all cases, it is the responsibility of the defense attorney to invoke the right to privileged communications and the challenge will be ruled upon by the trier of fact for the trial at hand.

In all cases, it is the responsibility of the defense attorney to invoke the right to privileged communications and the challenge will be ruled upon by the trier of fact for the trial at hand. If there is any question as the privilege of the communication, advice should be sought from the legal counsel of the agency.

CONSTITUTIONAL RIGHTS

If there comes a time where a person of interest becomes the focus of a criminal investigation, the interview will become much more pointed toward a quest to have the person confess to that crime. This is a precarious point in any interview and one that must be tendered with care to ensure the constitutional rights of the suspect are observed so as not to taint any confession the suspect may give. All interviews are a simple inquiry for information guided by the questions of the interviewer to bring out all the facts of the incident.

Miranda v. Arizona[3] is a decision of the U.S. Supreme Court that has been around since 1966 and is still a matter of controversy and litigation. *Miranda v. Arizona* set up guidelines for government **interrogation** of suspects that must be adhered to in order to admit into court the evidence gained from any statements made by the defendant and protect the accused under the Fifth and Sixth Amendments of the U.S. Constitution.

interrogation
formal questioning, same as interviewing but has a more formal connotation

The Fifth Amendment

> No person shall be held to answer for a capital, or otherwise infamous crime, unless on a presentment or indictment of a Grand Jury, except in cases arising in the land or naval forces, or in the Militia, when in actual service in time of War or public danger; nor shall any person be subject for the same offence to be twice put in jeopardy of life or limb; nor shall be compelled in any criminal case to be

a witness against himself, nor be deprived of life, liberty, or property, without due process of law; nor shall private property be taken for public use, without just compensation.

The Sixth Amendment

In all criminal prosecutions, the accused shall enjoy the right to a speedy and public trial, by an impartial jury of the State and district wherein the crime shall have been committed, which district shall have been previously ascertained by law, and to be informed of the nature and cause of the accusation; to be confronted with the witnesses against him; to have compulsory process for obtaining witnesses in his favor, and to have the Assistance of Counsel for his defense.

Note
The rules are simple; the Miranda Warning is required during any custodial interview of a suspect.

The rules are simple; the Miranda Warning is required during any custodial interview of a suspect. The two-prong test of Miranda is that the subject of the interview is in custody and the subject is asked questions about the crime.

Custody lies in the mind of the defendant: Does the person reasonably feel free to leave? This may be apart from actual arrest. The question of custody is a reasonable effect and can be subjectively looked at by the investigator. For example, if a homeowner is told by fire suppression personnel that she cannot leave, then she may reasonably assume she is in custody by government action. Therefore, it may be important to ensure that she realizes she is free to leave prior to any questioning. Likewise, if a person were to be surrounded by several investigators at once, perhaps displaying weapons, does that person feel he is free to leave or decline the interview? The U.S. Supreme Court defined *custody* in the case of *State v. Juarez-Godinez*[4] in 1997 with these words ". . . a police officer has seized a person, if that person believes that he or she has been seized and belief is objectively reasonable." Although the wording of the case indicates a *police officer,* that term means any government official. In the case of *Michigan v. Tyler,*[5] the court related that fire officials are indeed agents of the government and are bound to respect the rights of citizens. The opinion of Justice Stewart, regarding the U.S. Supreme Court decision in *Michigan v. Tyler,* explained, ". . . there is no diminution in a person's reasonable expectation of privacy nor in the protection of the Fourth Amendment simply because the official conducting the search wears the uniform of a firefighter rather than a policeman, or because his purpose is to ascertain the cause of a fire rather than to look for evidence of a crime."

A person who is questioned in a jail or an interrogation room is normally considered in custody and the Miranda Warning should be given. (See **Figure 19-3** and **Figure 19-4.**) The Miranda Warning is basically a two-part process where the suspect is warned and advised that he or she can remain silent under the Fifth Amendment and that he or she has a right to be assisted by an attorney as provided by the Sixth Amendment.

Peace Officers
Constitutional Pre-Interrogation Requirements

The following warnings must be given prior to questioning a person who is in custody or is deprived of his freedom of action in any significant way:

THE CONSTITUTION REQUIRES I INFORM YOU THAT:
1. YOU HAVE THE RIGHT TO REMAIN SILENT.
2. ANYTHING YOU SAY CAN AND WILL BE USED AGAINST YOU IN COURT.
3. YOU HAVE THE RIGHT TO TALK TO A LAWYER NOW AND HAVE HIM PRESENT NOW OR AT ANY TIME DURING QUESTIONING.
4. IF YOU CANNOT AFFORD A LAWYER, ONE WILL BE APPOINTED FOR YOU WITHOUT COST.

Waiver of Rights
The suspect may waive his rights, but the burden is on the officer to show the waiver is made voluntarily, knowingly and intelligently.
He must affirmatively respond to the following questions:
1. DO YOU UNDERSTAND EACH OF THESE RIGHTS I HAVE EXPLAINED TO YOU?
2. DO YOU WISH TO TALK TO US AT THIS TIME?

Election of Rights
A subject can avail himself of his rights at any time and interrogation must then cease.
If a subject will not waive his rights or during questioning elects to assert his rights, no testimony of that fact may ever be used against him at trial.

Figure 19-3 *The Miranda Warning should be read to the suspect from either a form or a card.*

It would be a monumental error to assume that just because the rights were read that the suspect has an understanding. It is wise for the investigator to note the person's condition as to alertness, mental capacity, and sobriety before moving on. Has the subject been an abnormally long time without sleep or food? Has the subject been in need of medical care or suffering from lack of medications? It should be determined if the subject has been taking intoxicating medications or has consumed alcohol. Mere consumption of alcohol is not a bar to interrogation, but if the subject is intoxicated to the point of not being able to understand his or her rights, that is a bar to interrogation. If there is any question as to the capacity of the suspect to knowingly and intelligently understand his or her rights, further questioning to determine the person's capacity needs to be done. It must be shown that interviewees were in control and aware of their surroundings, the seriousness of the situation, and that with this in mind they knowingly and intelligently submitted to the interview. In all cases make sure the suspect has a working knowledge of the English language. If an interpreter is used, ensure that he or she is certified and will be available for examination by the court if needed.

■ Note
Mere consumption of alcohol is not a bar to interrogation, but if the subject is intoxicated to the point of not being able to understand his or her rights, that is a bar to interrogation.

Age is not necessarily a barrier to interrogation. The investigator must ensure that if the accused is a juvenile, the juvenile fully understands his rights and the situation and can make a logical and intelligent decision. (See **Figure 19-5.**) It is advisable to have the parents present at the interview or

■ Note
Age is not necessarily a barrier to interrogation.

ANYWHERE POLICE DEPARTMENT

DEFENDANT Curtis Remke

INTERROGATION: ADVICE OF YOUR MIRANDA RIGHTS

Before we ask you any questions, you must understand your rights.

You have the right to remain silent . . . . CR Initials

If you give up your right to remain silent, anything you say can and will be used against you in a court of law. . . . . CR Initials

You have the right to speak with an attorney for advice before we ask you any questions and to have him with you during questioning . . . CR Initials

If you cannot afford an attorney, one will be appointed for you without charge before any questioning if you wish . . . . CR Initials

If you decide to answer questions now without an attorney present, you will still have the right to stop answering questions at any time . . . . CR Initials

Do you understand each of these rights I have read to you? . . . . CR Initials

Are you willing to answer questions and make a statement, knowing that you have these rights, and do you waive these rights freely and voluntarily with no threats or promises of any kind having been made to you? . . . . CR Initials

Charles Good
Witness's Signature

Curtis Remke
Signature of the defendant

Witness's Signature

Date 3-14-20__ TIME 1330 hrs D.R. # 97-860

Figure 19-4 *The "Advice of Miranda Warning Form" (waiver) is a good tool to use to make sure the individual knows and understands his or her rights.*

Figure 19-5 *Juveniles can be interviewed using certain precautions.*

someone to stand in for the parents and protect the rights of the child. This *in loco parentis* can be a responsible adult who will act in the best interest of the child.

Miranda is administered to a juvenile in the same manner as an adult with the parent acting on behalf or counseling the child for the decision to submit to the interview. Care should be taken that the accused can knowingly understand and willingly submit to the interview. In the case of a juvenile, the investigator must establish that the child has enough maturity to willingly and knowingly make the decision to be interviewed and that the child understands the rights that protect him or her. When in doubt about the subject's ability to consent to the interview, it is wise for the investigator to postpone the interview until advice can be sought from the prosecuting attorney.

■ Note
When in doubt about the subject's ability to consent to the interview, it is wise for the investigator to postpone the interview until advice can be sought from the prosecuting attorney.

In many cases, suspects ask the investigator for advice on what action they should take. The investigator should avoid this trap because the investigator cannot advise the suspect on any of the actions. Under no circumstances should the investigator attempt to advise the suspect on matters of Miranda. The investigator cannot speak for the prosecuting attorney or make promises or deals, and cannot offer any type of consideration in the criminal proceedings. The investigator will also need to avoid any intimidating statements as such may prejudice any information obtained from the interview. Such statements normally relate to the consequences of conviction. In such matters, simply state that *if* convicted what the maximum punishment *could* be, and ask the suspect, "With these rights in mind, do you wish to talk to me now?"

If the subject invokes the rights and refuses to be interviewed, then no interview can take place. Neither can the investigator initiate any contact with the suspect for the interview.

If the subject agrees to the interview after such warnings and the execution of a rights waiver, conduct the interview, mindful that at any time the subject can terminate the interview and invoke his or her constitutional rights. At that point, all questioning about the crime or facts of the incident must cease. Normal booking questions can continue, such as who the person is and other booking information.

PLANNING FOR THE CONFESSION

Note
The suspect should be interviewed privately without support or distractions.

Planning for an interview where the goal is to gain a confession to a crime requires the investigator to look at the logistics of the interview. The interview should take place where the investigator has the advantage whenever possible. The suspect should be interviewed privately without support or distractions. (See **Figure 19-6.**)

Interviews often fail when there is not enough time devoted to them. Typically, an interview has four parts. As with witness interviews, the prelude to an interview is a time to build a relationship with the interviewee. It should also establish that the interview is voluntary and consensual. Even in noncustodial settings where the Miranda Warning is not given there should be some declaration to the interviewees that they are there of their own free will without promise or duress of any type. This is easily done with an opening statement such as "Sir (or Ma'am), do you mind answering some questions about the fire at this time? If not, you are free to leave at any time."

Figure 19-6 *A typical governmental interview room provides minimal distractions for a successful event.*

The first part of a criminal interview is to confront the crime, or to establish that a crime was committed. Confront the suspect with the fact that the fire was incendiary in nature and you believe the fire was intentionally set. In addition, you may want to let the suspect know there is reason to believe that he or she was involved in some way. This is the first part of a calculated set of moves in which you will reveal why you believe the suspect committed the crime. As a point of caution, never release all your evidence at the beginning because this will leave you with little to use later in the interview. The majority of the time this early phase of the interview will be met with firm denials of the suspect's involvement. This is normal and should be expected. But this is a point where many interviews fail. Even if the suspect denies any involvement, continue to question. Always stay with the interview as long as the suspect will talk to you.

Inasmuch as you have worked the scene, questioning should be based on the facts gathered. If the suspect knows that you have evidence to support the accusation, the suspect will begin the process of considering whether an admission will be in his or her best interest or whether continuing to deny the accusation is best. You need to work diligently to overcome any denial of participation by asserting how the evidence points to the suspect as being the perpetrator of the crime.

■ Note
Any suspect interview involves questions that are accusatory in nature.

Any suspect interview involves questions that are accusatory in nature. At some point, there will be a confrontation, when the investigator asks the suspect if he or she committed the crime. To accomplish this, you should carefully prepare for the interview, knowing the information from the examination of the scene and the conclusions of witness interviews. Questions should be tempered with facts and allow for the suspect to fill in information. Patience and the expectation of evasiveness mean that questions will be asked time and again until proper information is attained.

Unfortunately, there is not a list of "sample questions" that always work. You will have to draw upon your communication skills, professional and life experiences, and people skills to accomplish a successful interview. Questions should be arranged so as to allow variations of the same questions in a different manner at strategic points in the interview. Suspect denials and variations on answers should be noted and used throughout the interview. It is well known that when a person lies or makes up information, it is difficult to maintain the story because additional information must be fabricated to support the first lie. That principle coupled with the stress of being interviewed by a person in authority makes it difficult for the suspect to keep on track and tell the same story time and time again because the person must make up the information quickly and under stress. When you challenge the suspect on the inconsistencies, he or she may come unraveled and confess. In any event, pointing out inconsistent information places the suspect at a disadvantage.

■ Note
Pointing out inconsistent information places the suspect at a disadvantage.

Word questions carefully, using professional tones and wording. Remember that the suspect expects to be treated with dignity. Firm statements and accusations should be presented professionally, without the "tough guy" approach. Use communication skills to ensure you are being understood by the suspect. Focus questions not on the crime itself, but on what motivated the suspect to commit the crime. This is often done by allowing the suspect to describe himself or herself, which can reveal points that may be useful in determining motive. This description can also be helpful in determining themes to use later; for example, if the suspect describes having financial difficulties, you may question the suspect later on whether the fire was set to gain money from the insurance on the property.

When the suspect denies participation in the crime, do not argue but firmly work on the principle that you have evidence that implicates the suspect in the arson. Inform the suspect of selected parts of evidentiary information that show his or her association with the fire scene. Let the suspect provide the defense or alibi information, which you can check on at a later time. Confront the suspect with your belief that he or she is a perpetrator and let the suspect try to convince you of innocence.

Firmly stand your ground and cut off efforts to deny by asking for reasons why the person committed the arson. This part is critical to the rest of the interview because you must take care to keep the suspect talking. If the suspect stops the denials and starts to offer justifications, you are making definite progress. It may be advisable to reinforce with the suspect that telling the truth is the best alternative. At this point, it is important to remember that you cannot make any deal with the suspect concerning prosecution. Only the prosecuting attorney can do so. You can relate to the suspect that you will make the suspect's cooperation known to the prosecution.

■ Note
You cannot make any deal with the suspect concerning prosecution.

In the third phase of interviewing, start to develop themes that minimize the crime and allow the suspect to maintain some dignity in confession. For example, if the target of the arson was a vacant house, you can tell the suspect, "That old house was about to fall down, you probably did the owner a favor by burning it." At this point, any justification is progress if the suspect admits any amount of involvement. Themes for the interview should focus on the reasons for the arson and can speculate on why the arson was committed. Present these themes as speculation for the suspect to comment on. Presenting a way for the suspect to confess while at the same time "saving face" can be a valuable incentive for the suspect. For example, presenting themes such as a prank that got out of hand may be a way to allow the suspect to maintain his or her dignity and confess, feeling that the confession will fall upon compassionate ears.

Themes should be relevant to the case and based on fact. You can use a certain amount of exaggeration or indulgence to make a theme more acceptable to the suspect; however, ensure that such themes are not **exculpatory** in nature. All themes presented to the suspect must have a central focus on the unlawful actions of the suspect, although they can be presented as mitigating factors to the unlawful act.

exculpatory
tends to clear or clears the suspect of guilt

Allow sufficient time to complete the interview process because in many cases interviews can last hours depending on the suspect's willingness to admit his or her involvement in the arson. In most cases, terminating the interview early results in failure, so it is advisable to allow up to 1 hour for each phase of the interview.

Investigators must conduct every interrogation with the belief that suspects, when presented with the proper avenue, will use it to confess their crimes. Research indicates that most guilty persons who confess are, from the outset, looking for the proper opening during the interrogation to communicate their guilt to the interrogators. Suspects confess when the internal anxiety caused by their deception outweighs their perceptions of the crime's consequences.[6]

■ Note
Suspects confess when the internal anxiety caused by their deception outweighs their perceptions of the crime's consequences.

The Breaking Point

The last part of the interview is the breakthrough point when the suspect actually admits his or her involvement in the arson. This is significant, although not the end of the interview. At this point, the suspect has acknowledged his or her participation in the arson to whatever degree, so there is no doubt as to the person's guilt. At this point, questioning should start to refine the details of the methods by which the arson was committed, allowing more latitude for the suspect to tell the story. It may be advisable to ask the suspect to write out the confession. The suspect should be questioned as to details and motives, with time to clarify each salient point of the situation. Once the suspect has admitted to the crime, it may be the desire of the investigator to arrest the suspect or cause the suspect to be arrested by law enforcement. If this is the case, the suspect will be in custody and the Miranda Warning must be administered.

■ Note
The investigator must ensure that all legal requirements are met.

SUMMARY

The first responder investigator has the opportunity to do preliminary interviews of victims, witnesses, and emergency responders. It is invaluable for first responders to know how the case will carry forward with additional interviews by the assigned investigator.

The assigned investigator needs compassion for the victims and to understand that sometimes witnesses are victims and have emotions that may come out in the interview. If there is one aspect of this chapter that must be remembered, it is that everyone involved in the event must be interviewed because each has a story to tell and that information may be exactly what the investigator needs when interviewing the suspect.

Certain steps taken in the interview of a suspect can provide the best opportunity for a confession. The most important rule is not to put a set limit on the amount of time needed for an interview. Always let the witness or suspect continue to talk as long as the person has something to say that is pertinent to the case.

The investigator must know how to listen; you need to hear the spoken word and the unspoken word—that is, what is not being said. This is especially true for victims who don't even refer to their loss. But be cautious about making assumptions about communication behavior because different people act differently in similar circumstances. The unspoken word also applies to the body language of the victim, witness, and suspect. These actions sometimes can speak volumes.

Legal issues of interviewing are very important to understand. The investigator must be fully versed in what is allowed or not allowed as per the law, which includes local laws and rules as well as decisions handed down by the U.S. Supreme Court.

The successful interview process can easily make the difference in proving a case for court. Moreover, even in accidental fires interviews can provide a key to confirm or dispel the investigator's hypothesis. All interviews become part of the case file through proper documentation and reports. Documentation must follow acceptable rules of evidence and be within the policy of the agency.

The investigator must ensure that all legal requirements are met. Not only will this protect the admissibility of any evidence that may be gathered via the interviews, but it will protect the investigator from liability. It is the investigator's responsibility to protect the rights of the citizen, including the person being interviewed.

KEY TERMS

Exculpatory Tends to clear or clears the suspect of guilt.

Interrogation Formal questioning, same as interviewing but has a more formal connotation.

REVIEW QUESTIONS

1. Why might some investigators like to interview the witnesses after they have examined the scene?
2. Why might some investigators prefer to interview the witnesses prior to examining the scene?
3. In what type of setting (give examples) would the investigator want to interview a suspect?
4. What authority do you as the investigator have to clandestinely record your conversation with a witness or suspect?
5. The Miranda Warning is based on which amendments of the Constitution of the United States?
6. When is the Miranda Warning required?
7. What must the investigator do if the suspect asks for an attorney?
8. List three instances when the principle of privileged communications can be invoked.
9. What must the investigator do to prepare for the interview of a suspect?
10. How much time should be taken to interview a suspect?

DISCUSSION QUESTIONS

1. The suspect is in custody, and he has asked for his attorney. When the attorney shows up, should you ask her whether you can interview the client?
2. Deep in your gut you know he started the fire. He refused to take the polygraph and refuses to do any more interviews. The prosecuting attorney has said that there is just

not enough evidence to convict without a confession. You approach him, and when the two of you are alone, you tell him that you know he set the fire but that you do not have enough evidence to arrest him. Then, you tell him that he is good, that he beat you. He looks you straight in the face and says, "Thank you." Is that a confession?

ACTIVITIES

1. Before there was the Miranda ruling there were requirements for federal investigators to give similar warnings. Search on the Internet to find the law that required federal law enforcement officials to give individuals their rights prior to the Miranda case. Also find out when that case took place.
2. Search on the Internet to find out who Miranda was and how he met his final demise.

NOTES

1. *United States v. White*, 401 U.S. 745 (1971).
2. Susan H. Adams, "Statement Analysis: What Do Words Really Reveal?" *FBI Law Enforcement Bulletin*, October 1996, 12–20.
3. *Miranda v. Arizona*, 384 U.S. 436 (1966).
4. *State v. Juarez-Godinez*, 326 Or 1, 942 P2d 772 (1997).
5. *Michigan v. Tyler*, 436 U.S. 499 (1978).
6. David D. Tousignant, "Why Suspects Confess," *FBI Law Enforcement Bulletin,* March 1991.

HUMAN BEHAVIOR

Learning Objectives

Upon completion of this chapter, you should be able to:

- Describe and understand the dynamics of an interview and the importance of starting with a base line before interpreting the physical and verbal reactions of suspects to certain questions.
- Describe the use of the polygraph examination in a fire investigation.
- Describe and understand actions people take or do not take at the fire scene based on their physical limitations or understanding of the emergency scene.
- Describe and understand the dynamics of a group and how group actions can be affected by gender, strength of organization, and knowledge of the structure or type of situation.

CASE STUDY

It was a typical police interrogation room: a table, three chairs, one wall that housed a mirror that took up most of the wall, and a huge bible sitting on the edge of the table. Built into the table was a polygraph machine. The operator could sit with his back to the mirror, and the person taking the polygraph examination sat on the other side of the table, not facing the examiner but at a right angle, facing the door.

With all the movies and TV police shows that are out, the person being interviewed should have known the mirror was two way and others were watching. He also should have known that those watching could also hear what was happening. He did not know the microphone was in the telephone wall jack adjacent to his chair. It would become obvious that he did not know that from the other side of the mirror, those watching could see every part of the room.

Let us refer to the person being interviewed as the suspect. Preliminary interview with this individual brought up more questions than answers. His body language was friendly and open upon first meeting, but as more questions were asked about the fire, he began to exhibit nervousness. He would not look the investigator in the face and always looked down when giving a specific answer. Furthermore, this individual had opportunity and motive to set the fire.

The polygraph was set up in advance with the investigator and polygraph examiner discussing the case so that a set of questions could be designed specifically for this individual. The investigator greeted the suspect, offered coffee and civil conversation while the equipment was being set up. The investigator thanked the suspect for his willingness to go through the process to eliminate him as potentially being involved, which would enable authorities to concentrate on finding who set the fire.

The suspect was left in the room for a while and observed. Again, everyone assumed he knew he was being watched. As the suspect took a few paces around the room, the examiner was quick to see that the suspect was walking in such a way as to put his weight on his heels rather than keeping his weight evenly dispersed on his entire foot.

The polygraph examiner went over the questions one at a time with the suspect before attaching him to the polygraph machine. Each question was adjusted to get a yes or no response, and then the suspect was hooked up to the machine. All this time, the large, 4-inch-thick bible sat on the corner of the table beside the suspect. The suspect glanced at it during the prequestion phase.

Remembering what the examiner had mentioned about the suspect walking on his heels, the investigator watched through the mirror as the exam was conducted. There did seem to be foot movement even though he was warned repeatedly to keep still by the examiner. During the examination, the suspect answered each question the same as he did in the prequestioning phase. The test was run a second time, and then a third time, possibly because of all that foot movement. Then, the examiner left the room and entered the adjacent room on the other side of the mirror.

While the examiner was sitting at the desk doing the evaluation for the results, there was the sound of a chair moving. The suspect stood up, looked at the mirror for a full 10 seconds, and then looked at the door. He then gently walked over to the wall

(Continued)

(Continued)

that contained the mirror and placed his back against the wall up against the corner. By moving up to the glass, the investigator and examiner could still see the suspect clearly. The suspect stood there for about a minute, and then lifting his foot, he took off his shoe and sock. He then pulled a tack out of his big toe; it had been buried all the way into the foot. He put his sock and shoe back on, and then did the same with the other foot. Sure enough, there was a tack in that toe as well, pushed all the way into the big toe. Blood could be seen on both of the toes, but he put his shoes on and went back and sat down in the chair.

Not surprisingly, he failed, really failed, the polygraph exam. The investigator took his file and entered the room. He took the third chair and pulled it up to the corner just in front of the suspect but to the side so that he was leaning against the table. He pulled the Bible up and rested his file on the book.

Silence followed as the investigator looked down at the suspect's shoes. Then, the investigator quietly asked if they could treat his wounds for him, stating an infection could get quite painful. Looking down at his feet, the suspect declined the treatment but stated that he watches TV and knows his rights. The investigator turned his head and looked at the door; looking back, he advised the suspect that he was free to leave at any time.

The suspect stayed. Glancing between his own feet and the Bible, he confessed that he read he could foil a polygraph test by inflicting pain by stepping on tacks. The problem he had was that just before he walked into the room he accidentally rocked onto his toes and both tacks embedded themselves. He went through the whole process with those tacks buried into each big toe.

Having confessed to trying to rig the polygraph test, it wasn't long before he confessed to setting the fire. Investigators must develop their skills of knowing what to say when, which questions to ask, how to establish the right atmosphere, all of which are important. Then, sometimes you can watch suspects set themselves up for failure and eventual confession.

INTRODUCTION

Note
Investigators try to be detectors of deception by evaluating what they hear from a person, how it is said, and the body language that goes with the conversation.

Interpreting human behavior is something we attempt to do our entire lives. Understanding human behavior can help investigators interview a person of interest to determine if further study of that individual is warranted. Investigators try to be detectors of deception by evaluating what they hear from a person, how it is said, and the body language that goes with the conversation. However, this is not always successful and is subjective at best. Although these clues are not admissible in court, they do give the experienced and trained investigator a direction to take in the investigation.

Understanding human behavior is critical in the investigation of fire injuries and fatalities, as well as understanding the actions of children that are behavioral versus criminal. Many studies have been completed to address why someone did not try to escape or why panic took hold or even in some cases why someone reacted heroically. These studies, if conducted scientifically, can be of great value in understanding what may have just occurred at the fire scene being investigated. These studies can help to explain how someone acts remarkably different when alone compared to how they act or react in a group. Group dynamics may take the individual on a path he or she would never have taken as an individual.

■ **Note**
The investigator not superimpose his or her own feelings and thoughts on anyone who has just suffered a loss.

It is imperative that the investigator not superimpose his or her own feelings and thoughts on anyone who has just suffered a loss. Crying or not crying, being angry or relaxed are all traits of individual personalities. Everyone handles tragedy differently. This does not mean that investigators ignore people's reactions. You document their actions, words, emotions, and then check these observations against what you learn through interviews and comments of others as to what they would have expected. Even then, any speculation is not definitive but simply an indication of the direction the investigation may take in the near future.

■ **Note**
Understanding human behavior is critical to the success of the investigation.

Whether investigating an individual for potential involvement in the setting of a fire or working a fatal fire and trying to determine why the victim did not escape, understanding human behavior is critical to the success of the investigation. This chapter is simply an introduction. Both the first responder investigator and the assigned investigator will benefit by understanding human behavior.

CRIMINAL INVESTIGATIVE INTERVIEWS

First responder investigators will rarely have the opportunity to get involved at this level in a criminal case, first, because they have not received the necessary training and, second, because their assignment is with their engine company and few departments have the resources to have the engine company do these types of functions.

As a general rule, we do not like to use the term *interrogation* to describe an interview. *Interrogation* is a proper term, a legal term in some respects. What is described as an interview of the suspect in many circles can be called an interrogation. Because of abuses in the past of using highhanded threats, using a single light shined on the face, or even the cliché of beating with a rubber hose has given the proper term a negative connotation. Thus, it may be better to use the phrase *conducting interviews*.

■ **Note**
A properly conducted interview can provide valuable information leading to a proper conclusion to the investigation.

A properly conducted interview can provide valuable information leading to a proper conclusion to the investigation. An improper, botched, or poorly prepared interview can at the least create a lost opportunity or a longer path of ever getting the truth. The worse-case scenario is accusing an innocent person, and just below that would be the lost opportunity to solve the case.

■ **Note**
The investigator must be prepared before conducting interviews.

The investigator must be prepared before conducting interviews. Preparation is not just having paper, pen, or a recorder; the investigator must be mentally prepared. The interview should start on a neutral tone and the investigator must stay in control through the end. No interview should be put on a time limit. If the investigator does not have time to conduct the interview, then it should be rescheduled.

The Suspect

In previous chapters, we talked about the advantage of investigators having a fire background, which enables them to interpret the burn patterns based on their firefighting experience. In this instance, the street experience of a police officer can provide the skills that will help the investigator to read people, to interpret body language. Also in previous chapters, we discussed legal issues about questioning and detaining individuals and about how the rules differ between representatives of the government versus someone representing the insurance entity.

■ Note **To make a fair assessment of an individual the investigator may need to observe the suspect in a more relaxed or neutral attitude.**

To make a fair assessment of an individual the investigator may need to observe the suspect in a more relaxed or neutral attitude. This can be accomplished by using standard pleasantries prior to the more in-depth interview. As more pressure is put on the potential suspect, there will be more and more indicators that the investigator can interpret such as body movement, posture, and eye movement, which are all referred to as *kinesics*. Sometimes these actions by themselves speak volumes.

The line of questions should take the suspect along a logical path. A generic list of questions on any fire situation can be a good leadoff with clarifying questions along the way as needed. Eventually, the specifics about the fire will need to be asked. As there is more evidence linking an individual to the fire, the questioning may want to cover the motive of the fire as well.

If there is not enough information for the government investigator to make an arrest, this is not the only opportunity to question the subject. Certain items might need to be researched to determine whether answers were truthful or deceptive. If the investigation reveals deceptive answers to the questions asked, then another interview may be in order, this time in a location where the investigator will have control, such as a station interview room.

The interview room is a controlled environment. The only thing in the room should be a small table and no more chairs than are necessary for that interview. The process should be videotaped, unobtrusively if possible. This will allow the recording of any confession or admission of fact that could lead to a conviction. Just as important, the videotaping can ensure that false accusations will not stand against anyone during such events. If video recording is not available, the conversations should be recorded on audio devices for the same reasons. Investigators must always follow the law and take the higher road in such situations, for the good of the people. It is critical that all recordings of the interview are kept, unedited, for review upon appropriate and proper request.

■ Note **There is no one solution for every situation. Sometimes the tactic being used just does not work, so tactics can be changed to see if another will yield results.**

There is no one solution for every situation. Sometimes the tactic being used just does not work, so tactics can be changed to see if another will yield results. Investigators' ability to speak their own body language back to the suspect is important to send messages, such as feigning surprise, disbelief, or even sympathy in some instances.

What formal college classes can help someone prepare to be an investigator? For fire scene investigation, a good scientific background in physics, chemistry, and engineering can be beneficial. Drama or acting classes can enhance interview skills, because an investigator may need to role play to carry out the task.

Witnesses

The interview of witnesses must be done to glean as much information as possible. In the case of a victim, it must be done with grace, compassion, and understanding, and yet sometimes be insistent enough to get the information necessary. Emotions can be quite high depending on the circumstances. Victims may have been closer to the crime and will have more information than others. Victims may even know the suspect. They may also have insight on the motive, why the individual committed the crime.

■ Note

The basic information needed from each witness is who, what, when, where, and how.

The basic information needed from each witness is who, what, when, where, and how. As mentioned, witnesses may also know why. They will use all their senses to provide such information. The obvious is their vision: what did they see? So often, witnesses say they didn't see anything important, but talking through the incident they may come up with something they did not think was important but very well could be of assistance. There could have been some sound that could provide clues as well, especially if they can narrow the sound down to a time. Some witnesses could even tell the type of car that drove by their house even if they never looked out the window, especially if they once owned that type of vehicle. Some sounds can be quite unique such as the sound of an older Volkswagen Beetle. The value of such information varies on a case-by-case situation.

Smell can be telling, especially if an odor reminded the witness of a volatile liquid. Smells may not always be fire-related. One witness of a Dumpster fire behind a high-rise apartment stated that the suspect ran past and had a very strong smell of menthol. Upon arrest, it was found that the suspect worked in the local cigarette manufacturing plant and worked in the menthol room.

Something Smells

An extreme pounding on the front door brought the Reverend jumping out of bed and running to the front door. There, out of breath, was his neighbor who stated his house was on fire. They called 9-1-1, and the neighbor stayed in the living room until the fire department arrived. The neighbor talked about how he had jumped

(Continued)

(Continued)

out of the second-floor window with flames around him. He was limping from the fall, but he was fully dressed even though it was 2 o'clock in the morning. While waiting, the adrenalin factor started to decrease and the Reverend noticed an unusual smell. He then realized it smelled like gasoline. As the victim was sitting in the chair, the pant legs of his trousers had risen up. He was wearing athletic socks, which were usually white, and he had on white sneakers. Both sneakers and socks were slightly yellow stained.

When the fire engine arrived, the reverend and the neighbor went out to meet the firefighters. Standing in the side yard, they watched as flames came out of both the first- and second-floor windows. The neighbor was also a deacon in the Reverend's church, and the Reverend found it hard to admit to himself that the deacon could have been involved in setting the fire.

The fire investigator showed up on the scene and interviewed the deacon. He found clear marks under the window that affirmed that the deacon had indeed jumped from the second-floor window. But the marks were from shoes and the deacon was clearly barefoot. The investigator then approached the Reverend and asked if he was the one who called 9-1-1. General questions were asked, and then the investigator asked if there was anything he noticed that might have been unusual about the victim. There was a pause, and then the Reverend explained how there was a smell of gasoline and the white socks and sneakers of the deacon appeared to have yellow stains.

Also during the interview, the Reverend mentioned that once the engine arrived, the deacon went to a shed in the backyard. The investigator asked permission to look in the shed and the deacon assented. There on the floor were two sneakers and two athletic socks, reeking of a gasoline odor and yellow stained. Recovered and properly sealed, they were taken as evidence. The deacon saw the investigator with the items and mentioned that he had spilled gasoline while filling his lawnmower the previous day and had taken off his shoes at that time.

The statement and eventual testimony of the Reverend, that he smelled what he believed to be gasoline and that the deacon was wearing shoes when he came to the Reverend's home, were enough to eventually elicit a confession. The deacon had poured gasoline throughout the first floor to burn down the house; with some fuel still in his container he walked up the stairs while pouring and entered the bedrooms, pouring more gasoline. He wanted to make sure the house went to the ground. However, vapors from the gasoline in the kitchen, mixed with the right percentage of air, eventually reached the gas hot water heater pilot light. The sound of the ignition of the vapors could be heard upstairs and the deacon, recognizing his plight, dove out the second-floor window he had left open to make sure the fire had sufficient air to burn. His quick thinking and the open window may well have saved his life, but for the next 10 years that life was behind bars. Thanks in part to the Reverend's sense of smell, this case was solved.

The sense of touch and taste may not come up as often in fire cases but have been known to assist in solving other crimes where the crime scene was set on fire to cover up the evidence. A person's senses are tools that can be used to eventually help identify the perpetrator of the crime.

HUMAN REACTION IN FIRES

An important function of any fire investigation is to determine why or how someone got hurt or died as a result of the fire. Understanding human behavior may help explain some of the actions taken by citizens, but a thorough investigation and follow-up are necessary to accurately determine the facts surrounding fire deaths. The first responder investigator can use human behavior to explain many aspects of people's actions in the initial fire report. The assigned investigator will also find that understanding human behavior is required in the follow-up and detailed investigation needed for every fatal fire.

■ Note
The first responder investigator can use human behavior to explain many aspects of people's actions in the initial fire report.

Some outstanding scientific reports explain how different people react in different fire scenarios. These reports are available from the National Fire Protection Association and the National Institute of Standards and Technology. They specify that many variables need to be explored to explain why some people act one way and others act another in the same situation.

It has always been known that a leader can affect the actions of a group. History books are filled with situations where an army on the verge of being routed were revitalized and led to victory by an able leader who could grab their attention and lead them on. The same can be said for a leader leading citizens to safety in a fire.

The concept of leadership was proved during fires at the Iroquois Theater and more recently at the Beverly Hills Supper Club, where actors Eddie Foy and John Davidson, respectively, were able to get the attention of patrons and pointed them to safety before heat and smoke forced the performers to leave the stage. In another situation in the early 1900's, a milkman rushed into a crowd and, because of his white uniform, gained the trust of the people so that he was able to calm a panic and lead some to safety; those rescued admitted they followed because of the uniform. Theaters in the mid- and late twentieth century allowed police officers and firefighters free entry, to badge in, at shows. The reasons were twofold: an appreciation for their service, and if something should happen, these public servants could lead patrons to safety rather than let panic set in.

FACTORS AFFECTING HUMAN BEHAVIOR IN FIRES

When working a fire fatality, there is always the question of why a person did or did not do something. Finding bodies just under windows or adjacent to doors is so distressing knowing they were so close to rescue. Many times, the effect of

A Sad Discovery

Fire was coming out of only the kitchen windows, but smoke permeated the entire structure when the fire department arrived. Even though the structure was occupied, a neighbor called 9-1-1 only after smelling smoke and discovering the house next door was on fire. Two of the young children were found in the backyard, but the baby and the babysitter were not outside. The first firefighting crew entered the structure to do a search. Turning right, they found the babysitter immediately. She was on the floor and the baby was in her arms. Both were pulled out and attempts were made to revive them, but both were pronounced dead at the hospital.

The fire was knocked down within minutes and ventilation was conducted. As the smoke cleared, the discovery adjacent to the front door was bone chilling. Approximately 3 feet off the floor were marks on the wall where the babysitter had presumably clawed away to escape. The doorknob was only 18 inches away.

carbon monoxide in their systems takes away the ability to reason and make an escape. Other times, a multitude of causes, some of which we address in this chapter, can prevent a person from escaping.

Ability to Recognize the Hazard

■ Note
The individual must be able to recognize that there is a hazard and that danger warrants action.

The first factor is that the individual must be able to recognize that there is a hazard and that danger warrants action. This is more a cognitive issue and the individual may be too young or too old to understand and comprehend the danger. The individual may be incapacitated or impaired by drug or alcohol use. Those with developmental disabilities, although they are the size and have the strength of an adult, just do not have the mental capacity to make a proper decision in a fire situation. Also, although it may be difficult to believe, lack of sleep can dampen or prevent lack of good judgment and prevent clear understanding of the hazard, leading to the wrong choice with dire consequences.

Carbon Monoxide

Another factor that contributes to not recognizing the hazard or more aptly not reacting properly to the hazard is being overcome by the smoke. Carbon monoxide is one of many by-products of incomplete combustion. Inhalation of this gas leads to oxygen deprivation. As the gas enters the lungs, it combines with blood cells more easily than oxygen does; thus, the blood does not carry sufficient oxygen to the brain, which leads to loss of reasoning ability.

Inability to React

Physical limitations can also prevent the escape of an occupant. Many of these conditions are the same as those listed in the section titled "Ability to

Recognize the Hazard." Whether too old or too young, the individual may recognize the hazard but is physically unable to react. Alcohol may not completely take away the ability to recognize the hazard, but its effects on the body may prevent escape.

Medical conditions can prevent escape. Those who are bed-ridden, infirmed, or of limited mobility are certainly at risk in a fire situation not only because of the speed of the fire, but also because of obstacles in any typical structure that limits their ability to escape.

Knowledge of Surroundings

■ Note **Not being familiar with the surroundings can be a factor preventing escape.**

Not being familiar with the surroundings can be a factor preventing escape. This is a common theme in motel fires. All too often, occupants take the longer path to an exit because they are not familiar with the closest exit. This is usually because they take the path they already traveled when they entered the structure.

Regretfully, the average person does not enter the room, look at the exit marked on the diagram on the back of the door or the safety section in the book in the room, and count the number of doors to the closest exit. Even if the person does this, it is doubtful whether he or she will remember this information at 2 in the morning. For this reason, many advocate having lit exit signs at floor level in hotel hallways. This way, when exiting through a smoke-filled corridor the occupants can see the way to the exit door. Some products on the marked can actually create a line of flashing lights that can be followed to the exit.

Group Dynamics

■ Note **People behave differently when they are around others than when they are alone.**

People behave differently when they are around others than when they are alone. A group is usually less likely to react quickly than is a lone individual in similar situations. The larger the group, the slower the reaction.

Consider how the "Station" fire in Rhode Island broke out. The crowd saw the flames, but it seems that each individual was thinking that if no one else was worried, then maybe I shouldn't either. Deep down inside, they began to recognize there was a problem. Another interesting observation about that fire was the fact that there were exits by the stage, but the patrons all rushed to the door they came in and not to a secondary exit. Band staff members were standing by a door, fire was at the stage, but all the patrons headed toward the front. This is typical: They were familiar with the door they came in, so they wanted to exit the same way.

An organized group in which everyone knows each other is more apt to react more quickly. In such groups, a hierarchy is already in place and the usual leaders will step up in the emergency and the group will react. The longer the group is together, the quicker the reaction in an emergency situation.

Gender

There are differences in how men and women react in a fire. Most dramatically, men tend to search for the fire, whereas women are more apt to call 9-1-1 immediately. Men are more apt to get extinguishers and fight the fire, whereas women will gather the family first and leave the structure. With a nighttime fire, women are twice as likely to get dressed than are men. Men have a tendency to go back into a burning building, but that is far less likely to be done by a woman.

Children

Most likely, the hardest thing to face at a fire scene is a child fatality. There are so many factors in dealing with children and fire that texts, classes, and programs have been specifically designed to address such issues.

■ Note
To see things through the eyes of a child may be difficult at times for some adults.

To see things through the eyes of a child may be difficult at times for some adults. To find a child under a window is such a loss especially when there is no fire damage, just smoke. But the window was jammed. You may ask yourself why the child didn't just break the window. Perhaps because the child knew that to break a window was against the rules and feared punishment.

Every investigator should take specialized training in dealing with children in the areas of fire curiosity, fire setting, and fire play. Many fire departments hire specialist public fire and life safety educators to deal with these types of problems. Commonly called public educators, these specialists can team up with the fire investigators to find solutions for these children. They also work with parents to help them understand why a child would play with fire or why a child got hurt as a result of a fire and that it does not necessarily make them bad parents.

This is an area in which the first responder investigator should receive some training as well. A fire officer with a uniform, badge, and turnout gear can make a big impression on a child. Sometimes visiting the fire station and receiving specialized fire safety training can make a difference in a child's life.

■ Note
By investigating the actions of victims in fires, we learn much more that enables us to prevent future injuries and loss of life.

COLLECTION OF HUMAN BEHAVIOR DATA

By seeking as many answers as possible from every fire, the fire service can identify what causes fires and take steps to prevent fires. Likewise, by investigating the actions of victims in fires, we learn much more that enables us to prevent future injuries and loss of life. If we could eventually discover how a majority of people would react to a certain situation, then we can create products and procedures to endeavor to keep them safe when an emergency arises.

SUMMARY

The investigative interview is an opportunity to find clues that can lead to a successful conclusion of an investigation. Interviews should be planned in advance and start off in a neutral, relaxed demeanor in hopes of getting a base line to evaluate future reactions to your inquiries. Sometimes a polygraph test can be a good tool to help in the interview of a strong suspect. It is also a good elimination tool but is an expensive, time-consuming process that should be limited in use.

However, the real study of human behavior comes from the research on each and every fire scene that resulted in a fire injury or fire death. Every death must be investigated to find out how the person died. Not just the medical examiner's report is needed, but an answer to why, for instance, that person never got out of bed to save himself when he was physically and mentally able to do so.

Today, fire and building codes mandate the use of smoke detectors because research was able to prove that they worked in saving lives. It has been proved that without them people die. Research has also shown that when danger arises inside a structure, people flee in the direction they entered the building, which can sometimes create a bottleneck and result in people dying.

For the future, the fire service could change the way we educate people and create new fire safety products and procedures—all this is within our grasp. But it will only happen when fires are thoroughly investigated, documented, and reported and when that information is used for future research.

REVIEW QUESTIONS

1. What is meant by getting a base line in an interview before starting to question the individual on details of the case?
2. What is the study of kinesics?
3. What is the benefit of using an interview room at a station? What should be in this room?
4. What are the five senses a witness or victim may use to provide information on a suspect?
5. Give three examples of factors that would prevent someone from recognizing the hazard.
6. Give three examples of factors that would prevent someone from reacting to the hazard.
7. Why is there a faster reaction time to an emergency when a group is organized?
8. Between men and women, who is more apt to look for the fire and attempt extinguishment?
9. What basic information does the investigator need from each witness?
10. Why is it important to collect human behavior data?

DISCUSSION QUESTIONS

1. Why not just call an interrogation exactly what it is, an interrogation?
2. When in an airplane, do you look at the emergency safety information card? Do you count the number of seats to the closest exit? Why don't people do this every time they fly?

ACTIVITIES

1. On the Internet, go to the National Institute of Standards and Technology (NIST) site at http://www.nist.gov/ and review recent fire investigative reports under the Building and Fire Research link.
2. On the Internet, go to the National Fire Protection Association (NFPA) site at http://www.nfpa.org/ and review recent reports on fire investigations and human behavior. (Search for "human behavior" on the site to access statistical reports.)

SOURCES OF INFORMATION

Learning Objectives

Upon completion of this chapter, you should be able to:

- Describe the freedom of information laws.
- Describe the various sources of information available at both municipal and state agencies.
- Describe various sources of information available from private entities such as insurance companies.
- Describe and understand that all reports are relevant to your case and admissible in any necessary legal proceeding.

CASE STUDY

In the scope of things, it was a relatively small explosion. Some living nearby remember hearing something but were not aware there was a problem until they heard the sirens in the distant. The explosion was in the attached garage, resulting in the destruction of the garage and the wife's car in the driveway, just in front of the garage. The detonation set off a fire that eventually spread to the house, causing some fire damage in the kitchen and smoke damage throughout the structure.

The homeowner stated he was not at home when the incident occurred. The first engine company officer, the first responder investigator, realized that this could have been an explosive device with the destruction that occurred near a work bench. He secured the scene and called for an assigned investigator.

There were no vehicles inside the garage and the lawnmower and spare tanks were in a shed in the backyard. No volatile liquid or propane containers could be found in the garage. A screwdriver was found embedded in a stud, and there were small holes in various areas of the exterior wall that gave the impression of shrapnel. Explosive experts were brought in before digging the scene. It was a joint venture. Evidence of an explosive device was recovered and sent to the laboratory for examination.

The homeowner was married at the time but living alone because his wife left him and flew home to her mother's in another state. When questioned about the device, he stated that he had no knowledge of such things. He admitted to being in the army but was just in the infantry and had no special skills. He claimed to have no enemies and got along with everyone in the neighborhood and at work.

Neighbors who were interviewed confirmed that there were no problems. But they also indicated they didn't know the homeowner and that he was hardly ever home. His employer said there was no problem with his work as a laborer in a warehouse, but they had heard stories about his drinking and about him being thrown out of a bar recently.

In a courthouse search, authorities were able to confirm the property ownership that he had been paying his taxes, that he had no misdemeanor convictions, and that he did not appear on any financial statements recorded at the courthouse. One file that had not been searched yet was a book with copies of all the military DD-214s, which are the discharge papers of anyone leaving military service. The forms list the person's time in service, rank, specialty, and tours of duty via medals. This file contained the records only of veterans who voluntarily brought in their DD-214. Almost everyone brought in their papers after World War II and Korea, fewer after Vietnam, and very seldom are these papers brought in today. However, the homeowner's DD-214 was in the file. He was not infantry; his military specialty was listed as EOD, explosive ordinance disposal.

The facts stood that he had lied about his military occupation. The laboratory confirmed that the material submitted had trace residue of explosive material. Multiple credible witnesses saw the bouncer throw the suspect out of the drinking establishment. These same witnesses also heard the suspect threaten to blow up the bouncer and the establishment the day before the device went off at the suspect's home. It took the jury just 2 hours to reach a decision of guilty. Ironically, the defense counsel pleaded mercy at the sentencing hearing based on the defendant's military record.

INTRODUCTION

■ **Note**
There is a wealth of information out there just for the taking.

■ **Note**
A lot more information can be obtained with a kind request than with a threat.

There is a wealth of information out there just for the taking. The search can be tedious and time consuming and may not always uncover exactly what is needed. But not to look may mean passing up great opportunities that could make a difference in the case you are working. Information can be obtained from governmental entities at the local, state, and federal levels. The private sector may have information available; getting it may result from how it is requested. A lot more information can be obtained with a kind request than with a threat. Never threaten, but give stern warnings. Save such warnings as a last resort, and before ever issuing a warning, make sure you have the legal ability to carry out the consequences.

The investigator must know all laws, rules, and regulations as they pertain to obtaining and giving out information pursuant to any request. Laws differ between states and the federal government and are subject to change each legislative year. State supreme courts and the United States Supreme Court can make rulings that may affect how you handle information requests. A local investigative association may provide training events so that you can stay on top of such changes. But the one authority who should be consulted on any information dealing with a criminal case is the prosecuting attorney, or the local jurisdiction counsel should be consulted if the subject is an administrative request that is not criminal court–related.

■ **Note**
The wallet badge will open doors without shouting your position to the world.

One final note: When doing a search at a physical location, it may not always be best to do so in uniform, which may bring unnecessary scrutiny from others in the facility. This may be one time when plain clothes are the better choice. Also, if in plain clothes, investigators should put their best foot forward by making themselves look professional by wearing a coat and tie or business attire. A wallet badge is appropriate because you can show it only when needed. The wallet badge will open doors without shouting your position to the world. One last suggestion: Always leave a business card, which is not only professional but may also provide future information from those who keep your card.

FIRST RESPONDER CLUES

The basic background information uncovered by the first responder investigator can be beneficial to the investigation. Items such as a homeowner's comment about a prior employer or information on a T-shirt that may indicate a previous or current affiliation with a job, organization, or the military can be important clues. Granted, many people wear T-shirts and have nothing to do with that organization. But don't let that ever deter you from taking the opportunity to seek information.

Some items noticed by the first responder investigator may also be noticed by the assigned investigator. This should not be a problem at all and should be discussed when the two meet for a briefing before the first responder investigator leaves the scene.

FREEDOM OF INFORMATION ACT

Most people do not fully understand the Freedom of Information Act (FOIA, or FOI Act). They take the title for granted and assume their rights from that alone. Federal requirements under FOIA are usually different from states' requirements. The Department of Justice mandates that the request be in writing, whereas some state FOIA requests need only be verbal. However, most do give a time limit for initial response from the government, from weeks to a month or more. However, all require a response and are allowed to charge the requestor for the items, provided that the cost is fair and equitable.

The first responder investigator needs only to know that he or she does not release any information, that any such request must go to the fire administration to be handled from that office. However, the first responder investigator can have an impact on any request received by the administration. There is a finite amount of time for response to a legitimate FOIA issue. For this reason, fire reports should be completed as soon as possible after the event to enable the administration to comply with any such requests.

States have their own FOIA that addresses how requests are to be handled. The assigned investigator should be completely familiar with the federal and state statutes and all they require to obtain information and to request information.

Much of what the investigator may need can be found at the local courthouse without putting in a FOIA request. Some federal records may be beneficial to the investigator's request, and investigators need to be familiar with how to obtain such documents. The local prosecuting attorney can provide guidance.

COURTHOUSE SEARCH, CLERK OF COURTS

As for that courthouse search, most courthouses have switched over to using electronic documents for most types of records. Deed history and plats may still be found in the large texts, but the rest of the information can usually be obtained from a data terminal. If you are doing an in-depth study of a particular individual, going back several years, it may be beneficial to review the old records. The clerk of courts may grant your request to review these records.

If you are fortunate enough to be allowed to peruse these older documents, be very careful. Do not have a pen hovered over any document because this makes the clerks nervous. Do not take the books apart to make copies; ask permission or have a clerk do it. Always put the books back where they belong even if you find them out on the counter. This respect will go a long way for any future searches you may request.

Be cautious when asking for assistance. Remember, the clerk may have lived in the community for years; more important, the clerk assisting could be related to your suspect. Or worse, the person you are investigating may be innocent, and the fact that you are checking on him or her may give some in earshot the impression that the person may have done something terribly wrong. Just remember, you can't go wrong if you always treat people as you would want to be treated.

■ **Note**
You can't go wrong if you always treat people as you would want to be treated.

GOVERNMENT RECORDS SEARCH

Depending on the state, some records are kept at a local government authority and others are kept at the state capital. Most are available for search. The decision to make is how much information you are going to reveal to get what you need. Under the premises that most clerks would not have admissible information on your particular case, it may be best not to talk about the case. As mentioned in the section titled "Courthouse Search, Clerk of Courts," the person you are talking to may be related to or may know the defendant, so why take that chance at this point.

■ **Note**
The person you are talking to may be related to or may know the defendant.

Any record obtained must be relevant and germane to the case at hand. Records must be authentic, and it may be worth the effort to have the agency create a document to prove their validity. Foremost, records obtained should be able to be admitted in a court of law.

Federal Records

Depending on circumstances, a multitude of records can be obtained through the federal government. If the individual is active military or a veteran, then there will be records on the person's expertise. Records can prove when the defendant was out of the country. Remember, as an investigator, you are a seeker of truth and information about someone's innocence is just as important as evidence of guilt. When working a fatality with a body whose identity cannot be confirmed, the military may have dental records that can help in the process.

If the fire involved a farm, the farmers may have been involved in several federal programs, records of which might have information on crop production, blight with a crop, or recent loss of a subsidy.

■ **Note**
The State Department may have information about canceled or prohibited contracts associated with the property in question.

If the property was a manufacturer and evidence at the scene indicates that the products were shipped overseas or raw materials came from overseas, the State Department may have information about canceled or prohibited contracts associated with the property in question. If there are indications that the fire was incendiary in nature and the property was involved in interstate commerce with a suspect that may be from another state, the Bureau of Alcohol, Tobacco, Firearms, and Explosives may have special agents who can assist with your case.

If there is any information that the individual or business had contact with any federal agency, then it would be worth the effort to explore this with that agency. Again, this may be a missed opportunity if you do not ask the questions or do the research.

State Agencies

Many state agencies may be of assistance. Birth records are primarily handled by the state now, whereas in the past they were a local issue. Most of those records were transferred to the state when states took over that responsibility. Birth records can help to confirm the identity of the suspect and give the name and information of the mother and father if a detailed background check is being conducted.

One thing that any new fire investigator may find interesting is the direction that an investigation may take. Motives don't always have to make sense. One individual was approached by his ex-girlfriend, who said she had proof that he was the father of her baby and was seeking child support. He set fire to the house that night, the house that had copies of the proof that came from the Bureau of Vital Statistics.

■ Note
Motives don't always have to make sense.

There are different names for similar state agencies, and a visit to the state's website will clarify which agency to contact to get information. Some examples of assistance from state agencies to verify motive are the state gaming commission, which pulls the license to run an individual's bingo hall; the alcohol control bureau may pull a restaurant's alcohol license; or the state health department may pull a restaurant's cooking permit for gross or frequent food safety violations. Even a beauty salon or barber shop needs to maintain a state license to stay in business. An agency pulling a license could cause a business to go into bankruptcy. On the other hand, the owner may rather collect the fire insurance money.

■ Note
An agency pulling a license could cause a business to go into bankruptcy. On the other hand, the owner may rather collect the fire insurance money.

The state agency responsible for creating or widening roads may be able to offer tax value to the landowner to buy the landowner out to make improved roads. That offer may be less than the market value of the property, but instead of fighting it, the owner may burn the property to get the insurance money, and then sell the property to the state. This is important because some state or local laws have allowed the highway department to condemn properties, giving a fair value for the land and property, and doing so for the good of the people.

Local Government

Very similar to the state, various local agencies may have information of interest to the case being worked. A local health inspector may enforce the state health regulations or the locality may create their own streets and roads very similar to the state highway department.

The economy can provide a motive for setting property on fire. The housing market may be in a slump with many homes going months or even years without selling. This can create an extreme fiscal burden. Then, the tax assessor increases the assessed value of the home, making an even larger fiscal burden with increased tax bills. Checking the assessor's office can give an indication of whether tax rates have risen or the home or business had recently been reappraised.

The building official will have information on building permits issues and those denied. That office also has information on buildings built in the past and may even have a set of working plans for when specific buildings were erected.

We would be remiss if we did not mention the obvious: Both the fire and police department records should be searched for a connection with the suspect. Also, do a search based on the address where the fire occurred. There may be a history of goings-on at this address, anything from illegally burning trash to domestic calls on a regular basis—both of which may be of interest to the investigator.

UTILITIES

Most, but not all, local utilities are in the private sector with some oversight by the government. However, it is not unusual for items such as providing gas service to be a function of the city. A quick search on the Internet can identify who to call to check on the status of the local utilities. For example, the power company can tell you whether a property recently defaulted or had a change in its electricity use.

■ Note
The power company can tell you whether a property recently defaulted or had a change in its electricity use.

One investigation indicated that the electricity use had tripled over the previous 2 months. There was some speculation that it was because the homeowner was growing illegal drugs and needed a large amount of grow lights. But, in fact, the fire was electrical in nature, and there had been a dead short to ground from the back of the electrical panel as a result of a defect.

■ Note
The gas company has records of visiting the property for prior problems with the gas equipment.

The gas company has records of visiting the property for prior problems with the gas equipment. The water department may provide information, for example, as in one recent case, that the water had been shut off for the previous 2 months, but the homeowner says they were living in the home and everything was fine. In fact, they were not living at the home, but the homeowner's policy would not cover a home not occupied at the time of the fire. Although the family did not set the fire, they lied to collect the insurance money. This is plain and simple fraud. The insurance company denied the claim but did not want to press charges.

INSURANCE COMPANIES

Throughout this text, we have discussed assistance from insurance companies. Further assistance may be from a third party that can be used with the cooperation of the insurance company involved with the fire.

The Property Insurance Loss Register (PILR) is a private corporation whose members are insurance companies. Member companies list their major losses with PILR so that information can be shared with other members. Should there be a large loss, a member can research whether the insured has experienced large-loss fires with other insurance companies.

Another private sector resource for information is the National Insurance Crime Bureau (NICB). This is a not-for-profit organization fighting insurance fraud and vehicle theft by providing analysis, investigations, training, and public awareness of this type of crime. Specifically, it has a wealth of knowledge on vehicle fires, especially if the fire involves the identification of the vehicle itself. If a vehicle identification number (VIN) is missing, NICB has knowledge of alternate locations where the VIN can be found. The training classes are outstanding and can be a resource for an investigative association training event.

PRIVATE SECTOR

With the proper search warrant, the investigator can seek the identification with known business associates, suppliers, and customers. Researching any one may reveal information that can be considered in the investigation. For example, it would be good to know if the business owner was current with payments to raw materials suppliers. Keep in mind that it is not unusual for a business to hold onto its capital as long as possible; in other words, to not pay their bills until absolutely necessary. Checking the history will tell whether this was a normal practice, or if the payments only fell behind during the months leading up to the fire.

> **■ Note**
> **For example, it would be good to know if the business owner was current with payments to raw materials suppliers.**

This same theory can be used with residential homes: Were they paying their bills late, or was this a normal practice for many years? Just because someone is behind in making payments does not mean that person set their house on fire. It is just one more piece of data that must be evaluated using scientific methodology.

Competitors may have insight but be careful about taking this at face value. Competitors can be friends or would do anything to eliminate present or future competition. The instinct of the investigator will help give weight to any data collected. Verification of that data may help as well.

> **■ Note**
> **Competitors may have insight but be careful about taking this at face value.**

PEER ASSOCIATIONS

One of the best resources for any investigator is an association with neighboring investigators where they meet on a regular basis to discuss cases, issues, problems, and solutions. This liaison with other interested professionals is critical for success in the region. This association can be an informal gathering at a local coffee shop or a formal association with bylaws and officers. Either way, it is an opportunity to share, educate, and cooperate.

These associations can set up the necessary training to allow the investigator to stay current on investigative issues such as legal decisions and advances in technology. Even more advantageous is the opportunity for the first responder investigator to be involved as well. This can strengthen the team and allow for better collaboration in the future.

This group would be remiss not to include their peers in other law enforcement agencies, local, state, and federal. Because many incendiary fires involve other crimes, it would be advantageous to establish a rapport with these other professionals, considering that there eventually will be opportunities to work together on future crime scenes.

SUMMARY

The documents you obtain can provide a wealth of information. Investigators must leave themselves enough time to do the search for information. New investigators may stumble along in doing these information searches. Doing the search is an education in itself. As investigators conduct more information searches they learn of more resources and will get better with the process as time passes.

If you are not familiar with who or what agency may have information, go to a state or the federal government's Web site and scroll down the various agencies. Each government agency may have its own Web page describing its duties and responsibilities. Then, you can arrange to go and meet with representatives to search for data for a better hypothesis.

Joining or creating a regional investigative group is advantageous for the entire area. This provides opportunities to share information on events in each locality as well as to have the potential for cooperative training events to enhance the investigators' knowledge and skills.

REVIEW QUESTIONS

1. What type of information may be found at an assessor's office?
2. What is a DD-214 and what type of information might it contain?
3. What reason would you have for not bringing up the suspect's name when doing a courthouse search?
4. What state or local agencies may have information that can validate a motive for burning a restaurant?
5. What state or local agency has the right to condemn property for the good of the public?
6. What local department may have a set of working plans for the property where the fire occurred?
7. What is NICB? What assistance can this organization provide to the investigator?
8. What type of benefits could be gained by belonging to an association of fellow investigators from the region?
9. What is FOIA? Please explain.
10. What condition at a local business may require an inquiry to the State Department?

DISCUSSION QUESTIONS

1. What attire should the fire investigator wear when conducting interviews and doing document searching: uniform, suit, or casual attire including jeans?
2. What type of information might the first responder investigator see the night of the fire that would be beneficial to pass on to the assigned investigator, who may not notice the next morning?

ACTIVITIES

1. Go to the local courthouse and do your own research on yourself or your parents' property and information about any family member. If your family is not from the area where you are now, pick a friend. You may want to take that person along for this eye-opening activity.
2. On the Internet, search on your local government and make a list of the various agencies it includes. Identify which may be of assistance in doing an investigation.

THE EXPERT WITNESS

Learning Objectives

Upon completion of this chapter, you should be able to:

- Describe the role of the first responder investigator and this investigator's possible involvement in a court trial.
- Describe the basic premises in the *Daubert, Frye,* and *Kumho* cases as they relate to the testimony of an expert witness.
- Describe Federal Rule 702 and its impact on the fire investigator.

This is an unusual subject that requires an unusual departure from the book format. The following cases affect the testimony of a fire investigator. Each is an abbreviated synopsis. It is the responsibility of each and every fire investigator to be aware of both *Frye* and *Daubert* along with other cases that have an impact on the fire investigator testifying as an expert witness.

A word of caution: These case descriptions are extremely brief but convey the basic elements to understand the expert's testimony. It is highly recommended if and when someone is to become an expert witness in any case that they should know these cases and completely understand all the rules of the court in which he or she will testify. Granted, Rule 702 may be for the federal courts, but most state courts have very similar rules.

CASE STUDY

Frye v. United States[1]

In 1920, James Alphonso Frye confessed to murdering Robert Brown, a Washington, D.C., physician. Frye later recanted his confession, and to prove his innocence his lawyer wanted to produce evidence that his client was innocent by use of an examination of systolic blood pressure to show he was telling the truth when he said he did not murder Dr. Brown. They even offered to run the test in front of the jury. The request to use the test was denied.

The case went to the Court of Appeals of the District of Columbia in 1923. The only question of error was the denial by the trial judge of letting the defendant use the results of a deception test. The basis of the test is that the systolic (upper) measurement of the blood pressure will rise when the person experiences nervousness. Based on this theory, people are always nervous when they tell a lie and the pressure will rise. Telling the truth, on the other hand, creates no rise in the blood pressure. A researcher, William Marston, who already had a law degree and was working on his Ph.D. in psychology gave the test to Frye, and based on the test alone said he was being truthful when he said he did not commit the murder.

One piece of trivia: The press came up with the phrase "lie detector test." This was not a term that Marston used, but he did state that it was "the end of man's long futile striving for a means to distinguishing truth-telling from deception."

The Court of Appeals in their decision stated the following:

> Just when a scientific principle or discovery crosses the line between the experimental and demonstrable stages is difficult to define. Somewhere in this twilight zone the evidential force of the principle must be recognized, and while courts will go a long way in admitting expert testimony deduced from a well-recognized scientific principle or discovery, the thing from which the deduction is made must be sufficiently established to have gained general acceptance in the particular field in which it belongs.

(Continued)

(Continued)

> We think the systolic blood pressure deception test has not yet gained such standing and scientific recognition among physiological and psychological authorities as would justify the courts in admitting expert testimony deduced from the discovery, development, and experiments thus far made.

This decision is not just about the lie detector. It is about science, in that before something can be admitted into court as an expert opinion it must have scientific recognition and general acceptance in that specific field.

CASE STUDY

Daubert et ux., individual and guardians ad litem for Daubert, et al. v. Merrell Dow Pharmaceuticals, Inc.[2]

In 1974, William and Joyce Daubert had a son, Jason Daubert, who was born with a congenital disorder. He had only two fingers on his right hand and was missing a lower bone in his right arm. During Joyce's pregnancy, she had taken the prescribed drug Bendectin for nausea. There was a second minor in this case: Eric Schuller was born without a left hand and one leg was shorter than the other leg.

The Dauberts and Schullers sued on behalf of their sons, saying Bendectin caused the deformities. Merrell Dow brought forth expert testimony from Dr. Stephen Lamm, who stated that he reviewed multiple published human studies and concluded that Bendectin used during the first trimester of pregnancy was not supposed to be a health risk.

The Dauberts' lawyer brought forth eight affidavits based on animal testing that showed a link between the use of Bendectin and birth defects. Because the animal testing had not been published and had not been peer reviewed it was not admissible as expert testimony. The courts found for the defendant, Merrell Dow.

As discussed in Chapter 5, "Legal Issues and the Right to Be There," the Supreme Court chooses cases that will best establish, reestablish, refine, and define rules that are consistent with the Constitution of the United States. In 1993, they chose to hear the *Daubert v. Merrell Dow Pharmaceuticals Inc* case. The reason to hear the case had nothing to do with the question of whether Bendectin caused the birth defects of the Dauberts' son. They chose that case to clarify the issue on who can or should testify as an expert witness.

The Supreme Court affirmed both the Trial Court and Court of Appeals that testimony cannot be allowed unless generally accepted in that relevant scientific community. However, they stated that the decision is based on the Federal Rules of Evidence, not *Frye*.

CASE STUDY

Kumho Tire Co. v. Carmichael[3]

On July 6, 1993, the right rear tire blew out on a minivan driven by Patrick Carmichael. As a result of the accident, one passenger died and the others were severely injured. In October, the Carmichaels sued the tire manufacturer and distributor. They claimed that the tire was defective, causing it to blow, which caused the accident and subsequent death and injuries. The Carmichaels obtained a tire expert to testify that the tire in question was defective.

The lower court had no problem with the expert being a tire engineer, especially because he had worked for Michelin for 10 years. The fact that the engineer actually looked and touched the tire was a reliable method. However, the expert based his conclusions on the fact that the tire was defective because he did not see any other causes. The lower court's concern was that they were to take this expert's opinion only by the *ipse dixit*, just because he said so. Based on *Daubert*, the testimony was excluded.

The Eleventh Circuit Court overruled the District Court on appeal saying that *Daubert* only applied to scientific testimony. They felt that *Daubert* was not an issue in this case. The Supreme Court reversed the Eleventh Circuit Court and upheld the District Court's decision that the testimony of the expert should have been excluded and it was proper to invoke *Daubert* because it applies to all experts, not just scientists. The District Court did as they should, act as the gatekeeper evaluating the reliability of the evidence to be heard.

INTRODUCTION

Daubert or *Frye* has affected almost every federal court case involving expert testimony for the past 80 years. However, which expert testimony may be allowed in court in one state may not be the same as what is allowed in another state. The differences are slowly disappearing but have not totally gone away. On the other hand, Congress and the Supreme Court have endeavored to create rules that will be used in all federal courts, creating consistency across the country.

■ Note
The expert witness is to testify if qualified by their knowledge, skill, experience, training, or education.

Under Federal Rule 702,[4] the expert witness is to testify if qualified by their knowledge, skill, experience, training, or education. A lifetime of reports is only as good as the paper they are written on if the author cannot testify in court as to their contents. Fire investigators must be knowledgeable in all of the rules of the court for which they are about to testify. There are federal rules that require presubmittal of reports and the expert witness's curriculum vitae. State courts may or may not have similar requirements.

The bottom line is that the courts have to find a way to ensure that the scientific material that will be introduced into evidence for a trial is sound and not based on junk science or no science. This is truer today than ever before. The Innocence Project[5] put together a Peer Review Panel of top fire investigative specialists in this field to review two cases. Their findings clearly exonerated both individuals wrongly convicted of arson. One was released from prison prior to the publication of the Innocence Project Report. For the other, it was too late. The state had already executed him as a result of his sentencing. He had been tried and convicted for setting the fire that killed his three children. What put him on death row and eventually took his life was junk science.

■ Note One was released from prison prior to the publication of the Innocence Project Report. For the other, it was too late.

THE FIRST RESPONDER INVESTIGATOR AND COURT

A completely filled out fire report such as used by the National Fire Incident Reporting System includes the identification of the area of origin and the cause of the fire. This documentation is usually the extent to which the first responder investigator will usually go on any case. Without the involvement of the assigned investigator, this will be the only government report available. In almost every circumstance, there will be no other notes or files on the investigation.

■ Note The first responder investigator at some time in his or her career may very well receive a subpoena to testify in a deposition for a civil case.

Based on that report, the first responder investigator at some time in his or her career may very well receive a subpoena to testify in a deposition for a civil case. When this occurs, the first call should go up the chain of command. There is little doubt that the assigned investigator will assist and no doubt that the jurisdiction's legal counsel will provide guidance from that point forward. Depending on the make-up of the organization and relationship between their offices, there may be involvement, as an advisor, from the prosecutor's office because the prosecutor deals with trial law more than the jurisdiction's legal counsel does.

■ Note The first responder's level of training, years of firefighting experience, and expertise will be sufficient to allow that investigator to testify in a court of law. However, this will be the decision of the judge hearing that case.

At this point, some first responder investigators might think that they will always call the assigned investigator no matter how the fire looks. This is not the solution. Doing a deposition, or testifying in court, is not as harrowing as it may seem. Preparation and pretrial actions will be dictated by the court hearing the case. There are different rules for most state versus federal courts.

There is no way to predict what will happen. Most likely, the first responder's level of training, years of firefighting experience, and expertise will be sufficient to allow that investigator to testify in a court of law. However, this will be the decision of the judge hearing that case. For the assigned investigator, as an expert in the field, he or she must be prepared to testify and meet the rules of both federal and state courts for criminal and civil cases.

COURT RULES

Federal and state courts have an abundance of rules for every aspect of the trial. They cover both civil and criminal trials. Of interest to the fire investigative field are the rules for federal civil trials. The Supreme Court has worked hard in choosing cases to facilitate the understanding of who is an expert and the limitations and reliability of the expert's testimony. A general rule of thumb is that if you can meet Federal Rule 702, you should have little problem testifying in most other courts, both criminal and civil.

Note
If you can meet Federal Rule 702, you should have little problem testifying in most other courts, both criminal and civil.

Federal Rule 702

In 2000, Rule 702 was updated based on the *Daubert/Kumho/Joiner* trilogy. This helped to solidify that the courts do need to act as gatekeepers. The synopsis of Rule 702 states:

> If scientific, technical, or other specialized knowledge will assist the trier of fact to understand the evidence or to determine a fact in issue, a witness qualified as an expert by knowledge, skill, experience, training, or education, may testify thereto in the form of an opinion or otherwise, if (1) the testimony is based upon sufficient facts or data, (2) the testimony is the product of reliable principles and methods, and (3) the witness has applied the principles and methods reliably to the facts of the case.

Note
All expert testimony, not just scientists', will be evaluated on the rules.

Note
The trial judge will decide admissibility based on the preceding and, in particular, on the witness's knowledge, skill, experience, training, or education.

All expert testimony, not just scientists', will be evaluated on the rules. The trial judge will decide admissibility based on the preceding and, in particular, on the witness's knowledge, skill, experience, training, or education. The evidence must be based on facts or data that are the product of reliable principles and methods and that were gathered using those reliable principles and methods.

The National Fire Protection Association in NFPA 921 has strived and worked endless hours with input from all disciplines on the committee to stress the use of scientific methodology and a systematic approach. Applying these scientific principles accepted by the fire investigative field should help to make accurate final determinations of the fire cause that will meet the evidentiary rules should the case go to court, state or federal, civil or criminal.

THE FUTURE OF COURTROOM TESTIMONY

The fire service and the fire investigative field have not heard the last of the *Daubert* issues. It will be imperative that the investigative community as a whole begin taking every step necessary to ensure that we get our own house in order. The first step is education, not just on investigating the fire scene, but education on the law, the courts, and the rules associated with courtroom testimony.

In the *Benfield* case, *Michigan Miller's Mutual Insurance Company v. Janelle R. Benfield*,[6] the court allowed the insurance fire investigator to testify based on his credentials, but following cross–examination the testimony was stricken. The court, in particular, the judge, found that the expert did not cite any scientific theory and applied no scientific method. Under cross-examination, the expert admitted that "no source or origin can be found on his personal visual examination and therefore the source and origin must be arson." The judge ruled that *Daubert* applies even though the investigator was testifying on his investigative experience and not as a scientist.

Ironically a new trial was eventually allowed and even with the exclusion of the insurance investigator's testimony the insurance company prevailed. In the second trial, the assigned fire department investigator testified as an expert in fire investigations, not fire science, and that testimony was heard.

The *Benfield* decision was not a Supreme Court decision and at the time of publication was an isolated case. However, it is a wake-up call for the fire investigative community to ensure that they are testifying as fire investigators and not as fire scientists. It is critical that investigators follow the scientific methodology in conducting their case, using logic and being systematic.

SUMMARY

The assigned investigator ends up in criminal court on a regular basis. There is little doubt that investigators may also end up in court on a civil case as well. Likewise, first responder investigators too may have to testify in court as to their involvement in a case. For this reason, both need to become extremely familiar with the rules of the courts and their ability to testify as expert witnesses.

The *Frye, Daubert,* and *Kumho* cases were briefed at the beginning of this chapter. Of interest is that fact that none of these cases involved a fire. Because they each involve the testimony of an expert witness, they affect the fire service and the investigator's testimony as to the fire area of origin and cause. The investigator should know that there are other cases that also affect the expert witness.

By using these cases, the Supreme Court has been able to define and describe more accurately what is necessary to be an expert witness and has solidified the reliability and relevance of what is to be included in the expert's testimony. Congress created the Federal Rules for court cases, and Rule 702 affirms that the trial court is the gatekeeper to ensure that any expert testimony meets the mandates of Rule 702. As such, allowing only proper testimony has a better chance of not only seeking the truth but making wiser decisions in the adjudication of cases.

For the future, fire investigators need to heed the *Benfield* decision as a wake-up call to make sure they testify as fire investigators and not as scientists and that they are systematic in their efforts, ensuring that scientific methodology is followed in each and every case they work.

REVIEW QUESTIONS

1. What was the expertise in question in the *Frye* case and why did the judge not allow the expert to testify?
2. What was the expert testimony that was not allowed in the *Daubert* case and why was it not allowed?
3. What was the expert testimony in the *Kumho* case and what was the ground for denying the expert to testify?
4. Under Rule 702, on which five items does the judge decide as to the expert witness's qualifications?
5. Rule 702 states that an expert may testify thereto in the form of an opinion if three conditions are met. What are these conditions?

DISCUSSION QUESTIONS

1. As a result of a mutual aid agreement, you responded to a neighboring jurisdiction to assist with a fatal fire investigation where you clearly saw a fire that went to flashover. Your expert opinion is that the fire was accidental in nature. However, the investigator for that jurisdiction has insisted it was arson and the homeowner is to blame. This investigator has been a close professional friend for many years. What do you do?
2. Reference the preceding question. The trial is about to start and you still know that, in your professional opinion, the fire was accidental and the accused is innocent. What do you do?

ACTIVITIES

1. On the Internet, on the Innocence Project page, download the "Report on the Peer Review of the Expert Testimony in the Cases of *State of Texas v. Cameron Todd Willingham* and *State of Texas v. Ernest Ray Willis*" (http://www.innocenceproject.org/docs/ArsonReviewReport.pdf). Print and review this document to find inaccuracies in the investigation.
2. On the Internet, download documents and writings on the *Michigan Miller's Mutual Insurance Company v. Janelle R. Benfield* (*Benfield* decision). Be sure to become conversant with the facts surrounding the case and the impact the case may have on the fire investigative community in the future.

NOTES

1. Peter Nordberg, "The *Frye* Opinion," http://www.daubertontheweb.com/frye_opinion.htm.
2. *Daubert v. Merrell Dow Pharmaceuticals* (92-102), 509 U.S. 579 (1993), http://supct.law.cornell.edu/supct/html/92-102.ZS.html.
3. *Kumho Tire Co. v. Carmichael* (97-1709) 526 U.S. 137 (1999) 131 F.3d 1433, reversed, http://supct.law.cornell.edu/supct/html/97-1709.ZS.html.
4. Federal Rules of Evidence, "Notes on Rule 702" (Jan. 2, 1975, P.L. 93-595, § 1, 88 Stat. 1937), http://www.law.cornell.edu/rules/fre/ACRule702.htm.
5. Innocence Project Arson Review Committee, "Report on the Peer Review of the Expert Testimony in the case of *State of Texas v. Cameron Todd Willingham* and *State of Texas v. Ernest Ray Willis,*" http://www.innocenceproject.org/docs/ArsonReviewReport.pdf.
6. Guy E. Burnette, Jr., "Fire Scene Investigation: The *Daubert* Challenge," InterFire Online, http://www.interfire.org/res_file/daubert.asp#benfield.

THE FINAL REPORT, TESTIFYING IN COURT, AND PIO

Learning Objectives

Upon completion of this chapter, you should be able to:

- Describe the process of developing a final analysis and hypothesis.
- Describe the process of preparing and completing a final, accurate, and concise report.
- Describe the process of preparing and delivering a verbal report in the allotted time.
- Describe the process of preparing for court and testifying in a legal proceeding.
- Describe the process and benefits of preparing a public information presentation on the investigation findings.

CASE STUDY

The judge had set aside 2 days for the trial with instructions to the attorneys that he saw no reason for it to take longer, with which the attorneys agreed. The judge got the witness lists from the two attorneys, looked at them, and handed them to the bailiff. The bailiff read off the names and instructed the witnesses to line up across the bar, which is the small fence-like separation between the audience and the activities of the court. Once lined up, they were asked by the bailiff to raise their right hand. He then read an oath asking if they were to going to tell the truth, the whole truth, and nothing but the truth in the case to be heard before this court on this day. All the witness said, "I do."

The judge addressed the witnesses, asking general questions about whether anyone had special needs and if they were here freely of their own accord, and then he gave one final instruction. No one was to discuss this case in any form, in any way, while they were **sequestered.** On that note, everyone was led off to the witness waiting room. Everyone was in civilian attire, even the two investigators. Before everyone had settled into their chairs two individuals started talking about their testimony, about what they would say and comparing notes.

sequestered
to be separated or isolated from the public during a trial

The investigator wrote a small note about what was happening, opened the door, and gave it to the bailiff. The bailiff asked which two, and then escorted them outside the room where another bailiff stayed with them.

The judge was in jury selection when he got the note. He cleared the courtroom except for the attorneys and court security. The two men were brought before the judge, and then were brought back to the waiting room. It is not known exactly what the judge said, but after their testimony, they were escorted to the hall adjacent to the courtroom, handcuffed, and led off. All we know is that they were in contempt of court, allowed to testify for the sake of justice, but would be guests of the state for a couple of nights.

INTRODUCTION

■ Note
The final report must contain all aspects of the facts surrounding the case.

The final report must contain all aspects of the facts surrounding the case. Every hypothesis must be included, even those discounted and why they were rejected. The report will need to stand on its own but the investigator will need to deliver the report to an intended audience as well as to a judge and jury at the time of the trial. The report must be made available during the discovery portion of the trial and civil courts may require an individual to testify in a deposition prior to the trial.

Regardless of whether the case goes to court there may be an opportunity to provide public information on the findings of the investigation. A public information officer for the fire department is a key element to make this happen. If there will be a court case, the information released may be minimal, with the full set of facts to be released after the trial. If the fire was accidental, the public information officer can put together messages to be provided to the news media advising the public about what to do to prevent similar occurrences in the future.

REPORTS

Putting the report together requires a review of the entire file to see what other avenues need to be explored prior to closing the case. Even though the investigator knows his or her investigation was systematic and followed the scientific methodology, it must be reflected in the report. The author of the report must keep in mind that anyone reading that report will not have someone to explain what is missing; the report will be judged on what is in, or not in, the final findings.

Any information that needs corroborative evidence or documentation before going to trial needs to be documented with a final worksheet in the back of the report to serve as a guide. This is necessary because someone else may be obtaining this information. For example, a suspect's alibi was that he was 600 miles away in another town. A fortunate thing about networking is that there may be another investigator you know in that area. Based on your report and the additional information sheet, the other investigator may be able to handle documenting whether the suspect was or was not at the location given.

■ Note Many times showing a motive is not necessary to get a conviction. If motive can be confirmed, then it should be in the report.

Review the report, and look for any discrepancies. If you find any, make immediate corrections; to delay means having to find that location in the report again. This would be unproductive and there may be a chance the discrepancy does not get corrected.

■ Note For the jury to not know the motive would be like turning off *Law and Order* 2 minutes before the end of the episode.

Motive

Many times showing a motive is not necessary to get a conviction. If motive can be confirmed, then it should be in the report. The report should reflect the relative surety of the investigator's opinion of the motive. Again, the reader should understand if the listed motive is absolute or an educated speculation; it must be based on data received and documented.

The jury is going to want to know why that individual committed that crime. With so many crime drama shows on television the public has come to expect to know why the person committed the crime. For the jury to not know the motive would be like turning off *Law and Order* 2 minutes before the end of the episode.

■ Note Accidental cases deserve the same scrutiny as criminal cases. You are just as apt to end up in court over a civil issue as you are in a criminal case.

Accidental Causes

Accidental cases deserve the same scrutiny as criminal cases. You are just as apt to end up in court over a civil issue as you are in a criminal case. The same attention to details and documentation is important. To be careless just because it is not a criminal case can jeopardize your credibility. In a state courtroom system the same judge may hear both criminal and civil cases. If you want that judge to have faith in your testimony, always do your best.

Final Report

The final written report must accurately reflect the investigator's findings, must be concise, and must include the investigator's expert opinion on fire (or explosion) area of origin, cause, and product or person responsible for the causation of the event. It should be written so that it could be easily understood by the judge or jury. Always keep in mind that some basic principles on chemistry of fire or engineering may need to be explained. The report does not need to be a novel with witty prose but needs to hold the interest of readers so that they do not skip the details. Word of warning: Most prosecuting attorneys will flip to the back of your report and look at your final findings, so don't be offended.

■ Note
Always keep in mind that some basic principles on chemistry of fire or engineering may need to be explained.

Verbal Report

This is just a minor notation about something that should never happen in the government sector. In the insurance industry, a claims manager or a special investigation unit chief may send out an investigator to just look at the scene to see if it warrants further investigation. When the scene looks accidental and it seems that the insured did not increase the hazard by storing drums of gasoline in the basement, the investigator's instructions may be to not write a report and move on to the next case.

This type of instruction for the insurance industry is appropriate and proper. It is not acceptable for the fire service or any law enforcement agency. A written report must be completed on each and every case. To suggest a report is not necessary is an invitation to future disasters. We have a responsibility to the public to document our activities and at least to put any pertinent information in the basic fire report.

■ Note
In a perfect world, every fire department would submit reports either directly to the National Fire Incident Reporting System (NFIRS) or there would be a state version of this system that would collect state fire statistics and forward them on to the national system.

NATIONAL FIRE INCIDENT REPORTING SYSTEM

In a perfect world, every fire department would submit reports either directly to the National Fire Incident Reporting System (NFIRS) or there would be a state version of this system that would collect state fire statistics and forward them on to the national system. To not collect these data means incomplete national or state reports on fires, which in turn means missed opportunities to educate and promote all the good things the fire service accomplishes.

A case in point is a state that only has 50 percent of its departments submitting electronic reports to the state for collection. These statistics from only half of the fire departments are compiled for a report on fire in that state. Even with special notation that indicates that the report reflects only half of the state's fire departments it can be a problem. Not everyone reads those notations. Should the fire service want to submit legislation to get training money for every fire department in the state, the legislators would only see half of the fires, injuries, and civilian deaths that occurred. Needless to say, that legislation may only have half a chance to succeed.

For those departments reporting to NFIRS, this is where the work of the first responder investigator can really pay off. The report will be complete as to the findings of cause and origin. But there is one problem. As the assigned investigator takes on the case and additional findings or different findings are discovered, the investigator must go back and make supplemental reports so that the fire report reflects exactly what happened.

■ Note
The investigator must go back and make supplemental reports so that the fire report reflects exactly what happened.

COURTROOM PREPARATION

So many things must be done prior to going to court. There should be pretrial meetings, creation of a multimedia presentation, and final review of the report. Here are a few things to think of before going to trial. However, there may be very little time for the first court appearances if this is a criminal trial. Many states require an almost immediate hearing for setting bail and to hear testimony by the arresting officer, or prosecuting attorney, on probable cause.

In many courts, the bail amount will be a request by the prosecuting attorney with an alternate lower amount from the defense attorney with a plea of releasing the client immediately. Also, for most criminal proceedings, the prosecuting attorney will question the arresting officer to establish whether probable cause to make the arrest was justified. The investigator should not provide any more information that is asked for by the attorney or judge. An interesting comment from most fire investigators from coast to coast is that they do not get to see the prosecuting attorney before this hearing and even in some cases they present their findings on their own without comment from the prosecuting attorney. Keep in mind this is the judge's courtroom, and things need to be run the way he or she says, period.

■ Note
A curriculum vitae (CV) is nothing more than a résumé that includes your personal history and professional qualifications.

Curriculum Vitae

A curriculum vitae (CV) is nothing more than a résumé that includes your personal history and professional qualifications. Unlike a job-search résumé where the shorter, the better, the CV should be thorough and complete with all your background information along with all classes you have taken as well as any state and national certifications. Awards that were specific to the job can be added as well. The last part of your CV should consist of a chronological list of every case where you had an opportunity to testify. The information should consist of the case name, date, and docket number; it has been mentioned by some investigators that it may be beneficial to include the name of the presiding judge as well as the attorney with whom you worked on the case.

Other information that can be beneficial includes listing your jobs other than those in public safety. If you were a licensed electrician, it may be beneficial to list that on your CV. If you served in the military and worked in any field that may have a connection with your position such as bomb disposal, electronic surveillance, military police, or intelligence, then this too should be on the CV.

As you can see, every time you take a new class or training event your CV will need to be updated and each time after you testify, it will need to be updated. As you can imagine, over the years this document will certainly grow.

Multimedia Presentations

Note
The investigator should identify what type of multimedia equipment may be available in the courtroom.

The investigator should identify what type of multimedia equipment may be available in the courtroom. The bailiff or prosecuting attorney may be able to assist. Most new courthouses have built-in screens, projectors, computers, and some judges actually have a built-in screen for them to see anything being projected. (See **Figure 23-1.**)

Other courtrooms don't even have a screen. If your presentation will be computer based with a multimedia projector, if possible have a backup plan just in

Figure 23-1 *Fire Marshal giving testimony from the witness stand. Note the computer monitor; modern courtroom design can include technology such as computer monitors and multimedia all built-in for use during the trial.*

case. Be sure to think of everything, including an extension cord. Again, the court bailiff or court security will know what you can and cannot do in that judge's courtroom. Do not delay in this aspect to make sure you have your best presentation possible.

If photos are an important part of your case, also bring at least 5 × 7 prints, if not 8 × 10s to submit as evidence and so marked when submitted. This means that the jury may get to take these photos during deliberation, which can be an advantage.

Attorney Conference

Most prosecuting offices are no different from most fire investigative units—short of personnel and all too often short of time to get things done. Don't be too critical when you have to set up the meeting, and you may have to do this several times.

■ Note
The file must be reviewed in its entirety by the prosecuting attorney and read again by the investigator who will be testifying as an expert witness.

The file must be reviewed in its entirety by the prosecuting attorney and read again by the investigator who will be testifying as an expert witness. The conference should cover all aspects of the case. If this is an important trial, there may even be an opportunity for a pretrial run through on the testimony. Regretfully, this does not happen often enough, primarily because of the caseload of the attorneys and the investigator. If possible, this is an outstanding tool, even more so for the less experienced investigator. Don't be fooled because even those who have testified hundreds of times will benefit from this process.

DISCOVERY

The discovery process should be basically the same in all courts but maybe not exactly the same in every state. Regardless, opposing counsel should ask for copies of your report and anything to which you will be attesting to in court. As seekers of truth, you know this is not a bad thing and only fair. The prosecuting attorney will do the same with the defense experts, requesting all information that will be presented. Surprises in the courtroom make for great drama on television but are bad for the justice system.

This process will usually come through the court with the judge's assent, order, to make it happen. As such, it will give clear instructions on what is to be given, and you should follow these instructions to the absolute letter. This is a court order.

interrogatories
a series of written questions for discovery, whose answers are given under oath

There may be a request to answer **interrogatories** from opposing counsel. Interrogatories are nothing more than a list of questions that you must answer relating to the case. Always let the prosecuting attorney review the questions first. If you have been subpoenaed to testify in a civil trial, let the attorney who subpoenaed you review the questions. The written answers must be complete and thorough, and they are required to be provided with an affidavit as to the truth of the answers. Again, answers should be shared with the attorney with whom you are working on the case.

Any reports, such as the forensic laboratory report, may be requested along with information about your expertise. This of course is your curriculum vitae. If a metro task force was activated to help with the event, the names and contact information of everyone who worked on the case may be requested. But with a thorough report you can simply state that the information is in the report and give the page number. To save a phone call, also state that the list is true and complete as written in the report, providing this is the absolute truth.

Special note: Consider the case in which you got a discovery document and you produced everything and signed that it was true and correct. Then, in a couple of weeks, you discover you made a mistake. Never hesitate to call the attorney you are working with and provide that information immediately. A supplemental report can be provided along with an explanation on why it was left off. Will you hear about this later? Yes, there is no doubt. To decide not to say anything out of embarrassment would be a tremendous mistake, and it may jeopardize the case and your career.

COURTROOM TESTIMONY

Note

If it is a civil case, you should advise the attorney requesting your testimony that a subpoena will be necessary. This subpoena should be kept with your file.

If this is a criminal case, you are an employee of the government; you do not need a subpoena to arrive in court. If it is a civil case, you should advise the attorney requesting your testimony that a subpoena will be necessary. This subpoena should be kept with your file.

Attire

As for what to wear on your day in court, that should be a local issue decided by your department. If they do not have any guidance, here are some recommendations. The first responder investigator is usually an engine company officer. As such, it may be best for this investigator to be in dress uniform. It sends the accurate message that the investigator is a line officer and as such one of the first people on the scene. Along this line of thought, this individual will be testifying as to what was discovered during or immediately after suppression activities. The reason for wearing a dress uniform is out of respect for the courts and portrays a professional image.

The assigned investigator can make a better impression if dressed in a suit and tie or business suit. The tie should be simple and professional. A single pin on the lapel is fine but not a lot of jewelry. In the past, it was general knowledge that a person looks trustworthy if he wore a blue suit, white shirt, and a maroon or red tie. That may or may not be true today, but it does seem to be the attire of many CEOs across the country. Regardless, look professional.

Demeanor and Your Testimony

The process of your testimony can change from state to state. There may also be a significant difference as to the flow of the case and the order of the proceedings from state to federal court. The following information is given as an example and may not match exactly how things are done in your state.

■ Note Demeanor is a critical aspect of the testimony.

Demeanor is a critical aspect of the testimony. Taking an oath is done differently in different states. Some courts have all witnesses step forward; they are all sworn in at the same time, and then marched off to be sequestered until they testify. *Sequestered* means that you will be separated from the trial and you will not hear what others say because it may influence your testimony. Instructions from the judge should be that you will not discuss the case. This means that while sequestered, isolated, witnesses are to say nothing of the case. If two or more witnesses are discussing the case, call for the bailiff and advise of such. There is nothing illegal about discussing issues of the case before trial, but once the trial starts there is to be no conversation about anyone's testimony until the judge has released you from the stand.

Be prepared to be in this room for a long time. Once you have been sequestered, the judge will hear any motions. These are requests to the court by attorneys to make a ruling on one thing or another on behalf of their client. There may be a motion to exclude someone's testimony based on one reason or another. Motions may be for summary judgment, asking the court to dismiss the charges based on the prosecution having no evidence, which is based on what was turned over in the discovery phase. While this is going on, the judge may have to review various aspects and this may take some time. Once that is complete, the jury selection will begin. It has not been unusual for investigators to sit for a day or more waiting to testify.

■ Note When taking an oath, stand up straight, place your right arm out straight and at a right angle with palm facing forward. Keep this pose until the swearing is complete.

When taking an oath, stand up straight, place your right arm out straight and at a right angle with palm facing forward. Keep this pose until the swearing is complete. From the very beginning show your respect for the court and be professional.

When called to come forward you will have already talked to the bailiff or court security prior to the trial so that you know what path to take and which seat is the witness chair. There are courts of every description and design. In just an 80-mile radius, there are courtrooms with dramatic differences. An older court could have the witness chair sitting all by itself, on a slightly raised dais, in the middle of the room facing the judge. Another courtroom looks like the spaceship *Enterprise:* The room is square, but the area for attorneys, jury, and judge is round, with the floor slightly below grade, the ceiling is slightly raised with recessed lighting all around the edge, and even the attorneys' desks are curved to match the flow of the room. Some are perfectly balanced with the seating areas on each side of the judge exactly alike, witness on one side, court clerk on the other. If the clerk were to step away, it would be hard to know where to sit.

Go straight to the seat, but if you have not taken your oath, remain standing. If allowed, it is a good idea to take a bottle of water along with your file from which you will testify. Everything you bring into the courtroom is open for discovery; leave all magazines or other work outside the courtroom.

Sit up straight. If addressed by the judge acknowledge him or her; this is not unusual when this is the judge whom you will testify to over several years. Go ahead and get comfortable, open your binder, and sit your bottle of water on the floor; otherwise, the bottle can become a focal point instead of you. Do not lean back in the chair, lounging. This may give the impression of too much of a lax attitude about the proceedings. Sit up straight with your notes in front of you so that you can see without too much movement.

No matter who asks the questions always look at that person while they are asking. If there is any doubt about the question, ask that it be repeated. If the question makes no sense whatsoever, say so politely and ask for clarification. This is a better solution than giving a wrong answer. Only do this when necessary. Remember, you also want to come across as intelligent as well as professional.

■ Note
When answering a question look at the jury and occasionally at the judge.

When answering a question look at the jury and occasionally at the judge. If this is a trial or hearing where there is no jury, then give the answers to the judge. Eye contact is important. Don't rush the answer, but talk at a conversational speed. Even experienced investigators can tend to talk fast sometimes. You should only answer the question and nothing more; don't editorialize.

■ Note
You should only answer the question and nothing more; don't editorialize.

Show no emotion whatsoever toward the defendant. Regardless of the question, show no animosity toward opposing counsel. On the other hand, don't be robotic and neutral in your tone. Remember, you want to be believed. Opposing counsel may have their own expert who may very well have a view completely opposite to your findings. If you believe in your report and your opinion, let it show.

■ Note
If you believe in your report and your opinion, let it show.

If for any reason an attorney on either side should state that they object, say nothing more, but wait for the judge to decide. Even if you are in the middle of a sentence, stop immediately. If the objection is sustained, you will not answer that question, even if you really want to answer, and your attorney should recognize the need for that answer and find a way to bring it out in cross examination. If the judge overrules the objection, then continue with your answer. If there has been an appreciable pause, you can start over with your answer.

The process should be the prosecuting attorney asking you a series of questions. The opposing attorney will then ask you a series of questions. If the prosecuting attorney wants clarification on topics brought up by opposing counsel's questions, he or she can redirect and ask you for clarification. Although unlikely, the judge may ask for clarification as well.

At this point you will be dismissed. This is important: If there are any comments that either side reserves the right to call you back, you must go back to the room where the rest of the witnesses are sequestered. If the judge releases you, then you may sit in the audience or leave.

One final word on professionalism in the courts. If you stay to watch the remainder of the trial and if you are present when the verdict is read, refrain from making overt outbursts, shaking hands with fellow investigators, and so forth. It is understood and acceptable that you are emotionally attached to your testimony, and you may have good reason to want this person found guilty and sent off to jail for a long time. However, remember that there may very well be family members of the accused in the room. These people, citizens, may be victims in a way and will experience pain in hearing that their loved one is going to jail. There is nothing to be gained by causing them more strain and consternation. Your glee may be misinterpreted as your taking a vendetta against their loved one. More important, no judge likes outbursts in their courtroom.

PUBLIC INFORMATION OFFICER

The public information officer (PIO) has a role to play from the time the fire is reported until the investigation is complete, and if an arrest is made in the case, the PIO continues this role all the way through the trial.

In many jurisdictions, the PIO may be the fire investigator. In small departments, the PIO may even be the fire chief. In contrast, larger departments may assign an administrative or line officer as the PIO, with that being his or her only assignment. Usually after a few years that officer is reassigned and another officer fills that role for a set period of time. Some departments have been known to hire media specialists such as television reporters or members of the press to fill this full-time career role. What better person to understand and work with reporters than someone who once did this job themselves? In addition, these individuals may have degrees in communications, which can also be a great tool for the department.

■ Note
The PIO prepares written press releases for the media in an attempt to ensure that the right message gets out to the public.

The PIO prepares written press releases for the media in an attempt to ensure that the right message gets out to the public. Most of those working as PIOs want to give their own briefings to the press. Others may provide scripted information for the chief or investigator to convey to the press. Notice that the previous sentence does not say *read*. *Reading* statements is not unusual for those solemn situations when a life has been lost; but on other occasions reading may come off as stiff and insincere. (See **Figure 23-2.**)

When working with the press, professional relationships may be forged that will allow the PIO to suggest to reporters questions to ask the chief or investigator. The press may even go along with questions not to ask as well. However, the press does not work for the fire department. They are looking for information for their own audience, and they have their own need for information and they may also be on a time line as well. The PIO should be able to understand this and work with them.

Figure 23-2 *Fire chief giving live interview at the incident scene. (Photo courtesy of Henrico Division of Fire.)*

Interview Attitude

There was no question it was an incendiary fire. There had also been a fatality and the investigation had just gotten under way. A firefighter approached the investigator, who was also the PIO, and advised him that a reporter and TV crew from a local station was on the scene. The investigator met with the reporter, and she asked him if he would go on live with her in 15 minutes for the 11 o'clock news. The investigator agreed, provided that she did not ask four specific questions because he did not have those answers yet. He said that when he did get those answers he would call her and release that information, and she agreed.

The appointed time was approaching, the lights went on, camera ready, and the red light started blinking. The reporter addressed her audience stating that a devastating fire had just occurred. She turned to the investigator and, sure enough, specifically asked those four questions she had just a few minutes ago agreed not to ask. After each question, the investigator said, "The fire is under investigation and we do not have any information on that issue to release at this time." Then, the reporter stepped in front of the investigator, the camera zoomed in on her face, and she told her audience that if the local authorities have any answers, they were certainly not sharing them with you, the public.

Figure 23-3 *PIO working with the press to get a public safety message on the evening news. (Photo courtesy of Tommy Herman.)*

The relationship between the press and the department should not be adversarial; for the good of the public, there should be a positive working relationship between all of the media in that jurisdiction. Reporters should feel that they can call when necessary and get information for various stories. Likewise, a good relationship allows for the PIO to call and get a media crew for various events throughout the year. This will allow safety messages to get out to the public on holiday safety and general fire safety if there is a specific hazard.

The reason we investigate fires is to prevent them from occurring again. This is a major role for the PIO to get this information out to the public. (See **Figure 23-3.**) Smoke detector messages are vital; we need to sell that message to the public by using our partners in the press. Some examples of events for which the PIO should seek media coverage are the following:

- Kerosene heater safety
- Backyard barbecue safety
- Propane gas safety
- Fireworks safety and a message about legal and illegal fireworks

- Open burning safety, in both the spring and the fall
- Halloween safety, costume flammability, and the use of candles
- Christmas safety with Christmas trees, candles, electric lights, and so forth

A good PIO is vital to the success of the department. The primary and most important role is fire safety for the public. With no shame, PIOs should sell the department as well. Good press makes for positive image, which in turn can make a difference in future growth of the department. Future growth provides for better protection of the public. A positive image can also make a difference in recruiting for both paid and volunteer firefighters.

This is accomplished by developing good relationships with all the media in the jurisdiction—not just the television stations and the newspapers, but the radio stations as well. Seek all the media in the jurisdiction. One particular jurisdiction had a growing Vietnamese population that was well in excess of 30,000 individuals. Unknown at the time was that they had a small weekly newspaper. What better way to get out safety messages to a culture of which many, in this particular instance, spoke very little English?

SUMMARY

The investigator's report must contain all facts pertinent to the case at hand. If it is not in the report, then it most likely cannot be testified to in court. Finalizing the report is an opportunity to ensure that all avenues have been taken to secure all necessary information; and to establish guilt or innocence in criminal cases. Remember, you are a seeker of truth.

Your case may culminate in court where you will testify as an expert witness, providing an expert opinion as to the area of origin and the cause of the fire. Other facts discovered during the detailed investigation will stand on their own merit, provided you documented the information properly.

Remember that the overall goal of investigating a fire is to prevent future similar occurrences. One way to make this happen is through the work of a public information officer. This person can establish a working relationship with all media to provide a means of publicizing fire safety messages. The ultimate goal of the investigator and the PIO is to provide a better environment for the citizens of that jurisdiction.

KEY TERMS

Interrogatories A series of written questions for discovery, whose answers are given under oath.

Sequestered To be separated or isolated from the public during a trial.

REVIEW QUESTIONS

1. Why is it important to provide a motive on the case being presented in court?
2. If the investigation was just a small fire with a simple cause, why can't the investigator just give a verbal report to the chief?
3. What is a CV?
4. What is discovery?
5. What is the role of the PIO?

DISCUSSION QUESTIONS

1. Regarding attire, why shouldn't the investigator be in uniform when testifying in court?
2. What type of individual is best for the position of PIO: a 20-year articulate veteran of the fire service or a roving reporter with a degree in communications and 5 years of experience at a local news station?

ACTIVITIES

1. Research your local courtroom to see what type of media equipment it has installed for expert witness testimony.
2. If you have cable TV, go on the Internet to the Web site for Court TV, which is now Tru TV. Search their shows and identify which shows cover live court action. Arrange to watch these shows to observe real expert witnesses give testimony in a court of law. If you do not have cable, check with your local courts to see when they will have a criminal trial scheduled so that you can visit and observe a courtroom in action.

GLOSSARY

Accelerant Usually considered to be an ignitable liquid that can be used to ignite the fire, spread fire, or increase the rate of fire growth. When chemical processes are used, an oxidizer can contribute in the same manner as the ignitable liquid. Combustibles can also be used in the same manner, especially to move the fire from one location to another.

Accountability tag An identification tag with the holder's name and sometimes unit number or assigned company. More sophisticated tags also include the individual's medical history, including blood type and medical allergies. Used in the accountability system to track the location of all personnel at an emergency scene.

The Act Occupational Safety and Health Act.

Administrative search warrant Warrant issued by a magistrate or judge allowing the investigator to be on the scene to determine the cause of the fire for the good of the people.

AL/CU A designation given to electrical connectors by the industry to indicate that the device is safe for use with aluminum and copper wires. However, no definitive test by a third party could validate that the devices were designed or safe for aluminum wire.

American National Standards Institute (ANSI) Coordinates the creation, development, promulgation, and use of voluntary consensus standards and guidelines.

American Society for Testing and Materials (ASTM) A trusted source for technical standards for materials, products, systems, and services.

American Wire Gauge The standard used by the industry to determine the size and designation for electrical wiring.

Amperage The strength of an electrical current measured in amperes.

Annealing The collapse of coil springs, loss of temper, as the result of heat.

Arc The discharge of electricity from one conductive surface to another, resulting in heat and light.

Arrow patterns As described in the NFPA 921, the patterns on wooden structural members that show a direction or path of fire travel.

Arson Intentionally and willfully setting a fire with malice.

Authority Having Jurisdiction (AHJ) An entity responsible for enforcing a code, procedure, standard, or law. Usually referring to a political entity such as city council, board of supervisors, town elders, and so on.

Autoignition temperature The minimum temperature at which a properly proportioned mixture of vapor and air will ignite with no external ignition source.

Backdraft An explosion resulting from the sudden introduction of air (oxygen) into a confined space containing oxygen-deficient superheated products of incomplete combustion.

Barricade tape Wide banner, 3 inches or more wide, brightly colored and with appropriate markings to convey a message for individuals to not cross the tape. Tape is used to create a barrier to keep out most individuals for their own safety and to protect the scene. Markings may say "CRIME SCENE—DO NOT CROSS" or "FIRE LINE—DO NOT CROSS."

Bead The melted end of a metal conductor that shows a globule of resolidified material with a sharp line of demarcation to the remainder of the wire.

Bernoulli effect As it applies to chimneys, the wind blowing across the opening of a chimney decreases the pressure in the flue, increasing the updraft.

Building official The person responsible for the enforcement of the local and state building code. The building official's office is responsible for issuing construction permits and certificates of occupancy as a means of ensuring compliance with the code.

Building Officials Office Empowered by the Authority Having Jurisdiction (AHJ) to enforce the provisions of a state or local building code, electrical code, plumbing code, and so forth. In some instances empowered to enforce the fire code.

Bureau of Alcohol, Tobacco, Firearms, and Explosives (ATF) A federal agency with fire investigative authority; it operates a National Response Team, which can provide almost every resource necessary at a fire scene.

Burglary The entry into a structure or part of a structure with the intent to commit a crime.

Burner assembly The assembly houses the electrodes in a furnace to ignite the oil that is pumped out of the jet nozzle. This unit also contains the fan, which blows air into the chamber, allowing for a bright and efficient flame.

Calcination The process of driving off the moisture in the gypsum, discoloring and softening it in the process. The process includes the driving off of moisture that is chemically bonded within the gypsum.

Canister A manufactured metal or plastic cylindrical device filled with filtering material. Canisters are designed to be attached to a filter mask with an airtight seal, forcing all air entering the mask to pass through the filtering medium, essentially removing all hazardous airborne particulates.

Carbon monoxide (CO) A colorless, odorless gas usually produced as a by-product of a fire.

Casement windows Windows with hinges on the side that usually open with a rotating crank assembly.

Catalytic converter A device installed as part of the exhaust system of an automobile; designed to reduce emissions by burning off pollutants.

Ceiling jet A thin layer of buoyant gases that moves rapidly just under the ceiling in all directions away from the plume.

Chain of custody A means of documenting who had control of the evidence from the time of collection through the trial and up to its release or destruction.

Chimney chase Decorative hollow covering around a cement block or metal chimney; designed to be more aesthetic; usually framed in wood and covered with the same material as the siding of the structure.

Chimney liner The covering inside the chimney that enables the flow of hot gases and smoke from the chimney and keeps them from seeping through the mortar into the structure.

Circuit breakers A protection device as part of an electrical system that will automatically cut off the electricity should there be an excessive overload.

Claims managers Within almost every insurance company is a division assigned to evaluate submitted insurance claims to assess the value of the loss and, if justified, to write the checks necessary to settle the claims.

Clinker A solid mass created during a spontaneous combustion event with hay; it is green to gray and may be shiny like glass. This same term is used to describe residue found in furnaces, but the mass appears more rock-like.

CO/ALR A designation given to electrical devices that can safely use aluminum wiring. This is only given to those devices that meet a predetermined test accepted by both Underwriters Laboratory and the industry.

Cold scenes The remains of a fire scene after the fire has been completely extinguished. Although the debris may still be warm, there is little or no chance of reignition. Cold scenes may be days old.

Comparison samples Can refer to contaminated materials as well as product items such as coffee pots, portable heaters, and so forth. When submitting contaminated samples, such as carpeting, to test for accelerants, another sample from an area known, or believed, to not have been contaminated by the accelerant can be submitted. Comparison samples enable the lab to distinguish typical components of the material compared to components not associated with the material.

Conductors Anything that is capable of allowing the flow of electricity. Commonly used to describe the wires used to provide electricity within a structure.

Conduit A tube or trough made from both plastic and metal that is designed to provide protection for electrical conductors.

Consent Permission given by the person responsible for or controlling the property, allowing the investigator to search the property.

Contamination Anything introduced into the fire scene or into the evidence that makes test results unreliable or places doubt on the results of the laboratory testing.

Corpus delecti The body of the crime; the physical proof that the crime did occur.

Crazing Glass with small cracks that go in all directions where the cracks do not go all the way through the glass.

Criminal act As stated in *Black's Law Dictionary,* an external manifestation of one's will which is prerequisite to criminal responsibility. There can be no crime without some act, affirmative or negative.

Criminal search warrant A warrant issued by a magistrate or judge based on the sworn (and written affidavit) testimony of an investigator that probable cause exists that a crime has been committed and the person or place to search and the items being sought.

Deductive reasoning Taking the facts of the case and doing a thorough and meticulous challenge to all facts known, using logic.

Demarcation and sharp line of demarcation A distinct line between a fire-damaged area and nondamaged areas; not a gradual charring but an abrupt end.

Deposition Obtaining testimony of witnesses that is recorded, authenticated, and reduced to writing and that can be used in future court testimony. Used as a discovery device by either side in a civil or criminal trial.

Discovery A pretrial device used by both (all) sides in a case to obtain all the facts about the case to prepare for the trial.

Discovery sanctions Penalty for failing to comply with discovery rules.

Dispatcher A person who works in an Emergency Communications Center (E 9-1-1 centers), whose duties are to receive calls from the public for emergency assistance, and then to send the appropriate emergency unit(s) to handle the situation.

Drying oils Organic oil used in paints and varnishes that when dried leave a hard finish.

Drywall Older walls, called plaster and lath, were created by application of wet plaster onto walls. Drywall is a new type of board that is made of gypsum, coated with a hard paper, and created in 4-foot widths and various lengths from 8 feet to 12 feet. Drywall is also known as plasterboard, sheetrock, or gypsum board in various parts of the country. Thus, when drywall is applied to the wall it is just that: dry.

Electric matches Devices used to ignite items remotely such as pyrotechnics. Usually consist of 22-gauge wire with a loop at the end that has a coating, referred to as the pyrogen, that ignites at low energy; it may require an outside electrical source to cause ignition.

Electrons Negatively charged particles that are a part of all atoms.

Emergency Communications Center (Dispatch) Sometimes referred to as the E 9-1-1 center; a place where calls from the public for emergency assistance are received and whose staff (dispatchers) alert and send the appropriate emergency unit(s) to handle the situation.

Emergency medical technicians (EMTs) Individuals with specialized medical skills, trained to offer basic first aid in the field under emergency circumstances, and then to assist in the transporting of the patient to medical facilities if necessary.

Empirical data Data that are based on observations, experiments, or experience.

Energy release rate (ERR) Amount of energy produced in a fire over a given period of time.

Evidence transmittal A form designed to identify the evidence to be submitted to the laboratory and the tests to be conducted on that evidence; can also serve as the chain of custody of the evidence.

Exculpatory Tends to clear or clears the suspect of guilt.

Exemplar A comparison sample of an electrical or mechanical device that enables the investigator to compare the damaged item to an undamaged version of the product.

Exigent circumstances For the good of the people, without permission, public safety personnel can enter property on fire to save lives and control the fire.

Exothermic reaction The release of heat from a chemical reaction when certain substances are combined.

Fairness test A process adopted by the American National Standards Institute that requires openness, balance with a lack of dominance, consideration of all views and objections, an appeals process, and audit requirements. These requirements produce a level playing field for all involved.

Federal Bureau of Investigation (FBI) The nation's law enforcement agency; it is available to assist on major fire incidents. In addition to an outstanding crime lab, the FBI has one of the best training academies in the world.

Fire Marshal The predominant title used to describe government employees who have the primary responsibility to investigate fire and explosions, enforce a fire code, and provide public fire and life safety education.

Fire point The lowest temperature at which a fuel, in an open container, gives off sufficient vapors to support combustion once ignited from an external source.

Fire tetrahedron Tetrahedron is a solid figure with four triangular faces. When each face represents the four items necessary for a fire (heat, fuel, oxygen, and uninhibited chemical reaction), the figure then becomes the fire tetrahedron.

First responders Any public safety responders who may be or may have the potential of being first on the scene as the result of being sent there by the Emergency Communications Department.

Flameover The point at which a flame propagates across the undersurface of a thermal layer. Same as rollover.

Flash point The temperature of a liquid at which point it gives off sufficient vapors that, when mixed with air in proper proportions, ignites from an exterior ignition source; because of limited vapors only a flash of fire across the surface occurs.

Flashover A transition phase of fire where the exposed surface of all combustibles within a compartment reach autoignition temperature and ignite nearly simultaneously.

Fraud A false statement of fact or intentionally misleading statement made to cause someone to give up something of value.

Fuses A device used as a screw-in plug or a cartridge that contains a wire or thin strip of metal designed to melt at a specific temperature that relates to certain overcurrent situations. The melting of the wire or metal strip stops the flow of electricity as designed.

Gas chromatography (GC) A laboratory test method that separates the recovered sample into its individual components; provides a graphic representation of each component along with the amounts of each component.

Goggles, nonvented Eyewear that makes a tight seal with the face, preventing the introduction of any airborne particulates. *Nonvented* implies that there are no vent holes so that the glasses are airtight.

Good science Science that can be verified by independent means that meet accepted norms of the scientific community.

Greenfield The manufacturer's name for a flexible conduit. The name has become a common term to describe any flexible conduit, both metal and plastic.

Ground fault circuit interrupter (GFCI) A circuit breaker that is more sensitive than the standard breaker, tripping at a slight ground fault with the intent of preventing electrocution.

Grounding block A block of metal within an electrical panel, affixed with holes and screws to allow for the insertion of wires where the tip of the screw will bind and hold the wire in place.

Guide A nonmandatory document giving advice and recommendations on how to carry out a task.

Heat release rate (HRR) The rate at which heat is generated from a burning fuel.

Hux bar One brand name of a firefighting multitool that comes in different sizes but that averages 36 inches in length. Different tool heads on each end can be used for prying, poking, striking, and opening hydrants or closing gas valves.

Hydrocarbon detector An electronic device intended to be used in the field to identify the location of ignitable liquids; used to improve the probability of obtaining the best sample to submit to the laboratory for testing.

Imminent danger to life and health (IDLH) An atmosphere in which anyone entering endangers themselves unless proper protection is taken, which could include full turnout gear with self-contained breathing apparatus.

Incendiary The willful and intentionally setting of a fire, with malice.

Incident Command System (ICS) A system of command used by both military and public safety agencies that is designed to simplify command interstructures and enable safe and efficient handling of an incident. A plan that promotes safety and accountability of all personnel on the scene.

Insulators Something that is a nonconductor of electricity; commonly thought of as glass or porcelain, but today includes plastic and rubber as in the coverings of electrical wire.

International Association of Arson Investigators (IAAI) A private, nonprofit organization dedicated to the professional development of fire and explosion investigators through training and research of new technology.

International Association of Special Investigation Units (IASIU) A nonprofit international organization with local chapters dedicated to combat insurance fraud through training, awareness, and legislation.

Interrogation Formal questioning, same as interviewing but has a more formal connotation.

Interrogatories For the fire investigator, this is a series of written questions about the case to be completed by the one giving testimony in a deposition. Usually prepared for the discovery phase of legal process. Interrogatories must be signed by the person answering the questions, stating the information given for each question is true and correct.

Jet nozzle An oil furnace has a nozzle with an extremely small opening. The oil is pumped up to this nozzle, where it squirts out and is ignited by the electrodes, providing a proper flame for the burn chamber. Nozzles come in varying sizes.

Jetties A projecting or overhanging part of a building.

Job performance requirements A statement that describes a specific job task, items necessary to complete the task, measurable and observable outcomes of the task.

Junk science Untested scientific theory or faulty scientific data sometimes used to promote a private agenda or theory.

K-9 accelerant dog A dog specifically trained by a reputable organization to identify the presence of an accelerant and give the appropriate signal of its location.

Kinetic energy Energy as the result of a body in motion.

Litigant Someone who is involved in a lawsuit.

Lockout In reference to an electrical disconnect box, when it is in the open position with no current flowing and a lock is in place to keep the circuit from being energized until the lock is removed.

Loss Object of value that was destroyed.

Malice Intentionally, willfully, and knowingly doing a wrongful act with the intent to inflict harm on another; recognized by law as an evil intent.

Mass spectrometry (MS) One of two tests used to identify the presence of an ignitable liquid in a submitted sample, the other test being the gas chromatography (GC). MS is used to find the composition of a physical sample by generating a mass spectrum representing the masses of sample components; this test is sometimes referred to as mass-spec or MS.

Meter base The receptacle for the electrical meter installed by the power company.

Methane gas Colorless, odorless, flammable gas created naturally by decomposing vegetation. It can also be created artificially.

Micro torch A miniature torch that uses butane as a fuel. The mechanism can inject air into the chamber, allowing for a high heat output. Usually used by hobbyists.

Molotov cocktail A bottle filled with an ignitable liquid (usually gasoline) with an igniter to initiate combustion when the bottle breaks and releases the contents. The igniter can be a lit rag stuffed in the top of the bottle or a chemical igniter. Chemical igniters are two reactive chemicals, one coating the outside of the bottle and the other mixed with the contents of the bottle, that when the bottle is broken come together, reacting and igniting contents.

Multigas detector An electronic device to monitor and measure oxygen levels, presence and amount of carbon monoxide in the air, as well as hydrogen sulfide. Units usually have interchangeable modules to monitor different gases. These devices also measure the presence of flammable gases as a percentage in the air, allowing the detector to determine and display the LEL.

Multimeter An electronic device used to measure AC/DC voltage, current, resistance, capacitance, and frequency. Also referred to as a multitester.

National Association of Fire Investigators (NAFI) An organization whose primary purpose is to increase knowledge and improve skills of persons involved in fire and explosion investigations.

National Electrical Code (NEC) A consensus code for the safe installation of electrical components and appurtenances.

National Fire Incident Reporting System (NFIRS) A national computer-based reporting system that fire departments use to report fires and other incidents to which they respond. This is a uniform system to collect data for both local and national use.

National Fire Protection Association (NFPA) A private, nonprofit organization dedicated to reducing the occurrence of fires and other hazards.

National Institute of Science and Technology (NIST) Founded in 1901, a nonregulatory federal agency in the U.S. Commerce Department's Technology Administration. Its mission is to promote U.S. innovation and industrial competitiveness by advancing measurement science, standards, and technology in ways that enhance economic security and improve people's quality of life. The Fire Research Division consists of several groups that can conduct research on fire incidents.

National Professional Qualifications Board The entity created to oversee the application of professional standards by training entities to ensure that they meet the professional qualifications.

NFPA 921, *Guide for Fire and Explosion Investigations* A document designed to assist individuals investigate fire and explosion incidents in a systematic and efficient manner.

NFPA 1037, *Standard for Professional Qualifications for Fire Marshal* A standard outlining the minimum professional qualifications for fire marshals or equivalent positions.

NM Stands for *nonmetallic* and refers to a covering over wire conductors intended for use within a structure to deliver electricity to various points within the structure.

Off-gassing The release of gas or vapors in the process of aging or decomposing, also refers to vapors or gases that were absorbed by fabric, carpet, and so forth during a fire and then released after the fire is extinguished.

Officer in charge (OIC) The fire officer who has ultimate charge and control of the overall fire or emergency scene. Usually but not always the senior officer on the scene.

OSHA Occupational Safety and Health Administration; a federal agency created by the Occupational Safety and Health Act whose function is to ensure safety in the workplace.

Overhaul The act of searching for hidden fire, ensuring that the fire is completely extinguished.

Oxidize The effect that heat has on a metal surface, in which it consumes any covering, resulting in the rusting of the metal surface.

Oxidizer Any material that forms with a fuel to support combustion.

Peer review A process of review of written documents by persons in similar fields or professions in which anyone can make comments for changes and improvements, making the document more reliable and credible.

Perpetrator The one who commits the criminal act.

Physical evidence policy A locality's policy that guides and directs the proper collection, storage, handling, use, and disposal of all evidence.

Piezoelectric ignition Certain crystals that generate voltage when subjected to pressure (or impact).

Plume The column of smoke, hot gases, and flames that rises above a fire.

Policyholder Person who owns the insurance policy, possibly the property owner or someone who has an interest in the property.

Potential evidence Something that may, or could, possibly be used to make something else evident; evidence that is not yet used or that has yet to be used to prove a point or issue or support a hypothesis.

Private investigator (For fire and explosion): An individual involved in providing professional investigations of fires and explosions on a contract basis or as an employee of a private enterprise or organization.

Pro bono Doing the work free of charge; at no cost.

Proof of loss Formal statement made by the insurance policyholder to the insurance company validating the loss. If the policy covers contents, this statement lists all contents in the loss at the time of the incident.

Public fire investigator An investigator working for the government representing the locality, state, or federal government. Usually denotes someone whose job requirements include the determination of the area of origin and cause of fires and explosions.

Pyromania The tendency or impulse to set fires.

Report on Comments (ROC) A report created showing the committee actions taken on each proposal.

Report on Proposals (ROP) A report published by NFPA and sent to its membership and other interested parties to solicit feedback on proposed documents.

Requisite knowledge Basic knowledge a person must possess to perform an assigned task.

Requisite skills The skills a person must have to perform an assigned task.

Respirators A mask (full or half face) designed to cover the mouth and nose, allowing the wearer to breathe through filters attached to the mask so as to prevent the inhalation of dangerous substances usually in the form of dusts or airborne particulates.

Risk The item covered by the insurance policy; the item that may be lost.

Robbery Feloniously taking any item of value from another directly or in their immediate presence and against their will while using force or fear.

Rocket motors Hobby propulsion kits that provide thrust through the burning of an ignitable mixture.

Rollover The point at which a flame will propagate across the undersurface of a thermal layer. Same as flashover.

Rotor In every electrical motor, there are two basic parts, the rotor and the stator. The rotor rotates in an electric motor.

Safety match A match consisting of a head containing a chemical to initiate the fire and tinder to keep the fire burning; designed to ignite only when struck against an area on the matchbook or box that contains red phosphorus and ground glass.

Salvage A suppression activity used to protect the contents of a property from smoke or water damage by removing objects from the structure, covering materials within the structure with tarps, and/or removing water with squeegees, pumps, water vacuums, and so forth.

Scale A flaking of metal as the result of oxidation.

Scientific method The systematic pursuit of knowledge involving the recognition and formulation of a problem, the collection of data through observation and experiment, and the formulation and testing of a hypothesis.

Self-contained breathing apparatus (SCBA) Specially designed breathing apparatus that, when worn according to manufacturer's instructions, provides the wearer sufficient air to allow entry into atmospheres with insufficient oxygen or atmospheres where the air is contaminated. Must be worn with proper clothing such as firefighting turnout gear or encapsulated suits.

Sequestered To be separated or isolated from the public during a trial.

Service drop Overhead wiring from the electrical company that is attached to the weather-head to deliver electricity to the structure.

Service lateral Underground wiring that comes from the electrical company that comes up to the meter base.

Service panel Electrical wire travels from the meter base to the service panel. The panel provides a means to provide overcurrent protection and the ability to distribute electricity throughout the structure through branch circuits.

Sheetrock Often called drywall, gypsum, or wallboard; a crumbly material called gypsum sandwiched between two layers of thick paper. Gypsum is fire resistive.

Sifting Process of using a screen to separate out small particles so that larger particles can be examined.

Sleeving The slight separation of the insulation from around the electrical conductor(s), producing an effect that allows the insulation to loosely slide back and forth on the conductors.

Spalling When the surface of concrete pops off as a result of water in the concrete reaching boiling temperature and turning to steam. When this happens, the steam expands 1,700 times in volume, creating the energy to pop off the concrete from the surface.

Spark As a general term, anything that is incandescent, burning, that is air bound. In terms of electricity, it is the molten bit of metal that has absorbed sufficient energy to be incandescent and that is air bound. Both terms indicate that the spark is capable of being a competent ignition source for many combustibles.

Special Investigations Unit (SIU) A division usually within an insurance company that oversees or conducts investigations involving a loss.

Spoliation The destruction of evidence; the destruction or the significant and meaningful alteration of a document or instrument. It constitutes an obstruction of justice.

Spontaneous combustion A combustion process in which a chemical reaction takes place internally, creating an exothermic reaction that builds until it reaches the ignition temperature of the material involved.

Spontaneous heating A process in which a material increases in temperature without drawing heat from the surrounding area.

Spontaneous ignition Initiation of combustion from within a chemical or biological reaction that produces enough heat to ignite the material.

Standards Model providing minimum mandatory training requirement to meet a professional competency in a form that can be referenced or adopted into law.

Static electricity The buildup of a charge, negative or positive, as the result of items coming in contact and then breaking that contact, either taking away or leaving electrons that results in the change of the static charge of the items involved.

Stator In every electrical motor, there are two basic parts, the rotor and the stator; the stator is the part of the motor that stays stationary, housing the rotor.

Steel sole boots Boots with a lightweight steel plate that provides a barrier between the foot and the surface being walked on. The intent is to keep the wearer safe from punctures through the boot.

Steel toe boots Work boots designed with a steel protective cap covering the toes to prevent crushing blows that would otherwise injure the wearer's toes.

Strike-anywhere match A head containing a chemical to initiate the fire and tinder to keep the fire burning. These matches use a mixture of chemicals, usually in two layers and impregnated with ground glass. The outer layer is designed to be more friction oriented, and the inner layer helps the head burn longer and ensures the ignition of the tinder. Match is designed to be dragged against any rough surface to create the friction for ignition.

Subpoena The command of the courts to have a person appear at a certain place at a specific time to give testimony.

Subrogation The substitution of one person in the place of another with reference to a lawful claim.

Systematic search A search based on a system that is used for each fire, each and every time.

Tagged In reference to an electrical disconnect switch, the action of placing a tag on the device warning others not to turn the handle to the on position due to a safety concern.

Tool mark impressions Indentations or marks left on structural surfaces or other locations from tools used to open a door, window, safe, and so forth. The impressions might be unique enough to identify the tool used. Sometimes particles from the tools such as chips of metal or paint may be left in the impression and these particles can also help identify the specific tool used.

Tyvek coveralls Coveralls made out of Tyvek, which the manufacturer DuPont describes as lightweight, strong, vapor-permeable, water-resistant, and chemical-resistant material that resists tears, punctures, and abrasions.

UF The type of insulation on the electrical wiring that is used as an underground feeder. The insulation tends to be thicker and solid up to the insulated conductor.

United States Supreme Court The highest court of the nation.

Vapor density Density of a gas or vapor in relation to air, with air having a designation of 1.

Voltage Electromotive force or pressure, difference in electrical potential; measured as a volt, which is ability to make 1 ampere flow through a resistance of 1 ohm.

Voltage detector Tester or probe that either glows or emits a sound when placed in proximity of an energized electric circuit (electrical wires or cords).

Watts Unit of power equal to 1 joule per second.

ACRONYMS

AC	alternating current
AFCI	arc fault circuit interrupters
AHJ	Authority Having Jurisdiction
ANSI	American National Standards Institute
ASTM	American Society of Testing and Materials
AWG	American wire gauge
ATF	Alcohol Tobacco and Firearms (same as BATF)
BATF	Bureau of Alcohol Tobacco and Firearms
BTU	British thermal unit
CISM	critical incident stress management
CISD	Critical Incident Stress Debriefing
CNG	compressed natural gas
CO	carbon monoxide
CPA	Certified Public Accountants
CV	curriculum vitae
DC	direct current
DoD	Department of Defense
DOJ	Department of Justice
EFI	Electronic Fuel Injection
EMS	Emergency Medical Services
EMT	Emergency Medical Technician
ERR	energy release rate
EUO	examination under oath
FAA	Federal Aviation Administration
FBI	Federal Bureau of Investigations
FEMA	Federal Emergency Management Agency
FM	Factory Mutual
FMANA	Fire Marshal Association of North America
FOIA	Freedom of Information Act
GC	gas chromatography
GC/MS	gas chromatography – mass spectrometry
GFCI	ground fault circuit interrupter
HRR	heat release rate
HVAC	heating, ventilation and air conditioning
IAAI	International Association of Arson Investigators
IABPF	International Association of Black Professional Firefighters
IAFC	International Association of Fire Chiefs
IAFF	International Association of Firefighters
IASIU	International Association of Special Investigation Units
IC	Incident Commander
ICS	Incident Command System
IDLH	imminent danger to life and health
IFSTA	International Fire Services Training Association
IRA	Irish Republican Army
IRI	Industrial Risk Insurers

ISFSI	International Society of Fire Service Instructors
JPR	Job Performance Requirements
kW	kilowatts
LEL	lower explosive limit
ME	Medical Examiner
MS	mass spectrometry
NAFI	National Association of Fire Investigators
NCAVC	National Center for the Analysis of Violent Crime
NCIC	National Crime Information Center
NCST	National Safety Construction Team Act
NEC	National Electric Code
NFA	National Fire Academy
NFIRS	National Fire Incident Reporting System
NFPA	National Fire Protection Association
NICB	National Insurance Crime Bureau
NIOSH	National Institute for Occupational Safety and Health
NIST	National Institute of Standards and Technology
NPQB	National Professional Qualifications Board
NRT	National Response Team (BATF)
NTSB	National Transportation Safety Board
NVFC	National Volunteer Fire Council
NWCG	National Wildland Coordination Group
OIC	Officer in Charge
OSHA	Occupational Safety and Health Act (Administration)
PILR	Property Insurance Loss Register
PIO	Public Information Officer
psi	pounds per square inch
ROC	Report on Comments
ROP	Report on Proposals
SCBA	self contained breathing apparatus
SIU	Special Investigation Unit
SLR	single lens reflex
UEL	upper explosive limit
VIN	vehicle identification number

INDEX

D

N

O

P

R

S